Computational Mathematics and Applied Statistics

Computational Mathematics and Applied Statistics

Editor

Sandra Ferreira

MDPI • Basel • Beijing • Wuhan • Barcelona • Belgrade • Manchester • Tokyo • Cluj • Tianjin

Editor
Sandra Ferreira
University of Beira Interior
Portugal

Editorial Office
MDPI
St. Alban-Anlage 66
4052 Basel, Switzerland

This is a reprint of articles from the Special Issue published online in the open access journal *Mathematical and Computational Applications* (ISSN 2297-8747) (available at: https://www.mdpi.com/journal/mca/special_issues/Appl_Stat).

For citation purposes, cite each article independently as indicated on the article page online and as indicated below:

LastName, A.A.; LastName, B.B.; LastName, C.C. Article Title. *Journal Name* **Year**, *Volume Number*, Page Range.

ISBN 978-3-0365-7546-9 (Hbk)
ISBN 978-3-0365-7547-6 (PDF)

© 2023 by the authors. Articles in this book are Open Access and distributed under the Creative Commons Attribution (CC BY) license, which allows users to download, copy and build upon published articles, as long as the author and publisher are properly credited, which ensures maximum dissemination and a wider impact of our publications.

The book as a whole is distributed by MDPI under the terms and conditions of the Creative Commons license CC BY-NC-ND.

Contents

About the Editor . . . . **vii**

Sandra Ferreira
Preface to Computational Mathematics and Applied Statistics
Reprinted from: *Math. Comput. Appl.* **2023**, 28, 31, doi:10.3390/mca28020031 . . . . **1**

Lishamol Tomy, Christophe Chesneau and Amritha K. Madhav
Statistical Techniques for Environmental Sciences: A Review
Reprinted from: *Math. Comput. Appl.* **2021**, 26, 74, doi:10.3390/mca26040074 . . . . **5**

Muhammed Rasheed Irshad, Christophe Chesneau, Veena D'cruz and Radhakumari Maya
Discrete Pseudo Lindley Distribution: Properties, Estimation and Application on
INAR(1) Process
Reprinted from: *Math. Comput. Appl.* **2021**, 26, 76, doi:10.3390/mca26040076 . . . . **37**

Lishamol Tomy, Veena G and Christophe Chesneau
The Sine Modified Lindley Distribution
Reprinted from: *Math. Comput. Appl.* **2021**, 26, 81, doi:10.3390/mca26040081 . . . . **59**

Farrukh Jamal, Ali H. Abuzaid, Muhammad H. Tahir, Muhammad Arslan Nasir, Sadaf Khan and Wali Khan Mashwani
New Modified Burr III Distribution, Properties and Applications
Reprinted from: *Math. Comput. Appl.* **2021**, 26, 82, doi:10.3390/mca26040082 . . . . **75**

Anuresha Krishna, Radhakumari Maya, Christophe Chesneau and Muhammed Rasheed Irshad
The Unit Teissier Distribution and Its Applications
Reprinted from: *Math. Comput. Appl.* **2022**, 27, 12, doi:10.3390/mca27010012 . . . . **93**

Sadaf Khan, Gholamhossein G. Hamedani, Hesham Mohamed Reyad, Farrukh Jamal, Shakaiba Shafiq and Soha Othman
The Minimum Lindley Lomax Distribution: Properties and Applications
Reprinted from: *Math. Comput. Appl.* **2022**, 27, 16, doi:10.3390/mca27010016 . . . . **113**

Vasili B. V. Nagarjuna, Rudravaram Vishnu Vardhan and Christophe Chesneau
Nadarajah–Haghighi Lomax Distribution and Its Applications
Reprinted from: *Math. Comput. Appl.* **2022**, 27, 30, doi:10.3390/mca27020030 . . . . **133**

Mansura Jasmine, Abdolmajid Mohammadian and Hossein Bonakdari
On the Prediction of Evaporation in Arid Climate Using Machine Learning Model
Reprinted from: *Math. Comput. Appl.* **2022**, 27, 32, doi:10.3390/mca27020032 . . . . **147**

Porntip Dechpichai, Nuttawadee Jinapang, Pariyakorn Yamphli, Sakulrat Polamnuay, Sittisak Injan and Usa Humphries
Multivariable Panel Data Cluster Analysis of Meteorological Stations in Thailand for
ENSO Phenomenon
Reprinted from: *Math. Comput. Appl.* **2022**, 27, 37, doi:10.3390/mca27030037 . . . . **161**

Lishamol Tomy, Veena G and Christophe Chesneau
Applications of the Sine Modified Lindley Distribution to Biomedical Data
Reprinted from: *Math. Comput. Appl.* **2022**, 27, 43, doi:10.3390/mca27030043 . . . . **179**

Kindie Fentahun Muchie, Anthony Kibira Wanjoya and Samuel Musili Mwalili
Small Area Estimation of Zone-Level Malnutrition among Children under Five in Ethiopia
Reprinted from: *Math. Comput. Appl.* **2022**, 27, 44, doi:10.3390mca27030044 . . . . **195**

Jean-Louis Pinault
Morlet Cross-Wavelet Analysis of Climatic State Variables Expressed as a Function of Latitude, Longitude, and Time: New Light on Extreme Events
Reprinted from: *Math. Comput. Appl.* **2022**, *27*, 50, doi:10.3390/mca27030050 . . . . **211**

Kokou Essiomle and Franck Adekambi
A Note on Gerber–Shiu Function with Delayed Claim Reporting under Constant Force of Interest
Reprinted from: *Math. Comput. Appl.* **2022**, *27*, 51, doi:10.3390/mca27030051 . . . . **233**

Farrukh Jamal, Laba Handique, Abdul Hadi N. Ahmed, Sadaf Khan, Shakaiba Shafiq and Waleed Marzouk
The Generalized Odd Linear Exponential Family of Distributions with Applications to Reliability Theory
Reprinted from: *Math. Comput. Appl.* **2022**, *27*, 55, doi:10.3390/mca27040055 . . . . **247**

Abdulmtalb Hussen and Wenjie He
Prony Method for Two-Generator Sparse Expansion Problem
Reprinted from: *Math. Comput. Appl.* **2022**, *27*, 60, doi:10.3390/mca27040060 . . . . **269**

Schalk W. Human, Andriette Bekker, Johannes T. Ferreira and Philip Albert Mijburgh
A Bivariate Beta from Gamma Ratios for Determining a Potential Variance Change Point: Inspired from a Process Control Scenario
Reprinted from: *Math. Comput. Appl.* **2022**, *27*, 61, doi:10.3390/mca27040061 . . . . **289**

Shakaiba Shafiq, Sadaf Khan, Waleed Marzouk, Jiju Gillariose and Farrukh Jamal
The Binomial–Natural Discrete Lindley Distribution: Properties and Application to Count Data
Reprinted from: *Math. Comput. Appl.* **2022**, *27*, 62, doi:10.3390/mca27040062 . . . . **307**

About the Editor

Sandra Ferreira

Sandra Ferreira is an assistant professor at the Department of Mathematics at the University of Beira Interior. She received a Master's degree in Applied Mathematics from Évora University (2000) and a Ph.D. in Mathematics at the University of Beira Interior (2006). She is a researcher at the Center of Mathematics and Applications, in the Physics and Mathematical Modeling research group. She is also a researcher (external collaborator) at the Center for Mathematics and Applications (CMA), NOVA School of Science and Technology | FCT NOVA, in the Statistics and Risk Management research group. Her research interests include mathematical statistics; statistical analysis, modeling, and inference; applied and computational statistics; R Statistical Package; sample size; mixed, linear, and linear mixed models; data analysis models; F-Tests; applied mathematics; applied probability and probability theory; random sampling; confidence intervals; linear regression; multivariate, descriptive, and nonparametric statistics; and regression analysis. Her publications and current research interests focus on statistical inference for estimable functions and variance components in linear mixed models with commutative orthogonal block structure (COBS). Her research group is currently working on the project "Estimation in Mixed Linear Models—MLM". She has authored 78 scientific papers, receiving more than 340 citations (H-index 10). She was also awarded the "Mathematical sciences sponsorship fund" (2018) by Elsevier.

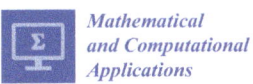

Editorial

Preface to Computational Mathematics and Applied Statistics

Sandra Ferreira

Center of Mathematics and Applications, Department of Mathematics, University of Beira Interior, 6201-001 Covilha, Portugal; sandraf@ubi.pt

Citation: Ferreira, S. Preface to Computational Mathematics and Applied Statistics. *Math. Comput. Appl.* **2023**, *28*, 31. https://doi.org/10.3390/mca28020031

Received: 22 February 2023
Accepted: 22 February 2023
Published: 27 February 2023

Copyright: © 2023 by the author. Licensee MDPI, Basel, Switzerland. This article is an open access article distributed under the terms and conditions of the Creative Commons Attribution (CC BY) license (https://creativecommons.org/licenses/by/4.0/).

The rapid advances in modeling research have created new challenges and opportunities for statisticians. These include statistical inferences in observational studies and many other emerging fields, which have motivated statisticians worldwide to develop cutting-edge methods and analytical strategies.

The main focus of this Special Issue is applications and methodological research in all fields of computational statistics. This Issue aims to provide a forum for computer scientists, mathematicians, and statisticians working in a variety of areas of statistics, including biometrics, econometrics, data analysis, graphics, and simulation. Computational Mathematics and Applied Statistics are two related fields that play a crucial and important role in the development and advancement of various scientific and technological domains.

Computational Mathematics involves developing and applying mathematical algorithms and models to solve complex problems using computers. This book integrates various mathematical disciplines, such as calculus, linear algebra, differential equations, optimization and numerical analysis, to develop efficient and accurate algorithms for scientific computation and data analysis. Computational Mathematics has numerous applications, including engineering, physics, biology, finance, and computer science.

Applied Statistics involves collecting, analyzing, and interpreting data to make informed decisions in various fields. It includes various statistical methods and tools, such as probability theory, hypothesis testing, regression analysis, and experimental design, to extract insights and patterns from data. Applied Statistics has applications in various fields, such as healthcare, finance, marketing, social sciences, and environmental studies.

In this book, the integration of these two fields has led to the development of data science, which has become a vital field for extracting insights from vast amounts of data. Furthermore, this book can be used as a textbook and/or reference book, which is especially suitable for undergraduates and graduates in computational mathematics, engineering, computer science, computational intelligence, and data science.

In [1], Tomy et al. explore the interdisciplinary partnership between Environmental Sciences and Statistics, highlighting the crucial role of statistical methods in solving environmental issues. By providing a clear roadmap, this paper facilitates collaborative learning between environmental scientists and quantitative researchers, enabling them to develop their analytical skills and knowledge base.

In [2], Irshad et al. present the discrete Pseudo Lindley (DPsL) distribution, which is a discrete version of the Pseudo Lindley (PsL) distribution. The authors conduct a systematic analysis of the mathematical properties of the DPsL distribution, including its probability-generating function, moments, skewness, kurtosis, and stress-strength reliability. The explicit forms of these properties make the distribution highly appealing. The practicality of the proposed distribution is demonstrated through its application to the first-order integer-valued autoregressive process, and its empirical relevance is validated through the analysis of three real-world datasets.

In [3], Tomy and Chesneau study a novel and flexible trigonometric extension of the modified Lindley distribution, known as the sine-modified Lindley distribution. This one-parameter survival distribution is created by incorporating features from the sine-generalized family of distributions, providing an attractive alternative to the Lindley and modified Lindley distributions, particularly for modeling lifetime phenomena with leptokurtic data. The results show that the sine-modified Lindley model outperforms important models such as the Lindley, modified Lindley, sine exponential, and sine Lindley models, based on various goodness-of-fit criteria.

In [4], Jamal et al. propose a new and improved functional form of the Burr III distribution, which enhances the classical distribution's flexibility and enables it to model the hazard rate functions of various shapes, including increasing, decreasing, bathtub, upside-down bathtub, and nearly constant. The article presents some of the distribution's fundamental properties, such as rth moments, sth incomplete moments, moment generating function, skewness, kurtosis, mode, ith order statistics, and stochastic ordering, in a clear and concise manner, and also demonstrates the effectiveness of the proposed model in three applications, consisting of both complete and censored samples.

In [5], Krishna et al. introduce the unit Teissier distribution, a bounded form of the Teissier distribution, and thoroughly examine its important properties. The analysis includes a shape analysis of the main functions, an analytical expression of moments based on the upper incomplete gamma function, incomplete moments, probability-weighted moments, and quantile function. The article also demonstrates the competency of the proposed model by analyzing two datasets from diverse fields.

In [6], Khan et al. utilize a new three-parameter continuous model, called the minimum Lindley Lomax distribution, by combining the Lindley and Lomax distributions. The article carefully examines various basic statistical aspects of the new distribution, including the quantile function, ordinary and incomplete moments, moment generating function, Lorenz and Bonferroni curves, order statistics, Rényi entropy, stress strength model, and stochastic sequencing. The article investigates the characterizations of the new model, estimates its parameters using maximum likelihood procedures and demonstrates the extensibility of the new model with two applications.

In [7], Nagarjuna et al. perform a hybrid distribution, the Nadarajah–Haghighi Lomax (NHLx) distribution, by combining the features of the Nadarajah–Haghighi and Lomax distributions. The NHLx distribution, with its four parameters, lower bounded support, and flexible distributional functions, including a unimodal probability density function and bathtub-shaped hazard rate function, provides an extension of the exponential Lomax distribution. Moreover, the distribution has the desirable statistical properties of moments and quantiles. The authors illustrate the statistical applicability of the NHLx distribution by conducting simulations and analyzing four real datasets.

In [8], Jasmine et al. conduct a study to manage hydrological resources such as reservoirs, rivers, and lakes. In recent years, data-driven techniques, such as the adaptive neuro-fuzzy inference system (ANFIS), have gained popularity in the hydrological field. This study explores the effective use of artificial intelligence for predicting evaporation in agricultural areas. Specifically, it examines ANFIS and hybridizes it with three optimizers: the genetic algorithm (GA), firefly algorithm (FFA), and particle swarm optimizer (PSO).

In [9], Dechpichai et al. aim to examine the spatial and temporal patterns of 124 meteorological stations in Thailand under ENSO. The research employs multivariate climate variables, including rainfall, relative humidity, temperature, max temperature, min temperature, solar downwelling and horizontal wind from the conformal cubic atmospheric model (CCAM), during the years of El Niño (1987, 2004, and 2015) and La Niña (1999, 2000, and 2011). This approach may be useful for planning and managing crop cultivation in different areas, using variables forecasted for the future and considering the effects of climate change.

In [10], Tomy and Chesneau examine the suitability of the sine-modified Lindley distribution, a relatively new statistical model, for analyzing biological data. The goodness-of-fit approach is employed to demonstrate the effectiveness of the model in estimating and modeling the lifespan of guinea pigs exposed to tubercle bacilli, the impact of growth hormone treatment on children, and the size of tumors in cancer patients. The researchers believe that this model has potential for the analysis of survival times related to cancer and other diseases. The paper also includes R codes for the figures and details on data processing.

In [11], Muchie et al. study small-area estimation methods to determine the prevalence of malnutrition among children under five in Ethiopia, specifically at the zonal level. To achieve this, the authors linked the most current survey data and census data available in Ethiopia. The findings revealed that stunting, wasting, and being underweight spatially varied across the zones, indicating that different regions face distinct challenges of varying degrees.

In [12], Pinault aims to advance the understanding of the mechanisms behind extreme climatic events, with the objective of improving forecasting methods and shedding light on the role of anthropogenic warming. The study utilizes wavelet analysis to identify the contribution of coherent Sea Surface Temperature (SST) anomalies produced from short-period oceanic Rossby waves to two case studies: a Marine Heatwave (MHW) in the northwestern Pacific with a significant impact in Japan, and a severe flood event in Germany. The study concludes by highlighting the need for further research to better understand how anthropogenic warming can modify key mechanisms in the evolution of dynamic systems that contribute to extreme events.

In [13], Essiomle and Adekambi focus their study on the Gerber–Shiu discounted penalty function with a constant interest rate, considering the delayed claim reporting times. The Poisson claim arrival scenario is used to derive the Laplace transform of the generalized Gerber-Shiu function, which leads to a second Volterra equation with a degenerated kernel. The study presents a closed-form expression for the Gerber–Shiu function in the case of an exponential claim distribution through sequence expansion. This expression enables the calculation of absolute and relative ruin probabilities.

In [14], Jamal et al. propose a generalized, odd, linear, exponential family of continuous distributions. The probability density and cumulative distribution function are represented as infinite linear combinations of the exponentiated-F distribution. Various statistical properties, including quantile function, moment-generating function, distribution of order statistics, moments, mean deviations, asymptotes, and the stress-strength model, are explored for the proposed family, and simulation studies are conducted for two sub-models to examine the asymptotic behavior of the maximum likelihood estimates.

In [15], Hussen and He discuss a generalized Prony method that can solve the sparse expansion problem for two generating functions, enabling the recovery of a wider range of function types using Prony-type methods. The two-generator sparse expansion problem has certain unique properties, such as the need to separate the two sets of frequencies from the zeros of the Prony polynomial. To address this issue, they propose a two-stage least-square detection method that effectively solves the problem.

In [16], Human et al. introduce a bivariate beta-type distribution that allows for users to detect a permanent upward or downward step shift in the process' variance without relying on parameter estimates. This approach offers an attractive and intuitive way to potentially identify the magnitude and time of shift occurrence. The paper derives certain statistical properties of this distribution, and simulations illustrate the theoretical results.

In [17], Shafiq et al. examine a novel discrete distribution, named the Binomial-Natural Discrete Lindley distribution, which is created by combining the binomial and natural discrete Lindley distributions. The distribution's properties, such as the moment-generating function, moments, hazard rate function, methods of moments, proportions, and maximum likelihood, are investigated.

Ultimately, this book is not intended to provide a comprehensive treatment of all the work in the field of Computational Mathematics and Applied Statistics development; rather, it is intended to help researchers focus on a few key strategies with the potential to have a high impact.

I would like to take this opportunity to offer my thanks for all the support offered by this journal, who invited me to guest-edit this Special Issue, and all the contributors and reviewers for their excellent work during the editing process.

Conflicts of Interest: The author declares no conflict of interest.

References

1. Tomy, L.; Chesneau, C.; Madhav, A.K. Statistical Techniques for Environmental Sciences: A Review. *Math. Comput. Appl.* **2021**, *26*, 74. [CrossRef]
2. Irshad, M.R.; Chesneau, C.; D'cruz, V.; Maya, R. Discrete Pseudo Lindley Distribution: Properties, Estimation and Application on INAR(1) Process. *Math. Comput. Appl.* **2021**, *26*, 76. [CrossRef]
3. Tomy, L.; Chesneau, C. The Sine Modified Lindley Distribution. *Math. Comput. Appl.* **2021**, *26*, 81. [CrossRef]
4. Jamal, F.; Abuzaid, A.H.; Tahir, M.H.; Nasir, M.A.; Khan, S.; Mashwani, W.K. New Modified Burr III Distribution, Properties and Applications. *Math. Comput. Appl.* **2021**, *26*, 82. [CrossRef]
5. Krishna, A.; Maya, R.; Chesneau, C.; Irshad, M.R. The Unit Teissier Distribution and Its Applications. *Math. Comput. Appl.* **2022**, *27*, 12. [CrossRef]
6. Khan, S.; Hamedani, G.G.; Reyad, H.M.; Jamal, F.; Shafiq, S.; Othman, S. The Minimum Lindley Lomax Distribution: Properties and Applications. *Math. Comput. Appl.* **2022**, *27*, 16. [CrossRef]
7. Nagarjuna, V.B.V.; Vardhan, R.V.; Chesneau, C. Nadarajah–Haghighi Lomax Distribution and Its Applications. *Math. Comput. Appl.* **2022**, *27*, 30. [CrossRef]
8. Jasmine, M.; Mohammadian, A.; Bonakdari, H. On the Prediction of Evaporation in Arid Climate Using Machine Learning Model. *Math. Comput. Appl.* **2022**, *27*, 32. [CrossRef]
9. Dechpichai, P.; Jinapang, N.; Yamphli, P.; Polamnuay, S.; Injan, S.; Humphries, U. Multivariable Panel Data Cluster Analysis of Meteorological Stations in Thailand for ENSO Phenomenon. *Math. Comput. Appl.* **2022**, *27*, 37. [CrossRef]
10. Tomy, L.; Chesneau, C. Applications of the Sine Modified Lindley Distribution to Biomedical Data. *Math. Comput. Appl.* **2022**, *27*, 43. [CrossRef]
11. Muchie, K.F.; Wanjoya, A.K.; Mwalili, S.M. Small Area Estimation of Zone-Level Malnutrition among Children under Five in Ethiopia. *Math. Comput. Appl.* **2022**, *27*, 44. [CrossRef]
12. Pinault, J.-L. Morlet Cross-Wavelet Analysis of Climatic State Variables Expressed as a Function of Latitude, Longitude, and Time: New Light on Extreme Events. *Math. Comput. Appl.* **2022**, *27*, 50. [CrossRef]
13. Essiomle, K.; Adekambi, F. A Note on Gerber–Shiu Function with Delayed Claim Reporting under Constant Force of Interest. *Math. Comput. Appl.* **2022**, *27*, 51. [CrossRef]
14. Jamal, F.; Handique, L.; Ahmed, A.H.N.; Khan, S.; Shafiq, S.; Marzouk, W. The Generalized Odd Linear Exponential Family of Distributions with Applications to Reliability Theory. *Math. Comput. Appl.* **2022**, *27*, 55. [CrossRef]
15. Hussen, A.; He, W. Prony Method for Two-Generator Sparse Expansion Problem. *Math. Comput. Appl.* **2022**, *27*, 60. [CrossRef]
16. Human, S.W.; Bekker, A.; Ferreira, J.T.; Mijburgh, P.A. A Bivariate Beta from Gamma Ratios for Determining a Potential Variance Change Point: Inspired from a Process Control Scenario. *Math. Comput. Appl.* **2022**, *27*, 61. [CrossRef]
17. Shafiq, S.; Khan, S.; Marzouk, W.; Gillariose, J.; Jamal, F. The Binomial–Natural Discrete Lindley Distribution: Properties and Application to Count Data. *Math. Comput. Appl.* **2022**, *27*, 62. [CrossRef]

Disclaimer/Publisher's Note: The statements, opinions and data contained in all publications are solely those of the individual author(s) and contributor(s) and not of MDPI and/or the editor(s). MDPI and/or the editor(s) disclaim responsibility for any injury to people or property resulting from any ideas, methods, instructions or products referred to in the content.

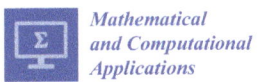

Review

Statistical Techniques for Environmental Sciences: A Review

Lishamol Tomy [1], Christophe Chesneau [2,*] and Amritha K. Madhav [3]

[1] Department of Statistics, Deva Matha College, Kuravilangad 686633, Kerala, India; lishatomy@gmail.com
[2] Laboratoire de Mathématiques Nicolas Oresme (LMNO), Université de Caen Normandie, Campus II, Science 3, 14032 Caen, France
[3] Department of Statistics, Nirmala College, Muvattupuzha 686661, Kerala, India; amrithakmadhav555@gmail.com
* Correspondence: christophe.chesneau@unicaen.fr

Abstract: This paper reviews the interdisciplinary collaboration between Environmental Sciences and Statistics. The usage of statistical methods as a problem-solving tool for handling environmental problems is the key element of this approach. This paper enhances a clear pavement for environmental scientists as well as quantitative researchers for their further collaborative learning with an analytical base.

Keywords: descriptive statistics; inferential statistics; species abundance data plots; abundance models; species richness indices; diversity measures; sampling; community comparisons; diversity in space (time); extreme value modeling; epidemiology; adaptive sampling; trend analysis; ecological modeling; detection limit

1. Introduction

In its simplest sense, the environment means the surrounding external conditions influencing the growth of people, animals or plants, living or working conditions, etc. Environmental Sciences (EVS) is an integrated multidisciplinary approach that studies the environment and solutions of environmental problems. In the present scenario, the environment has become a global agenda item, which has increased the scope and importance of EVS. In the development of different stages of civilization, humans were accompanied both by the environment and statistics. Since the early days, they were found to be knowingly accustomed to the environment and unknowingly played with statistics. Thus, both statistics and the environment have shared a long history of mutual reciprocation. In modern times, these two subjects are independently able to attract the academic attention of scholars throughout the world (see [1]).

The United Nations Statistics Division (UNSD) has an exclusive branch for environmental statistics, established in 1995. Its major area of work is data collection, methodology, capacity development, and coordination of environmental statistics and indicators. They have a dedicated newsletter called "ENVSTATS", which publishes the activities of UNSD in the area of environmental statistics. The Framework for the Development of Environmental Statistics (FDES 2013) is an updated version of the original FDES, which was published by UNSD in 1984. In India, the Ministry of Statistics and Programme Implementation has a specific publication report in the branch of environmental statistics called "EnviStats" which updates recent developments in the field of environmental statistics.

The extensive use of statistics in EVS led to the development of a new branch called Environmental Statistics. We all know that statistics are an inevitable context in any scientific arena. Even so, the motivation for conducting this specific review is that environmental statistics have an integrated multidisciplinary face, which will shed light on the pure biological field of modern science with its analytical nature. That undiscovered interconnection with statistics and environmental science will be revealed through this review, which will be an easy access point for future investigators. This review has been conducted in two parts, i.e., the pure statistical techniques and those specific techniques that have been exclusively invented for environmental science. A brief state of the art is presented below. The authors of [2] discussed different statistical techniques which are helpful to environmental engineers. It addresses different environmental problems with a solution-oriented approach that encourages students to view statistics as a problem-solving tool.

The use of statistical techniques to understand various environmental phenomena was explained in [3]. He examined different statistical tools, such as probabilistic and stochastic models, data collection, data analysis, inferential statistics, etc. In addition, he discussed principles and methods applicable to a wide range of environmental issues (including pollution, conservation, management, control, standards, sampling, monitoring, etc.) across all fields of interest and concern (including air and water quality, forestry, radiation, climate, food, noise, soil condition, fisheries and environmental standards). Accordingly, he considered sophisticated statistical techniques, such as extreme processes, stimulus response methodology, linear and generalized linear models, sampling principles and methods, time series, spatial models, multivariate techniques, design of experiments, etc.

This article is an attempt to describe some basic statistical concepts used in EVS, thereby establishing a link between the two subjects. It studies some basic statistical concepts relevant to environmental study. Illustrations are discussed on the basis of [4,5].

In this article, Section 2 presents the basic concepts in statistics, Section 3 describes the application of statistical tools in EVS, Section 4 is about the various illustrations regarding the topic, and Section 5 is the conclusion.

2. Basic Concepts

With the advent of the theory of probability and games of chance in the mid-seventeenth century, the concept of modern statistics was born. The name "statistics" appears to have come from the German word "Statistik," the Italian word "statista," or the Latin word "status," all of which mean "political state" or "state craft", respectively. The term statistics can be used in two different senses.

In the plural sense, it means "a collection of numerical facts". According to Horace Secrist, "Statistics may be defined as the aggregate of facts, affected to a marked extent by a multiplicity of causes, numerically expressed, enumerated or estimated according to a reasonable standard of accuracy, collected in a systematic manner, for a predetermined purpose and placed in relation to each other". This definition explains the characteristics of statistical data.

In its singular sense, it means "statistical methods for dealing with numerical data". According to Croxton and Cowden, "Statistics is the science of collection, presentation, analysis, and interpretation of numerical data". This definition points out different stages of statistical investigation. Hence, statistics is concerned with exploring, summarizing, and making inferences about the state of complex systems, for example, the state of a nation (social statistics), the state of people's health (medical and health statistics), the state of the environment (environmental statistics), as extensively described in [6].

In the midst of its wide range of applications and advantages, one important allegation about statistics is that the concerned parties may make misleading statements in their favor. However, the fact is that, as in the case of any science, only an expert can make use of statistical tools effectively. One should make sure that the statistical study is conducted by the right person. There are lots of good ways, many more bad and wrong ways too. So, be sure about the correctness of the tool used. The notorious allegation by Mark Twain citing the British Prime Minister Benjamin Disraeli that "there are three types of lies: lies, damned lies, and statistics" (but the phrase is nowhere in Disraeli's works, and the earliest known appearances were years after his death, so it is assumed to be by some anonymous writer in mid-1891) is just a lie, provided the precaution is served. In such a context, it is interesting that the author of [7] beautifully coined the title "Truth, Damn Truth and Statistics" for his article.

3. Application of Statistical Tools in EVS

In statistics, data analysis is divided into two sections: descriptive statistics and inferential statistics. The authors of [5] discussed the two in depth, and Sections 3.1 and 3.2 below present them in summary form.

3.1. Descriptive Statistics

Descriptive statistics are the initial stage of data analysis where exploration, visualization and summarization of data are done. We will look at the definitions of population and random sample in this section. Different types of data, viz. quantitative or qualitative, discrete or continuous, are helpful for studying the features of the data distribution, patterns, and associations. The frequency tables, bar charts, pie diagrams, histograms, etc., represent the data distribution, position, spread and shape efficiently. This descriptive statistical approach is useful for interpreting the information contained in the data and, hence, for drawing conclusions.

Further, different measures of central tendency viz. mean, median, etc., were calculated for analyzing environmental data. It is also useful to study dispersion measures, such as range, standard deviation, etc., to measure variability in small samples. One of the important measures of relative dispersion is the coefficient of variation, and it is useful for comparing the variability of data with different units. Skewness and kurtosis characterize the shape of the sample distribution. The concepts of association and correlation demonstrate the relationships between variables and are useful tools for a clear understanding of linear and non-linear relationships. Important measures of these fundamental characteristics are briefly discussed here in the following.

3.1.1. Central Tendency

The tendency of the observations to cluster around some central value is called central tendency. Any measure of central tendency is termed "average". The most commonly used averages are the following:

$$\text{Mean } \bar{x} = \sum_{i=1}^{n} \frac{x_i}{n},$$

where x_i denotes the ith observation and n is the number of observations.

Median is the middle-most observation when observations are arranged in ascending or descending order

and

Mode is the most frequently occurring observation.

3.1.2. Dispersion

The scattering of observations about the central value is called dispersion. Important measures of dispersion are range, quartile deviation, mean deviation, standard deviation and coefficient of variation. These four measures depend on the unit of measurement of the observations, hence, they are absolute measures. They can be defined as:

Range is the difference between largest and smallest observations.

Measure based on quartiles:
- Quartile deviation

$$QD = \frac{Q_3 - Q_2}{2}$$

where Q_3 and Q_2 are the third and first quartile in the frequency distribution, respectively;

- Mean deviation

$$MD = \sum_{i=1}^{n} \frac{|x_i - \bar{x}|}{n}$$

where $\bar{x}$ is mean of x_i (observed values);
- Standard deviation

$$SD = \sqrt{\sum_{i=1}^{n} \frac{(x_i - \bar{x})^2}{n}};$$

- Coefficient of variation

$$CV = \frac{SD}{Mean} \times 100.$$

Thus, CV is the relative measure (measure independent of unit) corresponding to SD.

3.1.3. Skewness

The lack of symmetry is termed as skewness or asymmetry. In a frequency curve, if both the sides of the mode are distributed in the same manner, the distribution is symmetric, otherwise it is skewed. When more area is on the right side of the mode, the distribution is positively skewed. If more area is on the left side of the mode, the distribution is negatively skewed. Figure 1 depicts the three situations. There are mainly two measures:

1. Pearson's measure

$$S = \frac{Mean - Mode}{SD}$$

If $S = 0$, the distribution is symmetric, if $S > 0$, positively skewed and if $S < 0$, negatively skewed.

2. Moment measure

$$\beta_1 = \frac{\mu_3}{\mu_2^{3/2}}$$

where

$$\mu_2 = SD^2$$

and

$$\mu_3 = \sum_{i=1}^{n} \frac{(x_i - \bar{x})^3}{n}.$$

If $\beta_1 = 0$, it is symmetric, if $\beta_1 > 0$, positively skewed and if $\beta_1 < 0$, negatively skewed.

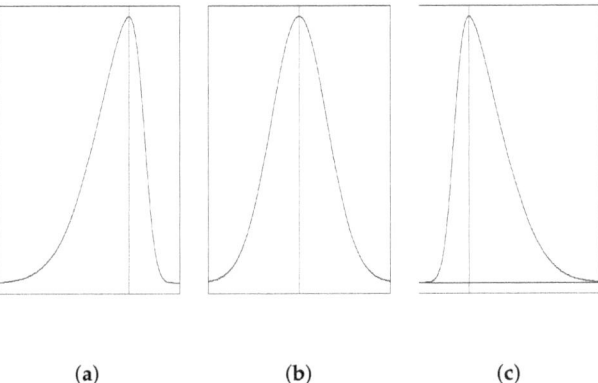

Figure 1. (a) Negative skewness, (b) symmetric, and (c) positive skewness.

3.1.4. Kurtosis

Kurtosis measures the degree of peakedness or flatness of a curve. The normal curve is called mesokurtic. If the curve is more peaked than normal, it is called leptokurtic. If it is flatter than normal, it is called platykurtic. Figure 2 illustrates the nature of different types of kurtosis.

The moment measure of kurtosis is

$$\beta_2 = \frac{\mu_4}{\mu_2^2}$$

where

$$\mu_4 = \sum_{i=1}^{n} \frac{(x_i - \bar{x})^4}{n}$$

If $\beta_2 = 3$, the distribution is mesokurtic, if $\beta_2 > 3$, the distribution is leptokurtic, and if $\beta_2 < 3$, the distribution is platykurtic.

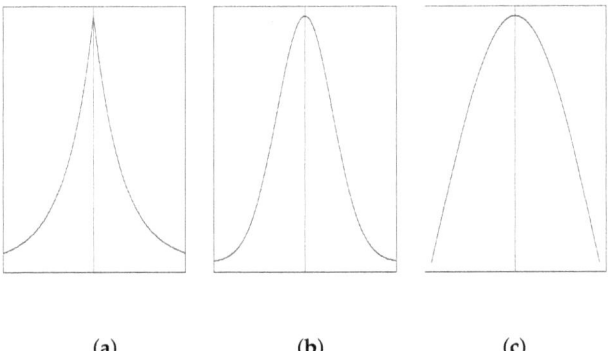

Figure 2. (a) Leptokurtic, (b) mesokurtic, and (c) platykurtic.

3.2. Inferential Statistics

In inferential statistics, the concept of probability is important for studying the uncertainties in the environment. For example, whether it will rain or not tomorrow can be best inferred by using probability. Several theoretical probability distributions, such as the Bernoulli distribution, the binomial distribution, the Poisson distribution, etc., are useful for modeling the probability distribution of real environmental data. For example, decisions, such as coin tossing, rain/no rain, yes/no, etc., are explained by Bernoulli variables since their outcomes are binary. In addition, if we are interested in counting the number of times floods occurred in the Dhemaji district of Assam, India, out of the total number of floods that occurred, because we are counting the number of times a flood (X), a Bernoulli event, occurs with a probability of p out of a total, i.e., out of n trials, the probability distribution of such variables is given by a binomial distribution. In addition, if we do not know the total number of flood occurrences but know the meaning of the flood occurrences, the distribution is modeled by the Poisson distribution. Statistical tools such as estimation, hypothesis testing, etc., play a vital role in analyzing environmental data. Some of the frequently used statistical tests in atmospheric and environmental science are the "Z-test," "T-test," "F-test," etc. Another statistical approach is time series analysis, which studies environmental quantities with respect to time. For example, the monthly/yearly mean temperature, rainfall, humidity, etc., are best studied by time series (see [8]).

4. Illustrations

In this section, we are discussing the available statistical techniques that are used in the field of environmental sciences along with some practical examples in the context of data based on environmental sciences. There are examples of how collaboration between environmental scientists and quantitative researchers has aided future learning in both fields, based primarily on two works: [4], which deals with statistical techniques, and [5], which deals with practical examples.

The statistical techniques available are given below, based on [4].

4.1. Methods of Plotting Species Abundance Data

4.1.1. Whittaker Plot

One of the best informative methods is the rank/abundance plot, or dominance diversity curve. Here, species are plotted from most to least abundant along the x axis and abundance in the y axis in $\log_{10}$ format (here, abundance of several orders of magnitude can be accommodated in the same graph). Proportional or percentage abundance are used in order to facilitate easy comparison.

The authors of [9] named this plot the Whittaker plot in remembrance of R. H. Whittaker for his famous contribution described in [10]. This plot has several advantages. Contrasting patterns of species richness are clearly displayed. If there are only a few of some species, all the information concerning their relative abundance is visible, as they are represented in their histogram format (see [11]). For following environmental impacts and succession, this plot is very effective. For that, we should plot a rank/abundance graph. The shape of the curve gives inference about which species abundance model best fits the data. The steep plot describes assemblages with high dominance, while the shallower plot symbolizes low dominance. High dominance plots are consistent with geometric or log series, while low dominance plots suit the log normal or broken stick model. However, the curves of different models are rarely fitted with empirical data (see [11]).

4.1.2. k-Dominance Plot

This kind of plot shows the relationship between percentage cumulative abundance (y axis) and species rank/log series rank (x axis). Here, the elevated curve represents the less diverse assemblages (see [12,13]).

4.1.3. Abundance/Biomass Comparison Curve or ABC Curve

A variant of the k-dominance plot was introduced by [14]. The related curve is constructed using two measures of abundance: the number of individuals and biomass. The level of disturbance, pollution-induced or otherwise, affecting the assemblage can be inferred from the resulting curve.

The method was developed for benthic macrofauna and has been used productively by a number of investigators in this context.

The ABC plot is used to study the entire species abundance distribution. The author of [15] has introduced a summary statistic specified as W (named after R. M. Warwick), and defined by

$$W = \sum_{i=1}^{S} \frac{B_i - A_i}{50(S-1)}$$

where B_i denotes the biomass value of each species rank (i) in the ABC curve, S represents the number of species, and A_i represents the abundance (individuals) value of each species rank (i). A_i and B_i do not necessarily refer to the same species, since species are ranked separately for each abundance measure. The result will be positive if the biomass curve is consistently above the individual curve. This symbolizes undisturbed abundance. In contrast, a grossly perturbed assemblage will give a negative value (consistently above the biomass curve). A curve that produces a value of W close to 0 and overlaps signifies moderate disturbance. W ranges usually from -1 to $+1$.

The W statistics are generally computed for each sample separately. ANOVA can be used to test for significant differences, if treatments have been replicated. Alternatively, graphing W values can be a very effective way of illustrating shifts in the composition of the assemblage if un-replicated samples have been taken along a transect or over a time series (such as before, during and after a pollution event). While considering ABC curves at discriminating samples, W statistics are most useful (see [16]).

4.2. Species Abundance Models

Statistical models were initially devised as the best empirical fits to the observed data (see [17]). They help the investigator to objectively compare different assemblages, which is one of its advantages. In some cases, a parameter of the distribution can be used as an index of diversity. Another set of models is biological or theoretical models.

4.2.1. Statistical Models

- log series model
  In this model, the number of species (y axis) is displayed in relation to the number of individuals per species (x axis), the abundance classes which are presented on log scale. This plot is typically used when the log normal distribution is chosen. This type of graph is sometimes dubbed the "Preston plot" (see [18]) in remembrance of Preston F., who pioneered the use of the log normal model in [19]. In the log series model, the mode will fall to the class with the lowest abundance, which represents a single individual, and in the case of this plot, it is more focused on rare species. In log transformation, the x axis has a tendency to shift a mode to the right so as to reveal a log normal pattern.
- Negative binomial model
  The author of [20] describes many applications of the negative binomial model in ecology. Particularly in estimating species richness (see [21]). However, the authors of [22] remarked that it is only rarely fitted to data of species abundance (one exception being [23]). Since it came from a stable log series model, it has some potential interest.
- Zipf-Mandelbrot model
  This model has its roots in linguistics and information theory. This model has several applications in environmental diversity, which are well described in [24–28]. The Zipf-Mandelbrot model is important for a rigorous sequence of colonists from the same species, always present at the same point in the succession in identical habitats. According to [29], this model is not better than the log series or log normal model. This model, however, has been successfully used in [28,30–32]. We also refer to [33–35] for the use of this model in terrestrial studies, and [36] for the use of this model in aquatic

systems. The author of [37] states that it can be used to test the performance of various diversity estimators. The Zipf-Mandelbrot model provided the best description of the cover data, while the biomass data are compatible with the log normal distribution.

4.2.2. Goodness of Fit Tests

A goodness of fit test, often called χ^2, is used to find the relationship between the observed and expected frequencies of a species in each abundance class [38]. To fit a deterministic model, the conventional method used is to assign the observed data to abundance classes. Classes based on $\log_2$ are usually used. According to the model used, the number of species expected in each abundance class is determined.

The model takes the S (number of species) as observed values and N (total abundance), and then determines how these N individuals should be distributed among the S species. If $p < 0.05$ (p-value), the model is rejected because it does not adequately describe the pattern of species abundances. If $p > 0.05$, the fit fails to be rejected or, ideally, $p >> 0.05$ is assumed to be a good fit. Tests of empirical data typically involve a very small number of abundance classes (10 or fewer). This causes a reduction in the degrees of freedom (d.f.) available. The more the degrees of freedom get the least value, the harder it becomes to reject a model.

The authors of [29] remarked that goodness of fit tests work most effectively with large assemblages (but might not be ecologically coherent units). Instead of χ^2 he recommends the Kolmogorov–Smirnov (K–S) goodness of fit (*GOF*) test, as said in [38,39]. Indeed, Tokeshi suggests adopting the K–S-*GOF* test, as the standard method of assessing the goodness of fit of deterministic models. He also suggests the K–S two-sample test can be used to compare two datasets directly to describe their abundance patterns.

The author of [11] reinforces that, if one model fits the data and another does not, it is not possible to conclude that the fit of the two is significantly different. His solution is to use replicated observations. The deviations can be log transformed, if necessary to achieve normality. A multiple comparison test, for example, Duncan's new multiple range test (see [38]) can then be used to infer which models are significantly different from one another.

4.2.3. Biological or Theoretical Models

- Deterministic and stochastic models

  Deterministic models assume that N individuals will be distributed amongst the S species in the assemblage. The geometric series is the only deterministic niche apportionment model. Stochastic models recognize that replicate communities structured according to the same set of rules will vary according to the relative abundances of species found there, and they try to capture the random elements inherent in natural processes. This makes biological sense. Perhaps not surprisingly, stochastic models are more challenging to fit than their deterministic counterparts. In a practical sense, it is necessary to know whether a model is deterministic or stochastic. Stochastic models have a complexity that requires replicated data, and this problem is solved in Tokeshi's refinements (see [40]).

- Geometric series

  Assume that the dominant species pre-empts a limiting resource percentage k, and the second most dominant species pre-empts the same k of the remaining part, and so on, until all S have been chosen. If the species abundance is proportional to the resource amount and the assumption stated above is fulfilled, the resulting pattern will follow a geometric series (or niche pre-emption hypothesis). Here, species abundance is ranked from most to least. Ratio of abundance of each species to abundance of predecessor is being a constant through the species and the ranked list is the reason. In addition, the series will appear as a straight line when plotted on log abundance/species rank graph. This plot helps identify whether the dataset is consistent or not with a geometric series. A full mathematical treatment of the geometric series can be found in [41], who also

presents the species abundance distribution corresponding to the rank/abundance series. In a geometric series, the abundances of species, ranked from the most to least abundant will be (see [41,42]):

$$n_i = NC_k k(1-k)^{i-1}$$

where n_i is the total number of individuals in the ith species, n is the total number of species, N is total number of individuals, k is the proportion of the remaining niche space occupied by each successively colonizing species (k is a constant), and $C_k = \left[1 - (1-k)^S\right]^{-1}$ is a constant that insures that $\sum_{i=1}^{n} n_i = N$. Because the ratio of the abundance of each species to the abundance of its predecessor is constant through the ranked list of species, the series will appear as a straight line when plotted on a log abundance/species rank graph.

- Broken stick model

  The broken stick model, alias the random niche boundary hypothesis, was proposed in [43]. This model plots relative species abundance in the y axis on a linear scale, and in the x axis, they plot the logged species sequence abundance, so as to represent it from most to least. Then, we will get a straight line. As [22] states, the model has a demerit in that it may be derived from more than one hypothesis. It provides evidence that some ecological factors are being shared more or less evenly between species (see [41]). It represents a group of S species with equal competitive ability vying for niche space, according to [29]. It is typically organized in the order of rank order abundance (see [11]). The authors of [44] prepared a program which estimates species abundance. This model is tricky enough to fit with empirical data (see [29]).

- Tokeshi's models

  Tokeshi has developed several niche apportionment models, including the dominance pre-emption, random fraction, power fraction, MacArthur fraction, and dominance decay models in [45,46]. They work with the assumption that abundance is proportional to the fraction of niche space occupied by a species. The model here assumes that the target niche selected is divided at random. The only difference between the models is how the target niche is selected. The larger the niche is, the more even the resulting species abundance distribution will be. Evenness ranges from least to most from the dominance pre-emption model, following the order of explanation. The random assortment model represents a random collection of niches of arbitrary sizes (see [45]).

  - Random fraction

    In this model, available niche space is divided at random into two pieces. Among these two, one is selected randomly for further subdivision, and so on, till all species are accommodated (see [40]). The sequential breakage model depicts a situation in which a new colonist competes for the niche of a species that is already in the community and takes over a random proportion of the previously existing niche. This model can be used to cover speciation events (see [40]). In addition, this is conceptually simple and found to be fit for a small community of freshwater *chironomids* (see [45,47]). The authors of [44] have created a Microsoft Excel program which can model the species abundance and distribution associated with it.

  - Power fraction model

    The Tokeshi model is applicable to species-rich assemblages, which is an exception to others (see [46]). In this case, the niche space is subdivided in the same way that a random fraction is. However, the probability of a niche splitting increases in this model, albeit only slightly in relation to size (x) via the power function (K). When K approaches 1, the largest niche is selected for fragmentation. When $K = 1$, the power fraction model resembles the MacArthur fraction model. Instead, when $K = 0$, niche fragmentation is done by random choice

and becomes a random fraction model. Usually K is set to 0.5 for the power fraction model (see [46]). Tokeshi accounts for virtually all assemblages. The author of [48] states that larger niches have a high fragmented probability or could occur either ecologically or evolutionarily.

- Dominance pre-emption model
  This model assumes that each species pre-empts more than half the niche space remaining. Because of this, it is dominant among combined species (see [45]). The proportion of available niche space is assigned between 0.5 and 1. When the number of replications increases (or $K = 0.75$, the same as the power fraction model), it becomes more similar to the geometric series (see [45]). It can also be applied to niche fragmentation (see [29,40]).
- MacArthur fraction model
  In the case of predicted species abundance distribution, the MacArthur fraction and the broken stick models paved the way to the same result. In this model, the probability of niche fragmentation is inversely proportional to size. This creates a very uniform distribution of species abundances and is only plausible in small communities of taxonomically related species. However, Tokeshi also reminds us that unreplicated data are not good for either the broken stick or the MacArthur fraction models.
- Dominance decay model
  Here, a more uniform pattern of species abundance is considered. At random, the niche space for fragmentation is selected at random. No empirical data indicate that communities as predicted by Tokeshi's dominance decay model can be found in nature till date. This can be due to insufficient investigations or due to the lower chance of finding an even distribution in nature.

4.2.4. Fitting Niche Apportionment Models to Empirical Data

The author of [45] found a new way of testing stochastic models. Species (S) are listed in decreasing order of abundance. The equation given below is used to fit a niche apportionment model if the mean observed abundance falls within the confidence limits of expected abundance.

$$R(x_i) = \mu_i \pm r\sigma_i \sqrt{n}$$

where

$x_{i=1}$ = mean abundance of most abundant;
$x_{i=2}$ = mean abundance of next most abundant;
.
.
.
$x_{i=S}$ = mean abundance of least abundant;
μ_i = mean of abundance ranked from $i = 1$ to S;
σ_i = standard deviation of abundance;
n = number of replicated samples;
r = breadth of confidence limit.

The mean abundance constitutes the observed distribution. For an assemblage of the same number of species (S), the expected abundance is estimated. For this model, we have to choose a large N, μ_i, σ_i and n. In addition, confidence limits are assigned to each rank of expected abundance by considering n rather than N (the number of times the model was simulated).

4.3. Species Richness Indices

There are two well-known species richness indices, which are easy to calculate too, which were introduced by [49,50], respectively.

- Margalef's diversity index (D_{Mg})

$$D_{Mg} = \frac{S-1}{\log N}$$

- Menhinick's index (D_{Mn})

$$D_{Mn} = \frac{S}{\sqrt{N}}$$

where S is the number of species recorded, N the total number of individuals in the sample.

Despite the attempt to correct for sample size, both measures remain strongly influenced by sampling effort. Nonetheless, they are intuitively meaningful indices that can be useful in biological diversity research.

Estimating Species Richness

There are two approaches to estimating species richness from samples, as cited in [51,52]. The first is the extrapolation of species accumulation or species–area curves. The second approach is to use a non-parametric estimator.

- Species accumulation curves
  Species accumulation curves, also known as collector curves, plot S, the total number of species, as a function of sampling effort (n) (see [51]). These curves are widely used in botanical research (see [53,54]). This is only a type of species accumulation curve. Curves that are S versus A for different areas (such as islands) and those used in increasingly larger parcels of the same region are the most common.
  The overall shape of species accumulation curves is determined by the order of samples (or individuals). By randomizing, the curve can be made smoother. It also helps deduce the mean and standard deviation of species richness. According to [55], these curves resemble rarefaction curves (see [56]). They usually move from left to right, as new species are added. However, rarefaction curves conventionally move from right to left. Many scientists have plotted species accumulation curves using linear scales on both axes. However, it is better to use a log-transformed x axis since semi-log plots make it easier to identify asymptotic curves from logarithmic curves (see [57]). To find an estimate of total species richness, the authors of [58] extrapolate the graph.
  Functions used in this kind of extrapolation can be classified into asymptotic or non-asymptotic. Both of their roles are to help the user predict an increase in species richness with additional sampling effort rather than to estimate total species richness.
  – Asymptotic curves
    They can be generated using two methods. The first is by using a negative exponential model (see [58]). The second is using the Michaelis–Menten equation (see [59]). The usual form of the equation is

    $$S(n) = \frac{nS_{\max}}{B+n}$$

    where $S(n)$ is the number of species observed in n samples, $S_{\max}$ is the total number of species in the assemblage, and B is the sampling effort required to detect 50% of $S_{\max}$ and n is the sample count.
  – Non-asymptotic curves
    These curves are used to estimate species richness. The authors of [60] proposed that the relationship between area and species be best described by a log linear model, extrapolated to a larger area. The authors of [61] imposed an asymptote on the log-log species area curve to avoid extremely high estimates of species richness.

* Parametric methods: log series and log normal distributions are the most potent two abundance models in this context (see [51]). Of these, the easiest fit is log series distribution, and it is also simple to apply. In addition, the log series model helps obtain a good estimate of total species richness if the number of individuals in the target area can be estimated. In this case, S will be underestimated where it should not be. Furthermore, this method is also used during rarefaction. Most people adopt the pragmatic approach when fitting continuous log normal distribution, which is inappropriate when observed data are in discrete form (see [22,51]). According to [21], this method has the unique property of generating a mode in the second or third class, giving the appearance of a log normal distribution even if it is not a log normal distribution. There is, however, no method for generating a confidence interval for any estimate of species richness found in a continuous log normal distribution (see [21,22,51,62]). An alternative to this is Poisson log normal (see [51]), which is rarely used as it is hard to fit. However, it produces higher estimates of species richness than any other method.
* Non-parametric methods:
  - Chao1

    It represents a simple estimator of the absolute number of species in an assemblage, which was introduced by Anne Chao (see [63]). The measure is named by [51] as Chao1 and it is based on the number of rare species in a sample. The following notation was provided by [52]:

    $$S_{Chao1} = S_{obs} + \frac{F_1^2}{2F_2}$$

    where S_{obs} denotes the number of species in the sample, F_1 is the number of observed species represented by a single individual (singletons) and F_2 denotes the number of observed species represented by two individuals (doubletons).

    The requirement for abundance data is an obvious disadvantage of Chao1. The abundance data should at least show whether they are singleton or doubleton. However, rather than presence/absence, they are often called incidence or occurrence data. The calculation of the variance of Chao1 is possible (see [64,65]).
  - Chao2

    Anne Chao was well aware that the number of species found in one sample is the only essential factor for calculation. For this, a new estimator, Chao2 was invented. It is as follows (see [51]):

    $$S_{Chao2} = S_{obs} + \frac{Q_1^2}{2Q_2}$$

    where Q_1 is the number of species that occur in one sample only (unique species) and Q_2 is the number of species that occurs in two samples.
  - Other estimators

    The author of [51] also invented another category of estimator called coverage estimators (see [66]). Coverage estimators are based on the assumption that widespread or abundant species can be included in any sample (see [67]). The abundance-based coverage estimator, alias ACE, is another estimator based on empirical data (see [68]). The partner incidence-based coverage estimator, ICE, focuses its eye on species found in <10 sampling units. Here, to estimate the true number of species, two estimators in this category are Jackknife and bootstrap estimators, which are described in the following sections. The estimators

are evaluated using some criteria, such as sample size, patchiness and overall abundance.

4.4. Diversity Measures

Species richness measures and estimators all fall into two categories: either parametric diversity indices or non-parametric diversity indices.

4.4.1. Parametric Measures of Diversity

- log series α

  The parameters of the log series model are x and α, where α is a diversity index. In addition, α is calculated during the fitting of a distribution. When S and N are known, the value of α can be easily calculated using the Williams monograph (see [69]) or appendix 4 of [70]. Here, x is estimated by iterating the following form:

$$\frac{S}{N} = -\log(1-x)\frac{1-x}{x}$$

  According to [70], until $x \geqslant 0.5$ and as $S > \alpha$, the log series distribution is not the best descriptor of species abundance pattern. In fact, for natural assemblages, usually $x > 0.9$ or close to 1 and $S > \alpha$. This implies that α is approximately the same as the number of species represented by a single individual.

- log normal λ

  The standard deviation (σ) of the log normal distribution would be a good measure of diversity. Although we can use it as an evenness measure and as an index for discriminating amongst samples, σ is not a good choice. It is also impossible to estimate for small sample sizes (see [71]). Then, S^* (S^* is the estimator of S, the number of species) is a good predictor of total species richness. However, the ratio of these two unsuitable parameters (S^*/σ) turns out to be an effective diversity measure (λ). It is effective in discriminating against assemblages (see [72]). Its ranking of sites suits well with α.

- The Q statistic

  The authors of [73,74] proposed the Q statistic, which is based on the distribution of species abundance data. For this measure, the user does not require a model to fit the empirical data. Hence, for empirical data, a cumulative species abundance curve is drawn and its inter-quartile slope is used to measure diversity. The author of [75] suggests that by restricting the measure to the inter-quartile region, the complete cumulative species abundance curve can be used to explain diversity as well as to remove the bias caused by the extremities (very rare and very abundant species). This is analogous to α and hence can be expressed in terms of a log series model, described by [76]. The following equation is estimated from empirical data:

$$Q = \frac{(1/2)n_{R_1} + \sum_{R_1+1}^{R_2-1} n_r + (1/2)n_{R_2}}{\log(R_2/R_1)}$$

  where n_r is the total number of species with abundance R, R_1 and R_2 are the 25% and 75% quartiles, n_{R_1} is the number of species in the class where R_1 falls, and n_{R_2} is the number of species in the class where R_2 falls.
  The quartiles are chosen so that:

$$\sum_{1}^{R_1-1} n_r < \frac{1}{4}S \leq \sum_{1}^{R_1} n_r$$

and

$$\sum_{1}^{R_2-1} n_r < \frac{3}{4} S \leq \sum_{1}^{R_2} n_r$$

where S is the total number of species in the sample, although the placement of R_1 and R_2 is not critical as the inter-quartile region of a cumulative species abundance curve, or indeed a rank/abundance plot, tends to be linear.

Because $Q = 0.371$ for the log normal model, it is not formally a parametric index. Thus, its performance is somewhat similar to that of parametric ones. However, for species which are censused >50%, Q may be biased (see [74]). The author of [77] has found an evenness measure which is similar to Q statistic, i.e., E_Q which will be discussed later.

4.4.2. Non-Parametric Measures of Diversity

Most diversity measures are not explicitly associated with named species abundance models, even though their performance is often governed by the underlying distribution of species abundances. They are non-parametric measures of diversity.

- Shannon Index (H')

  It was independently derived by Claude Shannon and Warren Weaver and is generally known as the Shannon index or Shannon information index. However, it is sometimes mistakenly referred to as the Shannon–Weaver index (see [9]). It is represented as

  $$H' = -\sum_{i=1}^{n} p_i \log p_i$$

  Usually, in samples, p_i will be unknown but it is estimated using the maximum likelihood estimator, n_i/N (see [78]), where n_i is the total number of individuals in the ith species and N is the total number of individuals. The ecological validity and computational easiness led Shannon to represent the index as logarithm of p_i. Historically, $\log_2$ is used for calculating the Shannon index, but this is without any biological reason. An increased trend in logarithm standardization is found in [79]. However, Shannon index does not have an unbiased estimate (see [80]).

  – A model using Shannon index: Caswell's neutral model

  Caswell's neutral model is very famous for its innovative approach to community structure analysis (see [81]). The model focuses on species abundance patterns when biological interactions are removed. It is represented by the deviation statistic defined by

  $$V = \frac{H' - E[H']}{SD(H')}$$

  where H' is the Shannon diversity index. It can be used to compare observed diversity (H') with the predicted neutral diversity $E[H']$. For values of $V > 2$ or $V < -2$, it depicts the departure from neutrality [82]. The author of [83] presented a computer program in PRIMER to calculate V which is termed a measure of environmental stress (see [84,85]) but is very rarely used. As richness and evenness are in complex relationships, V is probably useful only as a measure of disturbance. For large values of S and N, the expected values of H' are generated by a neutral model that closely resembles the predicted values in the log series model (see [70]), where S is the total number of species in the sample and N is the total number of individuals.

- **The Shannon evenness measure (J')**
  Assume a situation where all species have equal abundance. Then, the ratio of observed diversity will generate a new measure J' (see [22,78]). It is defined as

  $$J' = \frac{H'}{H_{\max}} = \frac{H'}{\log S}$$

  where S is the number of species and H' is the Shannon diversity index.
  To find $H_{\min}$, the author of [86] gives a simple method that can be utilized in other forms of the Shannon evenness (see [87]).

- **Heip's index of evenness (E_{Heip})**
  In [88], Heip notes that the evenness measure should not be based on species richness. So, according to this idea, he proposed the following new measure:

  $$E_{Heip} = \frac{e^{H'} - 1}{S - 1}$$

  Compared to J', E_{Heip} is least affected by species richness, it does not require sample size to be independent if there are only 10 species in 1 sample (see [89]). E_{Heip}'s minimum value is 0 and it usually goes to 0.006 when an extremely uneven community is considered.

- **SHE analysis**
  One of the main characteristics of the Shannon index is that it depends extremely on species richness and evenness. In [70,90], they identified that this property of the Shannon index can be utilized in another way. Consider a measure of evenness $E = e^{H'}/S$ (see [88]), such that

  $$H' = \log S + \log E$$

  This decomposition aids the user in interpreting changes in diversity.
  A decrease in diversity tends to cause pollution incidents due to loss of richness, evenness, or a combination of them.
  The essence of SHE analysis is the triangular relationship between S (species richness), H (diversity as measured by the Shannon index) and E (evenness). SHE analysis used by [91] in examining geographic patterns of body mass diversity in Mexican mammals found that evenness was high at intermediate spatial scales but low at the regional one.

- **The Brillouin index (HB)**
  The Brillouin index, abbreviated HB, is appropriate when sample randomness is not guaranteed, a community is completely censused, or every individual is accounted for (see [22,78]). It is given as

  $$HB = \frac{\log N! - \sum_{i=1}^{S} \log n_i!}{N}$$

  where N is the total number of individuals in the sample, n_i is the number of individuals from the ith species and S is the number of species. The HB value is rarely greater than 4.5. When compared to the Shannon Index, HB always yields a lower value, but they both provide similar or correlated estimates of diversity. The reason is that the Brillouin index describes a completely known collection without any uncertainty. Evenness (E) for the Brillouin diversity index is obtained from

  $$E = \frac{HB}{HB_{\max}}$$

where HB_{max} is calculated as

$$HB_{max} = \frac{1}{N} \log\left(\frac{N!}{N_S!^{S-r}(N_S+1)!^r}\right)$$

where N_S is the integer part of N/S and $r = N - S(N_S)$.
The index is unavailable with variance, and hence, no statistical test is needed to test significance. HB is mathematically speaking superior to the other two indices presented by [92]. However, some scientists state that it is more time-consuming and less familiar. Its over-dependence on sample size leads to unexpected results. This is unsuitable when abundance is measured as biomass or productivity (see [9,93]).

- Dominance and evenness measures
  A group of diversity indices is weighted by abundances of the commonest species and is usually referred to as either dominance or evenness measures.

  - Simpson's index (D)
    It is occasionally called the Yule index in remembrance of G. U. Yule (see [20]). The probability of any two individuals drawn at random from an infinitely large community being of the same species is given by [94] as

    $$D = \sum_{i=1}^{n} p_i^2$$

    where p_i denotes the proportion of individuals in the ith species, and n number of species. The form of the index appropriate for a finite community is:

    $$D = \sum_{i=1}^{n} \frac{n_i(n_i - 1)}{N(N-1)}$$

    where n_i is the number of individuals in the ith species and N is the total number of individuals in the sample.
    Simpson's index is expressed as $1 - D$ or $1/D$ because diversity decreases as D increases, and thus, it captures the variance of species abundance distribution. Simpsons' index, on the other hand, is less sensitive to species richness and more oriented toward species abundance. Confidence limits are applied using jackknifing. Simpson's index is the most meaningful and robust of all the measures. The reciprocal nature of the Simpson index was questioned by [95] and he recommends using $\log(D)$ instead of $(1 - D)$ or $(1/D)$, because this notation ensures severe variance problems. He also advises Kemp's transformation.
  - Simpson's measure of evenness ($E_{1/D}$)
    The Simpson measure of evenness, denoted by $E_{1/D}$ and stated in [9,89], is defined by

    $$E_{1/D} = \frac{1/D}{S}$$

    Here, $E_{1/D}$ usually ranges between 0 and 1 and is not so related to species richness. Because Simpson's index is a product of Simpson's evenness measure and S, multiplying S turns any good evenness index into a heterogeneity measure (see [96]).
  - McIntosh's measure of diversity (U)
    McIntosh postulated in 1967 that a community may be thought of as a point in a S-dimensional hyper volume, with the Euclidean distance between the assemblage and its origin serving as a measure of diversity (see [97]). The distance is known as U and is calculated as

$$U = \sqrt{\sum_{i=1}^{n} n_i^2}$$

where n_i is the number of individuals in the ith species and n number of species. The McIntosh U index is formally not a dominance index. However, a measure of diversity (D) or dominance that is independent of N can also be calculated as

$$D = \frac{N - U}{N - \sqrt{N}}$$

A further evenness measure can be obtained from the following formula (see [22]):

$$E = \frac{N - U}{N - N/\sqrt{S}}$$

- The Berger-Parker index (d)

  The Berger-Parker index, denoted by d, is an easy-to-calculate dominance measure (see [41,98]). The proportional abundance of the most abundant species is expressed by this index:

$$d = \frac{N_{max}}{N}$$

  where N_{max} is the number of individuals in the most abundant species. In this case, d denotes the relative importance of the most dominant species in the assemblage; both are considered equivalent. The reciprocal form of the Berger-Parker index is accepted because an increase in the value of the index accompanies an increase in diversity and a reduction in dominance, making it similar to Simpson's index. It is one of the most satisfactory diversity measures available because of its simplicity and biological significance (see [41]). In small assemblages, d is independent of S, and its value decreases with increasing species richness.

- Nee, Harvey and Cotgreave's evenness measure (E_{NHC})

  As an evenness measure, the authors of [77] proposed the slope b of a rank/abundance plot (with abundances log transformed). The resulting measure is

$$E_{NHC} = b$$

  E_{NHC} ranges from $-\infty$ and 0, where 0 is perfect evenness. This measure is difficult to interpret due to its range of values. It is more properly a measure of diversity than of evenness, and this is one of its demerits (see [73]). The authors of [89] therefore proposed a new form of the measure, which is

$$E_Q = -\frac{2}{\pi \arctan(b')}$$

  In this measure, the ranks are scaled before the regression is fitted, and b' denotes the corresponding slope. Thus, this is accomplished by dividing all ranks by the highest rank, such that the most abundant species receives a rank of 1.0 and the least abundant receives a rank of $1/S$. The transformation $(-2/[\pi \arctan(b')])$ places the measure in the 0 (no evenness) to 1 (perfect evenness) range.

- Camargo's evenness index (E_c)
  The author of [99] also introduced the following evenness measure:

  $$E_c = 1 - \sum_{i=1}^{S} \sum_{j=i+1}^{S} \frac{p_i - p_j}{S}$$

  where E_c is Camargo's index of evenness, p_i the proportion of species i in the sample, p_j the proportion of species j in the sample, and S the sample size.
  Although, the index is simple to calculate and relatively unaffected by rare species (see [100]). The authors of [37] found it to be biased, especially in comparison with the Simpson index.

- Smith and Wilson's evenness index (E_{var})
  The authors of [89] proposed a new index to provide an intuitive measure of evenness. This index takes the variation in species abundances and divides it by log abundance to produce proportional differences. This makes the index independent of measurement units. Smith and Wilson called their measure E_{var}. It is defined by

  $$E_{var} = 1 - \frac{2}{\pi \arctan\left\{\sum_{i=1}^{S}\left(\log n_i - \sum_{j=1}^{S} \frac{\log n_j}{S}\right)^2 / S\right\}}$$

  where n_i is the number of individuals in species i, n_j is the number of individuals in species j and S represents the total number of species. The conversion by $1 - 2/(\pi \arctan(x))$ ensures that the resulting measure falls between 0 (minimum evenness) and 1 (maximum evenness).

4.4.3. Taxonomic Diversity

If two assemblages have the same number of species and similar patterns of species abundance, but differ in the diversity of taxa to which the species belong, it seems intuitively reasonable that the assemblage with the most taxonomically diverse taxa is the more diversified assemblage. A taxonomic distinctness measure is one of the most recent developments in taxonomic diversity (see [101,102]).

- Clarke and Warwick's taxonomic distinctness index
  This measure gives the average taxonomic distance, or simply the path length between two randomly chosen organisms through phylogeny. Two forms can be taken by species in an assemblage. The first is taxonomic diversity (Δ), which considers taxonomic relatedness or species abundance. The two organisms may belong to the same species. The second form is taxonomic distinctness (Δ^*), a pure measure of taxonomic relatedness, which is equivalent to dividing Δ by the value it would take if all species belonged to the same genus, that is, in the absence of a taxonomic hierarchy. When presence/absence data are used, both measures reduce to the same statistic, Δ^+, which is the average taxonomic distance between two randomly selected species. It is calculated as follows:

  $$\Delta^+ = \frac{\sum_{i=1}^{S} \sum_{j=i+1, i<j}^{S} \omega_{ij}}{S(S-1)/2}$$

  where S denotes the number of species in the study and ω_{ij} is the taxonomic path length between species i and j.
  The taxonomic distinctness index is distinguished by its lack of reliance on sampling effort (see [103]).
  Using Δ^+, a significance test can be carried out. Here, the null hypothesis considered is "taxonomic distinctness of a locality is not significantly different from the global list". On the other hand, the author of [104] used multivariate methods during detection

of small variations in community structure and diversity. Multivariate analysis also helps find increased variability between samples (see [105]).

4.5. Sampling: An Essential Attribute

There are essentially two choices regarding sample size. The investigator may either adjust the sample size to cope-up with the situation or adopt a standard sample size. The second approach, which is also recommended by [70], is the best. If two samples with different sample sizes are drawn from the same assemblage, then this may lead to different conclusions about its diversity (see [22]). If samples are replicated several times, the curve obtained by plotting the measure of diversity (or evenness) against cumulative sample size may lead to a smooth curve.

- Replications
  The number of replications required is always an unanswerable question. Ideally, the available sample size and number of replications required to complete this are selected on the basis of the most diverse assemblage. In addition, it will be the same throughout the study. When sample size is not consistent, this becomes more true.
  One should be well aware of the difference between replication and pseudoreplication (see [106]). For more ideas in this context, users can refer to [107]. The primary condition is that all replicates must be independent (spatially).

4.6. Comparison of Communities

The manner in which the statistical comparison of communities or other ecological entities is achieved depends to some extent, though with significant overlaps, on the aspect of biodiversity that has been measured.

- Rarefaction—Sample data to common abundance level
  Rarefaction is a technique that reduces sample data to a common abundance level, which helps direct mapping between species richness in communities. During rarefaction, to estimate the richness of a small sample, complete information regarding all the collected species is required. Rarefaction curves converge when sample sizes are small (see [55,108]). Sampling should be enough to characterize the community, but there is a chance that estimates will be biased if the sample is insufficient.
  The author of [109] states that software can be used to create rarefaction curves. In [65], sample-based rarefaction curves were calculated using the EstimateS software. Confidence intervals can be incorporated into these curves. Rarefaction by the log series model is computationally simple. Indeed, it may even be used in circumstances where species abundances do not follow a log series distribution. However, if the sampling was inadequate in the first place, no method of rarefaction is going to compensate.
- Statistical tests
  Standard statistical techniques such as T-tests and ANOVA can be used to compare assemblages (see [38]). Alternatively, jackknifing or bootstrapping can be used to attach confidence intervals to a diversity statistic.
  – Jackknifing: a measure of diversity
    Jackknifing (see [110]) is a strategy for improving the estimate of almost any statistic. It can also be used to calculate the number of species present. It was first proposed by Quenouille in 1956, with Tukey making changes in 1958. The author of [111] was the first to apply the approach to diversity statistics. This application was further investigated by [112,113].
    Jackknifing does not require assumptions about the underlying distribution. Instead, it uses a set of "pseudo-values" which are artificially produced. These pseudo-values are (usually) normally distributed, their mean forms the best estimate of the statistic. Approximate confidence limits can also be attached to the estimate. The procedure is simple. The first step is to estimate the diversity of all n samples together. This produces St, the original diversity estimate. Next, the

diversity measure is recalculated n times, missing out each sample in turn. Each recalculation produces a new estimate, St_{-i}. The pseudo-value (ϕ_i) can then be calculated for each of the n samples as

$$\phi_i = nSt - (n-1)St_{-i}$$

The jackknifed estimate of the diversity statistic is simply the mean of these pseudo-values:

$$\bar{\phi} = \sum_{i=1}^{n} \frac{\phi_i}{n}$$

The approximate standard error of the jackknifed estimate is

$$SE_{\bar{\phi}} = \sqrt{\sum_{i=1}^{n} \frac{(\phi_i - \bar{\phi})^2}{n(n-1)}}$$

This standard error may be used to assign approximate confidence limits to the jackknifed diversity estimate. Confidence limits are set in the usual way, i.e.,

$$\bar{\phi} \pm t_{0.05(n-1)} SE_{\bar{\phi}}$$

Prior to jackknifing, the author of [38] recommended that statistics with a restricted range (such as those constrained between 0 and 1) should be modified. Following that, same methods were used to estimate species richness, with considerable success. They are called Jackknife 1, a first-order jackknife estimator that employs the number of species that occur only in a single sample (see [114,115]), and Jackknife 2, a second-order estimator which, like the *Chao*2 equation, takes both the number of species found in one sample only (Q_1) and in precisely two samples (Q_2) into account (see [116]). Both require incidence data.
In the following equations, m denotes the number of samples:

$$S_{Jack1} = S_{obs} + Q_1\left(\frac{m-1}{m}\right)$$

$$S_{Jack2} = S_{obs} + \left(\frac{Q_1(2m-3)}{m} - \frac{Q_2(m-2)^2}{m(m-1)}\right)$$

The variances of both estimators can be calculated.
- Bootstrapping
  A related method for producing standard errors and confidence bounds is bootstrapping. It is more computationally intensive than the jackknife, although it is regarded as an improvement. In essence, the original dataset is sampled numerous times to obtain a large number of different observations. These are then used to deduce the standard error. The authors of [20,38] provide more details. Bootstrapping, like jackknifing, can be used in species richness estimation.

- Null models
  In the last decade, there has been a rising use of null models in diversity measurement. Ecologists are becoming aware of the importance of developing testable null hypotheses (see [117]). The observed patterns are not due to the presumed causal explanation, according to the null hypothesis. It is based on the assumption that nothing significant has occurred (see [118]). Null models can also be used to determine whether perceived differences in diversity are simply an artifact of sampling. As [55] emphasizes, a null model does not presume that there is no structure in a community or that all processes are random. Instead, randomness is assumed only in respect of the mechanism being

tested. Null models are already used extensively to evaluate species co-occurrence patterns (see [119]).

4.7. Diversity in Space (and Time)

Till now, we have focused on the diversity of a defined assemblage or habitat, or α diversity. The author of [120] makes the distinction between α and β diversity where diversity increases as the similarity in species composition decreases. β diversity reflects biotic change or species replacement, whereas α diversity is a property of a specific spatial unit. The diversity of two or more spatial units differs. We can use β diversity. The relationship between α and β diversity is scale-dependent. The observation made by [80] is that

$$D_\gamma = \bar{D}_\alpha + D_\beta$$

When species richness is used to measure α and γ diversity, β diversity may be estimated as follows:

$$D_\beta = S_T - \bar{S}_j = \sum_{j=1}^{n} q_j (S_T - S_j)$$

where S_T is the species richness of the landscape (γ diversity), S_j denotes the richness of assemblage j and q_j is the proportional weight of assemblage j based on its sample size(n) or importance.

This approach is also used in the Shannon and Simpson diversity measurements. Low α and high β diversity will come from many small sampling units, but the opposite will be true if there are fewer but larger samples. If all other factors are equal, both sampling procedures yield the same conclusions concerning γ diversity.

- Indices of β diversity

  The majority of these indices use presence/absence data and, as such, focus on the species richness element of diversity.

  1. Whittaker's measure (β_W)

     One of the simplest, and most effective, measures of β diversity was devised by [120]:

     $$\beta_W = \frac{S}{\bar{\alpha}}$$

     where S is the total number of species recorded in the system (i.e., γ diversity) and α is the average sample diversity, where each sample is a standard size and diversity is measured as species richness. This is equivalent to:

     $$D_\beta = \frac{S_T}{\bar{S}_j}$$

     in Lande's notation.

     When Whittaker's measure is used to compute β_W, values of the measure will range from 1 (complete similarity) to 2 (no overlap in species composition). The author of [121] introduced a modification of Whittaker's measure. This allows the user to compare two transects (or samples) of different size. The related formula is

     $$\beta_{H1} = \frac{S/\alpha - 1}{N - 1} \times 100$$

     where S denotes the total number of species recorded, α means α diversity and N is the number of sites (or grid squares) along a transect. The measure ranges

from 0 (no turnover) to 100 (every sample has a unique set of species) and can be used to examine pairwise differentiation between sites. The author of [121] suggested a second modification which is insensitive to species richness trends. It is given by

$$\beta_{H2} = \frac{S/(\alpha_{max} - 1)}{N - 1} \times 100$$

Here, α_{max} is the maximum within-taxon richness per sample. The authors of [122] used β_{H2} to compare the turnover of various taxa in relation to disturbance in a Cameroon forest.

2. Cody's measure (β_C)
   The author of [123] proposed an index, which is easy to calculate and is a good measure of species turnover. It is given by

$$\beta_C = \frac{g(H) + l(H)}{2}$$

where $g(H)$ is the number of species gained and $l(H)$ is the number of species lost.

3. Routledge's measures (β_R, β_I and β_E)
   The author of [124] was concerned with how diversity measures can be partitioned into α and β components. His first index, denoted by β_R, takes overall species richness and the degree of species overlap into consideration. This index is defined by

$$\beta_R = \frac{S^2}{2r + S} - 1$$

where S is the total number of species in all samples and r is the number of species pairs with overlapping distributions.

β_I, the second index, stems from information theory and has been simplified for presence/absence data and equal sample size by [125]:

$$\beta_I = \log T - \frac{1}{T} \sum_{i=1}^{n} e_i \log e_i - \frac{1}{T} \sum_{j=1}^{n} S_j \log S_j$$

where e_i is the number of samples in the transect in which species i is present, S_j is the species richness of sample j, and $T = \sum_{i=1}^{n} e_i = \sum_{i=1}^{n} S_j$, and n the total number of samples.

The third index, β_E, is simply the exponential form of β_I. That is

$$\beta_E = e^{\beta_I}$$

4. Wilson and Shmida's index β_T
   The authors of [125] proposed a new measure of β diversity. It is given by

$$\beta_T = \frac{g(H) + l(H)}{2\bar{S}_j}$$

where $\bar{S}_j$ is the mean of S_j. Most measures of β diversity are sensitive to scale. Turnover decreases as progressively larger areas are investigated.

- Indices of complementarity and similarity
  The author of [126] coined the term complementarity to characterize the differences across locations in respect of the species they support. Complementarity is, of course, another name for the β variety. The larger the β diversity of two sites, the more

complimentary they are. Measures typically combine three variables: a, the total number of species present in both quadrants or samples, b the number of species present only in quadrant 1 and c the number of species present only in quadrant 2. There are mainly two indices.

1. Marczewski–Steinhaus (MS) distance

    Following [127], the author of [51] recommended the Marczewski–Steinhaus (MS) distance as a measure of complementarity. It is expressed as

    $$C_{MS} = 1 - \frac{a}{a+b+c}$$

    This measure is in fact the complement of the familiar [128] similarity index:

    $$C_J = \frac{a}{a+b+c}$$

    As suggested by Pielou, the statistic can also be adapted to give a single measure of complementarity across a set of samples or along a transect:

    $$C_T = \sum_{i=1}^{n} \sum_{j=1, i \neq j}^{n} \frac{U_{jk}}{n}$$

    where $U_{jk} = S_j + S_k - 2V_{jk}$ and is summed across all pairs of samples, V_{jk} is the number of species common to the two lists j and k (the same value as a in the formulae above), S_j and S_k are the number of species in samples j and k, respectively, and n is the number of samples.

    When n is large, C_T approaches a value of $nS_T/4$, where S_T is the species richness of all samples combined.

    A metric (as opposed to a nonmetric) measure is the Marczewski–Steinhaus dissimilarity measure (and hence the complement of the Jaccard similarity measure). This indicates that it meets specific geometric criteria. The significant result for the user is that it may now be used as a distance measure and in ordination (see [127]).

2. Sorensen's measure

    Another popular similarity measure was devised by [129]:

    $$C_S = \frac{2a}{2a+b+c}$$

    Sorensen's measure (see [20]) is widely recognized as one of the most effective presence/absence similarity metrics. The Bray-Curtis presence/absence coefficient is the same.

3. Lennon turnover measure

    Sorensen's measure will always be large. Therefore, they introduce a new turnover measure β_{sim}, that focuses more precisely on differences in composition:

    $$\beta_{sim} = 1 - \frac{a}{a + \min(b,c)}$$

    This is related to a measure derived by [130]. Any difference in species richness inflates either b or c. The consequence of using the smallest of these values in the denominator is thus to reduce the impact of any imbalance in species richness. The authors of [131] found that this measure performs well.

    One of the primary advantages of these measurements is that they are simple to calculate and comprehend. Furthermore, the coefficients do not take into consideration the relative abundance of species, which is a flaw.

4. Sorensen quantitative index or Bray-Curtis index
   Similarity/dissimilarity measures based on quantitative data. The author of [132] introduced a modified version of the Sorensen index. This is sometimes called the Sorensen quantitative index (see [133]). It is given by

$$C_N = \frac{2jN}{N_a + N_b}$$

where N_a is the total number of individuals in site A, N_b is the total number of individuals in site B, and $2jN$ is the sum of the lower of the two abundances for species found in both sites.

5. Other notable indices
   The authors of [134] looked into a number of quantitative similarity indices and discovered that, with the exception of the Morisita–Horn index, they were all heavily influenced by species richness and sample size. The Morisita–Horn index (MH) has the drawback of being extremely sensitive to the abundance of the most abundant species. Despite this, the author of [135] was able to measure β diversity in tropical cockroach assemblages using a modified version of the index. It is defined by

$$C_{MH} = \frac{2 \sum_{i=1}^{n} a_i b_i}{(d_a + d_b) \times N_a \times N_b}$$

where N_a is the total number of individuals at site A, N_b is the total number of individuals at site B, a_i is the number of individuals in the ith species in A, b_i is the number of individuals in the i species in B, n is the total number of species and d_a and d_b are calculated as follows:

$$d_a = \frac{\sum_{i=1}^{n} a_i^2}{N_a^2}$$

The Morisita–Horn measure is widely used (see [136,137]). The authors of [20] provided a version of Morisita's original index that is suitable for easy computation. A further simple measure is percentage similarity (see [20]):

$$P = 100 - 0.5 \sum_{i=1}^{S} |P_{ai} - P_{bi}|$$

where P_{ai} and P_{bi} is the percentage abundances of species i in samples a and b, respectively, and S is the total number of species.

Some practical applications are given below based on [5].

4.8. Extreme Values in Modeling Atmospheric Ozone

The traditional method of extreme value analysis popularized by [138] was the annual maximum method, in which one of the three classical types of extreme value distributions was fitted to, say, the annual maxima of a river or sea level series. Modified approaches to extreme value analysis which cope with time series dependence are discussed by [139,140]. The extreme value trend centered on the statistical features of insurance claims for environmental damage. The author of [141] suggested that exceedances over a high threshold can be modeled approximately by the generalized Pareto distribution (GPD).

4.9. Environmental Epidemiology

The study of associations between environmental pollutants and negative health consequences is a prominent topic in current environmental health science.

The authors of [142,143] have considered some methodological issues associated with detecting clusters in spatial point processes of disease. The authors of [144] extended the approach to the modeling of spatially aggregated data. Earlier, the authors of [145] proposed a non-parametric test for identifying disease clusters. However, as there are several sources for a disease, it has become impossible to associate the effect of each. Therefore, the cluster cannot be detected easily. In such cases, it is generally assumed that comparison of mortality or disease incidence with levels of counter-revolutionary spatial regions is subject to so much confounding with other environmental effects. To estimate the sequential mean and covariances, Zidek adapted Bayesian approach on spatial prediction of a multidimensional variable (see [146]).

4.10. Adaptive Sampling for Pollution 'Hot Spots'

The population mean concentration of the chemical pollutant will be estimated by identifying hotspots. Some clusters may be overlooked if basic random sampling is used. The sample mean, while unbiased as a population mean estimate, will have a substantial variance. In this circumstance, adaptive sampling is a viable alternative. In this case, the sampling procedure's direction at any stage is influenced at least in part by the information gathered in prior samplings.

The sampling procedure is as follows. Take a random sample of a certain size from the study area. Return and sample every unit adjacent to the contaminated unit if any of the selected units reveals contamination. If any neighboring units exhibit contamination, sample their neighbors, and so on, until each detected cluster has a clean boundary.

The total sample size is unknown in advance, however, the accuracy of the outcome will overcome this disadvantage. However, if the resulting data are evaluated naively, this strategy will produce erroneous estimates of population parameters. To avoid this, the authors of [147] outlined a sampling theory, i.e., employed a useful strategy for selecting the initial sample in clusters and stratifying those samples. Then, using modified Horvitz–Thompson or Hansen–Hurwitz estimators, unbiased estimators of the unknown population's mean can be obtained. These estimators, such as the mean of the initial sample, are unbiased, but they do not always have the lowest variance. The Rao–Blackwell theorem can be used to improve them.

4.11. Trend Analysis

Analysis of trends in environmental science leads to adjustments for autoregressive effects or other spatial-temporal correlations in the data. This is another important area of environmental trend analysis (see [148]). Any data that posses time-dependency will lead to auto-correlation and then to time series analysis.

4.12. Ecological Modeling

In building stochastic models of vertebrate populations, statistics have an useful interaction with fisheries and wildlife sciences. Analyzing the survival of the northern spotted owl after it experiences habitat loss and employing the well-known Leslie–Lefkovitch model suggested by [149] is an example. The model uses information about survival and fecundity in a matrix framework to predict future age structure based on past age structure information. After statistical analysis, it is found that the characteristic root was significantly less than zero, suggesting a decline in female owl populations due to habitat loss. However, other parameters, including vitality rates, do not show any negative trend. Here, they use an appropriate variance model, which is critical in stochastic modeling.

If single sampling is considered, the variance estimate computed will be misleading. However, if a number of sampling occasions are considered, then the process variance will give a better estimate.

4.13. Combining Environmental Information

Another increasingly important issue in the environmental sciences is the need to combine information from diverse sources that relate to a common endpoint. Combining information is a very active area of statistical and applied subject-matter research.

A common technique for combining independent results is Meta analysis (see [150]), which brings together the results of different studies, reanalyzes the disparate results within the concept of their common endpoints, and provides a quantitative analysis of the phenomenon of interest based on the combined data. In the case of environmental science, the effect of interest may be very small and therefore hard to detect. The limited sample sizes or data on many multiple endpoints lead to highly localized effects.

1. Combining p-values: Perhaps, the most known method of combining information is Fisher's inverse χ^2 method (see [151]), where individual p-values, (P_k), from K independent studies are combined. The resultant is combined p-value, which is compared to a χ^2 reference distribution with $2K$ degrees of freedom.
2. Hierarchical Bayesian method of combining information which leads to Bayesian or empirical Bayesian analysis.
3. Hierarchical regression model: The inclusion of factors that represented the various sources of variability was a key element. The odds ratio of exposure for responding patients (cases) versus non-responding, healthy subjects (controls) was the outcome of interest in each investigation. The hierarchical model was able to synthesize information across the ensemble of data, allowing more significant impacts to be investigated.

4.14. Space–Time Modeling with Applications to Atmospheric Pollution and Acid Rain

The space time autoregressive moving average (STARMA) approach was utilized by [152] (see [153]). For most latitudes, Niu and Tiao chose the STAR(2,1) model. The authors of [154] studied the logarithms of sulfate content in rainfall at 19 sites in the eastern and mid-western United States for 24 monthly measurements from 1982 to 1983, and came up with a substantially different atmospheric contaminant model. They calculated their estimator's variance. An empirical Bayesian approach was used to generate the sample spatial covariance matrix from the residuals of the fit.

4.15. Detection Limits

The authors of [155] illustrated a robust parametric method for quantifying non-detects, using a simple probability plot regression. A straight line is fitted through the observations displayed on normal (or lognormal) probability paper. The line offers estimations for the non-detected values when extrapolated back into the non-detect zone.

5. Conclusions

The statistical concepts act as a valuable tool for monitoring environmental systems. Agricultural activities, such as timing of cropping and harvesting, timing of chemical applications, type of crops planted, irrigation scheduling, etc., require knowledge of environmental statistics. The forestry activities of a country, such as extraction of timber, forestation, reforestation projects, etc., need statistical information. In addition, in measuring environmental diversity, statistics also plays a vital role. Thus, the application of statistics is important in environmental sciences for effective and innovative monitoring of environmental variables over time. To avoid mistakes at the end of the statistical analysis, it is very important to detect the actual distribution of the observed data (see [156,157]). For environment-related problems, the lack of sufficient data is a major problem. Given a small set of data, it is very difficult to correctly detect heavy-tailed distributions. Hence, the proportion of the middle and tails of the same set of data is taken for analysis. As a result, calculating the relative frequency of the outside values and the theoretical p-outside values is critical. In the particular case when $p = 0.25$, p-outside values coincide with extreme outliers and, at least, these outside values should be estimated from the sample (see [158,159]). They are useful in detecting the parameters that are used to find the tail of the distribution. They also help in finding probabilities of events. As *p*-outside values do not depend on moments, they can be easily applied to situations where moments are not essential or where they do not exist.

Many of the environmental difficulties discussed here are just a small sample of the wide range of challenging issues in quantitative environmental research, as well as the wide range of approaches to solving them. Based on [4,5], this review paper demonstrates that there are numerous viewpoints on the nature of Environmental Sciences and Statistics.

Funding: This research received no external funding.

Acknowledgments: We thank the two reviewers for the important remarks on the paper, completing the review in a thorough way.

Conflicts of Interest: The authors declare no conflict of interest.

References

1. Garfield, J. How students learn statistics. *Int. Stat. Rev. Int. Stat.* **1995**, *63*, 25–34. [CrossRef]
2. Brown, P.M.B.L.C.; Hambley, D.F. Statistics for environmental engineers. *Environ. Eng. Geosci.* **2002**, *8*, 244–245. [CrossRef]
3. Barnett, V. *Environmental Statistics: Methods and Applications*; John Wiley & Sons: Hoboken, NJ, USA, 2005; pp. 1–9.
4. Magurran, A.E. *Measuring Biological Diversity*; John Wiley & Sons: Hoboken, NJ, USA, 2013.
5. Piegorsch, W.W.; Smith, E.P.; Edwards, D.; Smith, R.L. Statistical advances in environmental science. *Stat. Sci.* **1998**, *13*, 186–208. [CrossRef]
6. Stephenson, D.B. *Statistical Concepts in Environmental Science*; Department of Meteorology, University of Reading: Reading, UK, 2003. Available online: https://met.rdg.ac.uk/cag/courses/Stats (accessed on 1 November 2021).
7. Velleman, P.F. Truth, damn truth, and statistics. *J. Stat. Educ.* **2008**, *16*. [CrossRef]
8. Bhagawati, B. Basic Statistical Concepts in Environmental Science: An Introduction. *IOSR J. Environ. Sci. Toxicol. Food Technol.* **2004**, *8*, 8–9. [CrossRef]
9. Krebs, C.J. *Ecological Methodology*; Benjamin Cummings: San Francisco, CA, USA, 1999.
10. Whittaker, R.H. Dominance and diversity in land plant communities: Numerical relations of species express the importance of competition in community function and evolution. *Science* **1965**, *147*, 250–260. [CrossRef]
11. Wilson, J.B. Methods for fitting dominance/diversity curves. *J. Veg. Sci.* **1991**, *2*, 35–46. [CrossRef]
12. Lambshead, P.J.D.; Platt, H.M.; Shaw, K.M. The detection of differences among assemblages of marine benthic species based on an assessment of dominance and diversity. *J. Nat. Hist.* **1983**, *17*, 859–874. [CrossRef]
13. Platt, H.M.; Shaw, K.M.; Lambshead, P.J.D. Nematode species abundance patterns and their use in the detection of environmental perturbations. *Hydrobiologia* **1984**, *118*, 59–66. [CrossRef]
14. Warwick, R. A new method for detecting pollution effects on marine macrobenthic communities. *Mar. Biol.* **1986**, *92*, 557–562. [CrossRef]
15. Clarke, K.R. Comparisons of dominance curves. *J. Exp. Mar. Biol. Ecol.* **1990**, *138*, 143–157. [CrossRef]
16. Roth, S.; Wilson, J.G. Functional analysis by trophic guilds of macrobenthic community structure in Dublin Bay, Ireland. *J. Exp. Mar. Biol. Ecol.* **1998**, *222*, 195–217. [CrossRef]
17. Fisher, R.A.; Corbet, A.S.; Williams, C.B. The relation between the number of species and the number of individuals in a random sample of an animal population. *J. Anim. Ecol.* **1943**, *12*, 42–58. [CrossRef]
18. Hubbell, S.P. *The Unified Neutral Theory of Biodiversity and Biogeography (MPB-32)*; Princeton University Press: Princeton, NJ, USA, 2001.
19. Preston, F.W. The commonness, and rarity, of species. *Ecology* **1948**, *29*, 254–283. [CrossRef]
20. Southwood, T.R.E.; Henderson, P.A. *Ecological Methods*; Oxford Blackwell Science: Oxford, UK, 2000; pp. 269–292.
21. Coddington, J.A.; Griswold, C.E.; Silva, D.; Peñaranda, E.; Larcher, S.F. Designing and testing sampling protocols to estimate biodiversity in tropical ecosystems. In *The Unity of Evolutionary Biology: Proceedings of the Fourth International Congress of Systematic and Evolutionary Biology*; Dioscorides Press: Portland, OR, USA, 1991; Volume 2.
22. Pielou, E. *Ecological Diversity*; Wiley Interscience: New York, NY, USA, 1975.
23. Brian, M.V. Species frequencies in random samples from animal populations. *J. Anim. Acol.* **1953**, *22*, 57–64. [CrossRef]
24. Zipf, G. *Human Behaviour and the Principle of Least Effort*; Hafner: New York, NY, USA, 1949.
25. Zipf, G. *Human Behaviour and the Principle of Least Effort*, 2nd ed.; Hafner: New York, NY, USA, 1965.
26. Mandelbrot, B.B. Fractals. Form, chance and dimension. *Encycl. Phys. Sci. Technol.* **1977**, *5*, 579–593. [CrossRef]
27. Mandelbrot, B.B. *The Fractal Geometry of Nature*, 1st ed.; WH freeman: New York, NY, USA, 1982.
28. Gray, J.S. Species-abundance patterns. In *Organization of Communities: Past and Present*; Gee, J.H.R., Giller, P.S., Eds.; Blackwell Scientific Publications: Oxford, UK, 1987; pp. 53–67.
29. Tokeshi, M. Species abundance patterns and community structure. *Adv. Ecol. Res.* **1993**, *24*, 111–186.

30. Reichelt, R.E.; Bradbury, R.H. Spatial patterns in coral reef benthos: Multiscale. *Mar. Ecol. Prog. Ser.* **1984**, *17*, 251–257. [CrossRef]
31. Frontier, S. Diversity and structure in aquatic ecosystems. *Oceanogr. Mar. Biol.* **1985**, *23*, 253–312.
32. Barange, M.; Campos, B. Models of Species Abundance: A Critique of and an Alternative to the Dynamics Model. *Mar. Ecol. Prog. Ser.* **1991**, *69*, 293–298. [CrossRef]
33. Watkins, A.J.; Wilson, J.B. Plant community structure and its relation to the vertical complexity of communities: Dominance/diversity and spatial rank consistency. *Oikos* **1994**, *70*, 91–98. [CrossRef]
34. Wilson, J.B.; Wells, T.C.; Trueman, I.C.; Jones, G.; Atkinson, M.D.; Crawley, M.J.; Dodd, M.E.; Silvertown, J. Are there assembly rules for plant species abundance? An investigation in relation to soil resources and successional trends. *J. Ecol.* **1996**, *84*, 527–538. [CrossRef]
35. Mouillot, D.; Lepretre, A. Introduction of relative abundance distribution (rad) indices, estimated from the rank-frequency diagrams (rfd), to assess changes in community diversity. *Environ. Monit. Assess.* **2000**, *63*, 279–295. [CrossRef]
36. Juhos, S.; Vörös, L. Structural changes during eutrophication of Lake Balaton, Hungary, as revealed by the Zipf-Mandelbrot model. *Hydrobiologia* **1998**, *369*, 237–242. [CrossRef]
37. Mouillot, D.; Lepretre, A. A comparison of species diversity estimators. *Res. Popul. Ecol.* **1999**, *41*, 203–215.
38. Sokal, R.R.; Rohlf, F.J. *Biometry: The Principles and Practice of Statistics in Biological Research*, 3rd ed.; Freeman: New York, NY, USA, 1995.
39. Siegel, S. *Nonparametric Statistics for the Behavioral Sciences*; McGraw-Hill: New York, NY, USA, 1956.
40. Tokeshi, M. *Species Coexistence: Ecological and Evolutionary Perspectives*; John Wiley & Sons: Hoboken, NJ, USA, 2009.
41. May, R.M. *Patterns of Species Abundance and Diversity*; Belknap Press of Harvard University Press: Cambridge, MA, USA, 1975; pp. 81–120.
42. Motomura, I. A statistical treatment of ecological communities. *Zool. Mag.* **1932**, *44*, 379–383.
43. MacArthur, R.H. On the relative abundance of bird species. *Proc. Natl. Acad. Sci. USA* **1957**, *43*, 293. [CrossRef] [PubMed]
44. Drozd, P.; Novotny, V. PowerNiche: Niche Division Models for Community Analysis. 2000. Available online: http://www.entu.cas.cz/png/powerniche/index.html (accessed on 1 November 2021).
45. Tokeshi, M. Niche apportionment or random assortment: Species abundance patterns revisited. *J. Anim. Ecol.* **1990**, *59*, 1129–1146. [CrossRef]
46. Tokeshi, M. Power fraction: A new explanation of relative abundance patterns in species-rich assemblages. *Oikos* **1996**, *75*, 543–550. [CrossRef]
47. Fesl, C. Niche-oriented species-abundance models: Different approaches of their application to larval chironomid (Diptera) assemblages in a large river. *J. Anim. Ecol.* **2002**, *71*, 1085–1094. [CrossRef]
48. Gaston, K.J.; Chown, S.L. Geographic range size and speciation. In *Evolution of Biological Diversity*; Magurran, A.E., May, R.M., Eds.; Oxford University Press: Oxford, UK, 1999; pp. 236–259.
49. Clifford, H.T.; Stephenson, W. *An Introduction to Numerical Classification*; Academic Press: New York, NY, USA, 1975. .
50. Whittaker, R.H. Evolution of species diversity in land communities. *Evol. Biol.* **1977**, *10*, 1–6.
51. Colwell, R.K.; Coddington, J.A. Estimating terrestrial biodiversity through extrapolation. *Philos. Trans. R. Soc. Lond. Ser. B Biol. Sci.* **1994**, *345*, 101–118.
52. Chazdon, R.L.; Colwell, R.K.; Denslow, J.S.; Guariguata, M.R. Statistical Methods for Estimating Species Richness of Woody Regeneration in Primary and Secondary Rain Forests of Northeastern Costa Rica. 1998. Available online: https://www.cifor.org/knowledge/publication/456 (accessed on 1 November 2021).
53. Arrhenius, O. Species and area. *J. Ecol.* **1921**, *9*, 95–99. [CrossRef]
54. Goldsmith, F.B.; Harrison, C.M. Description and analysis of vegetation. In *Methods in Plant Ecology*; Chapman, S.B., Ed.; John Wiley & Sons: New York, NY, USA, 1976.
55. Gotelli, N.J. Research frontiers in null model analysis. *Glob. Ecol. Biogeogr.* **2001**, *10*, 337–343. [CrossRef]
56. Sanders, H.L. Marine benthic diversity: A comparative study. *Am. Nat.* **1968**, *102*, 243–282. [CrossRef]
57. Longino, J.T.; Coddington, J.; Colwell, R.K. The ant fauna of a tropical rain forest: Estimating species richness three different ways. *Ecology* **2002**, *83*, 689–702. [CrossRef]
58. Holdridge, L.R.; Grenke, W.C. *Forest Environments in Tropical Life Zones: A Pilot Study*; Pergamon Press: Oxford, UK, 1971.
59. Michaelis, L.; Menten, M.L. Die kinetik der invertinwirkung. *Biochem. Z.* **1913**, *49*, 352.
60. Gleason, H.A. On the relation between species and area. *Ecology* **1922**, *3*, 158–162. [CrossRef]
61. Stout, J.; Vandermeer, J. Comparison of species richness for stream-inhabiting insects in tropical and mid-latitude streams. *Am. Nat.* **1975**, *109*, 263–280. [CrossRef]
62. Silva, D.; Coddington, J.A. Spiders of Pakitza (Madre de Dios, Perú): Species richness and notes on community structure. In *Manu: The Biodiversity of Southeastern Peru*; Smithsonian: Washington, DC, USA, 1996; pp. 253–311.
63. Chao, A. Nonparametric estimation of the number of classes in a population. *Scand. J. Stat.* **1984**, *11*, 265–270.
64. Chao, A. Estimating the population size for capture-recapture data with unequal catchability. *Biometrics* **1987**, *43*, 783–791. [CrossRef] [PubMed]
65. Colwell, R. Estimates: Statistical Estimation of Species Richness and Shared Species from Samples. 2000. Available online: http://purl.oclc.org/estimates (accessed on 1 November 2021).
66. Chao, A.; Lee, S.M. Estimating the number of classes via sample coverage. *J. Am. Stat. Assoc.* **1992**, *87*, 210–217. [CrossRef]

67. Chao, A.; Hwang, W.H.; Chen, Y.C.; Kuo, C.Y. Estimating the number of shared species in two communities. *Stat. Sin.* **2000**, *10*, 227–246.
68. Chao, A.; Yang, M.C. Stopping rules and estimation for recapture debugging with unequal failure rates. *Biometrika* **1993**, *80*, 193–201. [CrossRef]
69. Williams, C.B. *Patterns in the Balance of Nature and Related Problems of Quantitative Ecology*; Academic Press: London, UK, 1964.
70. Hayek, L.A.; Buzas, M. *Surveying Natural Populations*; Columbia University Press: New York, NY, USA, 1997.
71. Kempton, R.A.; Taylor, L.R. Log-series and log-normal parameters as diversity discriminants for the Lepidoptera. *J. Anim. Ecol.* **1974**, *43*, 381–399. [CrossRef]
72. Taylor, L.R. Bates, Williams, Hutchison—A variety of diversities. In *Diversity of Insect Fauna: 9th Symposium of the Royal Entomological Society*; Blackwell: Oxford, UK, 1978; pp. 1–18.
73. Kempton, R.A.; Taylor, L.R. Models and statistics for species diversity. *Nature* **1976**, *262*, 818–820. [CrossRef] [PubMed]
74. Kempton, R.A.; Taylor, L.R. The Q-statistic and the diversity of floras. *Nature* **1978**, *275*, 252–253. [CrossRef]
75. Whittaker, R.H. Evolution and measurement of species diversity. *Taxon* **1972**, *21*, 213–251. [CrossRef]
76. Kempton, R.A.; Wedderburn, R. A comparison of three measures of species diversity. *Biometrics* **1978**, *34*, 25–37. [CrossRef]
77. Nee, S.; Harvey, P.H.; Cotgreave, P. Population persistence and the natural relationship between body size and abundance. In *Conservation of Biodiversity for Sustainable Development*; Scandinavian University Press: Oslo, Norway, 1992; pp. 124–136.
78. Pielou, E.C. *An Introduction to Mathematical Ecology*; Wiley: New York, NY, USA, 1969.
79. Cronin, T.M.; Raymo, M.E. Orbital forcing of deep-sea benthic species diversity. *Nature* **1997**, *385*, 624–627. [CrossRef]
80. Lande, R. Statistics and partitioning of species diversity, and similarity among multiple communities. *Oikos* **1996**, *76*, 5–13. [CrossRef]
81. Caswell, H. Community structure: A neutral model analysis. *Ecol. Monogr.* **1976**, *46*, 327–354. [CrossRef]
82. Clarke, K.R.; Warwick, R.M. Change in marine communities. In *An Approach to Statistical Analysis and Interpretation*; PRIMER-E Ltd.: Plymouth, UK, 2001; Volume 2, pp. 1–168.
83. Goldman, N.; Lambshead, P.J.D. Optimization of the Ewens/Caswell neutral model program for community diversity analysis. *Mar. Ecol. Prog. Ser.* **1989**, *50*, 255–261. [CrossRef]
84. Platt, H.M.; Lambshead, P.J.D. Neutral model analysis of patterns of marine benthic species diversity. *Mar. Ecol. Prog. Ser.* **1985**, *24*, 75–81.
85. Lambshead, P.J.D.; Platt, H.M. Analysing disturbance with the Ewens/Caswell neutral model: Theoretical review and practical assessment. *Mar. Ecol. Prog. Ser.* **1988**, *43*, 31–41. [CrossRef]
86. Beisel, J.N.; Moreteau, J.C. A simple formula for calculating the lower limit of Shannon's diversity index. *Ecol. Model.* **1997**, *99*, 289–292. [CrossRef]
87. Hurlbert, S.H. The nonconcept of species diversity: A critique and alternative parameters. *Ecology* **1971**, *52*, 577–586. [CrossRef] [PubMed]
88. Heip, C. A new index measuring evenness. *J. Mar. Biol. Assoc. U. K.* **1974**, *54*, 555–557. [CrossRef]
89. Smith, B.; Wilson, J.B. A consumer's guide to evenness indices. *Oikos* **1996**, *76*, 70–82. [CrossRef]
90. Buzas, M.A.; Hayek, L.A.C. Biodiversity resolution: An integrated approach. *Biodivers. Lett.* **1996**, *3*, 40–43. [CrossRef]
91. Arita, H.T.; Figueroa, F. Geographic patterns of body-mass diversity in Mexican mammals. *Oikos* **1999**, *85*, 310–319. [CrossRef]
92. Laxton, R. The measure of diversity. *J. Theor. Biol.* **1978**, *70*, 51–67. [CrossRef]
93. Legendre, P. Numerical ecology: Developments and recent trends. In *Numerical Taxonomy*; Springer: Berlin/Heidelberg, Germany, 1983; pp. 505–523.
94. Simpson, E.H. Measurement of diversity. *Nature* **1949**, *163*, 688. [CrossRef]
95. Rosenzweig, M.L. *Species Diversity in Space and Time*; Cambridge University Press: Cambridge, UK, 1995.
96. Bulla, L. An index of evenness and its associated diversity measure. *Oikos* **1994**, *70*, 167–171. [CrossRef]
97. McIntosh, R.P. An index of diversity and the relation of certain concepts to diversity. *Ecology* **1967**, *48*, 392–404. [CrossRef]
98. Berger, W.H.; Parker, F.L. Diversity of planktonic foraminifera in deep-sea sediments. *Science* **1970**, *168*, 1345–1347. [CrossRef]
99. Camargo, J.A. Must dominance increase with the number of subordinate species in competitive interactions? *J. Theor. Biol.* **1993**, *161*, 537–542. [CrossRef]
100. Krebs, C. *Ecological Methodology*; Harper Collins Publishers: New York, NY, USA, 1989.
101. Clarke, K.R.; Warwick, R.M. A taxonomic distinctness index and its statistical properties. *J. Appl. Ecol.* **1998**, *35*, 523–531. [CrossRef]
102. Warwick, R.M.; Clarke, K.R. Taxonomic distinctness and environmental assessment. *J. Appl. Ecol.* **1998**, *35*, 532–543. [CrossRef]
103. Price, A.R.G.; Keeling, M.J.; O'callaghan, C.J. Ocean-scale patterns of 'biodiversity' of Atlantic asteroids determined from taxonomic distinctness and other measures. *Biol. J. Linn. Soc.* **1999**, *66*, 187–203.
104. Warwick, R.M.; Clarke, K.R. A comparison of some methods for analysing changes in benthic community structure. *J. Mar. Biol. Assoc. U. K.* **1991**, *71*, 225–244. [CrossRef]
105. Warwick, R.M.; Clarke, K.R. Increased variability as a symptom of stress in marine communities. *J. Exp. Mar. Biol. Ecol.* **1993**, *172*, 215–226. [CrossRef]
106. Hurlbert, S.H. Pseudoreplication and the design of ecological field experiments. *Ecol. Monogr.* **1984**, *54*, 187–211. [CrossRef]
107. Crawley, M.J. *GLIM for Ecologists*; Number 574.501519 C7; Blackwell Scientific: Hoboken, NJ, USA, 1993.

108. Tipper, J.C. Rarefaction and rarefiction—The use and abuse of a method in paleoecology. *Paleobiology* **1979**, *5*, 423–434. [CrossRef]
109. Gotelli, N.J.; Entsminger, G.L. *EcoSim: Null Models Software for Ecology*; Version 6.0; Acquired Intelligence and Kesey-Bear: Jericho, VT, USA, 2001.
110. Miller, R.G. The jackknife—A review. *Biometrika* **1974**, *61*, 1–15.
111. Zahl, S. Jackknifing an index of diversity. *Ecology* **1977**, *58*, 907–913. [CrossRef]
112. Adams, J.E.; McCune, E.D. Application of the generalized jackknife to Shannon's measure of information used as an index of diversity. In *Ecological Diversity in Theory and Practice*; International Cooperative Publishing House: Fairland, MD, USA, 1979; pp. 117–131.
113. Heltshe, J.F. Comparing diversity measures in sampled communities. In *Ecological Diversity in Theory and Practice*; International Cooperative Publishing House: Fairland, MD, USA, 1979; pp. 133–144.
114. Burnham, K.P.; Overton, W.S. Estimation of the size of a closed population when capture probabilities vary among animals. *Biometrika* **1978**, *65*, 625–633. [CrossRef]
115. Heltshe, J.F.; Forrester, N.E. Estimating species richness using the jackknife procedure. *Biometrics* **1983**, *39*, 1–11. [CrossRef] [PubMed]
116. Smith, E.P.; van Belle, G. Nonparametric estimation of species richness. *Biometrics* **1984**, *40*, 119–129. [CrossRef]
117. Gotelli, N.J.; Graves, G.R. *Null Models in Ecology*; Smithsonian Institution Press: Washington, DC, USA, 1996.
118. Strong, D.R. Null hypotheses in ecology. *Synthese* **1980**, *43*, 271–285. [CrossRef]
119. Gotelli, N.J. Null model analysis of species co-occurrence patterns. *Ecology* **2000**, *81*, 2606–2621. [CrossRef]
120. Whittaker, R.H. Vegetation of the Siskiyou mountains, Oregon and California. *Ecol. Monogr.* **1960**, *30*, 279–338. [CrossRef]
121. Harrison, S.; Ross, S.J.; Lawton, J.H. Beta diversity on geographic gradients in Britain. *J. Anim. Ecol.* **1992**, *61*, 151–158. [CrossRef]
122. Lawton, J.; Bignell, D.; Bolton, B.; Bloemers, G.F.; Eggleton, P.; Hammond, P.M.; Hodda, M.; Holt, R.D.; Larsen, T.B.; Mawdsley, N.A.; et al. Biodiversity inventories, indicator taxa and effects of habitat modification in tropical forest. *Nature* **1998**, *391*, 72–76. [CrossRef]
123. Cody, M.L.; MacArthur, R.H.; Diamond, J.M. *Ecology and Evolution of Communities*; Harvard University Press: Cambridge, MA, USA, 1975.
124. Routledge, R.D. On Whittaker's components of diversity. *Ecology* **1977**, *58*, 1120–1127. [CrossRef]
125. Wilson, M.V.; Shmida, A. Measuring beta diversity with presence-absence data. *J. Ecol.* **1984**, *72*, 1055–1064. [CrossRef]
126. Vane-Wright, R.I.; Humphries, C.J.; Williams, P.H. What to protect?—Systematics and the agony of choice. *Biol. Conserv.* **1991**, *55*, 235–254. [CrossRef]
127. Pielou, E.C. *The Interpretation of Ecological Data: A Primer on Classification and Ordination*; Wiley InterScience: New York, NY, USA, 1984.
128. Jaccard, P. Nouvelles recherches sur la distribution florale. *Bull. Soc. Vaud. Sci. Nat.* **1908**, *44*, 223–270.
129. Sorensen, T.A. A method of establishing groups of equal amplitude in plant sociology based on similarity of species content and its application to analyses of the vegetation on Danish commons. *Biol. Skar.* **1948**, *5*, 1–34.
130. Simpson, G.G. Mammals and the nature of continents. *Am. J. Sci.* **1943**, *241*, 1–31. [CrossRef]
131. Lennon, J.J.; Koleff, P.; Greenwood, J.J.D.; Gaston, K.J. The geographical structure of British bird distributions: Diversity, spatial turnover and scale. *J. Anim. Ecol.* **2001**, *70*, 966–979. [CrossRef]
132. Bray, J.R.; Curtis, J.T. An ordination of the upland forest communities of southern Wisconsin. *Ecol. Monogr.* **1957**, *27*, 326–349. [CrossRef]
133. Magurran, A.E. *Ecological Diversity and Its Measurement*; Princeton University Press: Princeton, NJ, USA, 1988.
134. Wolda, H. Similarity indices, sample size and diversity. *Oecologia* **1981**, *50*, 296–302. [CrossRef] [PubMed]
135. Wolda, H. Diversity, diversity indices and tropical cockroaches. *Oecologia* **1983**, *58*, 290–298. [CrossRef]
136. Arnold, A.E.; Maynard, Z.; Gilbert, G.S. Fungal endophytes in dicotyledonous neotropical trees: Patterns of abundance and diversity. *Mycol. Res.* **2001**, *105*, 1502–1507. [CrossRef]
137. Williams-Linera, G. Tree species richness complementarity, disturbance and fragmentation in a Mexican tropical montane cloud forest. *Biodivers. Conserv.* **2002**, *11*, 1825–1843. [CrossRef]
138. Gumbel, E.J. *Statistics of Extremes*; Columbia University Press: New York, NY, USA, 1958. [CrossRef]
139. Davison, A.C.; Smith, R.L. Models for exceedances over high thresholds. *J. R. Stat. Soc. Ser. B Methodol.* **1990**, *52*, 393–425. [CrossRef]
140. Gomes, M.I. On the estimation of parameters of rare events in environmental time series. *Stat. Environ.* **1993**, *2*, 225–241.
141. Pickands, J., III. Statistical inference using extreme order statistics. *Ann. Stat.* **1975**, *3*, 119–131.
142. Diggle, P.J. A point process modelling approach to raised incidence of a rare phenomenon in the vicinity of a prespecified point. *J. R. Stat. Soc. Ser. A Stat. Soc.* **1990**, *153*, 349–362. [CrossRef]
143. Diggle, P.J.; Rowlingson, B.S. A conditional approach to point process modelling of elevated risk. *J. R. Stat. Soc. Ser. A Stat. Soc.* **1994**, *157*, 433–440. [CrossRef]
144. Diggle, P.; Morris, S.; Elliott, P.; Shaddick, G. Regression modelling of disease risk in relation to point sources. *J. R. Stat. Soc. Ser. A Stat. Soc.* **1997**, *160*, 491–505. [CrossRef]
145. Stone, R.A. Investigations of excess environmental risks around putative sources: Statistical problems and a proposed test. *Stat. Med.* **1988**, *7*, 649–660. [CrossRef] [PubMed]

146. Zidek, J.V. Interpolating air pollution for health impact assessment. In *Statistics for the Environment, Volume 3, Pollution Assessment and Control*; Barnett, V., Feridun Turkman, K., Eds.; Wiley: Hoboken, NJ, USA, 1997; pp. 251–268.
147. Seber, G.A.; Thompson, S.K. 6 Environmental adaptive sampling. *Handb. Stat.* **1994**, *12*, 201–220.
148. Esterby, S.R. Review of methods for the detection and estimation of trends with emphasis on water quality applications. *Hydrol. Process.* **1996**, *10*, 127–149. [CrossRef]
149. Leslie, P.H. On the use of matrices in certain population mathematics. *Biometrika* **1945**, *33*, 183–212. [CrossRef] [PubMed]
150. Hedges, L.V.; Olkin, I. *Statistical Methods for Meta-Analysis*; Academic Press: New York, NY, USA, 2014.
151. Fisher, R.A. 224A: Answer to Question 14 on Combining independent tests of significance. *Am. Stat.* **1948**, *2*, 30.
152. Niu, X.; Tiao, G.C. Modeling satellite ozone data. *J. Am. Stat. Assoc.* **1995**, *90*, 969–983. [CrossRef]
153. Cliff, A.D.; Haggett, P.; Ord, J.K.; Bassett, K.A.; Davies, R.; Bassett, K.L. *Elements of Spatial Structure: A Quantative Approach*; Cambridge University Press: Cambridge, UK, 1975; Volume 6.
154. Loader, C.; Switzer, P. Spatial covariance estimation for monitoring data. In *Statistics in Environmental and Earth Sciences*; Walden, A., Guttorp, P., Eds.; Edward Arnold: London, UK, 1992; pp. 52–70.
155. Akritas, M.G.; Ruscitti, T.F.; Patil, G.P. 7 Statistical analysis of censored environmental data. *Handb. Stat.* **1994**, *12*, 221–242. [CrossRef]
156. Soza, L.N.; Jordanova, P.; Nicolis, O.; Střelec, L.; Stehlík, M. Small sample robust approach to outliers and correlation of atmospheric pollution and health effects in Santiago de Chile. *Chemom. Intell. Lab. Syst.* **2019**, *185*, 73–84. [CrossRef]
157. Stehlík, M.; Soza, L.N.; Fabián, Z.; Jiřina, M.; Jordanova, P.; Arancibia, S.C.; Kiseľák, J. On ecological aspects of dynamics for zero slope regression for water pollution in Chile. *Stoch. Anal. Appl.* **2019**, *37*, 574–601. [CrossRef]
158. Jordanova, P.K. Probabilities for *p*-outside values—General properties. *AIP Conf. Proc.* **2019**, *2164*, 020002.
159. Jordanova, P.K. Tails and probabilities for *p*-outside values. *arXiv* **2019**, arXiv:1902.03810.

Article

Discrete Pseudo Lindley Distribution: Properties, Estimation and Application on INAR(1) Process

Muhammed Rasheed Irshad [1], Christophe Chesneau [2,*], Veena D'cruz [1] and Radhakumari Maya [3]

[1] Department of Statistics, Cochin University of Science and Technology, Cochin 682022, Kerala, India; irshadmr@cusat.ac.in (M.R.I.); veenadicruz@gmail.com (V.D.)
[2] Laboratoire de Mathématiques Nicolas Oresme (LMNO), Université de Caen Normandie, Campus II, Science 3, 14032 Caen, France
[3] Department of Statistics, Government College for Women, Trivandrum 695014, Kerala, India; publicationsofmaya@gmail.com
* Correspondence: christophe.chesneau@unicaen.fr

Citation: Irshad, M.R.; Chesneau, C.; D'cruz, V.; Maya, R. Discrete Pseudo Lindley Distribution: Properties, Estimation and Application on INAR(1) Process. *Math. Comput. Appl.* **2021**, *26*, 76. https://doi.org/10.3390/mca26040076

Academic Editor: Paweł Olejnik

Received: 14 October 2021
Accepted: 9 November 2021
Published: 12 November 2021

Publisher's Note: MDPI stays neutral with regard to jurisdictional claims in published maps and institutional affiliations.

Copyright: © 2021 by the authors. Licensee MDPI, Basel, Switzerland. This article is an open access article distributed under the terms and conditions of the Creative Commons Attribution (CC BY) license (https://creativecommons.org/licenses/by/4.0/).

Abstract: In this paper, we introduce a discrete version of the Pseudo Lindley (PsL) distribution, namely, the discrete Pseudo Lindley (DPsL) distribution, and systematically study its mathematical properties. Explicit forms gathered for the properties such as the probability generating function, moments, skewness, kurtosis and stress–strength reliability made the distribution favourable. Two different methods are considered for the estimation of unknown parameters and, hence, compared with a broad simulation study. The practicality of the proposed distribution is illustrated in the first-order integer-valued autoregressive process. Its empirical importance is proved through three real datasets.

Keywords: Pseudo Lindley distribution; survival discretization method; over dispersion; moments; simulation; maximum likelihood estimation

MSC: 62E15; 62E20; 62E17

1. Introduction

Count data reflect the non-negative integers which represent the frequency of occurrence of a discrete event. Such datasets can be observed in numerous fields, such as actuarial science, finance, medical, sports, etc. For instance, the yearly number of destructive floods, the number of sports people injured in a month and the hourly number of COVID-19 vaccinations given are some examples of count data. Increasing the utilization of discrete distributions for modelling such datasets influenced researchers to propose more flexible distributions by reducing the estimation errors. Discretizing continuous distributions by survival discretization is one of the widely followed methods for introducing discrete distributions. The most famous discretization technique is described below. Assume that X is a continuous lifetime random variable with the survival function (sf) $S(x) = \Pr(X > x)$. Then, the probability mass function (pmf) dealing with X is given by:

$$\Pr(X = x) = S(x) - S(x+1), \quad x = 0, 1, 2, \ldots \quad (1)$$

Some of the recently introduced discrete distributions based on this survival discretization method are as follows: Discrete Lindley distribution by [1], discrete inverse Weibull distribution by [2], discrete Pareto distribution by [3], discrete Rayleigh distribution by [4], two-parameter discrete Lindley distribution by [5], exponentiated discrete Lindley distribution by [6], discrete Burr–Hatke distribution by [7], discrete Bilal distribution by [8], discrete three-parameter Lindley distribution by [9], etc. Recently, Ref. [10] proposed a discrete version of Ramos–Louzada distribution [11] for asymmetric and over-dispersed data with a leptokurtic shape.

Furthermore, count datasets arising in time series can be seen in many applied research areas. Examples include modelling and predicting the number of claims for next month for the insurance sector in a company, predicting the number of deaths from disasters, etc. The first-order integer-valued autoregressive process, or INAR(1), is appropriate for such cases. The authors of [12,13] independently developed the pioneer works of INAR(1) with Poisson innovations. Furthermore, since time series of counts mainly display over-dispersion (i.e., empirical mean is less than empirical variance), Poisson for innovation distribution is less efficient (since equi-dispersed). Hence, researchers have assembled many approaches concerning innovations in modelling over-dispersed time series count datasets. The INAR(1) process with geometric innovations (INAR(1)G) by [14], INAR(1) process with Poisson–Lindley innovations (INAR(1)PL) by [15], INAR(1) process with a new Poisson weighted exponential innovation ((INAR(1)NPWE)) by [16], INAR(1) process with discrete three-parameter Lindley as innovation by [9], INAR(1) process with discrete Bilal as innovation by [8], INAR(1) process with Poisson quasi Gamma innovations (INAR(1)PQX) by [17] and the INAR(1) process with Bell innovations (INAR(1)BL) by [18] are some of the recently developed over-dispersed INAR(1) processes.

Even though these processes provide better solutions to over-dispersed time series count datasets, they have some limitations that can sometimes cause computing difficulties. Even if a model has one parameter, the inclusion of special functions in the pmf, cumulative distribution function (cdf) and other statistical properties makes it difficult to obtain explicit expressions and, hence, for estimation procedures to generate them (see, e.g., [9,19]).

Hence, the main objective of the present work is to introduce a two-parameter discrete distribution, the discrete Pseudo Lindley (DPsL) distribution, which can serve as a model to analyse under as well as over-dispersed datasets, having a simple pmf and cdf. The main peculiarity of the proposed distribution is that it has closed-form expressions for its statistical properties such as a hazard rate function (hrf), probability-generating function (pmf), moments, skewness, kurtosis, mean past lifetime (mpl), mean residual lifetime (mrl), stress–strength reliability, etc. We embellish the importance of the DPsL distribution in the INAR(1) process by applying the DPsL distribution as an innovation process.

The remaining parts of the paper are organized as follows: Section 2 defines the proposed distribution and various properties such as moments, mean residual lifetime, mean past lifetime and stress–strength reliability,. Section 3 contains estimation methods and their simulation study. The INAR(1) process with DPsL innovations is developed in Section 4 with its parameter estimation and simulation study. In Section 5, three datasets are analysed by the DPsL distribution, and some other competitive and well-referenced distributions, in order to prove its applicability. Final remarks are provided in Section 6.

2. The Discrete Pseudo Lindley Distribution

2.1. Some Basics

A discrete analogue of the PsL distribution is derived in this section, namely, the DPsL distribution by using the survival discretization method. First of all, let us briefly present the work of [20], which introduced the Pseudo Lindley (PsL) distribution by mixing two independent random variables: one having the Exponential (θ) distribution, and the other having the Gamma (2,θ) distribution, with mixing probabilities $\frac{\beta-1}{\beta}$ and $\frac{1}{\beta}$, respectively. Assume that X is a continuous random variable having the PsL distribution; then, its probability density function (pdf) and sf are given by:

$$f_{\text{PsL}}(x;\theta,\beta) = \begin{cases} \dfrac{\theta(\beta - 1 + \theta x)e^{-\theta x}}{\beta}, & x > 0 \\ 0, & \text{otherwise} \end{cases}$$

and

$$S_{\text{PsL}}(x;\theta,\beta) = \begin{cases} \dfrac{(\beta + \theta x)e^{-\theta x}}{\beta}, & x > 0 \\ 1, & \text{otherwise} \end{cases} \tag{2}$$

respectively, where $\beta \geq 1$ and $\theta > 0$. Using the survival discretization technique as described in (1) by using (2), the pmf of the DPsL distribution can be derived as:

$$P_{\text{DPsL}}(x;\theta,\beta) = \frac{(\beta + \theta x)e^{-\theta x} - (\beta + \theta(x+1))e^{-\theta(x+1)}}{\beta}, \quad x = 0, 1, 2, \ldots. \qquad (3)$$

The parameter β can be considered as a shape parameter and θ as a scale parameter. The DPsL distribution can sometimes be denoted by the DPsL (θ, β) distribution to indicate the parameters.

The corresponding cdf and sf are given by:

$$F_{\text{DPsL}}(x;\theta,\beta) = 1 - \frac{e^{-\theta(1+x)}(\beta + (x+1)\theta)}{\beta}$$

and

$$S_{\text{DPsL}}(x;\theta,\beta) = \frac{e^{-\theta(1+x)}(\beta + (x+1)\theta)}{\beta}, \qquad (4)$$

respectively. As a first property, the pmf given in (3) is log concave, since:

$$\frac{P_{\text{DPsL}}(x+1;\theta,\beta)}{P_{\text{DPsL}}(x;\theta,\beta)} = \frac{\beta + \theta + x\theta - e^{-\theta}(\beta + (2+x)\theta)}{\beta(e^\theta - 1) + \theta((e^\theta - 1)x - 1)}$$

is a decreasing function in x for every possible value of the parameters.

The possible pmf shapes plotted for different values of the parameters of the DPsL distribution are displayed in Figure 1.

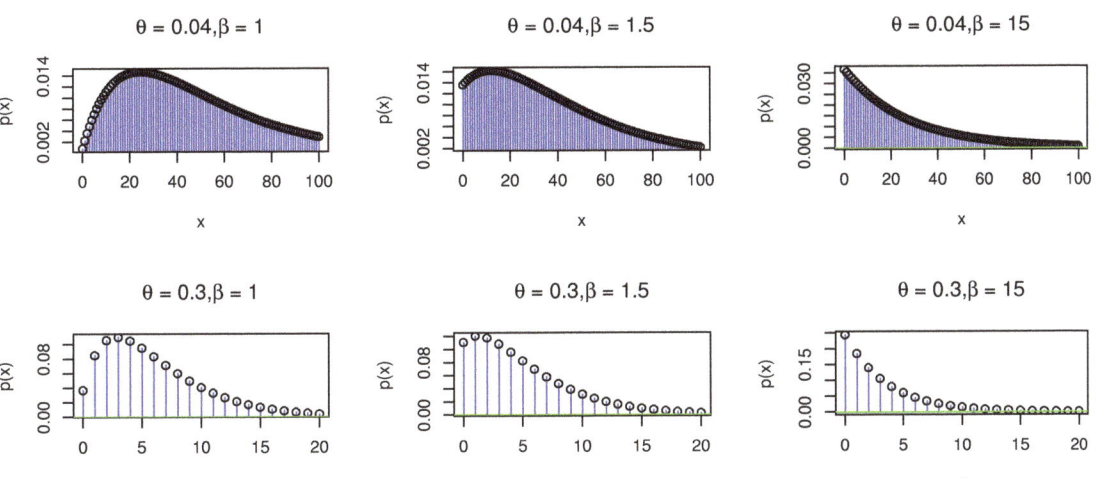

Figure 1. The pmf plots of the DPsL distribution for some set of values for θ and β.

The figure clearly indicates that the DPsL distribution is rightly skewed and has a longer right tail.

A mode of the DPsL distribution, e.g., x_m, is an integer value of x, for which the pmf $P_{\text{DPsL}}(x;\theta,\beta)$ is the maximum. That is $P_{\text{DPsL}}(x;\theta,\beta) \geq P_{\text{DPsL}}(x+1;\theta,\beta)$ and $P_{\text{DPsL}}(x;\theta,\beta) \geq P_{\text{DPsL}}(x-1;\theta,\beta)$, which is equivalent to:

$$\frac{\theta(1+e^\theta) - \beta(e^\theta - 1)}{\theta(e^\theta - 1)} - 1 \leq x_m \leq \frac{\theta(1+e^\theta) - \beta(e^\theta - 1)}{\theta(e^\theta - 1)}.$$

Hence, if $\frac{\theta(1+e^\theta)-\beta(e^\theta-1)}{\theta(e^\theta-1)} \geq 0$, and:

1. If $\frac{\theta(1+e^\theta)-\beta(e^\theta-1)}{\theta(e^\theta-1)}$ is not an integer, x_m is given as the integer part of $\frac{\theta(1+e^\theta)-\beta(e^\theta-1)}{\theta(e^\theta-1)}$;

2. If $\frac{\theta(1+e^\theta)-\beta(e^\theta-1)}{\theta(e^\theta-1)}$ is an integer, the DPsL distribution is bimodal, with the modes given by $x_m^{(1)} = \frac{\theta(1+e^\theta)-\beta(e^\theta-1)}{\theta(e^\theta-1)}$ and $x_m^{(2)} = \frac{\theta(1+e^\theta)-\beta(e^\theta-1)}{\theta(e^\theta-1)} - 1$.

If $\frac{\theta(1+e^\theta)-\beta(e^\theta-1)}{\theta(e^\theta-1)} < 0$, the mode of the DPsL distribution is $x_m = 0$.

The hrf of the DPsL distribution can be obtained as:

$$h_{\text{DPsL}}(x;\theta,\beta) = \frac{P_{\text{DPsL}}(x;\theta,\beta)}{1 - F_{\text{DPsL}}(x;\theta,\beta)}$$

$$= \frac{(\beta+\theta x)e^{-\theta x} - (\beta+\theta(x+1))e^{-\theta(x+1)}}{e^{-\theta(1+x)}(\beta+(x+1)\theta)}.$$

The hrf of the DPsL distribution was plotted for some set of values for θ and β in Figure 2.

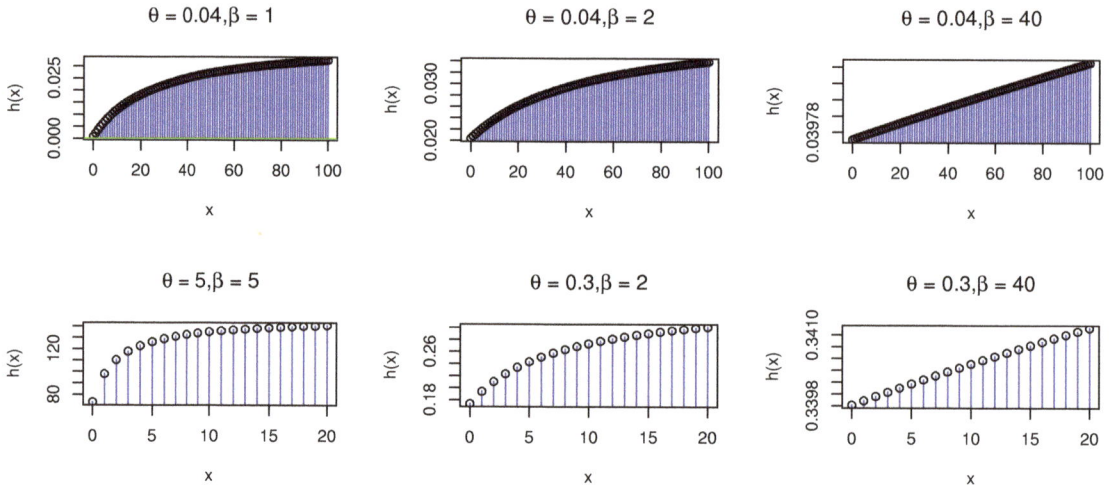

Figure 2. The pmf plots of the DPsL distribution for some set of values for θ and β.

Figure 2 clearly indicates that the hrf of the DPsL distribution is always increasing for different values of the parameters.

2.2. Identifiability

A set of unknown parameters of a model is stated to be identifiable if different sets of parameters give different distributions for a given x. Here, the identifiability property of the DPsL distribution is proved. Let $P_{\text{DPsL}}(x;\lambda_1)$ and $P_{\text{DPsL}}(x;\lambda_2)$ be different pmfs of the DPsL distribution indexed by $\lambda_1 = (\theta_1, \beta_1)$ and $\lambda_2 = (\theta_2, \beta_2)$, respectively. Then, the likelihood ratio is given by:

$$\begin{aligned} U &= \frac{P_{\text{DPsL}}(x; \lambda_1)}{P_{\text{DPsL}}(x; \lambda_2)} \\ &= \frac{\frac{(\beta_1 + \theta_1 x) e^{-\theta_1 x} - (\beta_1 + \theta_1 (x+1)) e^{-\theta_1 (x+1)}}{\beta_1}}{\frac{(\beta_2 + \theta_2 x) e^{-\theta_2 x} - (\beta_2 + \theta_2 (x+1)) e^{-\theta_2 (x+1)}}{\beta_2}} \\ &= \frac{\beta_2}{\beta_1} \frac{(\beta_1 + \theta_1 x) e^{-\theta_1 x} - (\beta_1 + \theta_1 (x+1)) e^{-\theta_1 (x+1)}}{(\beta_2 + \theta_2 x) e^{-\theta_2 x} - (\beta_2 + \theta_2 (x+1)) e^{-\theta_2 (x+1)}}. \end{aligned} \quad (5)$$

Taking logarithm of this ratio, we obtained:

$$\begin{aligned} \log U &= \log\left(\frac{\beta_2}{\beta_1}\right) + \log\left((\beta_1 + \theta_1 x) e^{-\theta_1 x} - (\beta_1 + \theta_1 (x+1)) e^{-\theta_1 (x+1)}\right) \\ &\quad - \log\left((\beta_2 + \theta_2 x) e^{-\theta_2 x} - (\beta_2 + \theta_2 (x+1)) e^{-\theta_2 (x+1)}\right). \end{aligned}$$

Now, by considering x as a continuous variable and taking the partial derivative of $\log U$ with respect to x and equating it to 0, we obtained:

$$\frac{\theta_1 (\theta_1 + \beta_1 - 1 + \theta_1 x - e^{\theta_1}(\theta_1 x + \beta_1 - 1))}{(\beta_1 + \theta_1 x) e^{-\theta_1 x} - (\beta_1 + \theta_1 (x+1)) e^{-\theta_1 (x+1)}} = \frac{\theta_2 (\theta_2 + \beta_2 - 1 + \theta_2 x - e^{\theta_2}(\theta_2 x + \beta_2 - 1))}{(\beta_2 + \theta_2 x) e^{-\theta_2 x} - (\beta_2 + \theta_2 (x+1)) e^{-\theta_2 (x+1)}},$$

which implies that:

$$e^{-(\theta_2 - \theta_1)x} \frac{(\beta_2 + \theta_2 x) - (\beta_2 + \theta_2(x+1))e^{-\theta_2}}{(\beta_1 + \theta_1 x) - (\beta_1 + \theta_1(x+1))e^{-\theta_1}} = \frac{\theta_2(\theta_2 + \beta_2 - 1 + \theta_2 x - e^{\theta_2}(\theta_2 x + \beta_2 - 1))}{\theta_1(\theta_1 + \beta_1 - 1 + \theta_1 x - e^{\theta_1}(\theta_1 x + \beta_1 - 1))}.$$

By performing $x \to +\infty$, we obtained $0 = \frac{\theta_2^2(1 - e^{\theta_2})}{\theta_1^2(1 - e^{\theta_1})}$ or $+\infty = \frac{\theta_2^2(1 - e^{\theta_2})}{\theta_1^2(1 - e^{\theta_1})}$ according to $\theta_2 > \theta_1$ or $\theta_2 < \theta_1$, respectively, which is impossible since $\theta_1 > 0$ and $\theta_2 > 0$. Therefore, $\theta_1 = \theta_2$. By taking into account this equality, by taking $x = 0$ in (5), we obtained $\frac{\beta_1 - (\beta_1 + \theta_1)e^{-\theta_1}}{\beta_2 - (\beta_2 + \theta_1)e^{-\theta_1}} = \frac{\beta_1}{\beta_2}$, which is possible if, and only if, $\beta_1 = \beta_2$. Therefore, we concluded that the DPsL model is identifiable and that the parameters uniquely determine the distribution, that is, $P_{\text{DPsL}}(x; \lambda_1) = P_{\text{DPsL}}(x; \lambda_2) \iff \lambda_1 = \lambda_2$.

2.3. Moments, Skewness and Kurtosis

In the rest of the study, X denotes a random variable that follows the DPsL distribution. Then, the probability generating function (pgf) of X can be derived as:

$$\begin{aligned} G(s) &= E(s^X) = \sum_{x=0}^{\infty} s^x P_{\text{DPsL}}(x; \theta, \beta) \\ &= \frac{e^{2\theta}\beta - e^{\theta}(\beta + s\beta + \theta - \theta s) + s\beta}{(e^{\theta} - s)^2 \beta}, \quad |s| < e^{\theta}. \end{aligned}$$

When s in pgf is substituted by e^t, the moment generating function (mgf) follows as:

$$M(t) = E(e^{tX}) = \frac{e^{2\theta}\beta - e^{\theta}(\beta + e^t\beta + \theta - \theta e^t) + e^t\beta}{(e^{\theta} - e^t)^2 \beta}, \quad t < \theta.$$

By using the well-known relationship between $M(t)$ and the (standard) moments of X, the first four moments of the DPsL distribution are:

$$E(X) = \frac{e^\theta(\beta+\theta) - \beta}{(e^\theta - 1)^2 \beta}, \quad (6)$$

$$E(X^2) = \frac{e^{2\theta}\beta + 3e^\theta\theta + e^{2\theta}\theta - \beta}{(e^\theta - 1)^3 \beta},$$

$$E(X^3) = \frac{-\beta - 3e^\theta\beta + 3e^{2\theta}\beta + e^{3\theta}\beta + 7e^\theta\theta + 10e^{2\theta}\theta + e^{3\theta}\theta}{(e^\theta - 1)^4 \beta}$$

and

$$E(X^4) = \frac{-\beta - 10e^\theta\beta + 10e^{3\theta}\beta + e^{4\theta}\beta + 15e^\theta\theta + 55e^{2\theta}\theta + 25e^{3\theta}\theta + e^{4\theta}\theta}{(e^\theta - 1)^5 \beta}.$$

Based on $E(X)$ and $E(X^2)$, the variance of X follows from the Koenig–Huygens formula as:

$$Var(X) = \frac{e^\theta[(e^\theta - 1)^2 \beta^2 + (e^{2\theta} - 1)\beta\theta - e^\theta \theta^2]}{(e^\theta - 1)^4 \beta^2}. \quad (7)$$

Expressions for skewness and kurtosis of the DPsL distribution can be derived explicitly by using the following formulas:

$$Skewness(X) = \frac{E(X^3) - 3E(X^2)E(X) + 2[E(X)]^3}{[Var(X)]^{3/2}}$$

and

$$Kurtosis(X) = \frac{E(X^4) - 4E(X^2)E(X) + 6E(X^2)[E(X)]^2 - 3[E(X)]^4}{[Var(X)]^2}.$$

2.4. Coefficient of Variation and Dispersion Index

The expressions of the coefficient of variation (CV) and dispersion index (DI) of X are given by:

$$CV(X) = \frac{\sqrt{Var(X)}}{E(X)} = \frac{\sqrt{(e^\theta - 1)^2 \beta^2 + (e^{2\theta} - 1)\beta\theta - e^\theta \theta^2}}{\sqrt{e^\theta}((\beta+\theta) - \beta e^{-\theta})}$$

and

$$DI(X) = \frac{Var(X)}{E(X)} = \frac{(e^\theta - 1)^2 \beta^2 + (e^{2\theta} - 1)\beta\theta - e^\theta \theta^2}{(e^\theta - 1)^2 \beta e^\theta (\beta + \theta)}, \quad (8)$$

respectively.

In full generality, when the DI is one, the distribution is equi-dispersed, and if DI is greater than (less than) one, the distribution is over-dispersed (under-dispersed). Some numerical values of the mean, variance, DI, skewness and kurtosis for the DPsL distribution for some values of the parameters are presented in Tables 1 and 2.

From the information contained in these tables, it is clear that the DPsL distribution would be an appropriate option for modelling under as well as over-dispersed and positively skewed datasets.

Table 1. Values for some moment measures for the DPsL distribution for $\beta = 1.5$ and different values of θ.

	θ				
Measures	4	5	6	7	8
Mean	0.06934	0.02955	0.01245	0.00518	0.00213
Variance	0.06901	0.02939	0.01241	0.00517	0.00212
DI	0.99525	0.99447	0.99649	0.99820	0.99911
Skewness	3.77540	5.77052	8.91577	13.86130	21.65950
Kurtosis	17.19030	35.95970	81.94370	194.42800	471.29900

Table 2. Values for some moment measures for the DPsL distribution for $\theta = 2$ and different values of β.

	β					
Measures	10	11	12	13	14	15
Mean	0.19272	0.18943	0.18669	0.18437	0.18238	0.18065
Variance	0.22724	0.22315	0.21972	0.21681	0.21430	0.21212
DI	1.17912	1.17799	1.17694	1.17595	1.17504	1.17420
Skewness	2.84454	2.86632	2.88457	2.90007	2.91340	2.92497
Kurtosis	12.9041	13.05360	13.17890	13.28540	13.3768	13.4562

2.5. Mean Residual Lifetime and Mean Past Lifetime

The mean residual lifetime (mrl) and mean past lifetime (mpl) of a component are two widely used measures to study the ageing behaviour of components. Both measures characterize the distribution uniquely. By assuming that the lifetime of a component is modelled by X, the mrl of X at $i = 0, 1, 2, \ldots$ is defined as:

$$\begin{aligned} \zeta(i) &= E(X - i \mid X \geq i) \\ &= \frac{1}{1 - F_{\text{DPsL}}(i - 1; \theta, \beta)} \sum_{j=i+1}^{\infty} (1 - F_{\text{DPsL}}(j - 1; \theta, \beta)). \end{aligned}$$

That is:

$$\begin{aligned} \zeta(i) &= \frac{1}{e^{-\theta i}(\beta + \theta i)} \sum_{j=i+1}^{\infty} e^{-\theta j}(\beta + \theta j) \\ &= \frac{e^{i\theta}\left((e^{\theta} - 1)\beta - i\theta + e^{\theta}(1 + i)\theta\right)}{e^{-\theta i}(\beta + \theta i)(e^{\theta} - 1)^2}. \end{aligned}$$

Furthermore, the mpl of X is another reliability measure that corresponds to the time elapsed since the failure of X given that the system has already failed before some i. Thus, the mpl of X at $i = 1, 2, \ldots$ is defined by:

$$\begin{aligned} \zeta^*(i) &= E(i - X \mid X < i) \\ &= \frac{1}{F_{\text{DPsL}}(i - 1; \theta, \beta)} \sum_{m=1}^{i} F_{\text{DPsL}}(m - 1; \theta, \beta), \end{aligned}$$

where $\zeta^*(0) = 0$. That is:

$$\zeta^*(i) = \frac{1}{\beta - e^{-\theta i}(\beta + i\theta)} \sum_{m=1}^{i} (\beta - e^{-m\theta}(\beta + m\theta))$$

$$= \frac{e^{-i\theta}}{\beta - e^{-\theta i}(\beta + i\theta)(e^\theta - 1)^2} \times$$
$$e^{-i\theta}\left\{\left[(e^\theta - 1)(1 + e^{\theta i}(1+i) - e^{\theta i}(1+i))\right]\beta - \left[e^{\theta(1+i)} + i - e^\theta(1+i)\right]\theta\right\}.$$

2.6. Stress–Strength Analysis

Stress–strength reliability has wide applications in almost all fields of engineering and machine learning. Let X_{stress} and $X_{strength}$ be random variables that model the stress and strength of a system, respectively. Then, the expected reliability can be calculated by the following formula:

$$Re_{Stress-Strength} = \Pr\left[X_{Stress} \leq X_{Strength}\right] = \sum_{x=0}^{\infty} P_{X_{Stress}}(x) S_{X_{Strength}}(x),$$

where $P_X(x)$ and $S_X(x)$ denote the pmf and sf, respectively, of a random variable X. Suppose that X_{stress} and $X_{strength}$ are two independent random variables following the DPsL (θ_1, β_1) and DPsL (θ_2, β_2) distributions, respectively. Then, from (3) and (4), the expected reliability is obtained in closed form as:

$$Re_{Stress-Strength} =$$
$$\frac{1}{\beta_1 \beta_2 (e^{\theta_1 + \theta_2} - 1)^3} \left\{ (e^{\theta_1} - 1)(e^{\theta_1+\theta_2} - 1)\beta_1 \left[(e^{\theta_1+\theta_2} - 1)\beta_2 + \theta_2 e^{\theta_1+\theta_2}\right] \right.$$
$$\left. - \theta_1 e^{\theta_1}\left[(e^{\theta_2} - 1)(e^{\theta_1+\theta_2} - 1)\beta_2 + e^{\theta_2}(1 - 2e^{\theta_1} + e^{\theta_1 \theta_2})\theta_2\right]\right\}.$$

Some numerical values for $Re_{Stress-Strength}$ for different values of the parameters are given in Tables 3–5.

From Tables 3 and 4, it is clear that the expected reliability increases (decreases) as $\beta_1 \to \infty$ ($\beta_2 \to \infty$). In addition, from Table 5, the expected reliability (decreases) as $\theta_1 \to \infty$ ($\theta_2 \to \infty$).

Table 3. Numerical values of $Re_{Stress-Strength}$ associated with the DPsL distribution at $\theta_1 = 0.3$, $\theta_2 = 0.1$ for different values of β_1 and β_2.

	$\theta_1 = 0.3, \theta_2 = 0.1$			
$\beta_1 \to$ $\beta_2 \downarrow$	1	2	3	7
1	0.82926	0.87819	0.89449	0.91314
2	0.6227	0.75075	0.77358	0.79967
3	0.63327	0.70827	0.73327	0.76184
7	0.57728	0.65972	0.68721	0.71862

Table 4. Numerical values of $Re_{Stress-Strength}$ associated with the DPsL distribution at $\theta_1 = 0.6$, $\theta_2 = 0.01$ for different values of β_1 and β_2.

	$\theta_1 = 0.6, \theta_2 = 0.01$			
$\beta_1 \rightarrow$ $\beta_2 \downarrow$	1	2	3	7
1	0.99903	0.99933	0.99943	0.99955
2	0.98084	0.98488	0.98623	0.98777
3	0.97478	0.98007	0.98183	0.98384
7	0.96785	0.97456	0.97679	0.97935

Table 5. Numerical values of $Re_{Stress-Strength}$ associated with the DPsL distribution at $\beta_1 = 1$, $\beta_2 = 1.5$ for different values of θ_1 and θ_2.

	$\beta_1 = 1, \beta_2 = 1.5$			
$\theta_1 \rightarrow$ $\theta_2 \downarrow$	0.1	0.5	0.7	0.9
0.1	0.40431	0.82936	0.87387	0.89879
0.5	0.04949	0.35792	0.45733	0.52947
0.7	0.02667	0.24765	0.33651	0.40671
0.9	0.01619	0.17754	0.25273	0.31061

2.7. Generating Random Values from the DPsL Distribution

Random values from the DPsL distribution can be generated by following the algorithm given below.

1. Generate u as a realization of a random variable U with the $U(0,1)$ distribution.
2. With the expression of the quantile function of the PsL distribution in mind, compute:

$$y = -\frac{\beta}{\theta} - \frac{1}{\theta}W_{-1}(e^{-\beta}\beta(u-1)),$$

where $W_{-1}(x)$ denotes the negative branch of Lambert–W function.

3. Then, $x = \lfloor y \rfloor$ represents a realization of a random variable with the DPsL distribution.

To generate a random sample of size n, repeat the algorithm n times.

3. Estimation Methods

The estimation of unknown parameters of a distribution is critical in accurately determining the behaviour of this distribution. Here, we use classical methods of estimation such as the method of maximum likelihood (mle) and weighted least square (wls) estimation for this purpose.

3.1. Maximum Likelihood Estimation

Let $X_1, X_2, \ldots, X_n$ be a random sample taken from the DPsL (θ, β) distribution, and $x_1, x_2, \ldots, x_n$ be observations of this random sample. The likelihood function is given by:

$$L = \left(\frac{1}{\beta}\right)^n \left\{\prod_{i=1}^{n}\left[(\beta + \theta x_i)e^{-\theta x_i} - (\beta + \theta(x_i + 1))e^{-\theta(x_i+1)}\right]\right\}$$

and the log likelihood function is given by:

$$\log L = -n \log \beta + \sum_{i=1}^{n} \log\left[(\beta + \theta x_i)e^{-\theta x_i} - (\beta + \theta(x_i+1))e^{-\theta(x_i+1)}\right].$$

Then, the maximum likelihood estimates (MLEs) of θ and β were obtained by maximizing L or log L with respect to these parameters. They can also be determined as the solutions of the normal equations given by:

$$\frac{\partial \log L}{\partial \theta} = 0 \implies$$

$$\sum_{i=1}^{n} \frac{e^{-\theta(2x_i+1)}\left[e^{\theta x_i}(x_i+1)(\theta x_i + \theta + \beta - 1) - e^{\theta(x_i+1)}x_i(\theta x_i + \beta - 1)\right]}{(\beta + \theta x_i)e^{-\theta x_i} - (\beta + \theta(x_i+1))e^{-\theta(x_i+1)}} = 0$$
(9)

and

$$\frac{\partial \log L}{\partial \beta} = 0 \implies$$

$$-\frac{n}{\beta} + \sum_{i=1}^{n} \frac{e^{-\theta x_i} - e^{\theta(x_i+1)}}{(\beta + \theta x_i)e^{-\theta x_i} - (\beta + \theta(x_i+1))e^{-\theta(x_i+1)}} = 0.$$
(10)

Equations (9) and (10) can be solved by numerical optimization techniques using mathematical software such as MATHEMATICA, MATHCAD and R.

3.2. Weighted Least Squares Estimation

Let $X_{(1)}, X_{(2)}, ..., X_{(n)}$ be the order statistics of a random sample taken from the DPsL (θ, β) distribution, and $x_{(1)}, x_{(2)}, \ldots, x_{(n)}$ be observations of these random variables. The weighted least squares estimates (WLEs) of the parameters θ and β of the DPsL distribution were obtained by maximizing the following function with respect to θ and β:

$$W = \sum_{i=1}^{n} \frac{(n+1)^2(n+2)}{i(n-i+1)}\left[F_{\text{DPsL}}\left(x_{(i)}; \theta, \beta\right) - \frac{i}{n+1}\right]^2.$$

3.3. Simulation Study

The current section deals with examining the efficiency of two estimation methods for estimating the parameters of the DPsL distribution using simulation. Estimates were calculated for different values of parameters (($\theta = 0.5, \beta = 1$) and ($\theta = 2.2, \beta = 1.5$)) for various sample sizes (25, 50, 75, 100) using the two estimation methods discussed and, thus, compared. Then, $N = 1000$ samples of values from the DPsL distribution using methods discussed in Section 2.7 were generated. The indices such as values of the estimates, mean square errors (MSEs), average absolute biases (Bias) and average mean relative estimates (MREs) were calculated in R software using the following formulas:

$$\text{MSE} = \frac{1}{N}\sum_{i=1}^{N}(\hat{\zeta}_i - \zeta)^2, \quad \text{Bias} = \frac{1}{N}\sum_{i=1}^{N}|\hat{\zeta}_i - \zeta|,$$

$$\text{MRE} = \frac{1}{N}\sum_{i=1}^{N}\frac{|\hat{\zeta}_i - \zeta|}{\zeta},$$

where $\zeta = \theta$ or β, and the index i refers to the ith sample. Simulation results, including values of estimates, Bias, MSEs and MREs for the two parameters θ and β of the DPsL distribution using the estimation approaches discussed, are reported in Tables 6 and 7.

Table 6. Simulation results of our estimation approaches for the DPsL distribution with $\theta = 0.5, \beta = 1$.

n	Indices	MLE		WLSE	
		θ	β	θ	β
25	Estimates	0.4902	1.1126	0.4289	1.0049
	Bias	0.0098	0.1126	0.0710	0.0049
	MSE	0.0069	0.1217	0.0448	7.1204×10^{-5}
	MRE	0.1319	0.1326	0.2599	0.0049
50	Estimates	0.4904	1.0808	0.4243	1.0035
	Bias	0.0096	0.0808	0.0757	0.0035
	MSE	0.0033	0.0379	0.0444	3.26×10^{-5}
	MRE	0.0908	0.0868	0.2444	0.0035
75	Estimates	0.4920	1.0614	0.4247	1.0030
	Bias	0.0079	0.0614	0.0753	0.0030
	MSE	0.0019	0.0160	0.0429	2.217×10^{-5}
	MRE	0.0704	0.0614	0.2328	0.0030
100	Estimates	0.4926	1.0553	0.4225	1.0028
	Bias	0.0074	0.0553	0.0775	0.0028
	MSE	0.0015	0.0119	0.0427	1.904×10^{-5}
	MRE	0.0634	0.0553	0.2350	0.0028

Table 7. Simulation results of our estimation approaches for the DPsL distribution with $\theta = 2.2$, $\beta = 1.5$.

n	Indices	MLE		WLSE	
		θ	β	θ	β
25	Estimates	2.3027	1.2005	1.7547	1.3939
	Bias	0.1027	0.2995	0.4452	0.1060
	MSE	1.5197	0.2979	0.2564	0.0154
	MRE	0.2509	0.3328	0.2079	0.0734
50	Estimates	2.1843	1.2621	1.8200	1.3993
	Bias	0.0157	0.2378	0.3799	0.1007
	MSE	0.2381	0.2774	0.1932	0.0134
	MRE	0.1829	0.3193	0.1749	0.0681
75	Estimates	2.1853	1.3217	1.8370	1.4066
	Bias	0.0147	0.1783	0.3629	0.0934
	MSE	0.1565	0.2519	0.1750	0.0118
	MRE	0.1457	0.2949	0.1689	0.0639
100	Estimates	2.2052	1.4245	1.8489	1.4133
	Bias	0.0052	0.0755	0.3511	0.0867
	MSE	0.0993	0.2468	0.1627	0.0105
	MRE	0.1154	0.2784	0.1642	0.0598

From the above tables, it is clear that, for estimating θ, the corresponding MLE performed well, and for β, the corresponding WLSE outperformed the MLE.

4. INAR(1) Process with DPsL Innovations

Numerous fields, such as agriculture, epidemiology, actuarial science, finance, etc., have come across certain time series of counts. Analysing these kinds of datasets using the INAR(1) process was first applied using Poisson innovations by [12,13]. Suppose that $\{\varepsilon_t\}_{t\in\mathbb{Z}}$ are the innovations, so are independent and identically distributed (iid) random variables, with $E(\varepsilon_t) = \mu_\varepsilon$ and variance $\text{Var}(\varepsilon_t) = \sigma_\varepsilon^2$. A stochastic process $\{X_t\}_{t\in\mathbb{Z}}$ defined as:

$$X_t = p \circ X_{t-1} + \varepsilon_t,$$

with $0 \leq p < 1$, is stated to be an INAR(1) process. The symbol $\circ$ is called as binomial thinning operator, which can be described as:

$$p \circ X_{t-1} = \sum_{j=1}^{X_{t-1}} U_j,$$

where $\{U_j\}_{j\in\mathbb{Z}}$ is a sequence of iid Bernoulli random variables with parameter p. The one step transition probability of the INAR(1) process is given by:

$$\Pr(X_t = k \mid X_{t-1} = l) = \sum_{i=1}^{\min(k,l)} \Pr(B = i)\Pr(\varepsilon_t = k-i), \ k, l \geq 0,$$

where B denotes a random variable following the Binomial (n,p) distribution. The mean, variance and dispersion index (DI) of $\{X_t\}_{t\in\mathbb{Z}}$ are given by [21]. They are:

$$E(X_t) = \frac{\mu_\varepsilon}{1-p}, \tag{11}$$

$$\text{Var}(X_t) = \frac{p\mu_\varepsilon + \sigma_\varepsilon^2}{1-p^2} \tag{12}$$

and

$$DI(X_t) = \frac{DI_\varepsilon + p}{1+p}, \tag{13}$$

where μ_ε, σ_ε^2 and DI_ε are the mean, variance and DI of the innovation distribution. The results of [12,13] influenced us to propose a new INAR(1) process with DPsL innovations, which are capable of modelling over as well as under-dispersed count datasets. Suppose that $\{\varepsilon_t\}_{t\in\mathbb{Z}}$ follow a DPsL distribution; then, the one step transition probability matrix of the corresponding process is:

$$\Pr(X_t = k \mid X_{t-1} = l) = $$
$$\sum_{i=1}^{\min(k,l)} \binom{l}{i} p^i (1-p)^{l-i}$$
$$\times \frac{(\beta + \theta(k-i))e^{-\theta(k-i)} - (\beta + \theta((k-i)+1))e^{-\theta((k-i)+1)}}{\beta},$$

which hereafter is called the INAR(1)DPsL process. By substituting μ_ε, σ_ε^2, and DI_ε in (11)–(13) with (6)–(8), the mean, variance and DI of the INAR(1)DPsL process could be attained. The conditional expectation and variance of the INAR(1)DPsL process are given by:

$$E(X_t \mid X_{t-1}) = pX_{t-1} + \mu_\varepsilon, \tag{14}$$

and
$$\text{Var}(X_t \mid X_{t-1}) = p(1-p)X_{t-1} + \sigma_\varepsilon^2, \tag{15}$$

respectively, where μ_ε and σ_ε^2 are given in (6) and (7), respectively (see [13,21]).

4.1. Estimation

Here, the inference of the INAR(1)DPsL process was examined using two estimation methods: the conditional maximum likelihood (CML) and Yule–Walker (YW) methods. A simulation study was performed to assess the efficiency of the two methods.

4.1.1. Conditional Maximum Likelihood

Let $X_1, X_2, \ldots, X_T$ be a random sample taken from the INAR(1)DPsL process, and $x_1, x_2, \ldots, x_T$ be observations of this random sample. Then, the conditional log likelihood function of the INAR(1)DPsL process is given by:

$$\begin{aligned} \ell(\Theta) &= \sum_{t=2}^{T} \log[\Pr(X_t = x_t \mid X_{t-1} = x_{t-1})] \\ &= \sum_{t=2}^{T} \log \left[\sum_{i=1}^{\min(x_t, x_{t-1})} \binom{x_{t-1}}{i} p^i (1-p)^{x_{t-1}-i} \right. \\ &\quad \left. \frac{(\beta + \theta(x_t - i))e^{-\theta(x_t-i)} - (\beta + \theta(x_t - i + 1))e^{-\theta(x_t-i+1)}}{\beta} \right], \end{aligned} \tag{16}$$

where $\Theta = (\theta, \beta, p)$ is the vector of unknown parameters to be estimated. Maximizing (16) with respect to Θ yields the CML estimates (CMLEs). In this regard, we used the optim-function in R software for the same. In addition, the fdHess function in R was used to obtain the observed information matrix and, hence, the standard errors (SE) of estimates of parameters in the INAR(1)DPsL process.

4.1.2. Yule–Walker

The YW estimates (YWEs) of the INAR(1)DPsL process were computed by solving simultaneous equations of sample and theoretical moments. Since the autocorrelation function (ACF) of the INAR(1) process at lag h was $\rho_X(h) = p^h$, the YWE of p is given by:

$$\hat{p}_{YW} = \frac{\sum\limits_{t=2}^{T}(x_t - \bar{x})(x_{t-1} - \bar{x})}{\sum\limits_{t=1}^{T}(x_t - \bar{x})^2}.$$

Now, the YWEs for θ and β were obtained by solving the equations of sample mean equals theoretical mean and sample dispersion equals theoretical dispersion of the process. Here, by denoting as $\hat{\theta}_{YW}$ and $\hat{\beta}_{YW}$ the YWEs of θ and β, respectively, the following relationship holds:

$$\hat{\beta}_{YW} = \frac{\hat{\theta}_{YW} e^{\hat{\theta}_{YW}}}{\bar{x}(1 - \hat{p}_{YW})(e^{\hat{\theta}_{YW}} - 1)^2 - (e^{\hat{\theta}_{YW}} - 1)}, \tag{17}$$

where $\bar{x} = \sum_{t=1}^{T} x_t / N$. Substituting $\hat{\beta}_{YW}$ with (17) in (13) and equating (13) to sample dispersion, we obtained $\hat{\theta}_{YW}$.

4.2. Simulation of INAR(1)DPsL Process

Here, a simulation study was conducted to comprehensively determine the performance of CMLEs and YWEs of the parameters of the INAR(1)DPsL process. In this regard, we generated $N = 1000$ samples each of sizes $n = 25, 50, 100$ from the proposed distribution for two sets of parameter values ($\theta = 0.1, \beta = 1.1$ and $\theta = 3, \beta = 4$). For each n, average absolute bias, MSE and MRE for the parameters were calculated for the two methods. The simulation results are presented in Table 8.

Table 8. Simulation results of the INAR(1)DPsL process.

		$\theta = 0.1, \beta = 1.1$					
Sample Size (n)	Parameters	CML			YW		
		Bias	MSE	MRE	Bias	MSE	MRE
25	θ	0.0183	0.0019	0.3271	0.0644	0.0047	0.6443
	β	0.2067	1.8986	0.9959	0.1305	0.2778	0.1186
	p	0.0449	0.0248	0.4289	0.6456	0.2627	2.1519
50	θ	0.0035	0.0007	0.1758	0.0633	0.0043	0.6330
	β	0.0916	0.3807	0.4131	0.0687	0.0881	0.0624
	p	0.0113	0.1187	0.2345	0.0232	0.0255	0.0773
100	θ	0.0014	0.0001	0.0841	0.0623	0.0040	0.6225
	β	0.0657	0.0178	0.0732	0.0369	0.0351	0.0336
	p	0.0096	0.0072	0.1812	0.0200	0.0019	0.0668
		$\theta = 3, \beta = 4$					
Sample Size (n)	Parameters	CML			YW		
		Bias	MSE	MRE	Bias	MSE	MRE
25	θ	0.7181	0.0853	1.6194	0.6708	1.1252	0.2236
	β	0.5259	0.2878	0.6254	0.1634	0.0276	0.0408
	p	0.0344	0.0502	0.2546	0.3809	0.5484	0.5441
50	θ	0.5244	0.0824	1.2841	0.5281	0.9221	0.1764
	β	0.0461	0.0434	0.4046	0.1609	0.0263	0.0402
	p	0.0054	0.0382	0.2157	0.2889	0.5318	0.4128
100	θ	0.0709	0.0816	0.3019	0.2791	0.1449	0.0930
	β	0.0363	0.0241	0.2953	0.1606	0.0260	0.0402
	p	0.0032	0.0282	0.1813	0.2553	0.0624	0.3647

From the above table, we observed that the average biases, MSEs and MREs of CMLEs tended to zero quicker than those of YWEs, making them efficient for small as well as large sample sizes. Therefore, the CML estimation was preferred to attain unknown parameters of the INAR(1)DPsL process.

5. Empirical Study

Three real datasets were used in this section to illustrate the performance of the DPsL distribution over some competitive distributions. The capability of the fitted distributions was compared using the goodness of fit criterion with its corresponding p-value.

5.1. Failure Times

The data of failure times for a sample of 15 electronic components in an acceleration life test (see [22]) were considered here. These data were based on the discretization concept. Adopting a data analysis setting, we compared the DPsL, discrete three-parameter Lindley (DTPL) (see [9]), discrete log-logistic (DLL) (see [23]), discrete inverse Weibull (DIW) (see [2]), discrete Burr–Hutke (DBH) (see [6]), discrete Pareto (DP) (see [3]), Poisson (P) and geometric (G) distributions. The MLEs with standard errors (SEs) and confidence intervals (CIs) for the parameter(s), estimated $-\log$ Likelihood ($-$L), Akaike information criterion (AIC), Bayesian information criterion (BIC) and goodness of fit statistic (Kolmogorov statistic (K-S) and p-value) of these distributions for this dataset are given in Table 9.

Table 9. The MLEs, CIs, $-$L, AIC, BIC, K-S and p-values of all the fitted distributions for the failure times data.

Statistic	Model			
	DPsL	DTPL	DLL	DIW
θ MLE (SE)	0.0623 (0.0043)	0.5084 (0.8277)	21.4627 (1.392)	0.0077 (0.0032)
CI	(0.0538, 0.0707)	($-$1.1139, 2.1307)	(18.7344, 24.1909)	(0.0013, 0.0140)
β MLE (SE)	1.3427 (0.1572)	0.0924 (0.1506)	1.7906 (0.1001)	0.7111 (0.0343)
CI	(1.0331, 1.6492)	(0.0629, 0.1219)	(1.5943, 1.9868)	(0.6439, 0.7782)
λ MLE (SE)	$-$	0.9397 (0.0040)	$-$	$-$
CI	$-$	(0.0845, 0.1003)	$-$	$-$
$-$L	64.2790	64.2790	65.6904	70.4214
AIC	132.558	134.558	135.3809	144.8427
BIC	133.9741	136.6822	136.797	146.2588
K-S value	0.1114	0.1116	0.1351	0.2194
p-value	0.9819	0.9816	0.9133	0.4068
Statistic	Model			
	DBH	DP	P	G
θ MLE (SE)	0.999 (0.0019)	0.7202 (0.0158)	27.535 (0.3498)	0.035 (0.0023)
CI	(0.9953, 1.0030)	(0.6893, 0.7511)	(26.8495, 28.2208)	(0.0305, 0.0395)
$-$L	91.3684	77.4023	151.2064	66.0001
AIC	184.7368	156.8047	304.4129	133.0002
BIC	185.4448	157.5127	305.1209	134.7083
K-S value	0.7912	0.4053	0.3815	0.1766
p-value	1.582×10^{-10}	0.0097	0.0179	0.6743

From Table 9, it is evident that, besides the DPsL distribution, the DTPL, G and DLL distributions also performed quite well, but it is clear that the DPsL distribution was the best among them, since it had the lowest K-S, AIC and BIC, with a higher p-value. In order to illustrate this claim, Figure 3 provides the probability–probability (P–P) plots, and Figure 4 displays the estimated cdfs of the fitted distributions.

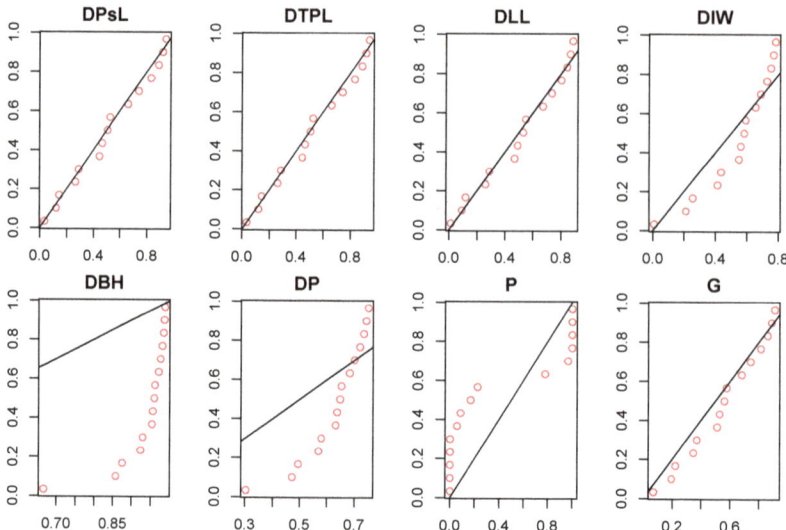

Figure 3. The P–P plots for the fitted distributions using the failure times data.

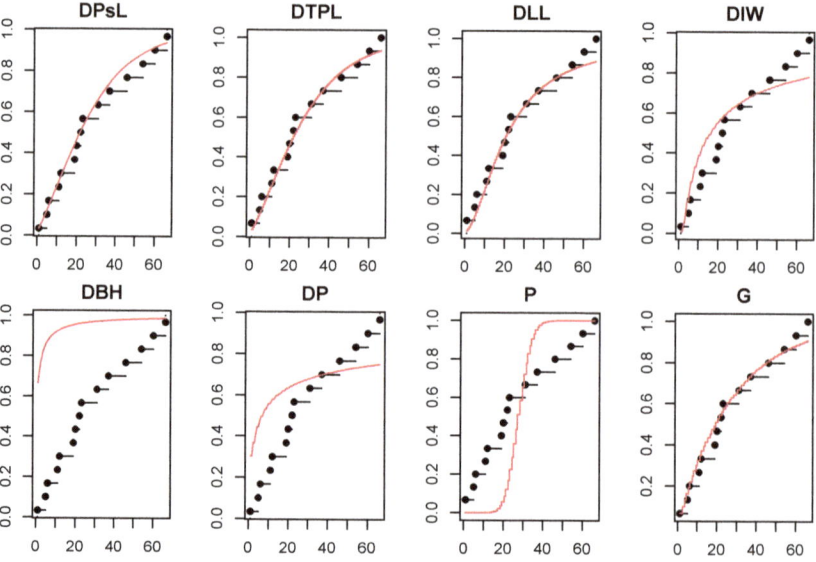

Figure 4. Estimated cdfs of the fitted distributions using the failure times data.

From the above figures, we could infer that the DPsL distribution yielded a better fit among other fitted distributions. Table 10 completes these results by presenting some descriptive measures of the fitted DPsL distribution. Hence, it is evident that the fitted DPsL distribution was over dispersed, moderately right skewed and leptokurtic.

Table 10. Values of some descriptive statistics of the DPsL distribution for the failure times data.

Mean	Variance	DI	Skewness	Kurtosis
27.8667	395.5822	14.1955	0.7020	2.3149

5.2. Numbers of Borers

The second dataset was the biological experiment data, which represented the number of European corn borer (No. ECB) larvae Pyrausta in the field (see [24]). It was an experiment conducted randomly on eight hills in 15 replications, and the experimenter counted the number of borers per hill of corn. The fits of the DPsL distribution were compared together with some competitive distributions which were the new Poisson weighted exponential (NPWE) (see [16]), DIW, discrete Burr-XII (DBXII) (see [23]), discrete Bilal (DBl) (see [8]), DP, DBH and Poisson (P) distributions. The MLEs with their corresponding SEs, CIs under the form (lower bound of the CI (LCI), upper bound of the CI (UCI)) for the parameter(s) and goodness of fit test for the numbers of borers dataset are reported in Table 11.

Table 11. The MLE, LCI, UCI, −L, AIC, BIC, χ^2 and p-values for the one parameter distributions considered using the number of borers dataset.

X	Observed Frequency	Expected Frequency							
		DPsL	NPWE	DIW	DBXII	DBl	DP	DBH	P
0	43	44.62	48.32	41.37	43.84	32.74	64.45	68.07	27.22
1	35	30.46	28.86	41.85	39.61	39.59	20.15	21.97	40.38
2	17	19.07	17.24	15.42	15.62	24.27	9.69	10.51	29.95
3	11	11.34	10.29	7.17	7.20	12.50	5.65	5.98	14.81
4	5	6.51	6.15	3.94	3.91	5.97	3.68	3.75	5.49
5	4	3.65	3.67	2.42	2.37	2.74	2.58	2.51	1.63
6	1	2.01	2.19	1.61	1.59	1.23	1.90	1.75	0.40
7	2	1.09	1.31	1.13	1.09	0.54	1.46	1.26	0.09
8	2	1.25	1.94	5.09	4.80	0.24	1.15	0.93	0.02
Total	120	120	120	120	120	120	120	120	120
θ MLE		0.7219	0.1434	0.345	0.519	0.6565	0.3292	0.8654	1.4834
θ SE		0.0122	0.2945	0.043	0.051	0.0017	0.0031	0.0035	0.0101
θ LCI		0.6980	0	0.261	0.419	0.6532	0.3232	0.8585	1.4635
θ UCI		0.7459	0.4339	0.429	0.619	0.6599	0.3352	0.8723	1.5033
β MLE		2.4635	0.5896	1.541	2.358				
β SE		0.1367	1.3706	0.156	0.3656				
β LCI		2.1956	0	1.235	1.641				
β UCI		2.7315	3.2760	1.847	3.074				
−L		200.4152	200.8774	204.812	204.293	204.6753	220.6182	214.0490	219.1879
AIC		404.8303	405.7548	413.624	412.586	411.3505	443.2363	430.0979	440.3759
BIC		410.4053	411.3297	419.199	418.161	414.138	446.0238	432.8854	443.1634
χ^2		1.4445	2.1591	5.511	4.664	10.0780	26.645	25.795	38.583
Degrees of freedom		3	3	3	3	4	4	4	4
p-value		0.9194	0.8267	0.138	0.198	0.0731	<0.001	<0.001	<0.001

From the above table, it is evident that, besides the DPsL distribution, the NPWE distribution also performed quite well, but it is clear that the DPsL distribution was the best among them, since it had the lowest −L, AIC, BIC and χ^2 value with the highest p-value.

From Figure 5, we could infer that the DPsL distribution yielded a better fit among other fitted distributions. To complete this, Table 12 contains some descriptive measures of the fitted DPsL distribution. Hence, here also, it is evident that the fitted DPsL distribution was over-dispersed, moderately right skewed and leptokurtic.

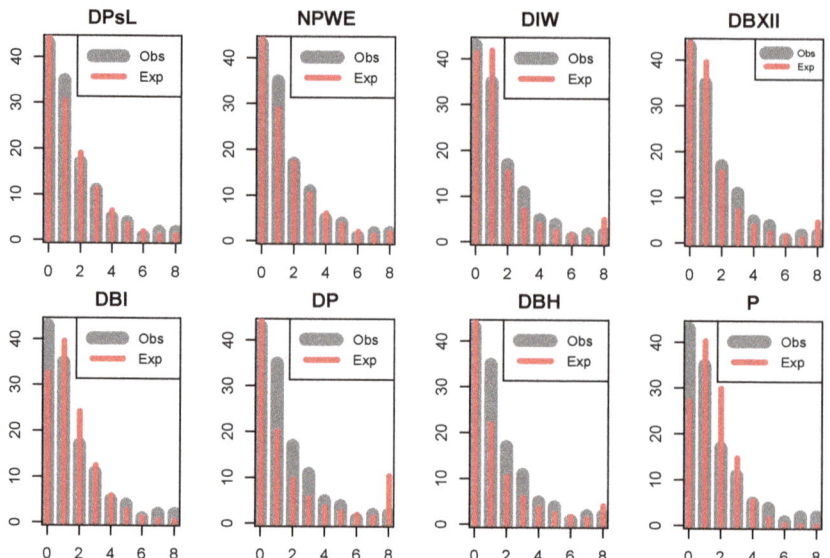

Figure 5. The estimated pmfs of the fitted distributions for the number of borers dataset.

Table 12. Values of some descriptive statistics of the DPsL distribution for the number of borers dataset.

Mean	Variance	DI	Skewness	Kurtosis
1.5917	2.6249	1.6491	0.8172	2.6435

5.3. Numbers of Claims

In this part, a comparison of the performance of the INAR(1)DPsL process with the INAR(1)DTPL (see [7]), INAR(1)NPWE (see [16]), INAR(1)DPLi (see [15]) and INAR(1)G (see [14]) processes was conducted. The one-step translation probabilities of the competitive INAR(1) processes were given as follows:

1. For the INAR(1)DPLi process:

$$\Pr(X_t = k \mid X_{t-1} = l) = \sum_{i=0}^{\min(k,l)} \binom{l}{i} p^i (1-p)^{l-i} \frac{\theta^2 (k-i+\theta+2)}{(\theta+1)^{k-i+3}}, \; \theta > 0.$$

2. For the INAR(1)DTPL process:

$$\Pr(X_t = k \mid X_{t-1} = l) =$$
$$\sum_{i=1}^{\min(k,l)} \binom{l}{i} p^i (1-p)^{l-i}$$
$$\times \frac{\lambda^{k-i} \{\beta(\lambda(\log(\lambda)-1)+1) + (\lambda-1)\log(\lambda)(\alpha+\beta(k-i))\}}{\beta - \alpha \log(\lambda)},$$
$$0 < \lambda < 1, \alpha\theta + \beta > 0, \theta = -\log(\lambda).$$

3. For the INAR(1)NPWE process:

$$\Pr(X_t = k \mid X_{t-1} = l) = \sum_{i=0}^{\min(k,l)} \binom{l}{i} p^i (1-p)^{l-i} \alpha(1+\theta)(1+\alpha+\alpha\theta)^{-(k-i)-1},$$

$\alpha > 0, \theta > 0$.

4. For the INAR(1)G process:

$$\Pr(X_t = k \mid X_{t-1} = l) = \sum_{i=1}^{\min(k,l)} \binom{l}{i} p^i (1-p)^{l-i} \left[\alpha(1-\alpha)^{k-i} \right],$$

$0 < \alpha < 1$.

The third data we used here were to illustrate the application of the DPsL distribution in the INAR(1) process. Originally, the data were studied by [25], which consisted of 67 monthly claims for short-term disability benefits made by injured workers to the B.C. Workers' Compensation Board (WCB). These data were reported from the BC Center, Richmond, for the period of 10 years from 1985 to 1994. The mean, variance, and DI of the dataset were 8.6042, 11.2392 and 1.3062, respectively. To check whether the data considered had statistically significant over-dispersion, the hypothesis test proposed by [26] was applied. The value test statistic was 51.971 with a p-value less than 0.001, which showed the data had significant over-dispersion. Figure 6 displays the plots of the autocorrelation function (ACF), partial ACF (PACF), histogram and time series plots, and in the PACF plot the unique first lag significance indicated that these data could be used for modelling the INAR(1) process.

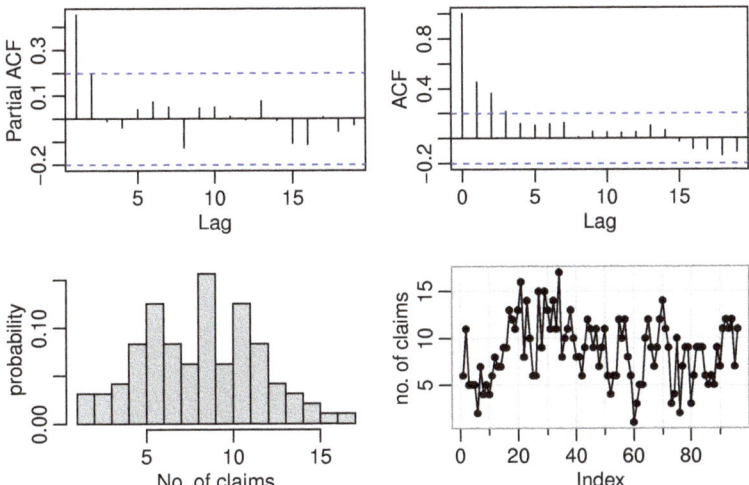

Figure 6. PACF, ACF, histogram and time series plot for the number of claims dataset.

The parameter estimates, modelling adequacy criteria, theoretical mean, variance and DI of the fitted INAR(1) process were recorded in Table 13. Since the INAR(1)DPsL process had lesser values for -L, AIC and BIC statistics than those of the INAR(1)DTPL, INAR(1)NPWE, INAR(1)PL and INAR(1)G processes, the INAR(1)DPsL process provided better fits than the competitors. Additionally, the obtained DI value of the INAR(1)DPsL process was very near the empirical one. It is conclusive that the INAR(1)DPsL process impressively explained the characteristics of the dataset.

Table 13. The estimates and modelling adequacy statistics of the fitted distributions for the number of claims dataset.

Model	Parameters	Estimates (SE)	$-L$	AIC	BIC	μ_x	σ_x^2	DI_x
INAR(1)DPsL	θ	0.4835(0.0526)	245.3344	496.6687	504.3618	8.7812	15.9626	1.8178
	β	1.9214(0.1254)						
	p	0.5620(0.0439)						
INAR(1)DTPL	θ	$-0.1211(0.3067)$	245.3344	498.6687	508.9261	8.7604	16.2473	1.8546
	β	0.4834(0.1903)						
	λ	0.7477(0.0324)						
	p	0.5619(0.0439)						
INAR(1)NPWE	θ	0.1729(0.8221)	252.3457	510.6913	518.3844	8.3542	18.4417	2.2075
	β	0.2738(0.1919)						
	p	0.6432(0.0338)						
INAR(1)DPL	θ	0.4938(0.0583)	248.6185	501.237	506.3657	9.375	23.1842	2.4729
	p	0.6139(0.0381)						
INAR(1)G	θ	0.2431(0.0263)	252.3457	508.6913	513.82	9.0417	31.4719	3.4808
	p	0.6432(0.0338)						
		Empirical				8.6042	11.2392	1.3062

The residual analysis was conducted to check whether the fitted INAR(1)DPsL process was accurate. For that, Pearson residuals for the INAR(1)DPsL process were calculated through the following formula:

$$r_t = \frac{x_t - E(X_t \mid X_{t-1} = x_{t-1})}{\text{Var}(X_t \mid X_{t-1} = x_{t-1})^{1/2}},$$

where $E(X_t \mid X_{t-1} = x_{t-1})$ and $\text{Var}(X_t \mid X_{t-1} = x_{t-1})$ were derived from (14) and (15), respectively. When the fitted INAR(1) process was statistically valid, the Pearson residual had to be uncorrelated and should have had zero mean and unit variance [27]. Here, we obtained the mean and variance of the Pearson residuals of the INAR(1)DPsL process as 0.035 and 0.967, respectively, which were very close to the desired values. According to the results of [28], the INAR(1)DPsL process for the data was

$$X_t = 0.5620 \circ X_{t-1} + \varepsilon_t,$$

where the innovation process was such that ε_t follows the DPsL (0.4835, 1.9214) distribution. Predicted values of the monthly number of claims dataset and the ACF plot of the Pearson residuals via this process were displayed in Figure 7.

Based on this figure, the ACF plot of the Pearson residuals specified that there was no presence of autocorrelation for the Pearson residuals.

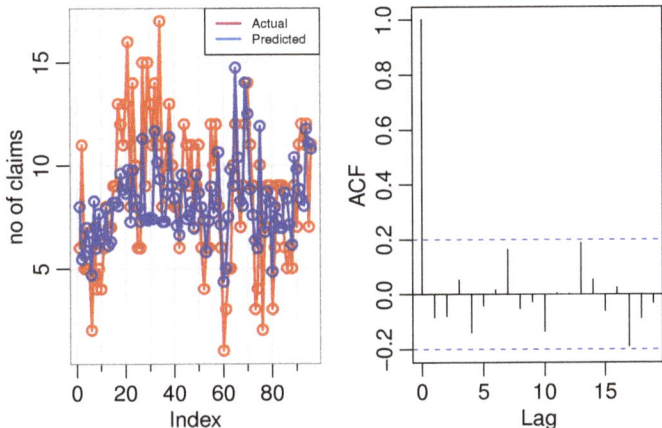

Figure 7. The predicted values of the number of claims dataset (**left**) and the ACF plot of the Pearson residuals (**right**).

6. Concluding Remarks

In this paper, a two-parameter discrete distribution, namely, the discrete Pseudo Lindley (DPsL) distribution, was proposed. Its primary motivation is the ability to model various phenomena with under- and over-dispersed observed values. Various statistical properties, almost all having a closed form, revealed the flexibility and simplicity of the distribution. The estimation of the unknown parameters was performed using two different methods. They conducted an extensive simulation study to reveal the finite sample performance of the distribution. Crucially, a new INAR(1) process with DPsL innovations was developed and studied in detail. Three real-life datasets were considered to prove the efficiency of the proposed distribution. As a future work, we could consider other methods of discretization for the PsL distribution, which would then provide better properties than the survival discretization method. Furthermore, we can attempt to extend it to bivariate models. We hope that the DPsL distribution, as well as the related modelling strategy, will be an interesting alternative to modelling count data, especially in modelling the over-dispersed count data.

Author Contributions: Conceptualization, M.R.I. and R.M.; methodology, M.R.I., C.C., V.D. and R.M.; software, V.D.; validation, M.R.I., C.C., V.D. and R.M.; software, V.D.; investigation, M.R.I., C.C., V.D. and R.M.; data curation, V.D.; writing—original draft preparation, V.D.; writing—review and editing, M.R.I., C.C., V.D. and R.M.; visualization, M.R.I., C.C., V.D. and R.M. All authors have read and agreed to the published version of the manuscript.

Funding: This research received no external funding.

Acknowledgments: We are grateful to the three reviewers for their helpful suggestions on the manuscript.

Conflicts of Interest: The authors declare no conflict of interest.

References

1. Gómez-Déniz, E.; Calderín-Ojeda, E. The discrete Lindley distribution: Properties and applications. *J. Stat. Comput. Simul.* **2011**, *81*, 1405–1416. [CrossRef]
2. Jazi, M.A.; Lai, C.D.; Alamatsaz, M.H. A discrete inverse Weibull distribution and estimation of its parameters. *Stat. Methodol.* **2010**, *7*, 121–132. [CrossRef]
3. Krishna, H.; Pundir, P.S. Discrete Burr and discrete Pareto distributions. *Stat. Methodol.* **2009**, *6*, 177–188. [CrossRef]
4. Roy, D. Discrete Rayleigh distribution. *IEEE Trans. Reliab.* **2004**, *53*, 255–260. [CrossRef]

5. Hussain, T.; Aslam, M.; Ahmad, M. A two parameter discrete Lindley distribution. *Rev. Colomb. Estadística* **2016**, *39*, 45–61. [CrossRef]
6. El-Morshedy, M.; Eliwa, M.; Nagy, H. A new two-parameter exponentiated discrete Lindley distribution: Properties, estimation and applications. *J. Appl. Stat.* **2020**, *47*, 354–375. [CrossRef]
7. El-Morshedy, M.; Eliwa, M.S.; Altun, E. Discrete Burr–Hatke distribution with properties, estimation methods and regression model. *IEEE Access* **2020**, *8*, 74359–74370. [CrossRef]
8. Altun, E.; El-Morshedy, M.; Eliwa, M. A study on discrete Bilal distribution with properties and applications on integer-valued autoregressive process. *Revstat. Stat. J* **2020**, *18*, 70–99.
9. Eliwa, M.S.; Altun, E.; El-Dawoody, M.; El-Morshedy, M. A new three-parameter discrete distribution with associated INAR (1) process and applications. *IEEE Access* **2020**, *8*, 91150–91162. [CrossRef]
10. Eldeeb, A.S.; Ahsan-ul Haq, M.; Eliwa, M.S. A discrete Ramos-Louzada distribution for asymmetric and over-dispersed data with leptokurtic-shaped: Properties and various estimation techniques with inference. *AIMS Math.* **2022**, *7*, 1726–1741. [CrossRef]
11. Ramos, P.L.; Louzada, F. A distribution for instantaneous failures. *Stats* **2019**, *2*, 247–258. [CrossRef]
12. McKenzie, E. Some simple models for discrete variate time series 1. *J. Am. Water Resour. Assoc.* **1985**, *21*, 645–650. [CrossRef]
13. Al-Osh, M.A.; Alzaid, A.A. First-order integer-valued autoregressive (INAR (1)) process. *J. Time Ser. Anal.* **1987**, *8*, 261–275. [CrossRef]
14. Aghababaei Jazi, M.; Jones, G.; Lai, C.D. Integer valued AR (1) with geometric innovations. *J. Iran. Stat. Soc.* **2012**, *11*, 173–190.
15. Lívio, T.; Khan, N.M.; Bourguignon, M.; Bakouch, H.S. An INAR (1) model with Poisson–Lindley innovations. *Econ Bull* **2018**, *38*, 1505–1513.
16. Altun, E. A new generalization of geometric distribution with properties and applications. *Commun. Stat. Simul. Comput.* **2020**, *49*, 793–807. [CrossRef]
17. Altun, E.; Bhati, D.; Khan, N.M. A new approach to model the counts of earthquakes: INARPQX (1) process. *SN Appl. Sci.* **2021**, *3*, 274. [CrossRef] [PubMed]
18. Huang, J.; Zhu, F. A New First-Order Integer-Valued Autoregressive Model with Bell Innovations. *Entropy* **2021**, *23*, 713. [CrossRef] [PubMed]
19. Winkelmann, R. Duration dependence and dispersion in count-data models. *J. Bus. Econ. Stat.* **1995**, *13*, 467–474.
20. Zeghdoudi, H.; Nedjar, S. A Pseudo Lindley distribution and its application. *Afr. Stat.* **2016**, *11*, 923–932. [CrossRef]
21. Weiß, C.H. *An Introduction to Discrete-Valued Time Series*; John Wiley & Sons: Hoboken, NJ, USA, 2018.
22. Lawless, J.F. *Statistical Models and Methods for Lifetime Data*; John Wiley & Sons: Hoboken, NJ, USA, 2011; Volume 362.
23. Para, B.A.; Jan, T.R. Discrete version of log-logistic distribution and its applications in genetics. *Int. J. Mod. Math. Sci.* **2016**, *14*, 407–422.
24. Bodhisuwan, W.; Sangpoom, S. The discrete weighted Lindley distribution. In Proceedings of the 2016 12th International Conference on Mathematics, Statistics, and Their Applications (ICMSA), Banda Aceh, Indonesia, 4–6 October 2016; pp. 99–103.
25. Freeland, R.K. Statistical Analysis of Discrete Time Series with Application to the Analysis of Workers' Compensation Claims Data. Ph.D. Thesis, University of British Columbia, Vancouver, BC, Canada, 1998.
26. Schweer, S.; Weiß, C.H. Compound Poisson INAR (1) processes: Stochastic properties and testing for overdispersion. *Comput. Stat. Data Anal.* **2014**, *77*, 267–284. [CrossRef]
27. Harvey, A.C.; Fernandes, C. Time series models for count or qualitative observations. *J. Bus. Econ. Stat.* **1989**, *7*, 407–417.
28. Jazi, M.A.; Jones, G.; Lai, C.D. First-order integer valued AR processes with zero inflated Poisson innovations. *J. Time Ser. Anal.* **2012**, *33*, 954–963. [CrossRef]

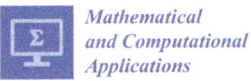

Article

The Sine Modified Lindley Distribution

Lishamol Tomy [1], Veena G [2] and Christophe Chesneau [3,*]

1. Department of Statistics, Deva Matha College, Kuravilangad 686633, Kerala, India; lishatomy@gmail.com
2. Department of Statistics, St. Thomas College, Palai 686574, Kerala, India; veenagpillai@hotmail.com
3. Laboratoire de Mathématiques Nicolas Oresme (LMNO), Université de Caen Normandie, Campus II, Science 3, 14032 Caen, France
* Correspondence: christophe.chesneau@gmail.com

Abstract: The paper contributes majorly in the development of a flexible trigonometric extension of the well-known modified Lindley distribution. More precisely, we use features from the sine generalized family of distributions to create an original one-parameter survival distribution, called the sine modified Lindley distribution. As the main motivational fact, it provides an attractive alternative to the Lindley and modified Lindley distributions; it may be better able to model lifetime phenomena presenting data of leptokurtic nature. In the first part of the paper, we introduce it conceptually and discuss its key characteristics, such as functional, reliability, and moment analysis. Then, an applied study is conducted. The usefulness, applicability, and agility of the sine modified Lindley distribution are illustrated through a detailed study using simulation. Two real data sets from the engineering and climate sectors are analyzed. As a result, the sine modified Lindley model is proven to have a superior match to important models, such as the Lindley, modified Lindley, sine exponential, and sine Lindley models, based on goodness-of-fit criteria of importance.

Keywords: goodness-of-fit; trigonometric distributions; modified Lindley distribution; engineering data; climate data; statistical analysis

1. Introduction

The last few years in applied sciences have been marked by the need and volume of data to be analyzed. To meet this need, new models have been proposed, and their improvement is a hot topic. These require, among other things, the underlying development of new (statistical or probabilistic) distributions. In this regard, one idea is to modify existing distributions in order to make the corresponding models more flexible and adaptable to several kinds of data. Hence, several modifications based on mathematical techniques have been proposed, generating distributions classified under "families of distributions". The readers are referred to [1] for a bird's-eye view. In recent times, the families described by "trigonometric transformations" have gained a lot of interest because of their applicability and working capability in a variety of situations. Related to this topic, Refs. [2–4] were among the first to study the sinusoidal transformation that leads to the sine generated (S-G) family. For this family, the cumulative distribution function (cdf) and probability density function (pdf) are given by

$$F_S(x;\eta) = \sin\left[\frac{\pi}{2}G(x;\eta)\right], x \in \mathbb{R}, \quad (1)$$

and

$$f_S(x;\eta) = \frac{\pi}{2}g(x;\eta)\cos\left[\frac{\pi}{2}G(x;\eta)\right], x \in \mathbb{R}, \quad (2)$$

respectively, where $G(x;\eta)$ and $g(x;\eta)$ represent the cdf and pdf of a certain continuous distribution with a parameter vector denoted by η. Thus, the functions $F_S(x;\eta)$ and $f_S(x;\eta)$ are linked to a baseline or parent distribution determined beforehand, relying on the purpose of study. It is worth noting that the baseline cdf has not been supplemented with any additional parameters. The S-G family was developed as a viable substitute for the parent distribution; we can see it from the following first-order stochastic ordering (FOSO) property:

$$G(x;\eta) \leq F_S(x;\eta) \tag{3}$$

for all $x \in \mathbb{R}$, as well as the possibility of creating versatile statistical distributions that can accept a wide range of data. To make the statement clearer, the exponential distribution is used as a parent distribution by [2] to define the sine exponential distribution. The inverse Weibull (IW) distribution proposed by [5] is used as the reference distribution by [4], thus creating the sine IW (SIW) distribution. The sine power Lomax distribution investigated by [6] is one of the most recent works highlighting the importance of the S-G family. It enhances the parental power Lomax distribution on several functional aspects. Among the trigonometric families of distributions, a few of them, including the C-S family by [7], SKum-G family by [8], STL-G family by [9], and T-G family by [10], were influenced by these efforts.

In this research, we contribute to the developments of the S-G family by linking it to a particular one-parameter distribution introduced by [11]: the modified Lindley (ML) distribution. The sine ML (S-ML) distribution is thus introduced. In order to comprehend the outlined approach, a review of the ML distribution is essential. As a first comment, the ML distribution presented by [11] is achieved by implementing the tuning exponential function $e^{-\theta x}$, with $\theta > 0$, to the Lindley distribution, with the motive of modifying its capabilities for new modeling perspectives. On the mathematical side, the cdf and pdf of the ML distribution are defined by

$$G_{ML}(x;\theta) = \begin{cases} 1 - \left[1 + e^{-\theta x}\dfrac{\theta x}{1+\theta}\right]e^{-\theta x}, & \text{if } x > 0 \\ 0, & \text{if } x \leq 0 \end{cases} \tag{4}$$

and

$$g_{ML}(x;\theta) = \begin{cases} \dfrac{\theta}{1+\theta}e^{-2\theta x}\left[(1+\theta)e^{\theta x} + 2\theta x - 1\right], & \text{if } x > 0 \\ 0, & \text{if } x \leq 0 \end{cases}, \tag{5}$$

respectively. Basically, the ML distribution satisfies the following FOSO property:

$$G_L(x;\theta) \leq G_{ML}(x;\theta) \leq G_E(x;\theta) \tag{6}$$

for all $x \in \mathbb{R}$, where $G_L(x;\theta)$ and $G_E(x;\theta)$ represent the cdfs of the Lindley and exponential distributions, respectively. In this sense, the ML distribution constitutes a real alternative to these two classical distributions. The ML distribution is also identified as a linear combination of the exponential distribution with parameter θ and the gamma distribution with parameters $(2, 2\theta)$, and it has an "increasing-reverse bathtub-constant" hazard rate function (hrf). The real benefit is quite noteworthy; the ML model is superior to the Lindley and exponential models for the three data sets seen in [11]. A few inspired distributions enhancing or generalising the ML distribution were proposed for the purpose of optimality. These include the Poisson ML distribution by [12], wrapped ML distribution by [13], and discrete ML distribution by [14].

The immediate aim of the S-ML distribution is to use the S-G technique to enhance the effectiveness of the ML distribution on diverse data sets. In particular, thanks to the FOSO properties in Equations (3) and (6), it is a real and attractive alternative to the Lindley and ML distributions. Further exploration in the following research will reveal deeper motives. To summarise, the S-ML model's utility and adaptability make it particularly appealing to fit data from various fields. Remarkably, the characterized pdf shows a variety of curve shapes, some of which have only one mode, are decreasing, and are asymmetrical to the right. In comparison to the pdf of the ML distribution, when it is unimodal, the pdf of the S-ML distribution has a more rounded peak, meaning that it is better adapted to fit a data histogram presenting a high kurtosis level. Furthermore, the S-ML distribution exhibits a non-monotonic hrf which is "increasing-reverse bathtub-constant" shaped. The hrf of the ML distribution has this feature as well. As with other competent models, the accuracy of the fits is persistent in the case of the S-ML model due to their characteristics. The claim is demonstrated by examining two published real-world data sets, primarily from engineering and climate data, against twelve competent models.

We prepare the rest of the paper in the following manner. The concept, quality, and key aspects of the S-ML distribution are covered in Section 2. A moment analysis is conducted in Section 3. The maximum likelihood estimation of the parameter θ is explained in Section 4. A simulation study is presented in Section 5. Section 6 assesses the proposed model's applicability to real-world data. Finally, in Section 7, the conclusions are provided.

2. The S-ML Distribution

The mathematical foundation for the S-ML distribution is first presented.

2.1. Functional Analysis

To begin, we perform a functional analysis of the S-ML distribution. By substituting Equations (4) and (5) in Equations (1) and (2), respectively, we derive the major functions of the S-ML distribution; the cdf and pdf are given as follows

$$F_{S-ML}(x;\theta) = \begin{cases} \cos\left[\dfrac{\pi}{2}\left(1 + e^{-\theta x}\dfrac{x\theta}{1+\theta}\right)e^{-\theta x}\right], & \text{if } x > 0 \\ 0, & \text{if } x \leq 0 \end{cases}$$

and

$$f_{S-ML}(x;\theta) = \begin{cases} \dfrac{\pi}{2}\dfrac{\theta}{1+\theta}e^{-2\theta x}\left[(1+\theta)e^{\theta x} + 2x\theta - 1\right]\sin\left[\dfrac{\pi}{2}\left(1 + e^{-\theta x}\dfrac{x\theta}{1+\theta}\right)e^{-\theta x}\right], & \text{if } x > 0 \\ 0, & \text{if } x \leq 0 \end{cases}, \quad (7)$$

with $\theta > 0$. As a primary result mentioned in the introduction section, the following FOSO property holds: $G_{ML}(x;\theta) \leq F_{S-ML}(x;\theta)$ for any $x \in \mathbb{R}$, making an immediate difference between the ML and S-ML modeling from the cdf viewpoint. Differences can also be observed on the respective pdfs, as discussed below. Naturally, variant forms of $f_{S-ML}(x;\theta)$ can be obtained by changing the value of θ. Due to the relative complexity of this function in the analytical sense, we propose a graphical study for shape analysis. The more representative shapes of this pdf are shown in Figure 1.

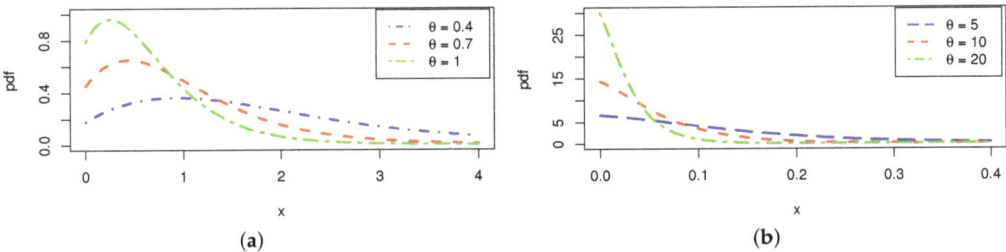

Figure 1. Plots of (a) unimodal shapes and (b) decreasing shapes for $f_{S-ML}(x;\theta)$.

We can observe from Figure 1, that, for smaller values of θ, the plot of $f_{S-ML}(x;\theta)$ is unimodal, and for larger values of θ, the plot of $f_{S-ML}(x;\theta)$ is decreasing. As a result, the S-ML distribution is suitable for modeling a vast majority of lifetime phenomena. Compared to the parent ML distribution, the following observations are made: When it is unimodal, we observe that the pdf of the S-ML distribution has a more rounded peak, meaning that it is better adapted to fit a data histogram presenting a high kurtosis level. In other words, the S-ML model is more able to analyze data of a leptokurtic nature.

2.2. Reliability Analysis

We complete the previous functional analysis by studying the complementary reliability functions, such as the survival function (sf), hrf (for hazard rate function), reversed hrf (rhrf), second rate of failure (srf), and the cumulative hrf (chrf) of the S-ML distribution. In a broader sense, the sf measures the probability that the life of an item will survive beyond any specified time. Mathematically, the sf of the S-ML distribution is given by

$$S_{S-ML}(x;\theta) = 1 - F_{S-ML}(x;\theta) = \begin{cases} 1 - \cos\left[\dfrac{\pi}{2}\left(1+e^{-\theta x}\dfrac{x\theta}{1+\theta}\right)e^{-\theta x}\right], & \text{if } x > 0 \\ 1, & \text{if } x \leq 0 \end{cases}.$$

The hrf measures the likelihood of an item deteriorating or expiring depending on its lifetime. As a direct consequence, it is critical in the classification of survival distributions. The hrf of the S-ML distribution is specified by

$$\begin{aligned} h_{S-ML}(x;\theta) &= \dfrac{f_{S-ML}(x;\theta)}{S_{S-ML}(x;\theta)} \\ &= \begin{cases} \dfrac{\pi}{2}\dfrac{\theta}{1+\theta}e^{-2\theta x}\left[(1+\theta)e^{\theta x}+2x\theta-1\right]\cot\left[\dfrac{\pi}{4}\left(1+e^{-\theta x}\dfrac{x\theta}{1+\theta}\right)e^{-\theta x}\right], & \text{if } x > 0 \\ 0, & \text{if } x \leq 0 \end{cases}. \end{aligned}$$

Further, Figure 2 displays the shapes of this hrf for various values of θ.

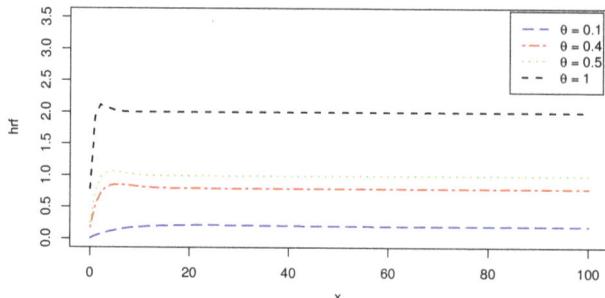

Figure 2. Plots of $h_{S-ML}(x;\theta)$ with selected values of θ.

Figure 2 emphasizes that the hrf of the S-ML distribution has "increasing-reverse bathtub-constant" shapes, which is also possessed by the hrf of the ML distribution. This makes a solid difference between the Lindley and exponential distributions. It is also a desirable property for modelling purposes.

The rhrf is the ratio between the pdf to its cdf and it plays a role in analyzing censored data. Analytically, it corresponds to

$$r_{S-ML}(x;\theta) = \begin{cases} \dfrac{\pi}{2}\dfrac{\theta}{1+\theta}e^{-2\theta x}\left[(1+\theta)e^{\theta x}+2x\theta-1\right]\tan\left[\dfrac{\pi}{2}\left(1+e^{-\theta x}\dfrac{x\theta}{1+\theta}\right)e^{-\theta x}\right], & \text{if } x > 0 \\ 0, & \text{if } x \leq 0 \end{cases}.$$

The srf is the logarithmic ratio of the sf at time x and $x+1$, and it is given by

$$r^*_{S-ML}(x;\theta) = \begin{cases} \ln\left(\dfrac{1-\cos\left[(\pi/2)(1+e^{-\theta x}x\theta/(1+\theta))e^{-\theta x}\right]}{1-\cos\left[(\pi/2)(1+e^{-\theta(x+1)}(x+1)\theta/(1+\theta))e^{-\theta(x+1)}\right]}\right), & \text{if } x > 0 \\ 0, & \text{if } x \leq 0 \end{cases}.$$

The chrf is the negative logarithm of sf and is given by

$$H_{S-ML}(x;\theta) = \begin{cases} -\ln\left(1 - \cos\left[\frac{\pi}{2}\left(1 + e^{-\theta x}\frac{x\theta}{1+\theta}\right)e^{-\theta x}\right]\right), & \text{if } x > 0 \\ 0, & \text{if } x \leq 0 \end{cases}.$$

With these functions, we conclude different reliability analysis in regard with the S-ML distribution.

3. Moment Analysis

For any lifetime distribution, a moment analysis is necessary to handle numerically its modeling capacities, identifying the behavior of various central and dispersion moment parameters, as well as moment skewness and kurtosis coefficients.

As a first notion, for any positive integer $r \geq 1$, and a random variable X with the S-ML distribution, the r-th moment of X exists. It can be expressed as

$$\text{mom}(r) = \mathbb{E}(X^r) = \int_0^{+\infty} x^r f_{S-ML}(x;\theta) dx$$
$$= \frac{\pi}{2}\frac{\theta}{1+\theta}\int_0^{+\infty} x^r e^{-2\theta x}\left[(1+\theta)e^{\theta x} + 2x\theta - 1\right]\sin\left[\frac{\pi}{2}\left(1 + e^{-\theta x}\frac{x\theta}{1+\theta}\right)e^{-\theta x}\right]dx. \quad (8)$$

Integral developments in the classical sense are limited. Computer software, on the other hand, can be used to quantitatively evaluate it for a given θ.

We propose a series development of mom(r) in the next result, which can be used for computational purposes in a less opaq method than a "ready to use but black box" computer program.

Proposition 1. *The r-th moment of X can be expanded as*

$$\text{mom}(r) = \frac{r}{\theta^r}\sum_{k=1}^{+\infty}\sum_{\ell=0}^{2k}\binom{2k}{\ell}\frac{(-1)^{k+1}}{(2k)!}\left(\frac{\pi}{2}\right)^{2k}(1+\theta)^{-\ell}\frac{(\ell+r-1)!}{(\ell+2k)^{r+\ell}}.$$

Proof. For the proof, we do not directly use the integral expression of mom(r) as described in (8). An integration by part gives

$$\text{mom}(r) = \int_0^{+\infty} x^r f_{S-ML}(x;\theta)dx = r\int_0^{+\infty} x^{r-1}S_{S-ML}(x;\theta)dx.$$

Now, by utilizing the series expansion of the cosine function and the classical binomial formula, we obtain

$$S_{S-ML}(x;\theta) = 1 - \sum_{k=0}^{+\infty}\frac{(-1)^k}{(2k)!}\left(\frac{\pi}{2}\right)^{2k}\left(1 + e^{-\theta x}\frac{x\theta}{1+\theta}\right)^{2k}e^{-2k\theta x}$$
$$= \sum_{k=1}^{+\infty}\frac{(-1)^{k+1}}{(2k)!}\left(\frac{\pi}{2}\right)^{2k}\left(1 + e^{-\theta x}\frac{x\theta}{1+\theta}\right)^{2k}e^{-2k\theta x} \quad (9)$$
$$= \sum_{k=1}^{+\infty}\sum_{\ell=0}^{2k}\binom{2k}{\ell}\frac{(-1)^{k+1}}{(2k)!}\left(\frac{\pi}{2}\right)^{2k}\left(\frac{\theta}{1+\theta}\right)^{\ell}x^{\ell}e^{-(\ell+2k)\theta x}.$$

Hence, after some developments including the change of variable $y = (\ell + 2k)\theta x$ (so that $dx = [1/((\ell + 2k)\theta)]dy$), and the calculus of gamma-type integral, we get

$$\text{mom}(r) = r \int_0^{+\infty} x^{r-1} \left[\sum_{k=1}^{+\infty} \sum_{\ell=0}^{2k} \binom{2k}{\ell} \frac{(-1)^{k+1}}{(2k)!} \left(\frac{\pi}{2}\right)^{2k} \left(\frac{\theta}{1+\theta}\right)^\ell x^\ell e^{-(\ell+2k)\theta x} \right] dx$$

$$= r \sum_{k=1}^{+\infty} \sum_{\ell=0}^{2k} \binom{2k}{\ell} \frac{(-1)^{k+1}}{(2k)!} \left(\frac{\pi}{2}\right)^{2k} \left(\frac{\theta}{1+\theta}\right)^\ell \int_0^{+\infty} x^{r+\ell-1} e^{-(\ell+2k)\theta x} dx$$

$$= \frac{r}{\theta^r} \sum_{k=1}^{+\infty} \sum_{\ell=0}^{2k} \binom{2k}{\ell} \frac{(-1)^{k+1}}{(2k)!} \left(\frac{\pi}{2}\right)^{2k} (1+\theta)^{-\ell} \frac{(\ell+r-1)!}{(\ell+2k)^{r+\ell}}.$$

Proposition 1 is proved. □

Then, based on Proposition 1, the following finite sum approximation remains acceptable:

$$\text{mom}(r) \approx \frac{r}{\theta^r} \sum_{k=1}^{U} \sum_{\ell=0}^{2k} \binom{2k}{\ell} \frac{(-1)^{k+1}}{(2k)!} \left(\frac{\pi}{2}\right)^{2k} (1+\theta)^{-\ell} \frac{(\ell+r-1)!}{(\ell+2k)^{r+\ell}},$$

where U represents any large integer.

From the above moment formulas, we can easily derive the mean, variance, moment skewness coefficient and moment kurtosis coefficient; the mean is given by $\text{mom}(1)$, the variance is obtained as $\mathbb{V}(X) = \mathbb{E}\left((X - \text{mom}(1))^2\right)$, the moment skewness coefficient can be derived as $\text{MS} = \mathbb{E}\left((X - \text{mom}(1))^3\right) / \mathbb{V}(X)^{3/2}$ and the moment kurtosis coefficient can be derived as $\text{MK} = \mathbb{E}\left((X - \text{mom}(1))^4\right) / \mathbb{V}(X)^2$.

Table 1 gives a glimpse of these values for different values of θ.

Table 1. Values of various moment measures of the S-ML distribution.

θ	mom(1)	$\mathbb{V}(X)$	MS	MK
0.5	1.448465	1.210367	1.449555	6.374828
1.0	0.6803329	0.3003946	1.520729	6.633623
1.5	0.4363871	0.1322356	1.573212	6.842306
2.0	0.318833	0.07375445	1.611987	7.005059
2.5	0.2502845	0.04687065	1.64148	7.133455
3.0	0.2056052	0.03235949	1.664569	7.236649

From Table 1, we can observe that, as the value of the parameter θ of the S-ML distribution increases, all the considered measures increase. Furthermore, since $\text{MS} > 0$, it is clear that the S-ML distribution is mainly right-skewed, and since $\text{MK} > 3$, it is mainly leptokurtic.

We can complete the previous moment results by investigating the incomplete moments. To begin, let $r \geq 1$ be an integer, $t \geq 0$, and X be a random variable with the S-ML distribution. Based on this variable, we define its incomplete version by $Y(t) = X$ if $X \leq t$ and $Y(t) = 0$ if $X > t$. Then, the r-th incomplete moment of X given at t exists, and it is defined by

$$\text{mom}(r,t) = \mathbb{E}(Y(t)^r) = \int_0^t x^r f_{S-ML}(x;\theta) dx.$$

It is involved in developments of important probabilistic objects, such as mean deviations, income curves, etc. More basically, it can be viewed as a truncated version of the standard r-moment. We may refer to [15] in this regard.

In the next results, we present a series expansion of $\text{mom}(r,t)$, which can be used for approximation purposes.

Proposition 2. *The r-th incomplete moment of X given at t exists and can be expanded as*

$$\mathrm{mom}(r,t) = -t^r \left\{ 1 - \cos\left[\frac{\pi}{2}\left(1 + e^{-\theta t}\frac{t\theta}{1+\theta}\right)e^{-\theta t}\right]\right\}$$
$$+ \frac{r}{\theta^r}\sum_{k=1}^{+\infty}\sum_{\ell=0}^{2k}\binom{2k}{\ell}\frac{(-1)^{k+1}}{(2k)!}\left(\frac{\pi}{2}\right)^{2k}(1+\theta)^{-\ell}\frac{1}{(\ell+2k)^{r+\ell}}\gamma(r+\ell,(\ell+2k)\theta t),$$

where $\gamma(a,t)$ *denotes the incomplete gamma function defined by* $\gamma(a,t) = \int_0^t x^{a-1}e^{-x}dx$, *where* $a > 0$ *and* $t \geq 0$.

Proof. The proof follows the lines of the one of Proposition 1. An integration by part gives

$$\mathrm{mom}(r,t) = \int_0^t x^r f_{S-ML}(x;\theta)dx = -t^r S_{S-ML}(t;\theta) + r\int_0^t x^{r-1} S_{S-ML}(x;\theta)dx.$$

It follows from the series expansion in Equation (9) and the change of variable $y = (\ell + 2k)\theta x$ that

$$\int_0^t x^{r-1} S_{S-ML}(x;\theta)dx = \sum_{k=1}^{+\infty}\sum_{\ell=0}^{2k}\binom{2k}{\ell}\frac{(-1)^{k+1}}{(2k)!}\left(\frac{\pi}{2}\right)^{2k}\left(\frac{\theta}{1+\theta}\right)^\ell \int_0^t x^{r+\ell-1}e^{-(\ell+2k)\theta x}dx$$
$$= \frac{1}{\theta^r}\sum_{k=1}^{+\infty}\sum_{\ell=0}^{2k}\binom{2k}{\ell}\frac{(-1)^{k+1}}{(2k)!}\left(\frac{\pi}{2}\right)^{2k}(1+\theta)^{-\ell}\frac{1}{(\ell+2k)^{r+\ell}}\gamma(r+\ell,(\ell+2k)\theta t).$$

Therefore

$$\mathrm{mom}(r,t) = -t^r\left\{1 - \cos\left[\frac{\pi}{2}\left(1 + e^{-\theta t}\frac{t\theta}{1+\theta}\right)e^{-\theta t}\right]\right\}$$
$$+ \frac{r}{\theta^r}\sum_{k=1}^{+\infty}\sum_{\ell=0}^{2k}\binom{2k}{\ell}\frac{(-1)^{k+1}}{(2k)!}\left(\frac{\pi}{2}\right)^{2k}(1+\theta)^{-\ell}\frac{1}{(\ell+2k)^{r+\ell}}\gamma(r+\ell,(\ell+2k)\theta t).$$

This concludes the proof of Proposition 2. □

In some sense, Proposition 2 generalizes Proposition 1; by taking $t \to +\infty$, Proposition 2 becomes Proposition 1.

The rest of the study is devoted to the applicability of the S-ML model, illustrated with concrete examples of data analysis.

4. Inferential Analysis

The inference of the S-ML model is covered in this section. The parameter θ is supposed to be unknown. In order to estimate it, the maximum likelihood estimation method is employed. We adopt the methodology as described in a broader context, as seen in [16].

Thus, the next is a mathematical representation of this methodology in the setting of the S-ML distribution. First, let n be a positive integer and $x_1, x_2, \ldots, x_n$ be observations drawn from a random variable X following the S-ML distribution. Then, the corresponding likelihood function and log-likelihood function are as follows

$$L = \prod_{i=1}^n f_{S-ML}(x_i;\theta) = \left(\frac{\pi}{2}\right)^n \left(\frac{\theta}{1+\theta}\right)^n e^{-2\theta\sum_{i=1}^n x_i} \prod_{i=1}^n \left[(1+\theta)e^{\theta x_i} + 2x_i\theta - 1\right]$$
$$\times \prod_{i=1}^n \sin\left[\frac{\pi}{2}\left(1 + e^{-\theta x_i}\frac{x_i\theta}{1+\theta}\right)e^{-\theta x_i}\right],$$

and

$$\ln L = n \ln \pi - n \ln 2 + n \ln \theta - n \ln(1+\theta) - 2\theta \sum_{i=1}^{n} x_i$$
$$+ \sum_{i=1}^{n} \ln\left[(1+\theta)e^{\theta x_i} + 2x_i\theta - 1\right] + \sum_{i=1}^{n} \ln\left\{\sin\left[\frac{\pi}{2}\left(1 + e^{-\theta x_i}\frac{x_i\theta}{1+\theta}\right)e^{-\theta x_i}\right]\right\},$$

respectively. The maximum likelihood estimate (MLE) of θ can be defined via the following argmax definition:

$$\hat{\theta} = \text{argmax} \ln_{\theta > 0} L. \tag{10}$$

This estimate can be formalized through the solution of the non-linear equations expressed as $d \ln L/d\theta = 0$, where

$$\frac{d}{d\theta} \ln L = \frac{n}{\theta} - \frac{n}{1+\theta} - 2\sum_{i=1}^{n} x_i + \sum_{i=1}^{n} \left[\frac{e^{\theta x_i}(\theta x_i + x_i + 1) + 2x_i}{(1+\theta)e^{\theta x_i} + 2x_i\theta - 1}\right] +$$
$$\sum_{i=1}^{n} \left[\frac{\pi}{2}e^{-\theta x_i}\left(-\frac{\theta x_i^2 e^{-\theta x_i}}{\theta+1} + \frac{x_i e^{-\theta x_i}}{\theta+1} - \frac{\theta x_i e^{-\theta x_i}}{(\theta+1)^2}\right) - \frac{\pi}{2}x_i e^{-\theta x_i}\left(1 + \frac{\theta x_i e^{-\theta x_i}}{\theta+1}\right)\right]$$
$$\times \cot\left[\frac{\pi}{2}\left(1 + e^{-\theta x_i}\frac{x_i\theta}{1+\theta}\right)e^{-\theta x_i}\right].$$

There is no analytical solution for this equation, but $\hat{\theta}$ can be determined at least numerically with any statistical software such as the R software (see [17]). Based on $\hat{\theta}$, the estimated pdf (epdf) of the S-ML model is given by $f_{S-ML}(x; \hat{\theta})$ and the estimated cdf (ecdf) of the S-ML model is given by $F_{S-ML}(x; \hat{\theta})$.

Let $I(\theta) = -E\left[d^2 \ln[f_{S-ML}(X; \theta)]/d\theta^2\right]$ be the expected Fisher information matrix. Then, the estimated standard error (SE) of θ is achieved by considering the value of the diagonal component of $I(\hat{\theta})^{-1}$ raised to half.

5. Simulation Study

In the framework of the S-ML model, a simulation study is carried out to study the performance of $\hat{\theta}$ given as Equation (10) in terms of their bias (bias) and mean squared error (MSE). The simulated procedure can be described as follows:

We generate samples of sizes $n = 20, 50, 100, 200, 500, 1000$ from the S-ML distribution with $\theta = (1.25, 1.50, 2.00, 2.50)$. For each sample, the MLE $\hat{\theta}$ is calculated. Here, 1000 such repetitions are made to calculate the standard mean MLE (MMLE), bias and MSE of these estimates using the formula:

$$\text{MMLE}(\hat{\theta}) = \frac{1}{1000}\sum_{i=1}^{1000} \hat{\theta}_i, \text{Bias}_\theta(\hat{\theta}) = \frac{1}{1000}\sum_{i=1}^{1000}(\hat{\theta}_i - \theta)$$

and

$$\text{MSE}_\theta(\hat{\theta}) = \frac{1}{1000}\sum_{i=1}^{1000}(\hat{\theta}_i - \theta)^2,$$

respectively, where $\hat{\theta}_i$ is the estimate of θ for each iteration in the simulation study; i is from 1 to 1000. The results of the study are reported in Table 2.

Table 2. Outcome of the simulation study.

θ	n	$\hat{\theta}$ Bias	$\hat{\theta}$ MSE
1.25	20	0.040548	0.052057
	50	0.017112	0.0194051
	100	0.007862	0.009083
	200	0.005018	0.0045803
	500	0.002377	0.001814
	1000	0.001318	0.000915
1.50	20	0.51523	0.072304
	50	0.021153	0.028311
	100	0.008343	0.0125480
	200	0.005195	0.006043
	500	0.0027303	0.002816
	1000	0.001619	0.001239
2.00	20	0.060237	0.135020
	50	0.029170	0.052147
	100	0.012338	0.024522
	200	0.009004	0.012285
	500	0.001464	0.004796
	1000	0.003824	0.002366
2.50	20	0.106934	0.246197
	50	0.033415	0.083405
	100	0.023273	0.044903
	200	0.007408	0.019908
	500	0.007856	0.008118
	1000	0.001431	0.004158

From Table 2, it is observed that as sample size n increases,

1. Bias decreases, which shows the accuracy of $\hat{\theta}$;
2. MSE decreases, which indicates the consistency (or preciseness) of $\hat{\theta}$.

6. Applications of the S-ML Model

We use the S-ML model on two data sets based on the maximum likelihood method as introduced previously. The data differ in size, traits, and background, but they are all of current interest in their areas.

6.1. Method

We proceed as follows for each data set:

1. The data are presented briefly, accompanied with their reference;
2. A table that encapsulates the basic statistical measures of the data is provided;
3. The goodness-of-fit measures of the models under consideration are evaluated and arranged in order of model performance in a table;
4. The MLE(s) of the model parameters is(are) shown, as well as the relevant SEs, as supplementary work;
5. It is concluded with a visual concept by presenting the histogram of the data and the epdf, as well as the empirical cdf plots and ecdf for the S-ML model exclusively in another graph.

The adequacy measures that are used for model fitting are provided here. Suppose $x_1, x_2, \ldots, x_n$ represent the data and $x_{(1)}, x_{(2)}, \ldots, x_{(n)}$ be their ordered values. As an initial step, we consider the Cramér von-Mises, Anderson Darling, and Kolmogorov–Smirnov statistics defined by

$$A^* = -n - \sum_{i=1}^{n} \tfrac{2i-1}{n} \left[\ln\left(F_{S-ML}\left(x_{(i)}; \hat{\theta}\right) \right) + \ln\left(S_{S-ML}\left(x_{(n+1-i)}; \hat{\theta}\right) \right) \right],$$

$$W^* = \tfrac{1}{12n} + \sum_{i=1}^{n} \left(F_{S-ML}\left(x_{(i)}; \hat{\theta}\right) - \tfrac{2i-1}{2n} \right)^2$$

and

$$D_n = \max_{i=1,2,\ldots,n} \left(F_{S-ML}\left(x_{(i)}; \hat{\theta}\right) - \tfrac{i-1}{n}, \tfrac{i}{n} - F_{S-ML}\left(x_{(i)}; \hat{\theta}\right) \right),$$

respectively. The p-value of the Kolmogorov–Smirnov test linked to D_n is also examined. Of course, the above definitions can be adapted to any other model than the S-ML model. The measures of adequacy are extensively employed to determine which model is best in terms of fitting the data set under study. The model having the least value for the W^* or A^*, and the highest p-value, is considered to give the best fit that is in correspondence with the data.

Furthermore, we consider the following goodness-of-fit measures: Akaike information criterion (AIC) and Bayesian information criterion (BIC), given as follows

$$\text{AIC} = 2k - 2\text{LL}, \text{BIC} = -2\text{LL} + k\ln(n),$$

respectively, where LL is the value of the log-likelihood function taken at $\hat{\theta}$ and k, being the number of parameters of the model, here $k = 1$ for the S-ML model. As it is widely understood, the model with the lowest value for AIC or BIC is selected as the greatest player of models that fits the data set compared to the other models. For more information on the usage and the underlying meaning of the measures W^*, A^*, D_n, AIC and BIC, we refer to [18].

In order to study the best fit of the S-ML model, we aim to compare it with some useful and competent models, which include the ML, Lindley, sine exponential and sine Lindley models listed in Table 3. It is worth noting that models with three parameters are also considered. The aim is to prove that our model can be efficient enough to outperform more complex models in the literature.

Table 3. Competent models with the S-ML model.

Models	Abbreviations	Cdfs	References
Lindley	Lindley	$1 - \left[1 + \tfrac{x\theta}{1+\theta}\right] e^{-x\theta}$	[19]
sine exponential	S-Expo	$\cos\left(\tfrac{\pi}{2} e^{-\theta x}\right)$	[2]
sine Lindley	S-Lindley	$\cos\left[\tfrac{\pi}{2}\left(1 + \tfrac{\theta x}{1+\theta}\right) e^{-\theta x}\right]$	[20]
modified Lindley	ML	$1 - \left[1 + e^{-\theta x} \tfrac{\theta x}{1+\theta}\right] e^{-\theta x}$	[11]
inverted modified Lindley	I-ML	$\left[1 + \tfrac{\theta}{1+\theta} \tfrac{1}{x} e^{-\theta/x}\right] e^{-\theta/x}$	[21]
inverted Lindley	IL	$\left[1 + \tfrac{\theta}{1+\theta} \tfrac{1}{x}\right] e^{-\theta/x}$	[22]
transmuted exponentiated inverse Rayleigh	TEIR	$e^{-\theta\alpha/x^2}\left[1 + \lambda - \lambda e^{-\theta\alpha/x^2}\right]$	[23]
transmuted inverse Rayleigh	TIR	$e^{-\theta/x^2}\left[1 + \lambda - \lambda e^{-\theta/x^2}\right]$	[24]
inverse Rayleigh	IR	$e^{-\alpha/x^2}$	[25]
Lomax	Lomax	$1 - \left[1 + \tfrac{x}{\lambda}\right]^{-\alpha}$	[26]
log normal	LNormal	$\phi\left[\tfrac{\ln x - \mu}{\sigma}\right]$	[27]
generalized beta type II	GB2	$\tfrac{|\alpha|}{\beta^{\alpha\theta} B(\theta,\delta)} \int_0^x \tfrac{y^{\alpha\theta-1}}{\left(1+(y/\beta)^\alpha\right)^{\theta+\delta}} dy$	[28]

6.2. Precipitation Data Set

The data set has thirty consecutive values of precipitation (in inches) in the month of March in Minneapolis, as provided by [29] and recently used by [30]. The data are: (0.77, 1.74, 0.81, 1.2, 1.95, 1.2, 0.47, 1.43, 3.37, 2.2, 3, 3.09, 1.51, 2.1, 0.52, 1.62, 1.31, 0.32, 0.59, 0.81, 2.81, 1.87, 1.18, 1.35, 4.75, 2.48, 0.96, 1.89, 0.9, 2.05). The descriptive statistical measures of these data are presented in Table 4.

Table 4. Descriptive statistical measures for the precipitation data set.

Mean	Median	Variance	Skewness	Kurtosis	Min	Max
1.68	1.47	1	1.086682	4.206884	0.32	4.75

Based on the information in Table 4, the data are right-skewed and leptokurtic. The MLE, SE, and goodness-of-fit measures of the S-ML model and those of the other models for precipitation data set are given in Tables 5 and 6.

Table 5. MLEs, SEs, and goodness-of-fit measures for the precipitation data set with one parameter models.

Model	MLE (SE)	AIC	BIC	A*	W*	D_n	p-Value
S-ML	$\hat{\theta} = 0.44$ (0.0551)	82.9109	84.3121	0.6796	0.0972	0.1273	0.7153
ML	$\hat{\theta} = 0.6644$ (0.0974)	85.8898	87.291	1.1278	0.1723	0.1566	0.4532
Lindley	$\hat{\theta} = 0.9096$ (0.1247)	88.2874	89.6886	1.5908	0.2618	0.1882	0.2383
S-Expo	$\hat{\theta} = 0.3396$ (0.0576)	90.7932	92.1944	2.1771	0.3873	0.2202	0.1088
S-Lindley	$\hat{\theta} = 0.6091$ (0.0729)	85.6414	87.0426	1.1566	0.1817	0.1637	0.3966
I-ML	$\hat{\theta} = 1.247$ (0.1906)	89.7366	91.1378	1.3909	0.2170	0.1975	0.1925
IL	$\hat{\theta} = 1.5833$ (0.2268)	92.4423	1.8266	0.3040	0.1904	0.2279	0.0887
IR	$\hat{\alpha} = 0.8588$ (0.1568)	92.292	92.674	2.1822	0.43077	0.2396	0.06369

Table 6. MLEs, SEs, and goodness-of-fit measures for the precipitation data set with models having more than one parameter.

Model	MLE (SE)	AIC	BIC	A*	W*	D_n	p-Value
TEIR	$\hat{\alpha} = 1.1878$ (5.739)	90.2022	94.4058	1.1359553	0.2117553	0.1817501	0.2749038
	$\hat{\lambda} = -0.67006$ (0.266)						
	$\hat{\theta} = 0.6362$ (4.778)						
TIR	$\hat{\lambda} = 0.0001$ (0.40171)	91.073	94.759	1.136	0.21177	0.1817	0.2748
	$\hat{\theta} = 0.8588$ (0.2136)						
Lomax	$\hat{\alpha} = 58619.76$ (52.96)	94.9	97.8	2.5139	0.4539	0.2352	0.0724
	$\hat{\lambda} = 98190$ (96.69)						
LNormal	$\hat{\mu} = 0.33737$ (0.11368)	81	83.8	0.19855	0.0311	0.0913	0.9640
	$\hat{\sigma} = 0.62263$ (0.08038)						
GB2	$\hat{\alpha} = 580.4141$ (1593.1699)	84.1	89.7	7.43	1.58	0.445	0.0000138
	$\hat{\beta} = 0.8125$ (0.6469)						
	$\hat{\theta} = 4.3731$ (6.6885)						
	$\hat{\delta} = 520.2863$ (584.0752)						

We can observe from Table 5 that the S-ML model has the lowest statistics with the highest p-value, implying that it delivers a better fit than the other models studied. Comparing the models in Table 6, we can see that the lognormal model gives a better fit, while the S-ML model takes the second place, but with less modeling complexity in terms of the number of parameters. Figure 3 depicts the epdf and ecdf plots of the S-ML model for the precipitation data set.

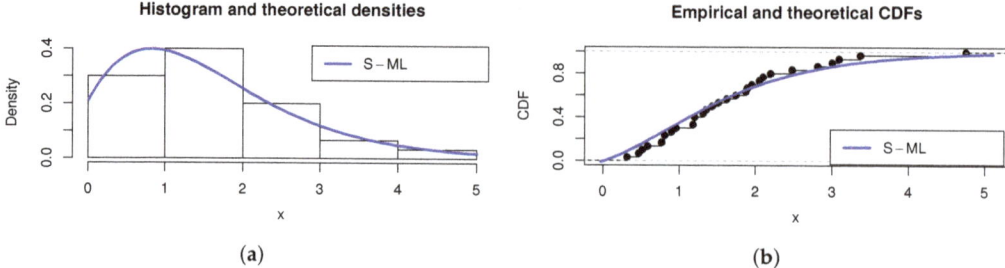

Figure 3. Plots of the (a) epdf and (b) ecdf of the S-ML model for the precipitation data set.

From Figure 3, it is obvious that the S-ML model captures the histogram's overall pattern and illustrates the comparison of the cdf with the empirical cdf of the S-ML model. The suitable behaviour of the S-ML model is further confirmed by these graphs. Apart from the lognormal model, the S-ML model clearly fits better than the Lindley, ML, S-Expo and S-Lindley, and other models.

6.3. Time between Failure Data Set

This data set refers to the time between failures for repairable items. It was obtained from [31]. The data are: (1.43, 0.11, 0.71, 0.77, 2.63, 1.49, 3.46, 2.46, 0.59, 0.74, 1.23, 0.94, 4.36, 0.40, 1.74, 4.73, 2.23, 0.45, 0.70, 1.06, 1.46, 0.30, 1.82, 2.37, 0.63, 1.23, 1.24, 1.97, 1.86, 1.17). The descriptive statistical measures of these data are presented in Table 7.

Table 7. Descriptive statistical measures for the failure time data set.

Mean	Median	Variance	Skewness	Kurtosis	Min	Max
1.542667	1.235000	1.127167	1.295462	4.319170	0.110000	4.730000

From Table 7, we can observe that the failure time data set is right-skewed and leptokurtic.

The MLE, SE, and goodness-of-fit measures of the S-ML model and those of the other models for the failure time data set are given in Tables 8 and 9.

Table 8. MLEs, SEs, and goodness-of-fit measures for the failure time data set with one parameter models.

Model	MLE (SE)	AIC	BIC	A*	W*	D_n	p-Value
S-ML	$\hat{\theta} = 0.47420$ (0.06039)	82.3276	83.7288	0.22656158	0.02897235	0.07805138	0.99313719
ML	$\hat{\theta} = 0.7297$ (0.1082)	83.5051	84.9063	0.4401	0.0650	0.1112	0.8514
Lindley	$\hat{\theta} = 0.9762$ (0.1345)	85.0946	86.4958	0.7265	0.1138	0.1406	0.5929
S-Expo	$\hat{\theta} = 0.3662$ (0.0625)	86.5547	87.9559	1.0582	0.17981	0.16722	0.3711
S-Lindley	$\hat{\theta} = 0.64690$ (0.0783)	83.7386	85.1398	0.4599	0.0654	0.1139	0.8310
I-ML	$\hat{\theta} = 0.9222$ (0.1361)	92.6416	94.0426	0.9582	0.1430	0.1394	0.6043
IL	$\hat{\theta} = 1.1603$ (0.1619)	95.8658	97.2670	1.261	0.1904	0.1411	0.5879
IR	$\hat{\alpha} = 0.237$ (0.043)	135.289	136.690	1.423	2.410	0.442	0.00001

Table 9. MLEs, SEs, and goodness-of-fit measures for the failure time data set with models having more than one parameter.

Model	MLE (SE)	AIC	BIC	A*	W*	D_n	p-Value
TEIR	$\hat{\alpha} = 0.022\ (0.065)$	122.99	127.19	8.373	1.47	0.37	0.000410
	$\hat{\lambda} = -0.880\ (0.114)$						
	$\hat{\theta} = 8.211\ (23.735)$						
TIR	$\hat{\lambda} = -0.880\ (0.114)$	120.990	123.792	8.369	1.471	0.3761	0.0004
	$\hat{\theta} = 0.185\ (0.035)$						
Lomax	$\hat{\alpha} = 19793.12\ (81.02)$	90	92.8	1.33	0.232	0.184	0.259
	$\hat{\lambda} = 305\ (51.04))$						
LNormal	$\hat{\mu} = 0.1597\ (0.1464)$	85.5	88.3	0.2577	0.0369	0.0987	0.9322
	$\hat{\sigma} = 0.801\ (0.1035)$						
GB2	$\hat{\alpha} = 655.80\ (2342.40)$	87.2	92.8	7.53	1.64	0.042	0.00003
	$\hat{\beta} = 0.907\ (0.77)$						
	$\hat{\theta} = 2.351\ (3.56)$						
	$\hat{\delta} = 582.259\ (519.98)$						

Tables 8 and 9 show that, for the failure time data set, the S-ML model has the lowest statistics and the highest *p*-value, meaning that it provides a better match than the other models investigated.

Figure 4 depicts the epdf and ecdf plots of the S-ML model for the failure time data set.

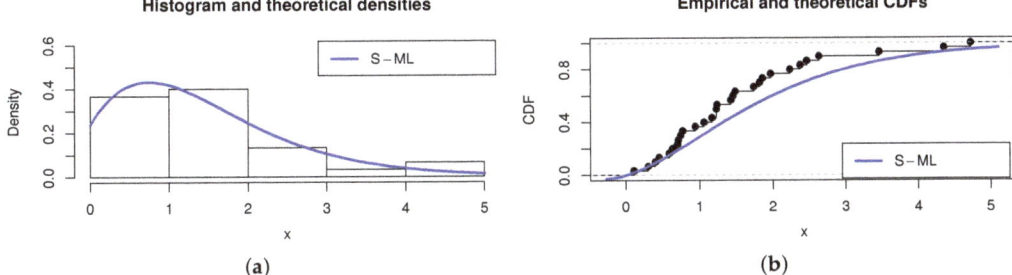

Figure 4. Plots of the (a) epdf and (b) ecdf of the S-ML model for the failure time data set.

From Figure 4, it is obvious that the S-ML model captures the histogram's overall pattern and illustrates the comparison of the cdf with the empirical cdf of the S-ML model. The suitable behaviour of the S-ML model is further confirmed by these graphs.

7. Conclusions

The article's major contribution is a flexible trigonometric extension of the well-known modified Lindley model that proposes a novel efficient statistical modelling technique. We employ the features of the sine generalized family of distributions in this regard, and develop the sine modified Lindley distribution. We have displayed a few of its more noteworthy attributes, with a focus on the shape properties of the corresponding probability density and hazard rate functions, as well as discussing moments. Simulation studies and applications demonstrate the utility of the model under consideration. In particular, we compared it to the primary current models derived from the Lindley, exponential and other models with one or more parameters, using two real-world data sets. As a result, the obtained findings are really satisfactory, demonstrating that the novel distribution has a wide range of applications that could be the subject of additional research in a variety of scientific fields.

Author Contributions: Conceptualization, L.T., V.G. and C.C.; methodology, L.T., V.G. and C.C.; software, L.T., V.G. and C.C.; validation, L.T., V.G. and C.C.; formal analysis, L.T., V.G. and C.C.; investigation, L.T., V.G. and C.C.; writing—original draft preparation, L.T., V.G. and C.C.; writing—review and editing, L.T., V.G. and C.C. All authors have read and agreed to the published version of the manuscript.

Funding: This research received no external funding.

Acknowledgments: We would like to express our gratitude to the four reviewers for their useful comments on the work's first draft.

Conflicts of Interest: The authors declare no conflict of interest.

References

1. Brito, C.R.; Rêgo, L.C.; Oliveira, W.R.; Gomes-Silva, F. Method for generating distributions and classes of probability distributions: The univariate case. *Hacet. J. Math. Stat.* **2019**, *48*, 897–930.
2. Kumar, D.; Singh, U.; Singh, S.K. A new distribution using sine function: Its application to bladder cancer patients data. *J. Stat. Appl. Probab.* **2015**, *4*, 417–427.
3. Souza, L. New Trigonometric Classes of Probabilistic Distributions. Ph.D. Thesis, Universidade Federal Rural de Pernambuco, Recife, Brazil, 2015.
4. Souza, L.; Junior, W.R.O.; de Brito, C.C.R.; Chesneau, C.; Ferreira, T.A.E.; Soares, L. On the Sin-G class of distributions: Theory, model and application. *J. Math. Model* **2019**, *7*, 357–379. [CrossRef]
5. Nelson, W. *Applied Life Data Analysis*; John Wiley & Sons: New York, NY, USA, 1982.
6. Nagarjuna, V.B.V.; Vardhan, R.V.; Chesneau, C. On the Accuracy of the Sine Power Lomax Model for Data Fitting. *Modelling* **2021**, *2*, 5. [CrossRef]
7. Souza, L.; Junior, W.R.O.; de Brito, C.C.R.; Chesneau, C.; Ferreira, T.A.E.; Soares, L. General properties for the Cos-G class of distributions with applications. *Eurasian Bull. Math.* **2019**, *2*, 63–79.
8. Chesneau, C.; Jamal, F. The sine Kumaraswamy-G family of distributions. *J. Math. Ext.* **2021**, *15*, 1–33.
9. Al-Babtain, A.A.; Elbatal, I.; Chesneau, C.; Elgarhy, M. Sine Topp-Leone-G family of distributions: Theory and applications. *Open Phys.* **2020**, *18*, 574–593. [CrossRef]
10. Souza, L.; Junior, W.R.O.; de Brito, C.C.R.; Chesneau, C.; Ferreira, T.A.E.; Fernandes, L.R. Tan-G class of trigonometric distributions and its applications. *Cubo* **2021**, *23*, 1–20. [CrossRef]
11. Chesneau, C.; Tomy, L.; Gillariose, J. A new modified Lindley distribution with properties and applications. *J. Stat. Manag. Syst.* **2021**. [CrossRef]
12. Chesneau, C.; Tomy, L.; G, V. The Poisson-modified Lindley distribution. *Appl. Math. E Notes.* in press.
13. Chesneau, C.; Tomy, L.; Jose, M. Wrapped modified Lindley distribution. *J. Stat. Manag. Syst.* **2021**. [CrossRef]
14. Tomy, L.; G, V.; Chesneau, C. The discrete modified Lindley distribution. 2021; unpublished work.
15. Cordeiro, G.M.; Silva, R.B.; Nascimento, A.D.C. *Recent Advances in Lifetime and Reliability Models*; Bentham Books: Sharjah, United Arab Emirates, 2020. [CrossRef]
16. Casella, G.; Berger, R.L. *Statistical Inference*; Duxbury Advanced Series Thomson Learning: Pacific Grove, CA, USA, 2002.
17. R Development Core Team. *R: A Language and Environment for Statistical Computing*; R Foundation for Statistical Computing: Vienna, Austria, 2005.
18. Konishi, S.; Kitagawa, G. *Information Criteria and Statistical Modeling*; Springer: New York, NY, USA, 2007.
19. Lindley, D.V. Fiducial distributions and Bayes theorem. *J. R. Stat. Soc. A* **1958**, *20*, 102–107. [CrossRef]
20. Kumar, D.; Singh, U.; Singh, S.K.; Chaurasia, P.K. Statistical properties and application of a Lifetime model using sine function. *Int. J. Creat. Res. Thoughts* **2018**, *6*, 993–1002.
21. Chesneau, C.; Tomy, L.; Gillariose, J.; Jamal, F. The inverted modified Lindley distribution. *J. Stat. Theory Pract.* **2020**, *14*, 1–17. [CrossRef]
22. Sharma, V.; Singh, S.; Singh, U.; Agiwal, V. The inverse Lindley distribution: A stress-strength reliability model with applications to head and neck cancer data. *J. Ind. Prod. Eng.* **2015**, *32*, 162–173. [CrossRef]
23. Haq, M.A. Transmuted exponentiated inverse Rayleigh distribution. *J. Stat. Appl. Prob.* **2016**, *5*, 337–343. [CrossRef]
24. Ahmad, A.; Ahmad, S.; Ahmed, A. Transmuted Inverse Rayleigh Distribution: A Generalization of the Inverse Rayleigh Distribution. *Math. Theory Model.* **2014**, *4*, 90–98.
25. Voda, V.G. On the inverse Rayleigh distributed random variable. *Rep. Statist. App. Res.* **1972**, *19*, 13–21.
26. Lomax, K.S. Business Failures, Another example of the analysis of failure data. *J. Am. Stat. Assosciation* **1954**, *49*, 847–852. [CrossRef]
27. Aitchinson, J.; Brown, J.A.C. *The Lognormal Distribution*; Cambridge University Press: Cambridge, UK, 1957.
28. Kalbfleisch, J.D.; Prentice, R.L. *The Statistical Analysis of Failure Time Data*; Wiley: New York, NY, USA, 1980.
29. Hinkley, D. On quick choice of power transformation. *Appl. Stat.* **1977**, *26*, 67–69. [CrossRef]

30. Yusuf, A.; Mikail, B.B.; Aliyu, A.I.; Sulaiman, A.L. The Inverse Burr Negative Binomial Distribution with Application to Real Data. *Stat. J. Theor. Appl. Stat.* **2016**, *5*, 53–65. [CrossRef]
31. Murthy, D.P.; Xie, M.; Jiang, R. *Weibull Models*; Wiley: Hoboken, NJ, USA, 2004.

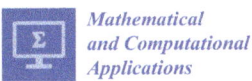

Article

New Modified Burr III Distribution, Properties and Applications

Farrukh Jamal [1,*], Ali H. Abuzaid [2], Muhammad H. Tahir [1], Muhammad Arslan Nasir [3], Sadaf Khan [1] and Wali Khan Mashwani [4]

1. Department of Statistics, The Islamia University of Bahawalpur, Bahawalpur 63100, Pakistan; mtahir.stat@gmail.com (M.H.T.); smkhan6022@gmail.com (S.K.)
2. Department of Mathematics, Al-Azhar University-Gaza, Gaza P.O. Box 1277, Palestine; a.abuzaid@alazhar.edu.ps
3. Department of Statistics, Govt. Sadiq Egerton College Bahawalpur, Bahawalpur 63100, Pakistan; arslannasir147@gmail.com
4. Institute of Numerical Sciences, Kohat University of Science and Technology, Kohat 26000, Pakistan; mashwanigr8@gmail.com
* Correspondence: farrukh.jamal@iub.edu.pk

Abstract: In this article, Burr III distribution is proposed with a significantly improved functional form. This new modification has enhanced the flexibility of the classical distribution with the ability to model all shapes of hazard rate function including increasing, decreasing, bathtub, upside-down bathtub, and nearly constant. Some of its elementary properties, such as rth moments, sth incomplete moments, moment generating function, skewness, kurtosis, mode, ith order statistics, and stochastic ordering, are presented in a clear and concise manner. The well-established technique of maximum likelihood is employed to estimate model parameters. Middle-censoring is considered as a modern general scheme of censoring. The efficacy of the proposed model is asserted through three applications consisting of complete and censored samples.

Keywords: Burr III distribution; stochastic ordering; middle-censoring; order statistics

MSC: 60E05; 62N05; 62F10

1. Introduction

Burr devised a dynamic family of probability distributions based on the Pearson differential equations. The Burr XII (BXII) and Burr III (BIII) distributions are widely used models from the system of Burr distributions. On the contrary, according to [1], the Burr X (BX) model has also gained much attention from applied statisticians along with the BXII and BIII models. The prime reason is that these densities exists in simpler forms and can yield a range of shapes to model a variety of scenarios in diverse scientific fields. The authors in [2] are of the view that the most adaptable of these three is BIII, especially in environmental, reliability, and survival sciences. The BIII distribution is also called the Dagum distribution in studies of income, wage, and wealth distribution [3]. In the actuarial literature, it is known as the inverse Burr distribution [4] and the kappa distribution in the meteorological literature [5]. As per [4], it is a prime case of the four-parameter generalised Beta-II distribution. In order to follow the ambit regarding the scope of this provision, we now shift our attention to the BIII distribution. For a random variable X defined on a positive real line, the cumulative distribution function (cdf) and probability density function (pdf) of two-parameter BIII distribution, respectively, are given below:

$$F(x;c,k) = \left(1 + x^{-c}\right)^{-k} \tag{1}$$

and
$$f(x;c,k) = ckx^{-c-1}(1+x^{-c})^{-k-1}, \quad (2)$$

where $c, k > 0$ are the shape parameters.

The shape parameter plays a significant role in yielding the hazard rate of BIII distribution, which can be decreasing or unimodal. Thus, it cannot be used to model lifetime data with a bathtub-shaped hazard function, such as human mortality and deterioration modelling. For the last few decades, statisticians have been developing various extensions and modifications in Weibull distribution due to its simple functional form. The two-parameter flexible Weibull extension of [6] has a hazard function that can be increasing, decreasing, or bathtub shaped. Zhang and Xie [7] studied the characteristics and application of the truncated Weibull distribution, which has a bathtub-shaped hazard function. A three-parameter model, called exponentiated Weibull distribution, was introduced by [8]. Another three-parameter model is referred to as the extended Weibull distribution by [9]. Xie et al. [10] proposed a three-parameter modified Weibull extension with a bathtub-shaped hazard function. A new modified Weibull distribution by the authors in [11] has been presented with increasing and a bathtub-shaped hazard function.

Various extensions of BIII distribution have been studied in the literature. In reference [12], the authors studied low-flow frequency analysis in hydrology with three-parameter-modified BIII distribution with supreme interest in the lower tail of a distribution. Çankaya et al. [13] extended the BIII model by adding a skew parameter with an epsilon skew extension approach. Modi and Gill [14] introduced the unit BIII model. Haq et al. [15] introduced the unit-modified BIII model. Ali et al. [16] re-parameterized BIII distribution and proposed the modified BIII (MBIII) distribution with the following cdf:

$$F(x) = \left(1 + \mu x^{-c}\right)^{\frac{-k}{\mu}} \quad x > 0, \quad (3)$$

where c, k, and μ are the shape parameters. The authors claimed that the newly structured model is a limiting case of generalized inverse Weibull, BIII, and log-logistic distribution. Still, the density of the improved model can only model positively skewed data, which greatly dented the proposition of the model in the first place. Other extensions are mostly based on the generalized families of distributions that sare complex in nature. Some of them are mentioned as: Beta Dagum by [17], Modified BIII by [18], Marshall Olkin BIII by [19], Gamma BIII by [20], and Gamma BIII by [21]. However, we feel that a flexible model with computationally simpler functional forms is still presently needed. Motivated by a lack of availability of literature related to the modified BIII distribution, we present a much more flexible new modification of BIII distribution. The cdf of the new, modified BIII (NMBIII) distribution is defined as

$$F(x;c,k,\lambda) = \left(1 + x^{-c}e^{-\lambda x}\right)^{-k} \quad x > 0, \quad (4)$$

where the $e^{-\lambda x}$ is the additional factor, with λ as the rate parameter and c, k are power parameters of the baseline model.

It is worth mentioning that when we use the additional term to add flexibility in the model, we specifically refer to the ability of the proposed model to fit a diverse range of real life phenomena. Additionally, flexibility may also be associated with the instantaneous failure rate or hazard rate, and is more commonly known as risk function. By selecting precise values for the shape parameters, the hazard rate function of the NMBIII distribution can take on a variety of appealing shapes. Generally speaking, the classical models deal with normal extreme observations. A new modification of BIII distribution will also enable us to observe the tail behaviour of the distribution, which is skewed in nature. Further, the BIII distribution has a monotonic decreasing and unimodal hazard rate function, but due to its modification, NMBIII has monotonic, decreasing, increasing, unimodal, bathtub, and approximately constant hazard-rate shapes. Moreover, many standard distributions are nested models or limiting cases of the Burr system of distributions, which include the

Weibull, exponential, logistic, generalised logistic, Gompertz, normal, extreme value, and uniform distributions. The NMBIII distribution outperforms most of these competitive existing models. When $\lambda = 0$, NMBIII distribution reduces to BIII distribution. When $\lambda = 0$ and $k = 1$, then NMBIII distribution gives us log-logistic distribution. When $k = 1$, then NMBIII distribution gives us modified log-logistic distribution (new). When $c = 0$ and $k = 1$, the NMBIII distribution reduces to logistic distribution. When $c = 1$, it reduces to modified skew logistic distribution (new). When $c = 0$ and $\lambda = 1$, it reduces to generalized logistic distribution type I or Burr type II, or this type has also been called the "skew-logistic" distribution (see [22]). In a nutshell, with the proposed NMBIII, we seek and hope to attract applied researchers from all scientific community to utilize it in the significant modelling of real-life scenarios.

The article is structured as follows: In Section 2, we focus our attention on the idea behind the new modification. {In Section 3, we acquaint the readers with some of the structural properties including the linear expansion, moments, mode, moment-generating functions, order statistics, and stochastic ordering of NMBIII distribution. In Section 4, model parameters are estimated by maximum likelihood method, and the Fisher information matrix is derived. Section 5 gives the simulation method based on complete and incomplete samples (middle censored). In Section 6, three data sets on complete and middle-censored data sets have been employed to established the authenticity of the proposed model to the readers. Section 7 consists of the concluding remarks and discussions.

2. The New Modified BIII Model

The modified Weibull (MW) distribution (see [23]) has the cumulative survival function that is the product of the Weibull cumulative hazard function αx^β and $e^{\lambda x}$. Hence, the distribution function was found to be

$$F(x) = \left(1 - e^{-\alpha x^\beta} e^{\lambda x}\right),$$

which was later generalized to exponentiated form by [24] using Lehmann alternative-I.

In the same vein, Equation (4) has been modified. The pdf corresponding to (4) is given as:

$$f(x; c, k, \lambda) = \frac{k\left(\lambda + \frac{c}{x}\right)}{x^c e^{\lambda x}} \left(1 + x^{-c} e^{-\lambda x}\right)^{-k-1}. \tag{5}$$

The corresponding survival and hazard functions of NMBIII are, respectively, given by:

$$S(x; c, k, \lambda) = 1 - \left(1 + x^{-c} e^{-\lambda x}\right)^{-k} \tag{6}$$

and

$$h(x; c, k, \lambda) = \frac{k\left(\lambda + \frac{c}{x}\right)}{x^c e^{\lambda x}} \frac{\left(1 + x^{-c} e^{-\lambda x}\right)^{-k-1}}{1 - \left(1 + x^{-c} e^{-\lambda x}\right)^{-k}}. \tag{7}$$

If a new random variable y is defined as $y = \frac{1}{x}$ in Equation (4), then we obtain the following model, referred to as modified Burr XII distribution, with cdf and pdf, respectively, as under

$$G(y) = 1 - \left(1 + \frac{y^c}{e^{\frac{\lambda}{y}}}\right)^{-k} \tag{8}$$

and

$$g(y) = \frac{k\left(c + \frac{1}{y}\right)}{e^{\frac{\lambda}{y}}} y^{c-1} \left(1 + \frac{y^c}{e^{\frac{\lambda}{y}}}\right)^{-k-1}. \tag{9}$$

As far as we can tell, Equations (4) and (8) are first modifications of BIII distribution and BXII distributions, respectively. Thus, the proposed distribution in (4) is more flexible and has tractable tail properties than its parent BIII distribution as well as MBIII distributions. The shapes of pdf and hrf are presented in Figures 1 and 2, respectively.

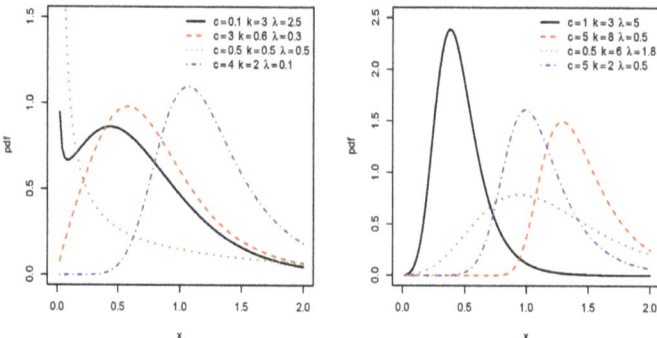

Figure 1. Density function of NMBIII distribution.

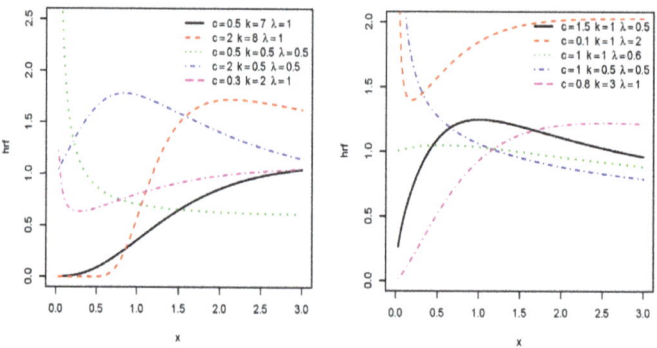

Figure 2. Hazard function for NMBIII distribution.

Figure 1 represents the different shapes of the proposed model, i.e., bimodal, reversed-J, right skewed, approximate left-skewed, and symmetrical shapes for different parameter values. Figure 2 reflects the different shapes of hazard function, which are increasing, decreasing, bathtub, upside-down bathtub, and nearly constant for different parameter values. The proposed distribution is more flexible and tractable than its parent BIII distribution, as well as MBIII distributions (see in Table 1).

Table 1. Sub models of NMBIII distributions.

Model	λ	c	k	$G(x)$	Reference
Burr III	0	-	-	$(1+x^{-c})^{-k}$	Standard
Log-Logistic	0	-	1	$\frac{x^c}{1+x^c}$	Standard
Modified Log-Logistic	-	-	1	$\frac{x^c e^{-\lambda x}}{1+x^c e^{-\lambda x}}$	New
Logistic	-	0	1	$\frac{e^{-\lambda x}}{1+e^{-\lambda x}}$	Standard
Modified skew logistic	-	1	-	$\frac{x e^{-\lambda x}}{1+x e^{-\lambda x}}$	New
Generalized logistic Type-I or Burr II or skew logistic	1	0	-	$(1+e^{-x})^{-k}$	Johnson et al. [22] and Aljouiee et al. [25]

3. Some Properties of NMBIII

In this section, we will provide some significant properties of the NMBIII distribution such as rth moment, sth incomplete moment, moment generating function, skewness, kurtosis, mode, and order statistics.

3.1. Useful Expansion

The generalized binomial theorem or power series is given by:

$$(1+z)^{-b-1} = \sum_{i=0}^{\infty} \binom{b+i}{i} (-1)^i z^i. \tag{10}$$

Using series expansion in (10), Equation (4) becomes

$$f(x;c,k,\lambda) = \sum_{i=0}^{\infty} \binom{k+i}{i} (-1)^i \frac{k\left(\lambda + \frac{c}{x}\right)}{x^{c(i+1)} e^{\lambda x(i+1)}}. \tag{11}$$

This expression can be used to obtain the following properties of the NMBIII distribution.

3.2. Moments

The rth moment of NMBIII distribution is given by:

$$m'_r = E(X^r) = \int_0^{\infty} x^r f(x)\, dx$$

$$= \sum_{i=0}^{\infty} \binom{k+i}{i} (-1)^i \int_0^{\infty} x^{r-c(i+1)} \left(\lambda + \frac{c}{x}\right) e^{-\lambda(i+1)x}\, dx$$

$$= \lambda \sum_{i=0}^{\infty} a_i \int_0^{\infty} x^{r-c(i+1)} e^{-\lambda(i+1)x}\, dx + c \sum_{i=0}^{\infty} a_i \int_0^{\infty} x^{r-c(i+1)-1} e^{-\lambda(i+1)x}\, dx$$

$$= \lambda \sum_{i=0}^{\infty} a_i \Gamma(r-c(i+1)-1) \left[\frac{1}{\lambda(i+1)}\right]^{r-c(i+1)-1} \tag{12}$$

$$+ c \sum_{i=0}^{\infty} a_i \Gamma(r-c(i+1)) \left(\frac{1}{\lambda(i+1)}\right)^{r-c(i+1)}$$

$$= \lambda \sum_{i=0}^{\infty} a_i \frac{\Gamma(r-c(i+1)-1)}{(\lambda(i+1))^{r-c(i+1)}} \left(\frac{1}{i+1} + c(r-c(i+1)-1)\right),$$

where $a_i = \binom{k+i}{i} (-1)^i$ and $\Gamma(a)\, b^a = \int_0^{\infty} x^{a-1} e^{-bx}\, dx$ is gamma function.

Remark 1. *By submitting $r = 1$ in Equation (13), one can find mean of the NMBIII distribution.*

The sth incomplete moment of NMBIII distribution is

$$T'_s(x) = \lambda \sum_{i=0}^{\infty} a_i \gamma\left(r-c(i+1)-1, \frac{x}{\lambda(i+1)}\right) \left(\frac{1}{\lambda(i+1)}\right)^{r-c(i+1)-1}$$

$$+ c \sum_{i=0}^{\infty} a_i \gamma\left(r-c(i+1), \frac{x}{\lambda(i+1)}\right) \left(\frac{1}{\lambda(i+1)}\right)^{r-c(i+1)}. \tag{13}$$

The application of incomplete moment refers to the mean deviations and Bonferroni and Lorenz curves. These curves are useful in economics reliability, demography, insurance, and medicine, to mention few.

3.3. Moment-Generating Function

The moment-generating function of NMBIII distribution is given by:

$$M_0(t) = E(e^{tx}) = \int_{i=0}^{\infty} e^{tx} f(x)\, dx$$

$$= \sum_{i=0}^{\infty} \binom{k+i}{i}(-1)^i \int_{i=0}^{\infty} x^{-c(i+1)}\left(\lambda + \frac{c}{x}\right) e^{(t-\lambda(i+1))x}\, dx$$

$$= \sum_{i=0}^{\infty} a_i \int_{i=0}^{\infty} x^{-c(i+1)}\left(\lambda + \frac{c}{x}\right) e^{(t-\lambda(i+1))x}\, dx \tag{14}$$

$$= \sum_{i=0}^{\infty} a_i \left(\lambda \int_0^{\infty} x^{-c(i+1)} e^{(t-\lambda(i+1))x}\, dx + c \int_0^{\infty} x^{-c(i+1)-1} e^{(t-\lambda(i+1))x}\, dx \right)$$

$$= \sum_{i=0}^{\infty} a_i \left(\lambda \frac{\Gamma(1-c(i+1))}{(\lambda(i+1)-t)^{1-c(i+1)}} + c \frac{\Gamma(-c(i+1))}{(\lambda(i+1)-t)^{-c(i+1)}} \right).$$

The skewness and kurtosis of the NMBIII distribution can be obtained numerically by the following expression.

$$\alpha = \frac{m_3' - 3\, m_2'\, m_1' + 2\, m_1'}{\{m_2' - (m_2')^2\}^{3/2}} \tag{15}$$

and

$$\beta = \frac{m_4' - 4\, m_3' m_1' + 6\, m_2'\, (m_1')^2 - 3\, (m_1')^4}{\{m_2' - (m_2')^2\}^2}, \tag{16}$$

where m_r' is the rth moment can be obtained form Equation (13).

Remark 2. The mode of the NMBIII distribution can be obtained as follows: taking the log of Equation (5), one obtains

$$\log f(x) = \log k + \log\left(\lambda + \frac{c}{x}\right) - c\, \log x - \lambda x - (k+1)\log\left(1 + x^{-c} e^{-\lambda x}\right), \tag{17}$$

Taking derivative with respect to x, we get

$$\frac{d}{dx}\log f(x) = \frac{-\frac{1}{x^2}}{\lambda + \frac{c}{x}} - \frac{c}{x} - \lambda + (k+1)\frac{x^{-c} e^{-\lambda x}\left(\lambda + \frac{c}{x}\right)}{1 + x^{-c} e^{-\lambda x}}, \tag{18}$$

by setting the above expression equal to zero and solving for x, one can find the mode. The numerical values of the first four moments are given in Table 2.

Table 2. The numerical values of the first four moments (m_r', $r = 1, 2, 3, 4$), skewness (α) and kurtosis (β) of the NMBIII for some parameter values.

c, k, λ	m_1'	m_2'	m_3'	m_4'	α	β
(0.5, 0.5, 0.5)	0.6754	2.0695	10.4250	72.6365	3.3418	20.5484
(1.5, 0.5, 0.5)	0.6662	1.0760	3.1293	14.2983	3.1239	27.1548
(1.5, 1.5, 0.5)	1.2849	2.6939	8.7612	41.8399	2.4599	23.6830
(1.5, 1.5, 1.5)	0.8024	0.8745	1.2564	2.3394	1.6650	70.6890
(2.0, 0.5, 0.5)	0.6814	0.9031	2.0319	7.3171	2.8155	31.7670
(2.0, 2.0, 0.5)	1.3695	2.6073	7.0943	27.8595	2.4280	39.1836
(2.0, 2.0, 2.0)	0.8041	0.7682	0.8775	1.2098	1.5101	226.2743

3.4. Order Statistics

The density function $f_{i:n}(x)$ of the i-th order statistic, for $i = 1,\ldots,n$, from i.i.d. random variables $X_1,\ldots,X_2$ following MBIII distribution is simply given by:

$$F_{i:n}(x) = \frac{n!}{(i-1)!(n-i)!} \sum_{j=0}^{n-i} \binom{n-i}{j} \frac{(-1)^j}{j+i} F(x)^{j+i}. \quad (19)$$

The corresponding pdf is

$$f_{i:n}(x) = \frac{n!}{(i-1)!(n-i)!} \sum_{j=0}^{n-i} \binom{n-i}{j} (-1)^j f(x) F(x)^{j+i-1}. \quad (20)$$

Using the pdf and cdf of NMBIII in Equations (4) and (5), we obtain

$$F_{i:n}(x) = \frac{n!}{(i-1)!(n-i)!} \sum_{j=0}^{n-i} \binom{n-i}{j} \frac{(-1)^j}{j+i} \left[1 + x^{-c} e^{-\lambda x}\right]^{-k(j+i)}. \quad (21)$$

Using series expansion in (10), we obtain

$$F_{i:n}(x) = \sum_{j=0}^{n-i} b_j \sum_{l=0}^{\infty} \binom{k(j+i)+l}{l} (-1)^l x^{-cl} e^{-\lambda l x}, \quad (22)$$

where $b_j = \frac{n!}{(i-1)!(n-i)!} \binom{n-i}{j} \frac{(-1)^j}{j+i}$. Similarly, following the above algebra, we have

$$f_{i:n}(x) = \sum_{j=0}^{n-i} a_j \sum_{l=0}^{\infty} \binom{j+i+l}{l} (-1)^l x^{-c(l+1)} \left(\lambda + \frac{c}{x}\right) e^{-\lambda(l+1)x}, \quad (23)$$

where $a_j = k \frac{n!}{(i-1)!(n-i)!} \binom{n-i}{j} (-1)^j$.

3.5. Stochastic Ordering

The concept of stochastic ordering is frequently used to show the ordering mechanism in life-time distributions. For more details about stochastic ordering, see [26]. A random variable is said to be stochastically greater ($X \leq_{st} Y$) than Y if $F_X(x) \leq F_Y(x)$ for all x. In the similar way, X is said to be stochastically lower ($X \leq_{st} Y$) than Y in the

1. Stochastic order ($X \leq_{st} Y$) if $F_X(x) \geq F_Y(x)$ for all x.
2. Hazard rate order ($X \leq_{hr} Y$) if $h_X(x) \geq h_Y(x)$ for all x.
3. Mean residual order ($X \leq_{mrl} Y$) if $m_X(x) \geq m_Y(x)$ for all x.
4. Likelihood ratio order ($X \leq_{lr} Y$) if $f_X(x) \geq f_Y(x)$ for all x.
5. Reversed hazard rate order ($X \leq_{rhr} Y$) if $\frac{F_X(x)}{F_Y(x)}$ is decreasing for all x.

The stochastic orders defined above are related to each other, as the following implications.

$$X \leq_{rhr} Y \Leftarrow X \leq_{lr} Y \Rightarrow X \leq_{hr} Y \Rightarrow X \leq_{st} Y \Rightarrow X \leq_{mrl} Y. \quad (24)$$

Let $X_1 \sim NMBIII(c_1, k_1, \lambda_1)$ and $X_2 \sim NMBIII(c_2, k_2, \lambda_2)$. Then, according to the definition of likelihood ratio ordering $\left[\frac{f(x)}{g(x)}\right]$,

$$f(x) = \frac{k_1\left(\lambda_1 + \frac{c_1}{x}\right)}{x^{c_1} e^{\lambda_1 x}} \left(1 + x^{-c_1} e^{-\lambda_1 x}\right)^{-k_1 - 1}, \qquad (25)$$

$$g(x) = \frac{k_2\left(\lambda_2 + \frac{c_2}{x}\right)}{x^{c_2} e^{\lambda_2 x}} \left(1 + x^{-c_2} e^{-\lambda_2 x}\right)^{-k_2 - 1}, \qquad (26)$$

and

$$\frac{f(x)}{g(x)} = \frac{k_1}{k_2} \frac{\left(\lambda_1 + \frac{c_1}{x}\right)}{\left(\lambda_2 + \frac{c_2}{x}\right)} \frac{x^{c_1} e^{\lambda_1 x}}{x^{c_1} e^{\lambda_1 x}} \frac{\left(1 + x^{-c_1} e^{-\lambda_1 x}\right)^{-k_1 - 1}}{\left(1 + x^{-c_2} e^{-\lambda_2 x}\right)^{-k_2 - 1}}. \qquad (27)$$

Taking log on both sides and taking the derivative with respect to x, we obtain

$$\frac{d}{dx}\left(\frac{f(x)}{g(x)}\right) = \frac{c_1}{x_i^2 \left(\lambda_1 + \frac{c_1}{x}\right)} - \frac{c_2}{x_i^2 \left(\lambda_2 + \frac{c_2}{x}\right)} + \frac{c_2 - c_1}{x_i} + (\lambda_2 - \lambda_1)$$
$$+ (k_2 + 1)\frac{x^{-c_2} e^{\lambda_2 x}\left(\lambda_2 + \frac{c_2}{x}\right)}{1 + x^{-c_2} e^{\lambda_2 x}} - (k_1 + 1)\frac{x^{-c_1} e^{\lambda_1 x}\left(\lambda_1 + \frac{c_1}{x}\right)}{1 + x^{-c_1} e^{\lambda_1 x}}, \qquad (28)$$

if $c_1 = c_2 = c$ and $\lambda_1 = \lambda_2 = \lambda$, then $\frac{d}{dx}\frac{f(x)}{g(x)} < 0$ if $(k_2 < k_1)$ and then $X <_{lr} Y$.

4. Maximum Likelihood Estimation

In this section, we will use the maximum-likelihood method to estimate the unknown parameters of the proposed model from complete samples only. Let $x_1, x_2, \ldots, x_n$ be a random sample of size n from the NMBIII family given in Equation (4) distribution. The log-likelihood function for the vector of parameter $\Theta = (c, k, \lambda)^T$ can be expressed as

$$l(\Theta) = n \log k - c \sum_{i=1}^{n} \log x_i - \lambda \sum_{i=1}^{n} x_i + \sum_{i=1}^{n} \log\left(\lambda + \frac{c}{x}\right)$$
$$- (k+1) \sum_{i=1}^{n} \log\left[1 + x^{-c} e^{-\lambda x}\right]$$

Taking the derivative with respect to λ, c, k, respectively, we get

$$U_k = \frac{\partial l(\Theta)}{\partial k} = \frac{n}{k} - \sum_{i=1}^{n} \log\left(1 + x^{-c} e^{-\lambda x}\right)$$

$$U_\lambda = \frac{\partial l(\Theta)}{\partial \lambda} = -\sum_{i=1}^{n} x_i + \sum_{i=1}^{n}\left(\lambda + \frac{c}{x}\right)^{-1} + (k+1)\sum_{i=1}^{n}\left(\frac{x^{-c} e^{-\lambda x} x_i}{1 + x^{-c} e^{-\lambda x}}\right)$$

$$U_c = \frac{\partial l(\Theta)}{\partial c} = -\sum_{i=1}^{n} \log x_i + \sum_{i=1}^{n}\left(\frac{1}{x_i \left(\lambda + \frac{c}{x}\right)}\right) + (k+1)\sum_{i=1}^{n}\left(\frac{x^{-c} e^{-\lambda x} \log x_i}{1 + x^{-c} e^{-\lambda x}}\right)$$

Setting U_k, U_λ, and U_k equal zero and solving these equations simultaneously yields the maximum likelihood estimates.

The observed information matrix for the parameter vector is given by

$$\begin{pmatrix} U_{kk} & U_{K\lambda} & U_{kc} \\ - & U_{\lambda\lambda} & U_{\lambda c} \\ - & - & U_{cc} \end{pmatrix}$$

whose elements are given below

$$U_{kk} = -\frac{n}{k^2}$$

$$U_{k\lambda} = \sum_{i=1}^{n} \left(\frac{x_i^{-c-1} e^{\lambda x_i}}{1 + x^{-c} e^{-\lambda x_i}} \right)$$

$$U_{kc} = -\sum_{i=1}^{n} \left(\frac{x_i^{-c} e^{\lambda x_i} \log x_i}{1 + x^{-c} e^{-\lambda x_i}} \right)$$

$$U_{\lambda c} = -\sum_{i=1}^{n} \frac{1}{x_i \left(\lambda + \frac{c}{x_i}\right)^2} + (k+1) \sum_{i=1}^{n} \left(\frac{x^{1-2c} e^{-2\lambda x_i} \log x_i}{(1 + x^{-c} e^{-\lambda x_i})^2} + \frac{e^{-\lambda x_i} x_i^{1-c} \log x_i}{1 + x^{-c} e^{-\lambda x_i}} \right)$$

$$U_{\lambda \lambda} = -\sum_{i=1}^{n} \frac{1}{\left(\lambda + \frac{c}{x_i}\right)^2} - (k+1) \sum_{i=1}^{n} \left(\frac{x^{2-2c} e^{-2\lambda x_i}}{(1 + x^{-c} e^{-\lambda x_i})^2} + \frac{e^{-\lambda x_i} x_i^{2-c}}{1 + x^{-c} e^{-\lambda x_i}} \right)$$

$$U_{cc} = -\sum_{i=1}^{n} \frac{1}{x_i^2 \left(\lambda + \frac{c}{x_i}\right)^2} + (k+1) \sum_{i=1}^{n} \left(\frac{x^{-2c} e^{-2\lambda x_i} (\log x_i)^2}{(1 + x^{-c} e^{-\lambda x_i})^2} + \frac{e^{-\lambda x_i} x_i^{-c} \log x_i}{1 + x^{-c} e^{-\lambda x_i}} \right)$$

5. Middle-Censoring

The middle-censoring scheme is a non-parametric general censoring mechanism proposed by [27], where other censoring schemes can be obtained as special cases of this middle-censoring scheme (see [28]).

For n identical lifetimes $T_1, \ldots, T_n$ with a random censoring interval $(L_i \leq R_i)$ at the ith item with some unknown bivariate distribution. Then, the exact value of T_i is observable only if $T_i \notin [L_i \leq R_i]$; otherwise, the interval $(L_i \leq R_i)$ is observed.

Middle-censoring had previously been applied to exponential and Burr XII lifetime distributions (see [28,29]). Furthermore, it was extended to parametric models with covariates [30], and its robustness was investigated by [31].

In this section, we analyse the NMBIII lifetime data when they are middle-censored. Assume that $T_1, \ldots, T_n$ are i.i.d. NMBIII (c, λ, k) random variable and let $Z_i = R_i - L_i$, $i = 1, \ldots, n$ be another random variable that defines the length of the censoring interval with exponential distribution with mean γ^{-1}, where the left-censoring point for each individual L_i is assumed to also be an exponential random variable with mean θ^{-1}. Moreover, the $T_i's$, $L_i's$, and $Z_i's$ are all independent of each other and the observed data, and $X_i's$ are given by $X_i = \begin{cases} T_i & \text{if } T_i \notin (L_i \leq R_i), \\ (L_i \leq R_i) & \text{otherwise.} \end{cases}$

5.1. Estimation

For n randomly selected units from the NMBIII (c, λ, k) population, where c, λ, and k are unknown, were tested under middle-censoring scheme. In this setting, there are $n_1 > 0$ uncensored observations and $n_2 > 0$ censored observations. Then, by re-ordering the observed data into the uncensored and censored observations, we therefore have the following data

$$\{T_1, \ldots, T_{n_1}, (L_{n_1+1}, R_{n_1+1}), \ldots, (L_{n_1+n_2}, R_{n_1+n_2})\},$$

where $n_1 + n_2 = n$.

The likelihood function of the observed data is given by:

$$L(c, \lambda, k|x) = \omega(k)^{n_1} \prod_{i=1}^{n_1} (\lambda + \frac{c}{x_i}) \prod_{i=1}^{n_1} (x_i^{-c} e^{-\lambda x_i}) \prod_{i=1}^{n_1} (1 + x_i^{-c} e^{-\lambda x_i})^{-k-1}$$

$$\times \prod_{i=n_1+1}^{n_1+n_1} [(1 + r_i^{-c} e^{-\lambda r_i})^{-k} - (1 + l_i^{-c} e^{-\lambda l_i})^{-k}],$$

where ω is a normalizing constant depending on γ and θ, and the estimation of them is not of interest and this is left as a constant. The log-likelihood function is given by

$$l(c,\lambda,k|x) = \log \omega + n_1 \log k + \sum_{i=1}^{n_1} \log(\lambda + \frac{c}{x_i}) + n \sum_{i=1}^{n_1} \log(x_i^{-c} e^{-\lambda x_i}) - (k+1) \sum_{i=1}^{n_1} \log(1 + x_i^{-c} e^{-\lambda x_i})$$
$$+ \sum_{i=n_1+1}^{n_1+n_1} \log[(1 + r_i^{-c} e^{-\lambda r_i})^{-k} - (1 + l_i^{-c} e^{-\lambda l_i})^{-k}].$$

The maximum-likelihood estimation (MLE) of c, λ, and k, denoted by $\widehat{c}_M$, $\widehat{\lambda}_M$, and $\widehat{k}_M$, can be derived by solving the following equations:

$$\frac{\partial l(c,\lambda,k|x)}{\partial c} = \sum_{i=1}^{n_1} (\lambda x_i + c)^{-1} - \sum_{i=1}^{n_1} \log x_i + (k+1) \sum_{i=1}^{n_1} \frac{(x_i^{-c} e^{-\lambda x_i}) \log x_i}{1 + x_i^{-c} e^{-\lambda x_i}}$$
$$+ \sum_{i=n_1+1}^{n_1+n_1} \frac{k(1 + r_i^{-c} e^{-\lambda r_i})^{-k-1} (r_i^{-c} e^{-\lambda r_i}) \log(r_i) - k(1 + l_i^{-c} e^{-\lambda l_i})^{-k-1} (l_i^{-c} e^{-\lambda l_i}) \log(l_i)}{[(1 + r_i^{-c} e^{-\lambda r_i})^{-k} - (1 + l_i^{-c} e^{-\lambda l_i})^{-k}]},$$

$$\frac{\partial l(c,\lambda,k|x)}{\partial \lambda} = \sum_{i=1}^{n_1} \frac{1}{\lambda + \frac{c}{x_i}} - \sum_{i=1}^{n_1} x_i - (k+1) \sum_{i=1}^{n_1} \frac{x_i^{-c+1} e^{-\lambda x_i}}{1 + x_i^{-c} e^{-\lambda x_i}}$$
$$- \sum_{i=n_1+1}^{n_1+n_1} \frac{k(1 + r_i^{-c} e^{-\lambda r_i})^{-k-1} (r_i^{-c+1} e^{-\lambda r_i}) - k(1 + l_i^{-c} e^{-\lambda l_i})^{-k-1} (l_i^{-c+1} e^{-\lambda l_i})}{[(1 + r_i^{-c} e^{-\lambda r_i})^{-k} - (1 + l_i^{-c} e^{-\lambda l_i})^{-k}]}$$

and

$$\frac{\partial l(c,\lambda,k|x)}{\partial k} = -\sum_{i=n_1+1}^{n_1+n_1} \frac{(1 + r_i^{-c} e^{-\lambda r_i})^{-k} \log(1 + r_i^{-c} e^{-\lambda r_i}) - k(1 + l_i^{-c} e^{-\lambda l_i})^{-k} \log(1 + l_i^{-c} e^{-\lambda l_i})}{[(1 + r_i^{-c} e^{-\lambda r_i})^{-k} - (1 + l_i^{-c} e^{-\lambda l_i})^{-k}]}$$
$$+ \frac{n_1}{k} - \sum_{i=1}^{n_1} \log(1 + x_i^{-c} e^{-\lambda x_i}).$$

It is obvious that the MLE of c, λ, and k cannot be solved explicitly. Therefore, the solutions can be obtained using Newton–Raphson method or numerically using the solve systems of nonlinear equations "nleqslv" package in R.

Since the MLE is asymptotically normal, the approximate confidence intervals for the parameters c, λ and k can be computed as follows: $\hat{c}_M \pm z_{\frac{\alpha}{2}} \sqrt{\hat{\sigma}_c^2}$, $\hat{\lambda}_M \pm z_{\frac{\alpha}{2}} \sqrt{\hat{\sigma}_\lambda^2}$ and $\hat{k}_M \pm z_{\frac{\alpha}{2}} \sqrt{\hat{\sigma}_k^2}$, where $\hat{\sigma}_{(.)}^2$ are the variances of the respective parameters c, k, and λ, and $z_{\frac{\alpha}{2}}$ is the value of the standard normal curve and α is the level of significance.

5.2. Simulation Results

We conducted Monte Carlo simulation studies to assess the finite sample behaviour of the MLEs of the parameters c, k and λ based on two settings; the first is the random variable generated from the NMBIII distribution, while the other considers the case where the NMBIII lifetime data were middle-censored.

The random samples for both settings were generated from distribution NMBIII(c, k, λ) based on accept-reject approach. Without loss of generality, random samples were used with five different sizes viz n = 10, 30, 50, 70, and 100 from NMBIII(c, k, λ) distribution with parameters $c = 1, k = 2$, and $\lambda = 0.5$.

The middle censoring settings considered three combinations of the censoring schemes $(\gamma^{-1}, \theta^{-1}) = (0.25, 0.25)$, $(1, 0.75)$, and $(1.25, 0.5)$.

The results were obtained from 1000 Monte Carlo replications from simulations carried out using the software R, and the average estimates and the mean squared error (MSE) are obtained and reported in Table 3.

Results in Table 3 show that the ML estimates for both settings behave similarly. In general, there is a decreasing function between the sample size and the mean squared error, which verifies the consistency property of the derived estimators. The average estimates are insignificantly effected by the censoring status.

Table 3. Average MLE estimates and the corresponding MSE (within brackets).

Distribution	n	Un-Censored			Middle-Censored								
					(0.25, 0.25)			(1, 0.75)			(1.25, 0.5)		
(c, k, λ)		c	k	λ	c	k	λ	c	k	λ	c	k	λ
(1, 2, 0.5)	10	1.114 (0.130)	2.079 (0.102)	0.397 (0.122)	1.123 (0.141)	2.233 (0.163)	0.447 (0.096)	1.087 (0.111)	2.130 (0.159)	0.524 (0.108)	1.196 (0.121)	2.088 (0.099)	0.561 (0.125)
	30	1.039 (0.034)	2.036 (0.039)	0.464 (0.080)	1.082 (0.096)	2.170 (0.072)	0.452 (0.043)	1.072 (0.036)	2.080 (0.082)	0.519 (0.046)	1.127 (0.052)	2.080 (0.093)	0.547 (0.037)
	50	1.036 (0.03)	2.032 (0.031)	0.484 (0.029)	1.071 (0.033)	2.096 (0.031)	0.536 (0.032)	1.066 (0.028)	2.071 (0.032)	0.508 (0.028)	1.103 (0.022)	2.022 (0.025)	0.529 (0.027)
	70	1.015 (0.016)	1.984 (0.015)	0.511 (0.019)	1.035 (0.017)	2.018 (0.021)	0.510 (0.022)	1.042 (0.015)	2.053 (0.021)	0.496 (0.020)	1.042 (0.021)	1.985 (0.016)	0.476 (0.017)
	100	1.001 (0.012)	1.991 (0.013)	0.502 (0.011)	1.019 (0.013)	1.998 (0.015)	0.495 (0.014)	0.980 (0.015)	2.020 (0.016)	0.498 (0.017)	0.981 (0.016)	1.907 (0.013)	0.491 (0.013)
(0.5, 2, 0.5)	10	0.621 (0.052)	2.074 (0.040)	0.427 (0.151)	0.582 (0.127)	2.325 (0.086)	0.522 (0.063)	0.534 (0.056)	2.135 (0.084)	0.524 (0.088)	1.196 (0.080)	2.098 (0.105)	0.530 (0.096)
	30	0.613 (0.034)	2.057 (0.036)	0.464 (0.032)	0.531 (0.038)	2.264 (0.039)	0.513 (0.040)	0.529 (0.034)	2.104 (0.033)	0.516 (0.037)	1.127 (0.030)	2.087 (0.033)	0.521 (0.037)
	50	0.538 (0.026)	2.010 (0.012)	0.484 (0.044)	0.519 (0.094)	2.125 (0.067)	0.489 (0.032)	0.518 (0.019)	2.014 (0.064)	0.505 (0.037)	1.103 (0.031)	2.054 (0.016)	0.518 (0.031)
	70	0.5017 (0.012)	1.928 (0.009)	0.511 (0.041)	0.491 (0.057)	2.020 (0.037)	0.490 (0.021)	0.506 (0.012)	1.982 (0.035)	0.501 (0.013)	1.042 (0.017)	2.010 (0.012)	0.509 (0.014)
	100	0.492 (0.002)	2.003 (0.001)	0.502 (0.027)	0.504 (0.046)	2.003 (0.027)	0.507 (0.007)	0.492 (0.006)	2.004 (0.026)	0.499 (0.005)	0.981 (0.011)	1.923 (0.010)	0.495 (0.012)
(2, 2, 2)	10	2.212 (0.063)	2.452 (0.127)	2.517 (0.096)	2.298 (0.105)	2.571 (0.056)	2.322 (0.151)	2.331 (0.040)	2.280 (0.088)	2.371 (0.052)	2.102 (0.086)	2.493 (0.084)	2.256 (0.080)
	30	2.176 (0.043)	2.420 (0.096)	2.161 (0.037)	2.179 (0.093)	2.552 (0.036)	2.291 (0.080)	2.238 (0.039)	2.222 (0.046)	2.328 (0.034)	2.045 (0.072)	2.258 (0.082)	2.173 (0.052)
	50	1.962 (0.032)	2.013 (0.094)	2.008 (0.031)	2.057 (0.016)	2.150 (0.019)	2.171 (0.044)	2.061 (0.012)	2.064 (0.037)	2.091 (0.026)	1.959 (0.067)	2.041 (0.064)	1.901 (0.031)
	70	1.953 (0.021)	1.875 (0.057)	1.949 (0.014)	1.809 (0.012)	1.823 (0.012)	1.956 (0.041)	1.864 (0.009)	2.054 (0.013)	1.903 (0.012)	1.953 (0.037)	2.004 (0.035)	1.825 (0.017)
	100	2.045 (0.007)	2.113 (0.046)	2.160 (0.012)	2.070 (0.010)	2.503 (0.006)	2.207 (0.027)	2.183 (0.001)	2.145 (0.005)	2.143 (0.002)	2.026 (0.027)	2.125 (0.026)	2.144 (0.011)

6. Applications

This section provides three applications for complete data sets to show how the NM-BIII distribution can be applied in practice. We compare NMBIII distribution to MBIII, BIII, Weibull (W), Gamma (Ga), Lognormal (LN), Generalised Weibull (EW), and Generalised Extreme value type-II (GEV-II) distributions. In these applications, the model parameters are estimated by the method of maximum likelihood. The Akaike information criterion (AIC), Bayesian information criterion (BIC), A*(Anderson Darling), and W*(Cramer–von Mises) are computed to compare the fitted models. In general, the smaller the values of these statistics, the better the fit to the data. Additionally, the asymptotic variance-covariance matrices of the NMBIII parameters are also provided. The plots of the fitted PDFs, CDFs,

Probability–Probabibility (PP), and Quantile–Quantile (QQ) of NMBIII are displayed for visual comparison. The required computations are carried out in the R software.

The first data set consists of 119 observations on fracture toughness of Alumina (Al_2O_3) (in the units of MPa m$^{1/2}$). These data were studied by [32]. The second data set refers to the material thickness of hole (12 mm) and sheet (3.15 mm), comprising 50 observations, as reported by authors in [33]. The third data set was first analysed by [34] and represents the survival times, in weeks, of 33 patients suffering from Acute Myelogenous Leukaemia.

Tables 4–6 list the MLEs, standard errors, AIC, BIC, A*, and W* of the model for the data sets 1–3. The results in Tables 4–6 indicate that the NMBIII model provides the best fit as compared to all the other models. Figures 3–5 also support the results of Tables 4–6.

Table 4. Data set 1.

Model	Parameters	MLE	Standard Error	AIC	BIC	A*	W*
NMBII	c	2.543	0.507	362.159	370.497	1.888	0.296
	k	25.243	5.185				
	λ	1.703	0.179				
MBIII	c	1111.230	461.820	379.380	387.718	3.515	0.583
	k	4.943	0.281				
	μ	770.050	398.963				
BIII	c	3.058	0.180	423.535	429.094	7.658	1.365
	k	51.879	11.180				
W	α	0.002	0.0002	394.821	405.379	1.955	0.422
	β	3.984	0.0773				
Ga	α	15.521	1.991	385.737	374.295	2.745	0.457
	β	3.588	0.468				
LN	μ	1.432	0.025	428.845	434.403	3.374	0.568
	σ	0.269	0.0174				
EW	α	0.0114	0.006	374.644	386.981	1.945	0.315
	β	3.2126	0.278				
	θ	2.0077	0.388				
GEV-II	α	48.447	10.816	425.796	431.354	7.875	1.408
	β	3.022	0.185				

The variance–covariance matrix of the MLEs of the NMBIII distribution for data set 1 is

$$\begin{pmatrix} 0.25674663 & 0.2275027 & -0.08608337 \\ 0.22750269 & 26.8853417 & 0.18880701 \\ -0.08608337 & 0.1888070 & 0.03215706 \end{pmatrix}$$

Table 5. Data set 2.

Model	Parameters	MLE	Standard Error	AIC	BIC	A*	W*
NMBII	c	2.802	1.620	−106.358	−100.622	0.524	0.090
	k	0.317	0.219				
	λ	17.274	5.605				
MBIII	c	0.0020	0.0002	−99.778	−94.042	0.988	0.159
	k	3.466	0.205				
	μ	0.0039	0.0007				
BIII	c	7.788	26.572	−26.027	−22.202	1.056	0.177
	k	0.065	0.221				
W	α	36.141	14.390	−101.784	−93.960	0.644	0.105
	β	2.118	0.246				
Ga	α	3.029	0.576	−102.743	−98.919	1.636	0.279
	β	18.561	3.836				
LN	μ	1.987	0.095	105.700	109.524	1.922	0.331
	σ	0.670	0.067				
EW	α	819.305	2409.321	−106.069	−100.333	0.535	0.093
	β	4.982	2.636				
	θ	0.297	0.200				
GEV-II	α	0.054	0.020	−70.449	−66.625	3.567	0.634
	β	1.236	0.118				

The variance–covariance matrix of the MLEs of the NMBIII distribution for data set 2 is

$$\begin{pmatrix} 2.6257440 & 0.34637888 & 8.616552 \\ 0.3463789 & 0.04803978 & -1.104049 \\ 8.6165520 & -1.10404897 & 31.417567 \end{pmatrix}$$

Table 6. Data set 3.

Model	Parameters	MLE	Standard Error	AIC	BIC	A*	W*
NMBII	c	0.521	0.121	303.703	308.101	0.440	0.064
	k	4.734	1.065				
	λ	0.012	0.005				
MBIII	c	153.592	319.615	309.465	313.863	0.672	0.098
	k	1.494	0.464				
	μ	0.201	796.017				
BIII	c	0.755	0.092	309.714	312.645	0.919	0.151
	k	5.705	1.228				
W	α	0.057	0.028	304.302	307.234	0.552	0.079
	β	0.792	0.112				
Ga	α	0.706	0.150	304.357	309.288	0.459	0.085
	β	0.017	0.005				
LN	μ	2.884	0.266	320.9177	323.8491	0.648	0.102
	σ	1.504	0.188				
EW	α	0.0431	0.186	306.296	310.693	0.554	0.079
	β	0.844	0.794				
	θ	0.901	1.352				
GEV-II	α	4.259	0.933	310.463	313.395	0.983	0.160
	β	0.685	0.091				

The variance–covariance matrix of the MLEs of the NMBIII distribution for data set 3 is

$$\begin{pmatrix} 0.014574568 & 0.075708071 & -0.0003995590 \\ 0.075708071 & 1.134424017 & -0.001232737 \\ -0.000399559 & -0.001232737 & 0.00002838794 \end{pmatrix}$$

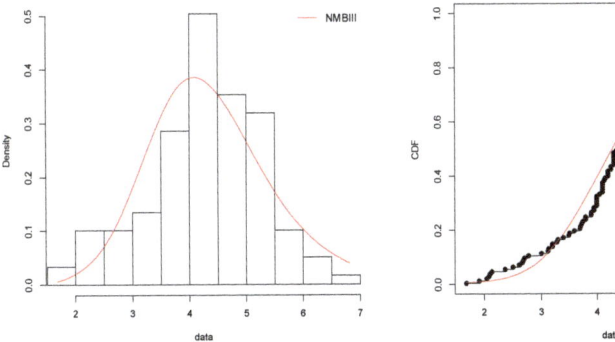

Figure 3. *Cont.*

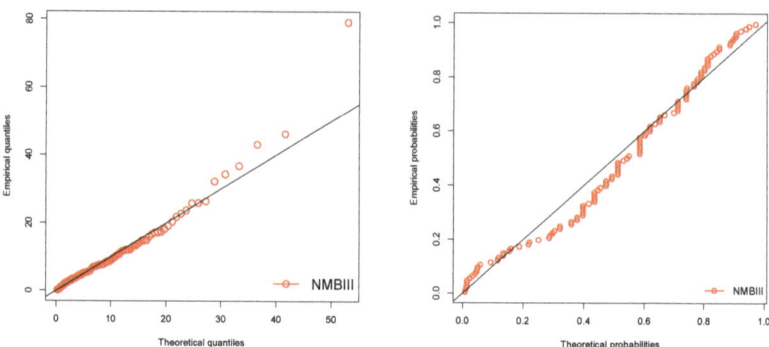

Figure 3. Estimated density (**top left**), cdf (**top right**), QQ-plot (**bottom left**), and PP-plot (**bottom right**) for data set 1.

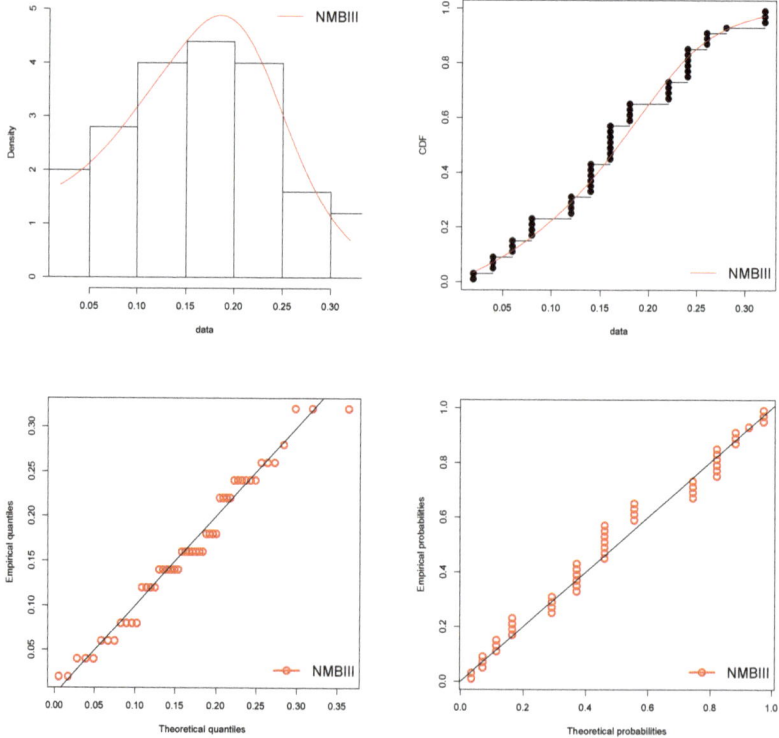

Figure 4. Estimated density (**top left**), cdf (**top right**), QQ-plot (**bottom left**) and PP-plot (**bottom right**) for data set 2.

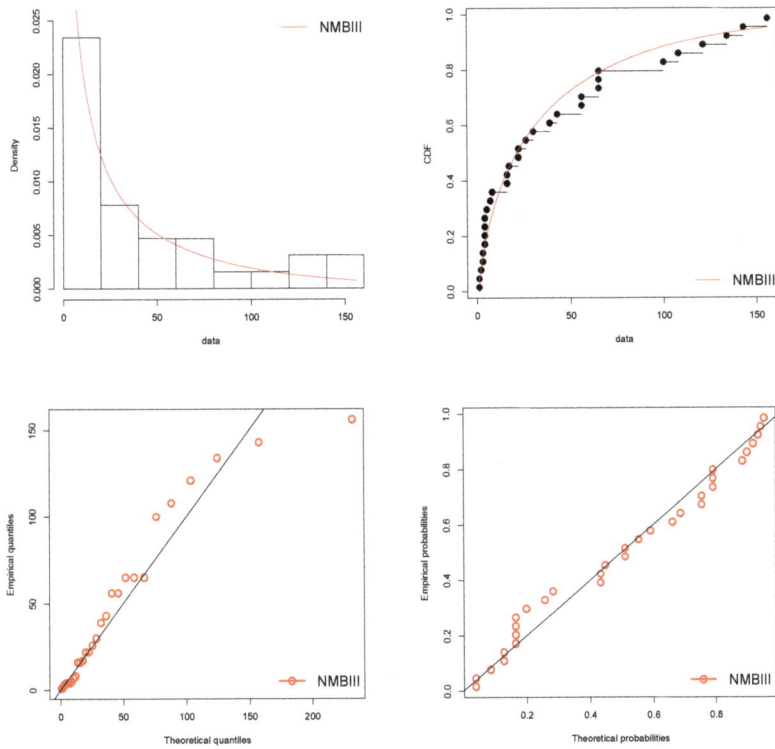

Figure 5. Estimated density (**top left**), cdf (**top right**), QQ-plot (**bottom left**), and PP-plot (**bottom right**) for data set 3.

7. Conclusions

A good theory should seek out the most concise explanation for the facts. With this in mind, a new modified form of BIII distribution has been introduced that can model well-specified forms of hazard rate shapes, including increasing, decreasing, bathtub, upside-down bathtub, and nearly constant. Some of its statistical properties, such as, rth moment, sth incomplete moment, moment generating function, skewness, kurtosis, mode, ith order statistics, and stochastic ordering have been derived. The maximum likelihood estimation is employed to estimate the model parameters. The usefulness of this model is demonstrated by applications on complete and censored samples. Simulation study is also performed. A future effort would include the contributions of new regression models, Bayesian parameter estimations, and research into diversified fields of data sets.

Author Contributions: F.J. gave the initial idea and proposed the new model. M.A.N. investigated mathematical properties. A.H.A. added the Middle-Censoring part of the paper. W.K.M. and M.H.T. wrote and structured the manuscript. S.K. investigated the numerical calculations applied to the data set and improved the overall presentation of the current draft. All authors have read and agreed to the published version of the manuscript.

Funding: This study did not receive any funds in any aspects.

Data Availability Statement: The data used in the article are given therein.

Acknowledgments: The pre-print version of this paper is also freely available here https://hal.archives-ouvertes.fr/hal-01902854/document (accessed on 14 December 2021).

Conflicts of Interest: No conflict of interest was declared by the authors.

References

1. Yousof, H.M.; Afify, A.Z.; Hamedani, G.G.; Aryal, G. The Burr X Generator of Distributions for Lifetime Data. *J. Stat. Theory Appl.* **2017**, *16*, 288–305. [CrossRef]
2. Gove, J.H.; Ducey, M.J.; Leak, W.B.; Zhang, L. Rotated sigmoid structures in managed uneven-aged northern hardwood stands: A look at the Burr Type-III distribution. *Forestry* **2008**, *81*, 161–176. [CrossRef]
3. Burr, I.W. Cumulative frequency distributions. *Ann. Math. Stat.* **1942**, *13*, 215–232. [CrossRef]
4. Kleiber, C.; Kotz, S. *Statistical Size Distributions in Economics and Actuarial Sciences*; John Wiley & Sons, Inc.: Hoboken, NJ, USA, 2003.
5. Mielke, P.W. Another family of distributions for describing and analyzing precipitation data. *J. Appl. Meteorol.* **1973**, *12*, 275–280. [CrossRef]
6. Bebbington, M.S.; Lai, C.D.; Zitikis, R. A flexible Weibull extension. *Reliab. Eng. Syst. Saf.* **2007**, *92*, 719–726. [CrossRef]
7. Zhang, T.; Xie, M. On the upper truncated Weibull distribution and its reliability implications. *Reliab. Eng. Syst. Saf.* **2011**, *96*, 194–200. [CrossRef]
8. Mudholkar, G.S.; Srivastava, D.K. Exponentiated Weibull family for analysing bathtub failure rate data. *IEEE Trans. Reliab.* **1993**, *42*, 299–302. [CrossRef]
9. Marshall, A.W.; Olkin, I. A new method for adding a parameter to a family of distributions with application to the exponential and Weibull families. *Biometrika* **1997**, *84*, 641–652. [CrossRef]
10. Xie, M.; Tang, Y.; Goh, T.N. A modified Weibull extension with bathtub-shaped failure rate function. *Reliab. Eng. Syst. Saf.* **2002**, *76*, 279–285. [CrossRef]
11. Almalki, S.J.; Yuan, J. A new modified Weibull distribution. *Reliab. Eng. Syst. Saf.* **2013**, *111*, 164–170. [CrossRef]
12. Shao, Q.; Chen, Y.D.; Zhang, L. An extension of three-parameter Burr III distribution for low-flow frequency analysis. *Comput. Stat. Data Anal.* **2008**, *52*, 1304–1314. [CrossRef]
13. Çankaya, M.N.; Yalçınkaya, A.; Altındağ, Ö.; Arslan, O. On the robustness of an epsilon skew extension for Burr III distribution on the real line. *Comput. Stat.* **2019**, *34*, 1247–1273. [CrossRef]
14. Modi, K.; Gill, V. Unit Burr-III distribution with application. *J. Stat. Manag. Syst.* **2019**, *23*, 579–592. [CrossRef]
15. Haq, M.A.U.; Hashmi, S.; Aidi, K.; Ramos, P.L.; Louzada, F. Unit modified Burr-III distribution: Estimation, characterizations and validation test. *Ann. Data Sci.* **2020**, 1–26. [CrossRef]
16. Ali, A.; Hasnain, S.A.; Ahmad, M. The modified Burr III distribution, properties and applications. *Pak. J. Stat.* **2015**, *31*, 697–708.
17. Domma, F.; Condino, F. The Beta-Dagum Distribution: Definition and Properties. *Commun. Stat. Theory Methods* **2013**, *42*, 4070–4090. [CrossRef]
18. Arifa, S.; Zafar Yab, M.; Ali, A. The Modified Burr III G Family of Distributions. *J. Data Sci.* **2017**, *15*, 41–60. [CrossRef]
19. Bhatti, F.A.; Hamedani, G.G.; Korkmaz, M.C.; Cordeiro, G.M.; Yousof, H.M.; Ahmad, M. On Burr III Marshal Olkin family: Development, properties, characterizations and applications. *J. Stat. Distrib. Appl.* **2019**, *6*, 1–21. [CrossRef]
20. Cordeiro, G.M.; Yousof, H.M.; Ramires, T.G.; Ortega, E.M.M. The Burr XII System of Densities: Properties, Regression Model and Applications. *J. Stat. Comput. Simul.* **2017**, *87*, 1–25. [CrossRef]
21. Oluyede, B.O.; Huang, S.; Pararai, M. A New Class of Generalized Dagum Distribution with Applications to Income and Lifetime Data. *J. Stat. Econom. Methods* **2014**, *3*, 125–151.
22. Johnson, N.L.; Kotz, S.; Balakrishnan, N. *Continuous Univariate Distributions*; Wiley: Hoboken, NJ, USA, 1995; Volume 2, pp. 140–142.
23. Lai, C.D.; Xie, M.; Murthy, D.N.P. A modified Weibull distribution. *IEEE Trans. Reliab.* **2003**, *52*, 33–37. [CrossRef]
24. Carrasco, M.; Ortega, E.M.; Cordeiro, G.M. A generalized modified Weibull distribution for lifetime modeling. *Comput. Stat. Data Anal.* **2008**, *53*, 450–462. [CrossRef]
25. Aljouiee, A.; Elbatal, I.; Al-Mofleh, H. A New Five-Parameter Lifetime Model: Theory and Applications. *Pak. J. Stat. Oper. Res.* **2018**, *14*, 403–420. [CrossRef]
26. Shaked, M.; Shanthikumar, J.G. *Stochastic Orders and Their Applications*; Wiley: New York, NY, USA, 1994.
27. Jammalamadaka, S.R.; Mangalam, V. Non-parametric estimation for middle censored data. *J. Nonparametr. Stat.* **2003**, *15*, 253–265. [CrossRef]
28. Iyer, S.K.; Jammalamadaka, S.R.; Kundu, D. Analysis of middle-censored data with exponential lifetime distributions. *J. Stat. Plan. Inference* **2008**, *138*, 3550–3560. [CrossRef]
29. Abuzaid A.H. The estimation of the Burr-XII parameters with middle-censored data. *SpringerPlus* **2015**, *4*, 1–10. [CrossRef]
30. Bennett, N.A. Some Contributions to Middle-Censoring. Ph.D. Thesis, Department of Statistics and Applied Probability, University of California Santa Barbara, Santa Barbara, CA, USA, 2011.
31. Abuzaid, A.H.; Abu El-Qumsan, M.K.; El-Habil, A.M. On the robustness of right and middle censoring schemes in parametric survival models. *Commun. Stat.-Simul. Comput.* **2017**, *46*, 1771–1780. [CrossRef]
32. Nadarajah, S.; Kotz, S. On the alternative to the Weibull function. *Eng. Fract. Mech.* **2007**, *74*, 451–456. [CrossRef]

33. Dasgupta, R. On the distribution of Burr with applications. *Sankhya* **2011**, *B73*, 1–19. [CrossRef]
34. Feigl, P.; Zelen, M. Estimation of exponential probabilities with concomitant information. *Biometrics* **1965**, *21*, 826–838. [CrossRef]

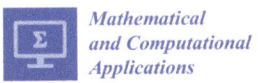

Article

The Unit Teissier Distribution and Its Applications

Anuresha Krishna [1], Radhakumari Maya [2], Christophe Chesneau [3,*] and Muhammed Rasheed Irshad [1]

[1] Department of Statistics, Cochin University of Science and Technology, Cochin 682022, Kerala, India; anuresha.stat@gmail.com (A.K.); irshadmr@cusat.ac.in (M.R.I.)
[2] Department of Statistics, Government College for Women, Trivandrum 695014, Kerala, India; publicationsofmaya@gmail.com
[3] Laboratoire de Mathématiques Nicolas Oresme (LMNO), Université de Caen Normandie, Campus II, Science 3, 14032 Caen, France
* Correspondence: christophe.chesneau@unicaen.fr

Abstract: A bounded form of the Teissier distribution, namely the unit Teissier distribution, is introduced. It is subjected to a thorough examination of its important properties, including shape analysis of the main functions, analytical expression for moments based on upper incomplete gamma function, incomplete moments, probability-weighted moments, and quantile function. The uncertainty measures Shannon entropy and extropy are also performed. The maximum likelihood estimation, least square estimation, weighted least square estimation, and Bayesian estimation methods are used to estimate the parameters of the model, and their respective performances are assessed via a simulation study. Finally, the competency of the proposed model is illustrated by using two data sets from diverse fields.

Keywords: Teissier distribution; unit Teissier distribution; Lambert W function; entropy; extropy; simulation; estimation

MSC: 60E05; 62F10; 33B30

1. Introduction

The introduction of new statistical distributions, defined on both the whole real line and the positive real line, is required for the interpretation of real-world occurrences. In both of these lines, a large number of probability distributions for real data sets have been presented recently. Several distributions that fall within these two categories have demonstrated their importance from both a theoretical and practical standpoint. It is true, however, that in many practical disciplines such as economics, medicine, biology, and others, there is a problem of bounded phenomena, or uncertainty resulting from factors such as rates, indices, proportions, and test scores. It is also important to note that, in comparing distributions with unbounded support to those with bounded support, one can find notable scarceness. As a result, in the recent past, some authors have focused on developing distributions that are defined on the bounded interval using any one of the parent distribution transformation techniques. For better understanding, we refer to recent papers in [1,2].

The beta distribution is definitely the most famous distribution among distributions defined in the $(0,1)$ interval. While the beta distribution is useful for modeling data on the unit interval, many other distributions have been proposed and studied over time. The readers may refer to the Topp–Leone distribution (see [3]), Kumaraswamy distribution (see [4]) and transformed Leipnik distribution (see [5]). However, in recent times, there has been a growing interest by statisticians in proposing distributions defined by the unit interval corresponding to any continuous distribution. The readers may refer to the log-Lindley distribution (see [6]), exponentiated Topp–Leone distribution (see [7]), unit inverse Gaussian distribution (see [8]), unit Lindley distribution (see [9]), unit Weibull distribution (see [10]), unit Burr-III distribution (see [11]), and unit half normal distribution (see [12]), among others. As a result, a lot of contributions are made to the existing literature. This recent leap in the number of research papers devoted to proposing new distributions in the unit interval exhibits their growing relevance. Despite the fact that many distributions have been proposed and studied as alternatives, there is still no agreement on which distribution is preferred.

In this article, a new bounded distribution is introduced by considering the baseline distribution as the Teissier distribution (TD) which was first introduced by the French biologist Georges Teissier in modeling the mortality of domestic animal species as a result of pure aging (see [13]). A continuous random variable Y is said to have the TD with parameter $\theta > 0$ if its probability density function (pdf) is given by

$$f_Y(y;\theta) = \theta\left(e^{\theta y} - 1\right)\exp\left(\theta y - e^{\theta y} + 1\right), \quad y > 0. \tag{1}$$

The cumulative distribution function (cdf) of Y is given by

$$F_Y(y;\theta) = 1 - \exp\left(\theta y - e^{\theta y} + 1\right), \quad y > 0.$$

Despite its importance, the TD has been overlooked in the statistical literature. Ref. [14] examined the location version of this model and explored a characterization based on life expectancy as part of a demographic study, four decades after it was introduced. In a later study, Muth [15] found that a TD-derived distribution has a heavier tail than well-known lifetime distributions such as the gamma, lognormal, and Weibull distributions. This distribution was later used by [16] to estimate the lifetime distribution for a German dataset based on used car prices. The TD and its location version have been forgotten since [16], and no further references appear in the literature until 2008. Ref. [17] showed that as the parameter decreases to zero, the model approaches the classical exponential distribution. The authors of [18] reintroduced this distribution and its scaled version, and studied its important properties using a generalized integro-exponential function [18,19] can be considered as the two-parameter extensions of the TD.

The study of the TD has gained momentum in both theoretical and applied perspectives after the work of [19–23] substantiate this claim. Meanwhile, the TD and its various variants proved their signature compared to other models for various real-life data sets. This is the chief motivation for looking into the TD defined on the unit interval, so-called the unit Teissier distribution (UTD), and its competency compared to other popular distributions defined on the unit interval. Finally, it can be concluded that it performs well compared to other distributions for the real-life data sets considered here.

As in the case of real-life data sets that vary over positive real lines and possess a bathtub-shaped hazard rate function (hrf), some real-life data sets that are defined on the unit interval also possess a bathtub-shaped hrf. In the former case, there is a tremendous amount of work in the literature on various models, such as [24–26]. As far as the latter case is considered, the papers in [8,27] showed that the unit inverse Gaussian and logit slash distributions, respectively, provide a bathtub-shaped hrf. These two distributions are more intricate because of the presence of the log function in their pdfs, and so their cdfs are not obtained in closed form. Hence, simulation experiments from these two distributions are very difficult. For some real data sets, the logit slash distribution does not yield a numerical estimation of parameters. Moreover, the hrfs of both of these distributions are not analytically tractable, even though they sketch the graph of the hrf and provide its various shapes. In addition, Korkmaz [27] proved that the logit slash distribution possesses a N-shaped (modified bathtub shape) and w-shaped bathtub shaped hrf. Another distribution defined on the unit interval, which also possesses a non-monotone shaped hrf, is the unit Modified Burr III distribution (see [28]), which has three parameters. Models with a lower number of parameters are preferable for various reasons, such as ease of simulation drawing and less difficulty in inferential aspects.

The UTD solves these two problems by providing closed form expressions for pdf, cdf, moments, inequality measures, and entropy measures. Secondly, it possesses monotone (increasing and decreasing) as well as non-monotone functions, such as bathtub-shaped and N-shaped hrf (modified bathtub shaped) with a smaller number of parameters. Based on [29], N-shaped hrfs appear in mortality among breast cancer patients.

The rest of the paper is presented as follows. In Section 2, the UTD is introduced and the statistical properties such as moments, incomplete moments, probability-weighted moments, and quantile function are studied. The Shannon entropy and extropy are derived in Section 3. Discussion on different methods of estimation, such as maximum likelihood (ML), ordinary, weighted least square and Bayesian estimation are carried out in Section 4. In Section 5, a simulation study is performed to evaluate the performance of the model parameter estimates. In Section 6, the proposed distribution is elucidated with two real data sets. Finally, the conclusions are presented in Section 7.

2. The Unit Teissier Distribution

In this Section, the UTD is presented and we determine some of its statistical properties.

2.1. Presentation

Mathematically, the UTD is derived from the transformation $X = e^{-Y}$ by taking into account the definition of the pdf given by (1). Its definition is formalized below.

Definition 1. *A random variable X is said to follow the UTD with parameter $\theta > 0$, if its pdf is of the following form:*

$$f(x;\theta) = \theta\left(x^{-\theta} - 1\right) x^{-(\theta+1)} e^{-x^{-\theta}+1}, \qquad x \in (0,1). \qquad (2)$$

The corresponding cdf is given by

$$F(x;\theta) = x^{-\theta} e^{-x^{-\theta}+1}. \qquad (3)$$

From (2) and (3), the survival function $S(x;\theta)$, hrf $h(x;\theta)$ and reversed hrf $\tau(x;\theta)$ of the UTD are obtained as follows (for $x \in (0,1)$):

$$S(x;\theta) = 1 - x^{-\theta} e^{-x^{-\theta}+1},$$

$$h(x;\theta) = \frac{f(x;\theta)}{S(x;\theta)} = \frac{\theta(x^{-\theta}-1)x^{-(\theta+1)}e^{-x^{-\theta}+1}}{1 - x^{-\theta}e^{-x^{-\theta}+1}} \qquad (4)$$

and

$$\tau(x;\theta) = \frac{f(x;\theta)}{F(x;\theta)} = \frac{\theta(x^{-\theta}-1)}{x}.$$

2.2. Shapes of the pdf and hrf

As immediate shape properties for the pdf and hrf, we can say that

$$\lim_{x \to 0} f(x;\theta) = \lim_{x \to 0} h(x;\theta) = 0,$$

and that the UTD is unimodal; the mode is the value of x maximizing $f(x;\theta)$, which is given by

$$x_{mode} = 2^{-\frac{1}{\theta}} \left(\frac{1 + 3\theta - \sqrt{5\theta^2 + 2\theta + 1}}{1 + \theta} \right)^{\frac{1}{\theta}}.$$

In order to obtain an overview of the shapes of the pdf (2) and hrf (4), the corresponding plots for different choices of the parameter θ are given in Figures 1 and 2, respectively.

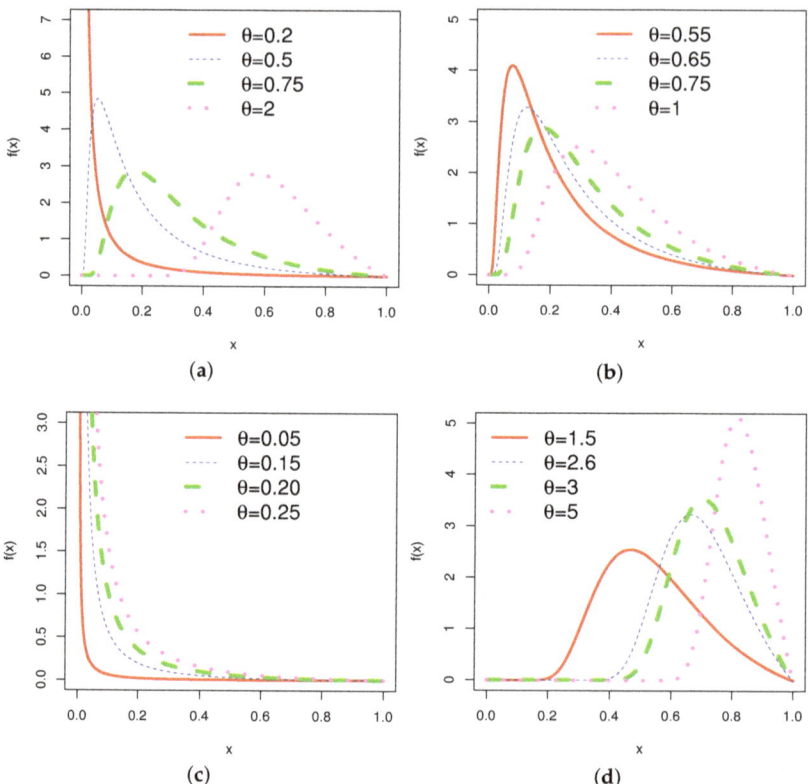

Figure 1. Plots of various shapes of the pdf of the *UTD* for varying values of the parameter θ.

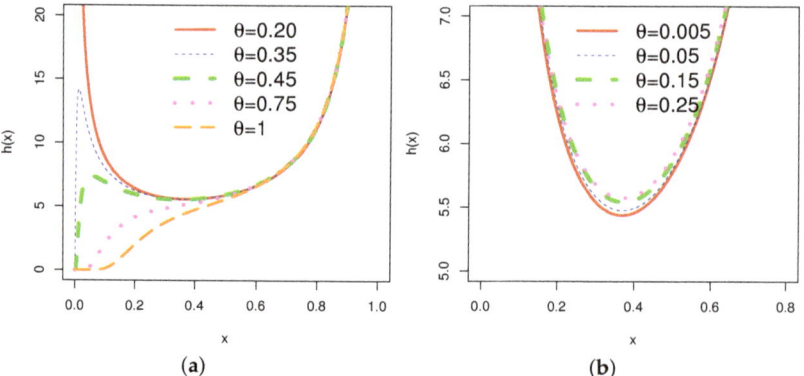

Figure 2. *Cont.*

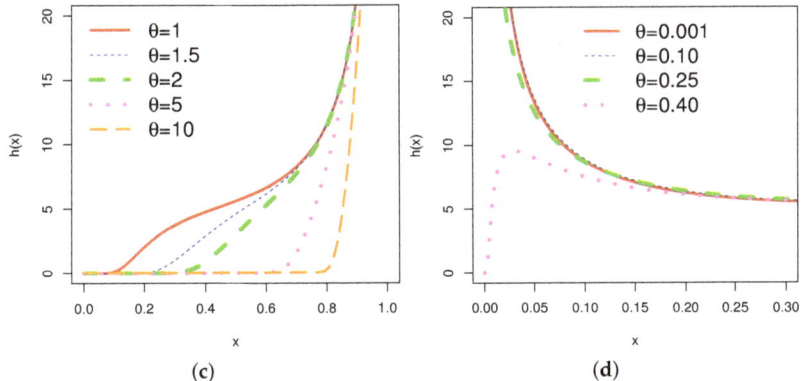

Figure 2. Plots of various shapes of the hrf of the UTD for varying values of the parameter θ.

The graphs of the hrf for various combinations of parameters show various shapes, including increasing, decreasing, bathtub, and N-shaped (modified bathtub shape). According to [29], mortality among breast cancer patients is characterized by an N-shaped hrf. This is one of the prominent properties of the UTD.

2.3. Moments

Analytically, the UTD's non-central moments can be expressed in terms of the upper incomplete gamma function. This is specified in the result below.

Result 1. *For any non-negative integer r, the rth non-central moment of a random variable X with the UTD is*

$$\mu'_r = E(X^r) = e\left\{\Gamma\left(-\frac{r}{\theta}+2,1\right) - \Gamma\left(-\frac{r}{\theta}+1,1\right)\right\},$$

where E denotes the expectation operator, $e \approx 2.718282$ (the exponential function taken at the value 1) and $\Gamma(a,b) = \int_b^\infty t^{a-1}e^{-t}dt$ is the upper incomplete gamma function.

Proof. Based on (2) and the definition of non-central moment, as well as the change of variable $u = x^{-\theta}$, and considering the definition of the upper incomplete gamma function, we have

$$\mu'_r = \int_0^1 x^r f(x;\theta)dx = \int_1^\infty u^{-\frac{r}{\theta}}(u-1)e^{-u+1}du$$
$$= e\left\{\Gamma\left(-\frac{r}{\theta}+2,1\right) - \Gamma\left(-\frac{r}{\theta}+1,1\right)\right\}.$$

This ends the proof. □

The statistical measures such as the mean (μ) and variance (σ^2) can be calculated numerically by using R software. Figure 3a,b illustrate the behavior of the mean and variance for varying values of the parameter θ, respectively.

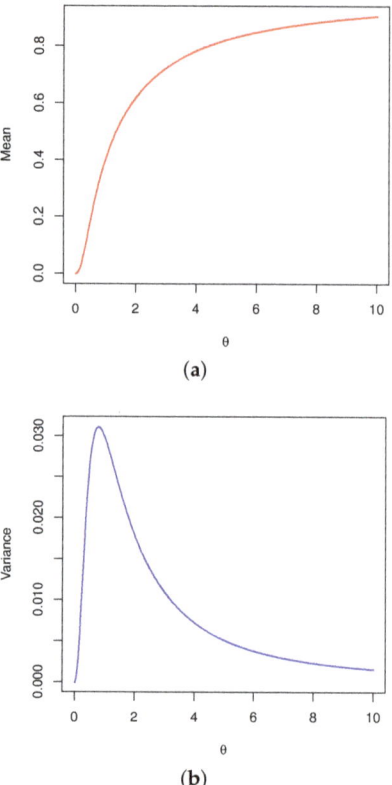

Figure 3. Plots of the (**a**) mean and (**b**) variance of the UTD for varying values of the parameter θ.

2.4. Incomplete Moments

As a generalized version of the non-central moments, the UTD's incomplete moments can be also expressed in terms of the upper incomplete gamma function.

Result 2. *For any non-negative integer k and $x \in (0,1)$, the kth incomplete moment at x of a random variable X with the UTD is*

$$m_k(x;\theta) = E(X^k 1_{X \leq x}) = e\left\{\Gamma\left(-\frac{k}{\theta}+2, x^{-\theta}\right) - \Gamma\left(-\frac{k}{\theta}+1, x^{-\theta}\right)\right\}.$$

Proof. By the definition of the kth incomplete moment, we have

$$m_k(x;\theta) = \int_0^x t^k f(t;\theta)dt.$$

In the above integral expression, after making the change of variable $u = t^{-\theta}$, the proof is similar to the one of Result 1, so the details are excluded here. □

There is an interesting aspect of the fact that the first incomplete moment $m_1(x;\theta)$ can be used to compute the mean deviation from the mean (μ) given by $E[|X - \mu|] = 2\mu F(\mu;\theta) - 2m_1(\mu;\theta)$.

2.5. Probability-Weighted Moments

The probability-weighted moments of the UTD are now under investigation.

Result 3. For any non-negative integers r and s, the $(r,s)^{th}$ probability-weighted moment of a random variable X with the UTD is given as

$$m_{r,s}(x;\theta) = E(X^r F(X;\theta)^s)$$
$$= \frac{e^{s+1}}{(s+1)^{-\frac{r}{\theta}+s+2}}\left\{\Gamma\left(-\frac{r}{\theta}+s+2,s+1\right) - (s+1)\Gamma\left(-\frac{r}{\theta}+s+1,s+1\right)\right\}.$$

Proof. By the definition of the $(r,s)^{th}$ probability-weighted moment and by making the change of variable $u = (s+1)x^{-\theta}$, we obtain

$$m_{r,s}(x;\theta) = \int_0^1 x^r (F(x;\theta))^s f(x;\theta) dx$$
$$= e^{s+1}\theta \int_0^1 x^{r-\theta(s+1)} x^{-(\theta+1)} e^{-(s+1)x^{-\theta}} dx - e^{s+1}\theta \int_0^1 x^{r-\theta s} x^{-(\theta+1)} e^{-(s+1)x^{-\theta}} dx$$
$$= \frac{e^{s+1}}{(s+1)^{-\frac{r}{\theta}+s+2}} \int_{s+1}^\infty u^{-\frac{r}{\theta}+s+1} e^{-u} du - \frac{e^{s+1}}{(s+1)^{-\frac{r}{\theta}+s+1}} \int_{s+1}^\infty u^{-\frac{r}{\theta}+s} e^{-u} du$$
$$= \frac{e^{s+1}}{(s+1)^{-\frac{r}{\theta}+s+2}}\left\{\Gamma\left(-\frac{r}{\theta}+s+2,s+1\right) - (s+1)\Gamma\left(-\frac{r}{\theta}+s+1,s+1\right)\right\}.$$

The desired result is obtained. □

2.6. Mean Residual Life Function

Result 4. For any $t \in (0,1)$, the mean residual life function at t of a random variable X with the UTD is

$$r(t;\theta) = E(X-t|X>t)$$
$$= \frac{1}{1-t^{-\theta}e^{-t^{-\theta}+1}}\left\{1-t-\frac{e}{\theta}\left[\Gamma\left(1-\frac{1}{\theta},1\right)-\Gamma\left(1-\frac{1}{\theta},t^{-\theta}\right)\right]\right\}.$$

Proof. By the definition of the mean residual life function and by making the change of variable $u = x^{-\theta}$, we obtain

$$r(t;\theta) = \frac{1}{S(t;\theta)} \int_t^1 S(x;\theta) dx$$
$$= \frac{1}{S(t;\theta)}\left\{1-t-\int_t^1 x^{-\theta} e^{-x^{-\theta}+1} dx\right\}$$
$$= \frac{1}{S(t;\theta)}\left\{1-t-\frac{e}{\theta}\int_1^{t^{-\theta}} u^{-\frac{1}{\theta}} e^{-u} du\right\}$$
$$= \frac{1}{1-t^{-\theta}e^{-t^{-\theta}+1}}\left\{1-t-\frac{e}{\theta}\left[\Gamma\left(1-\frac{1}{\theta},1\right)-\Gamma\left(1-\frac{1}{\theta},t^{-\theta}\right)\right]\right\}.$$

The claimed result is achieved. □

2.7. Quantile Function

In addition to its remarkable property of being expressed in closed form, the UTD is also capable of representing the quantile function in terms of the negative branch of the Lambert W function. We recall that the Lambert W function is a multivalued complex function defined as the solution of the equation $W(z)e^{W(z)} = z$, where z is a complex number. For any real numbers $z \geq -1/e$, $W(z)$ has two real branches. The real branch taking on values in $[-1,\infty)$ is called the principal branch and is denoted by W_0, and the one taking on values in $(-\infty, -1]$ is called the negative branch and is denoted by W_{-1}. For a comprehensive review of this special function, readers are referred to [30]. From a computational perspective, the Lambert W function is available in computer algebra systems such as Maple (function `LambertW`), Mathematica (function `ProductLog`), and Matlab (function `lambertW`), and also in programming languages such as R [31] (functions `lambert_W0` and `lambert_Wm1` for W_0 and W_{-1}, respectively, in the package gsl).

Result 5. *The quantile function of the UTD is given by*

$$Q(p;\theta) = \left[-W_{-1}\left(-\frac{p}{e}\right)\right]^{-\frac{1}{\theta}}, \quad (5)$$

where $p \in (0,1)$ and W_{-1} denotes the negative branch of the Lambert W function.

Proof. The cdf $F(x;\theta)$ should be inverted to obtain the quantile function of UTD. Thus, we need to solve $F(x;\theta) = p$ according to x, so

$$-x^{-\theta}e^{-x^{-\theta}} = -\frac{p}{e} \Leftrightarrow -x^{-\theta} = W_{-1}\left(-\frac{p}{e}\right) \Leftrightarrow x = \left[-W_{-1}\left(-\frac{p}{e}\right)\right]^{-\frac{1}{\theta}}.$$

The obtained x corresponds to the quantile function with respect to p. The desired result is achieved. □

The quantile function in (5) can be used to obtain the following quantities:

- the median given as $M = Q(0.5;\theta)$;
- the Galton coefficient of skewness specified by

$$S = \frac{Q(0.25;\theta) + Q(0.75;\theta) - 2M}{Q(0.75;\theta) - Q(0.25;\theta)};$$

- the Moors coefficient of kurtosis (with correction) defined by

$$T = \frac{Q(0.875;\theta) - Q(0.625;\theta) + Q(0.375;\theta) - Q(0.125;\theta)}{Q(0.75;\theta) - Q(0.25;\theta)} - 1.23.$$

3. Shannon Entropy and Extropy

This section discusses the Shannon entropy and extropy of the UTD.

3.1. Shannon Entropy

Conceptually, the Shannon entropy is the amount of information into a random variable. For a random variable T with pdf $f(t)$ with support $[0,1]$, its Shannon entropy is defined as

$$H(T) = -\int_0^1 f(t)\log f(t)dt. \quad (6)$$

The Shannon entropy for the UTD is obtained in terms of the following quantities:

- The upper incomplete gamma function already introduced;
- The nth derivative of the gamma function given as $\Gamma^n(a) = \int_0^\infty t^{a-1}(\log t)^n e^{-t}dt$;
- The nth derivative of the incomplete gamma function given as $\Gamma^n(a,z) = \int_z^\infty t^{a-1}(\log t)^n e^{-t}dt$;
- The exponential integral defined by $E_1(t) = \int_t^\infty (e^{-u}/u)du$.

The next result is about the expression of the Shannon entropy.

Result 6. *The Shannon entropy of a random variable T with the UTD has the following form*

$$H(T) = -(\log\theta)(e\Gamma(2,1)-1) - \Gamma^1(2) - \frac{e(\theta+1)}{\theta}\left(\Gamma^1(2,1) - E_1(1)\right) \qquad (7)$$
$$- 2e\Gamma(2,1) + e\Gamma(3,1) + 1.$$

Proof. By considering the change of variables, $u = t^{-\theta}$ and $v = t^{-\theta}-1$ and taking into account (2), we obtain

$$\int_0^1 f(t)\log f(t)dt = e(\log\theta)\left(\int_1^\infty ue^{-u}du - \int_1^\infty e^{-u}du\right) + \int_0^\infty ve^{-v}(\log v)dv$$
$$+ \frac{e(\theta+1)}{\theta}\left(\int_1^\infty u\log ue^{-u}du - \int_1^\infty e^{-u}(\log u)du\right) \qquad (8)$$
$$+ e\int_1^\infty (2u - u^2 - 1)e^{-u}du.$$

Substituting (8) in (6) and by using the quantities listed, (7) is obtained. □

3.2. Extropy

Extropy is a complementary dual of Shannon entropy that has a variety of interesting applications, including appropriate scoring of forecasting distributions, comparing the uncertainty of two random variables, and astronomical measurements of heat distributions in galaxies, among others. For a random variable T with pdf $f(t)$ with support $[0,1]$, its extropy is defined as

$$J(T) = -\frac{1}{2}\int_0^1 f^2(t)dt. \qquad (9)$$

Result 7. *The extropy of a random variable T with the UTD has the following form*

$$J(T) = \frac{\theta e^2}{2^{\frac{1}{\theta}+3}}\left\{\Gamma\left(\frac{1}{\theta}+3,2\right) - \frac{1}{4}\Gamma\left(\frac{1}{\theta}+4,2\right) - \Gamma\left(\frac{1}{\theta}+2,2\right)\right\}. \qquad (10)$$

Proof. By considering the change of variable $u = 2t^{-\theta}$ and taking into account (2), the numerator of (9) can be obtained as

$$\int_0^1 f^2(t;\theta)dt = \theta^2 e^2 \int_0^1 t^{-2(\theta+1)}t^{-2\theta}e^{-2t^{-\theta}}dt + \theta^2 e^2 \int_0^1 t^{-2(\theta+1)}e^{-2t^{-\theta}}dt$$
$$- \theta^2 e^2 \int_0^1 t^{-2(\theta+1)}2t^{-\theta}e^{-2t^{-\theta}}dt$$
$$= \frac{\theta e^2}{2^{\frac{1}{\theta}+4}}\int_2^\infty u^{\frac{1}{\theta}+3}e^{-u}du + \frac{\theta e^2}{2^{\frac{1}{\theta}+2}}\int_2^\infty u^{\frac{1}{\theta}+1}e^{-u}du - \frac{\theta e^2}{2^{\frac{1}{\theta}+2}}\int_2^\infty u^{\frac{1}{\theta}+2}e^{-u}du \qquad (11)$$
$$= \frac{\theta e^2}{2^{\frac{1}{\theta}+4}}\Gamma\left(\frac{1}{\theta}+4,2\right) + \frac{\theta e^2}{2^{\frac{1}{\theta}+2}}\Gamma\left(\frac{1}{\theta}+2,2\right) - \frac{\theta e^2}{2^{\frac{1}{\theta}+2}}\Gamma\left(\frac{1}{\theta}+3,2\right).$$

Substituting (11) in (9), (10) is obtained. □

4. Estimation and Inference

4.1. Maximum Likelihood Estimation

Let $x_1, x_2, \ldots, x_n$ be a random sample of values of size n from a random variable X with the UTD of unknown parameter θ. Then the likelihood function of θ is given by

$$l(\theta) = \theta^n \prod_{i=1}^n x_i^{-(\theta+1)} \prod_{i=1}^n \left(x_i^{-\theta}-1\right)e^{-\sum_{i=1}^n (x_i^{-\theta}-1)}. \qquad (12)$$

Then the ML estimate (MLE) of θ, say $\hat{\theta}$, is obtained by maximizing $l(\theta)$ with respect to θ. Thus, for any $\theta > 0$, we have $l(\theta) \leq l(\hat{\theta})$. Practically, one can use the derivative technique; the partial derivative of $\log l(\theta)$ with respect to the parameter is

$$\frac{\partial \log l(\theta)}{\partial \theta} = \frac{n}{\theta} - \sum_{i=1}^{n} \frac{x_i^{-\theta} \log x_i}{x_i^{-\theta} - 1} - \sum_{i=1}^{n} \log x_i + \sum_{i=1}^{n} x_i^{-\theta} \log x_i,$$

and $\hat{\theta}$ is obtained by solving the equation $\frac{\partial \log l(\theta)}{\partial \theta} = 0$ according to θ. Numerical optimization approaches employing mathematical tools such as R and Mathematica are the only way to achieve this.

Fisher Information Matrix and Asymptotic Confidence Interval

In order to carry out statistical inference on the parameters of the UTD, the 1×1 expected Fisher information matrix is needed. It can, however, be efficiently approximated by the observed Fisher information matrix $J(\hat{\theta})$ given by

$$J(\hat{\theta}) = -\left. \frac{\partial^2 \log l(\theta)}{\partial \theta^2} \right|_{\theta = \hat{\theta}}$$

and we can approximate variance of $\hat{\theta}$ as

$$Var(\hat{\theta}) \approx J^{-1}(\hat{\theta}).$$

Then the approximate $100(1 - \phi)\%$ two-sided normal confidence interval for θ is given by $\hat{\theta} \pm Z_{\frac{\phi}{2}} \sqrt{Var(\hat{\theta})}$, where $Z_{\frac{\phi}{2}}$ is the upper $\frac{\phi}{2}^{th}$ percentile of a standard normal distribution and ϕ is the significance level.

4.2. Ordinary and Weighted Least-Squares Estimation

The ordinary least square (LS) estimation and the weighted least square (WLS) estimation were proposed by [32] to estimate the parameters of the beta distribution. Let $x_{1:n}, x_{2:n}, \ldots, x_{n:n}$ be the ordered values of $x_1, x_2, \ldots, x_n$. Let us set

$$R(\theta) = \sum_{i=1}^{n} \left[F(x_{i:n}; \theta) - \frac{i}{n+1} \right]^2.$$

Then the LS estimate (LSE) of θ, say $\tilde{\theta}$, is obtained by minimizing $R(\theta)$ with respect to θ. Thus, for any $\theta > 0$, we have $R(\hat{\theta}) \leq R(\theta)$. Practically, the LSE can also be obtained by solving the following equation:

$$\frac{\partial R(\theta)}{\partial \theta} = 2 \sum_{i=1}^{n} \left[F(x_{i:n}; \theta) - \frac{i}{n+1} \right] D(x_{i:n}; \theta) = 0,$$

where

$$D(x; \theta) = \frac{\partial F(x; \theta)}{\partial \theta} = -e^{-x^{-\theta}+1} x^{-2\theta} (x^{\theta} - 1) \log(x),$$

according to θ. Similarly, the WLS estimate (WLSE) of θ, say $\breve{\theta}$, is obtained by minimizing the non-linear function

$$W(\theta) = \sum_{i=1}^{n} \frac{(n+1)^2 (n+2)}{i(n-i+1)} \left[F(x_{i:n}; \theta) - \frac{i}{n+1} \right]^2,$$

and it can also be obtained by solving the following equation:

$$\frac{\partial W(\theta)}{\partial \theta} = 2 \sum_{i=1}^{n} \frac{(n+1)^2 (n+2)}{i(n-i+1)} \left[F(x_{i:n}; \theta) - \frac{i}{n+1} \right] D(x_{i:n}; \theta) = 0.$$

4.3. Bayesian Estimation

In this subsection, the estimate of the UTD parameter is calculated by using Bayesian analysis. With this approach, prior knowledge about the problem can be incorporated. Here, the parameter should be given a prior density, and two types of priors are used as random priors: the half-Cauchy (HC) and the normal (N) priors. In the numerical integration, prior distributions that are not completely flat provide enough information to allow the numerical approximation algorithm to continue exploring the target posterior density. The HC distribution features such shapes, and its mean and variance do not exist, but its mode is equal to zero. Regarding the HC distribution, when the parameter becomes 25, the pdf is flat, but not entirely. According to [33], depending upon the information necessary, uniform or HC is a better choice of prior. Thus, for the parameter θ, $HC(25)$ and $N(0, 1000)$ are used as prior distributions.

- **Case 1:** $\theta \sim HC(25)$, then the posterior pdf is given by

$$\pi(\theta/x) \propto l(\theta) \times \frac{2 \times 25}{\pi(\theta^2 + 25^2)}. \tag{13}$$

- **Case 2:** $\theta \sim N(0, 1000)$, then the posterior pdf is given by

$$\pi(\theta/x) \propto l(\theta) \times \frac{1}{\sqrt{2\pi \times 10^3}} e^{-\frac{\theta^2}{2 \times 10^3}}, \tag{14}$$

where $l(\theta)$ is derived from (12).

Since (13) and (14) are not in closed form, one may use numerical integration or MCMC methods.

5. Simulation Study

5.1. Simulation for ML, LS, and WLS Estimates

In the simulation study, the Monte Carlo simulation was done in order to prove the efficiency of the model by using different estimation methods such as ML, LS, and WLS. The estimates were calculated for true values of parameters for $N = 1000$ samples of sizes 25, 50, 75, 100, 150, 200, and 500 and the following quantities were computed:

1. Mean of the estimates: $\text{Mean}(v) = \frac{1}{N} \sum_{i=1}^{N} v_i$;
2. Average bias of the estimates: $\text{Bias}(v) = \frac{1}{N} \sum_{i=1}^{N} (v_i - \theta)$;
3. Mean square error (MSE) of the estimates: $\text{MSE}(v) = \frac{1}{N} \sum_{i=1}^{N} (v_i - \theta)^2$,

where $v \in \{\hat{\theta}, \tilde{\theta}, \check{\theta}\}$, and the index i indicates the considered number of the sample. Tables 1–3 show the simulation results corresponding to the ML, LS, and WLS estimation methods. A graphical comparison of the MSEs obtained from the three methods is presented in Figure 4. Among these methods, the ML estimation provides accurate estimates of the parameter with the least MSE. As in all the estimation methods, the MSE decreases as sample size n increases, as expected.

Table 1. Simulation results: MLEs, bias, and MSE.

	$\theta = 0.26$			$\theta = 0.45$			$\theta = 0.60$		
n	$\hat{\theta}$	Bias($\hat{\theta}$)	MSE($\hat{\theta}$)	$\hat{\theta}$	Bias($\hat{\theta}$)	MSE($\hat{\theta}$)	$\hat{\theta}$	Bias($\hat{\theta}$)	MSE($\hat{\theta}$)
25	0.26352	0.00352	0.00045	0.45610	0.00610	0.00135	0.60813	0.00813	0.00241
50	0.26164	0.00164	0.00021	0.45284	0.00284	0.00063	0.60379	0.00379	0.00112
75	0.26116	0.00116	0.00014	0.45200	0.00200	0.00041	0.60267	0.00267	0.00072
100	0.26074	0.00074	0.00010	0.45129	0.00129	0.00031	0.60172	0.00172	0.00055
150	0.26042	0.00042	0.00006	0.45074	0.00074	0.00019	0.60098	0.00098	0.00034
200	0.26028	0.00028	0.00005	0.45049	0.00049	0.00015	0.60065	0.00065	0.00027
500	0.26011	0.00011	0.00002	0.45019	0.00019	0.00006	0.60025	0.00025	0.00011

Table 1. Cont.

	$\theta = 1$			$\theta = 1.5$			$\theta = 2$		
n	$\hat{\theta}$	Bias($\hat{\theta}$)	MSE($\hat{\theta}$)	$\hat{\theta}$	Bias($\hat{\theta}$)	MSE($\hat{\theta}$)	$\hat{\theta}$	Bias($\hat{\theta}$)	MSE($\hat{\theta}$)
25	1.01355	0.01355	0.00669	1.52033	0.02033	0.01505	2.02711	0.02711	0.02676
50	1.00632	0.00632	0.00312	1.50948	0.00948	0.00702	2.01264	0.01264	0.01248
75	1.00446	0.00446	0.00200	1.50668	0.00668	0.00450	2.00891	0.00891	0.00800
100	1.00286	0.00286	0.00153	1.50429	0.00429	0.00344	2.00572	0.00572	0.00612
150	1.00163	0.00163	0.00094	1.50245	0.00245	0.00213	2.00327	0.00327	0.00378
200	1.00109	0.00109	0.00074	1.50164	0.00164	0.00166	2.00218	0.00218	0.00295
500	1.00041	0.00041	0.00030	1.50062	0.00062	0.00068	2.00082	0.00082	0.00121

Table 2. Simulation results: LS estimates, bias, and MSE.

	$\theta = 0.26$			$\theta = 0.45$			$\theta = 0.60$		
n	$\tilde{\theta}$	Bias($\tilde{\theta}$)	MSE($\tilde{\theta}$)	$\tilde{\theta}$	Bias($\tilde{\theta}$)	MSE($\tilde{\theta}$)	$\tilde{\theta}$	Bias($\tilde{\theta}$)	MSE($\tilde{\theta}$)
25	0.26122	0.00122	0.00069	0.45210	0.00210	0.00207	0.60280	0.00280	0.00368
50	0.25998	−0.00002	0.00030	0.44997	−0.00003	0.00089	0.59996	−0.00004	0.00159
75	0.25997	−0.00003	0.00019	0.44994	−0.00006	0.00058	0.59992	−0.00008	0.00103
100	0.25990	−0.00010	0.00016	0.44983	−0.00017	0.00047	0.59977	−0.00023	0.00084
150	0.25954	−0.00046	0.00010	0.44920	−0.00080	0.00029	0.59893	−0.00107	0.00051
200	0.26057	0.00057	0.00008	0.44914	−0.00086	0.00022	0.59886	−0.00114	0.00039
500	0.25980	−0.00020	0.00003	0.44964	−0.00036	0.00009	0.59953	−0.00047	0.00016
	$\theta = 1$			$\theta = 1.5$			$\theta = 2$		
n	$\tilde{\theta}$	Bias($\tilde{\theta}$)	MSE($\tilde{\theta}$)	$\tilde{\theta}$	Bias($\tilde{\theta}$)	MSE($\tilde{\theta}$)	$\tilde{\theta}$	Bias($\tilde{\theta}$)	MSE($\tilde{\theta}$)
25	1.00467	0.00467	0.01021	1.50701	0.00701	0.02297	2.00935	0.00935	0.04083
50	0.99993	−0.00007	0.00442	1.49990	−0.00010	0.00994	1.99986	−0.00014	0.01768
75	0.99986	−0.00014	0.00287	1.49979	−0.00021	0.00646	1.99972	−0.00028	0.01148
100	0.99961	−0.00039	0.00233	1.49942	−0.00058	0.00523	1.99923	−0.00077	0.00930
150	0.99821	−0.00179	0.00142	1.49732	−0.00268	0.00319	1.99643	−0.00357	0.00567
200	0.99809	−0.00191	0.00109	1.49714	−0.00286	0.00244	1.99619	−0.00381	0.00434
500	0.99921	−0.00079	0.00045	1.49881	−0.00119	0.00101	1.99842	−0.00158	0.00180

Table 3. Simulation results: WLS estimates, bias, and MSE.

	$\theta = 0.26$			$\theta = 0.45$			$\theta = 0.60$		
n	$\breve{\theta}$	Bias($\breve{\theta}$)	MSE($\breve{\theta}$)	$\breve{\theta}$	Bias($\breve{\theta}$)	MSE($\breve{\theta}$)	$\breve{\theta}$	Bias($\breve{\theta}$)	MSE($\breve{\theta}$)
25	0.26141	0.00141	0.00062	0.45243	0.00243	0.00186	0.60324	0.00324	0.00330
50	0.26024	0.00024	0.00026	0.45041	0.00041	0.00078	0.60055	0.00055	0.00140
75	0.26018	0.00018	0.00017	0.45030	0.00030	0.00051	0.60041	0.00041	0.00090
100	0.26007	0.00007	0.00014	0.45012	0.00012	0.00041	0.60016	0.00016	0.00072
150	0.25971	−0.00029	0.00008	0.44950	−0.00050	0.00025	0.59934	−0.00066	0.00044
200	0.25968	−0.00032	0.00006	0.44944	−0.00056	0.00019	0.59925	−0.00075	0.00034
500	0.25986	0.00014	0.00001	0.44976	−0.00024	0.00008	0.59968	−0.00032	0.00014
	$\theta = 1$			$\theta = 1.5$			$\theta = 2$		
n	$\breve{\theta}$	Bias($\breve{\theta}$)	MSE($\breve{\theta}$)	$\breve{\theta}$	Bias($\breve{\theta}$)	MSE($\breve{\theta}$)	$\breve{\theta}$	Bias($\breve{\theta}$)	MSE($\breve{\theta}$)
25	1.00540	0.00540	0.00917	1.50809	0.00809	0.02063	2.01079	0.01079	0.03668
50	1.00091	0.00091	0.00388	1.50137	0.00137	0.00872	2.00182	0.00182	0.01550
75	1.00067	0.00067	0.00250	1.50101	0.00101	0.00562	2.00135	0.00135	0.00999
100	1.00026	0.00026	0.00201	1.50039	0.00039	0.00451	2.00052	0.00052	0.00802
150	0.99889	−0.00111	0.00122	1.49834	−0.00166	0.00275	1.99778	−0.00222	0.00490
200	0.99876	−0.00124	0.00093	1.49813	−0.00187	0.00210	1.99751	−0.00249	0.00373
500	0.99946	−0.00054	0.00038	1.49919	−0.00081	0.00087	1.99892	−0.00108	0.00154

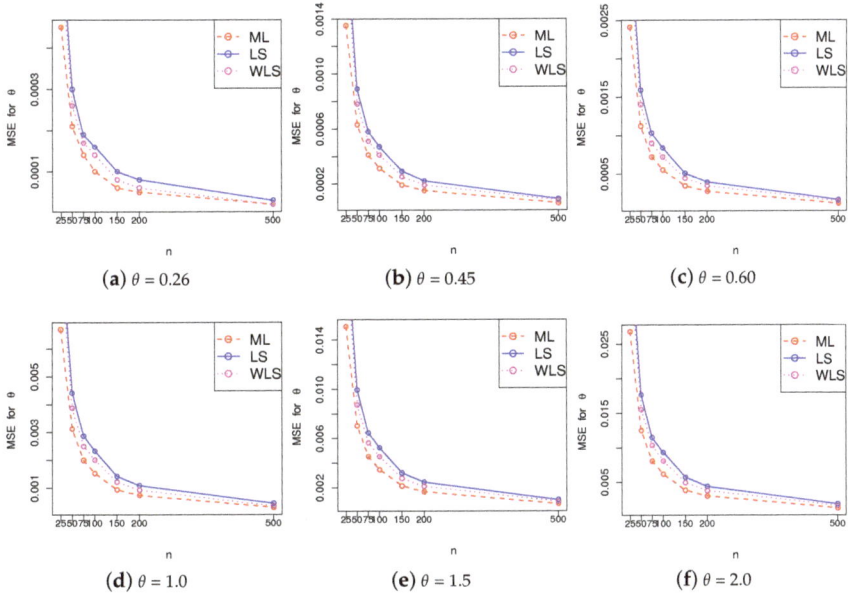

Figure 4. Graphical comparison of the MSEs obtained from ML, LS, and WLS estimation methods for (**a**) $\theta = 0.26$, (**b**) $\theta = 0.45$, (**c**) $\theta = 0.60$, (**d**) $\theta = 1.0$, (**e**) $\theta = 1.5$ and (**f**) $\theta = 2.0$.

5.2. Simulation for Bayesian Estimates

For the *UTD* parameter θ, in Section 4.3, the HC distribution (Case 1) and N distribution (Case 2) were motivated as prior distributions. As a result, the posterior summary results, such as means, standard deviations (SDs), Monte Carlo errors (MCEs), lower bounds (LBs), and upper bounds (UBs) of the 95% confidence intervals and medians are summarized in Tables 4 and 5 for Case 1 and Case 2, respectively. In both cases, increasing sample size leads to a decrease in SD and MCE, which predicts a consistency in the Bayesian estimates.

Table 4. Posterior summary results (Case 1: $\theta \sim HC(25)$).

	$\theta = 0.26$					
n	Mean	SD	MCE	LB	UB	Median
25	0.26217	0.01971	0.00300	0.22331	0.28030	0.26770
50	0.26834	0.01649	0.00213	0.24361	0.29064	0.26831
75	0.27247	0.01599	0.00206	0.25154	0.29064	0.27118
100	0.26770	0.01057	0.00175	0.25917	0.28695	0.26957
150	0.27098	0.00938	0.00154	0.25339	0.28218	0.26928
200	0.26300	0.00638	0.00123	0.25714	0.26781	0.26407
500	0.27236	0.00264	0.00109	0.26706	0.27597	0.27425
	$\theta = 0.45$					
n	Mean	SD	MCE	LB	UB	Median
25	0.45493	0.04530	0.00889	0.36841	0.52298	0.45676
50	0.47560	0.02125	0.00716	0.43660	0.51318	0.46884
75	0.46305	0.02069	0.00598	0.41964	0.50306	0.46555
100	0.47071	0.01648	0.00517	0.42378	0.49429	0.47592
150	0.46769	0.01489	0.00393	0.42507	0.49429	0.46646
200	0.45651	0.00991	0.00262	0.44271	0.47886	0.45687
500	0.46907	0.00830	0.00231	0.44839	0.48607	0.46841

Table 4. *Cont.*

			$\theta = 0.60$			
n	Mean	SD	MCE	LB	UB	Median
25	0.62836	0.05583	0.01269	0.51515	0.69347	0.65291
50	0.63488	0.03203	0.00815	0.55716	0.68275	0.63661
75	0.63130	0.02958	0.00740	0.54930	0.68275	0.63682
100	0.64362	0.02506	0.00657	0.59327	0.68275	0.65125
150	0.61670	0.02001	0.00473	0.59286	0.65175	0.60516
200	0.58433	0.01135	0.00361	0.56540	0.60849	0.58477
500	0.61527	0.00864	0.00333	0.60791	0.63575	0.61179
			$\theta = 1$			
n	Mean	SD	MCE	LB	UB	Median
25	1.02150	0.08400	0.01689	0.84895	1.13786	1.02060
50	1.02859	0.08394	0.01243	0.94449	1.15295	1.02883
75	1.04385	0.04133	0.01163	0.92371	1.09369	1.04292
100	1.06278	0.04029	0.01001	1.02172	1.22162	1.05589
150	1.03099	0.03427	0.00846	0.98819	1.07907	1.03083
200	1.02172	0.02549	0.00573	0.99575	1.05466	1.02620
500	0.97150	0.01249	0.00484	0.94968	0.98967	0.97079
			$\theta = 1.5$			
n	Mean	SD	MCE	LB	UB	Median
25	1.52918	0.14300	0.04025	1.31918	1.78305	1.50069
50	1.51870	0.11159	0.03641	1.17142	1.67897	1.55316
75	1.40491	0.09922	0.03165	1.20423	1.52690	1.42708
100	1.56450	0.07915	0.02093	1.25097	1.64389	1.57598
150	1.54486	0.05005	0.01196	1.47530	1.61097	1.55532
200	1.52040	0.02288	0.00591	1.50104	1.56919	1.52027
500	1.55005	0.02201	0.00390	1.52843	1.58571	1.55540
			$\theta = 2$			
n	Mean	SD	MCE	LB	UB	Median
25	2.06736	0.15240	0.03757	1.74925	2.30008	2.13126
50	2.07023	0.14915	0.02789	1.75671	2.29423	2.09929
75	2.07862	0.11586	0.02615	1.92777	2.22704	2.06513
100	2.09001	0.07422	0.02596	1.99647	2.20813	2.07646
150	2.03891	0.07322	0.02149	1.75963	2.14525	2.06189
200	2.06265	0.05250	0.01654	1.99405	2.26134	2.05505
500	2.09538	0.04211	0.01636	2.06500	2.21627	2.06596

Table 5. Posterior summary results (Case 2: $\theta \sim N(0, 1000)$).

			$\theta = 0.26$			
n	Mean	SD	MCE	LB	UB	Median
25	0.25141	0.05134	0.01312	0.04355	0.30548	0.26122
50	0.25790	0.04318	0.01151	0.15933	0.29794	0.26140
75	0.26324	0.03247	0.00592	0.24319	0.28917	0.26559
100	0.27617	0.01091	0.00322	0.25228	0.29550	0.27556
150	0.24699	0.01081	0.00208	0.23512	0.26737	0.24183
200	0.26654	0.01409	0.00285	0.25307	0.30800	0.26631
500	0.26818	0.01401	0.00159	0.26320	0.28228	0.26949

Table 5. *Cont.*

			$\theta = 0.45$			
n	Mean	SD	MCE	LB	UB	Median
25	0.44924	0.04779	0.01029	0.37730	0.52804	0.44985
50	0.46277	0.03141	0.00797	0.41231	0.51942	0.46657
75	0.46092	0.02074	0.00545	0.41995	0.49405	0.45864
100	0.47382	0.01515	0.00498	0.44194	0.50921	0.47630
150	0.43418	0.01103	0.00326	0.41547	0.44902	0.43349
200	0.46069	0.01011	0.00263	0.44098	0.48238	0.46259
500	0.46553	0.00599	0.00182	0.45494	0.47738	0.46694
			$\theta = 0.60$			
n	Mean	SD	MCE	LB	UB	Median
25	0.60442	0.05548	0.01444	0.49541	0.71972	0.60913
50	0.54888	0.03104	0.01281	0.49527	0.62074	0.55447
75	0.61557	0.02836	0.00759	0.57167	0.65640	0.61174
100	0.62561	0.01481	0.00538	0.60462	0.65368	0.62487
150	0.58561	0.01322	0.00402	0.55367	0.60185	0.58913
200	0.59753	0.01138	0.00449	0.58214	0.61691	0.59483
500	0.61042	0.01113	0.00361	0.59462	0.63647	0.61434
			$\theta = 1$			
n	Mean	SD	MCE	LB	UB	Median
25	1.02150	0.08400	0.01689	0.84895	1.13786	1.02060
50	1.03794	0.06691	0.01658	0.89983	1.14142	1.04985
75	1.04394	0.05572	0.01330	0.90487	1.14142	1.04432
100	1.04440	0.04849	0.01302	0.92824	1.14142	1.03943
150	1.05009	0.04125	0.01278	0.98320	1.13767	1.05431
200	1.00869	0.07225	0.01869	0.90478	1.07876	1.01865
500	1.01614	0.06391	0.00458	1.00317	1.04602	1.01625
			$\theta = 1.5$			
n	Mean	SD	MCE	LB	UB	Median
25	1.52918	0.14300	0.04025	1.31918	1.78305	1.50069
50	1.54627	0.07872	0.02170	1.38232	1.64742	1.55280
75	1.55881	0.06692	0.01959	1.44383	1.66958	1.55559
100	1.55876	0.04834	0.01311	1.45950	1.62751	1.55994
150	1.55805	0.03772	0.01221	1.48935	1.63206	1.55871
200	1.52328	0.10483	0.00849	1.45942	1.66891	1.52085
500	1.53107	0.09938	0.00683	1.51895	1.62075	1.51895
			$\theta = 2$			
n	Mean	SD	MCE	LB	UB	Median
25	1.93826	0.19074	0.03847	1.58829	2.30034	1.93391
50	2.06358	0.14649	0.03532	1.62112	2.28823	2.10545
75	2.02426	0.13764	0.02461	1.86401	2.29424	2.01176
100	2.11171	0.10124	0.02076	1.85800	2.22242	2.11979
150	2.05937	0.10096	0.01815	1.97012	2.21942	2.07596
200	2.03326	0.10071	0.01678	1.94723	2.23003	2.03772
500	2.05469	0.06821	0.01188	1.98674	2.26369	2.06343

6. Application

6.1. Methodology

With the help of two real-life data sets, the superiority of the UTD is illustrated. The first data set is from [34]. The data are the maximum flood level (in millions of cubic feet per second) for the Susquehanna River at Harrisburg, Pennsylvania. The second data set represents times between failures of secondary reactor pumps (see [35]).

An analysis of the total time on test (TTT) plot is used to identify the shape of the underlying hrf of the data. We specify that the hrf is decreasing, increasing, bathtub-shaped, and upside-down bathtub-shaped if the empirical TTT transform is convex, concave, convex then concave, and concave then convex, respectively. Thus, it will be shown that the first data set has an increasing hrf, while the second data set has a bathtub-shaped hrf. For the sake of comparison, the following lifetime distributions were considered: Log-Lindley distribution (LLD) (see [6]), unit Lindley distribution (ULD) (see [9]), Topp–Leone distribution (TLD) (see [3]), one-parameter Kumaraswamy distribution (OKD), power distribution (PD), transmuted distribution (TD), two-parameter Kumaraswamy distribution (KD), beta distribution (BD), exponentiated Topp–Leone distribution ($ETLD$), and unit Burr-III distribution (UBD). Using R software, the MLEs of all these distributions' parameters are computed along with information criteria, which are listed below.

- Akaike information criterion defined by $AIC = 2p - 2\log l$;
- Akaike information criterion corrected given as $AICc = AIC + \frac{2p(p+1)}{n-p-1}$;
- Consistent Akaike information criterion specified by $CAIC = -2\log l + p(\log n + 1)$;
- Bayesian information criterion defined by $BIC = p \log n - 2 \log l$.

Here, $\log l$ denotes the estimated value of the maximum log-likelihood, p denotes the number of parameters and n denotes the number of observations. The Kolmogorov–Smirnov (KS) test is also used to test the goodness of fit for all the data sets of the UTD and other distributions. Nonparametrically, this test measures how close the empirical distribution and the fitted distribution are. The AIC, AICc, CAIC, and BIC measure the adequacy, while the KS test measures the fit of each distribution. All the computations were performed using R software.

6.2. Flood Level Data

The data set is from [34]. The data represent the maximum flood level (in millions of cubic feet per second) for the Susquehanna River at Harrisburg, Pennsylvania, over 20 four-year periods from 1890 to 1969. Table 6 provides the measurements of the data set.

Table 6. Flood level data.

0.654	0.613	0.315	0.449	0.297	0.402	0.379	0.423
0.379	0.324	0.269	0.740	0.418	0.412	0.494	0.416
0.338	0.392	0.484	0.265				

In Table 7, the MLEs of the parameters of every distribution used and the observed KS test statistics of each distribution are given. We conclude that the UTD gives the best fit because it has the smallest KS value and the largest p-value. Table 8 shows that UTD has the largest maximum log-likelihood value and the smallest AIC, AICc, CAIC, and BIC values among other models. As a result, it can be concluded that the UTD model is more successful than the other models for the flood level data.

Table 7. The MLEs of the parameters with their standard errors, KS values, and the associated p-values.

Distribution	MLE of the Parameters (Standard Error)	KS	p-Value
UTD	$\hat{\theta} = 1.2184(0.1056)$	0.2527	0.1555
LLD	$\hat{\lambda} = 6.39 \times 10^{-10}(0.0959)$, $\hat{\sigma} = 2.2280(0.2597)$	0.3157	0.0372
ULD	$\hat{\theta} = 1.6205(0.2849)$	0.3182	0.0349
TLD	$\hat{\theta} = 2.2450(0.5019)$	0.3352	0.0223
OKD	$\hat{\theta} = 1.7290(0.3866)$	0.4128	0.0022
PD	$\hat{\theta} = 1.1140(0.2491)$	0.3941	0.0040
TD	$\hat{\theta} = 1(0.7359)$	0.4598	0.0004

Table 8. Log-likelihood, AIC, AICc, CAIC, and BIC values for flood level data.

Distribution	Log-Likelihood	AIC	AICc	CAIC	BIC
UTD	13.5818	−25.1635	−24.9414	−23.1679	−24.1678
LLD	6.6716	−9.3431	−8.6373	−5.3517	−7.3517
ULD	7.1410	−12.2821	−12.0598	−10.2863	−11.2864
TLD	7.3682	−12.7365	−12.5142	−10.7407	−11.7407
OKD	2.5110	−3.0220	−2.7998	−1.0263	−2.0263
PD	0.1124	1.7751	1.9974	3.7709	2.7708
TD	2.2856	−2.5713	−2.3489	−0.5755	−1.5756

Figure 5a,b represent the TTT plot and histogram of flood level data, respectively. The TTT plot indicates an increasing hrf, case covered by the UTD, and histogram depicts how well the proposed model fits the data, compared to other models.

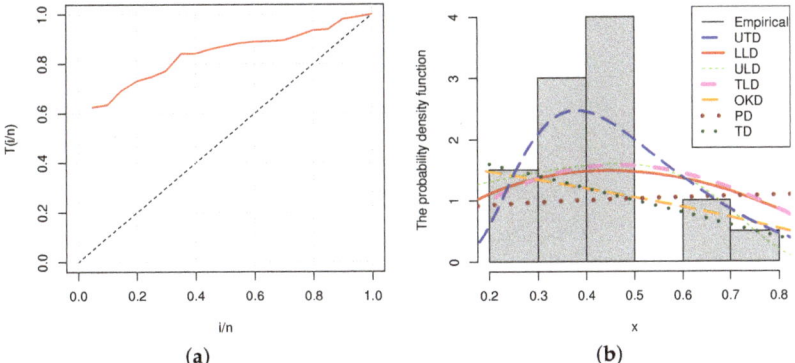

Figure 5. (a) TTT plot and (b) histogram for the flood level data.

The estimated variance of the MLE $\hat{\theta}$ of the UTD parameter θ for the flood level data is given by
$$J^{-1}(\hat{\theta}) = 0.0112.$$

Therefore, an approximate 95% confidence interval for θ is $[1.0114, 1.4254]$. Table 9 gives the median of bootstrap estimates and bootstrap confidence intervals.

Table 9. The median of bootstrap estimators and bootstrap confidence intervals.

Median of Estimates	Lower Limit	Upper Limit
$\hat{\theta} = 1.2239$	1.0815	1.4176

6.3. Times between Failures of Secondary Reactor Pumps Data

The data represent times between failures of secondary reactor pumps (see [35]). Here, a normalization operation is carried out by dividing the original data by 10, in order to obtain data between 0 and 1. Table 10 provides the measurements of the transformed data.

Table 10. Secondary reactor pumps data.

0.2160	0.0150	0.4082	0.0746	0.0358	0.0199	0.0402	0.0101
0.0605	0.0954	0.1359	0.0273	0.0491	0.3465	0.0070	0.6560
0.1060	0.0062	0.4992	0.0614	0.5320	0.0347	0.1921	

Listed in Table 11 are the MLEs of the parameters and the observed KS test statistic for each distribution. In terms of KS value and p-value, UTD is the best model. On the basis of Table 12, it can be seen that the UTD has the largest log-likelihood value, while the AIC, AICc, CAIC, and BIC values are the smallest. Therefore, the UTD model fits the secondary reactor pump failure data better than the other models.

Table 11. The MLEs of the parameters and their standard errors, KS values and the associated p-values.

Distribution	MLE of the Parameters (Standard Error)	KS	p-Value
UTD	$\hat{\theta} = 0.3625(0.0290)$	0.1366	0.7341
LLD	$\hat{\lambda} = 2.12 \times 10^{-9}(0.4578)$, $\hat{\sigma} = 0.7568(0.0680)$	0.1584	0.5573
ULD	$\hat{\theta} = 4.1497(0.7445)$	0.3274	0.0107
TLD	$\hat{\theta} = 0.4891(0.1020)$	0.1962	0.2982
OKD	$\hat{\theta} = 4.8569(1.0127)$	0.2568	0.0796
PD	$\hat{\theta} = 0.3773(0.0787)$	0.2247	0.1678
TD	$\hat{\theta} = 1(0.4902)$	0.4514	0.00008
KD	$\hat{\alpha} = 0.6766(0.1407)$, $\hat{\beta} = 2.9360(0.9558)$	0.1393	0.7123
BD	$\hat{\alpha} = 0.6307(0.1575)$, $\hat{\beta} = 3.2317(1.0648)$	0.1542	0.5919
$ETLD$	$\hat{\alpha} = 0.6567(0.1532)$, $\hat{\beta} = 1.6566(0.4805)$	0.1465	0.6536
UBD	$\hat{\lambda} = 0.1639(0.0873)$, $\hat{b} = 2.4273(1.2196)$	0.2243	0.1688

Table 12. Log-likelihood, AIC, AICc, CAIC, and BIC values for secondary reactor pumps data.

Distribution	Log-Likelihood	AIC	AICc	CAIC	BIC
UTD	21.7499	−41.4998	−41.3093	−39.3643	−40.3643
LLD	20.0761	−36.1522	−35.5522	−31.8812	−33.8812
ULD	14.5035	−27.0070	−26.8165	−24.8715	−25.8715
TLD	18.7827	−35.5653	−35.3749	−33.4299	−34.4298
OKD	18.0840	−34.1679	−33.9775	−32.0325	−33.0324
PD	15.5307	−29.0615	−28.8709	−26.9259	−27.9259
TD	11.2067	−20.2229	−20.2229	−18.2779	−19.2778
KD	20.3296	−36.6592	−36.0592	−32.3882	−34.3883
BD	20.0285	−36.0571	−35.4570	−31.7860	−33.7861
$ETLD$	20.1709	−36.3418	−35.7418	−32.0708	−34.0708
UBD	17.5294	−31.0588	−30.4588	−26.7878	−28.7878

Figure 6a,b represent the empirical TTT plot and histogram of the secondary reactor pumps data. A bathtub hrf is indicated by the TTT plot for the secondary reactor pumps data. From the histogram, it can be seen that the empirical line is closer to the fitted line of the UTD model than other models. The estimated variance of the MLE $\hat{\theta}$ of the UTD parameter θ for the times between failures of secondary reactor pumps data data is calculated by

$$J^{-1}(\hat{\theta}) = 0.0008$$

Therefore, an approximate 95% confidence interval for θ is given by $[0.3057, 0.4193]$. Table 13 provides the median of bootstrap estimates and bootstrap confidence intervals.

Table 13. The median of bootstrap estimators and bootstrap confidence intervals.

Median of Estimates	Lower Limit	Upper Limit
$\hat{\theta} = 0.3605$	0.3154	0.4216

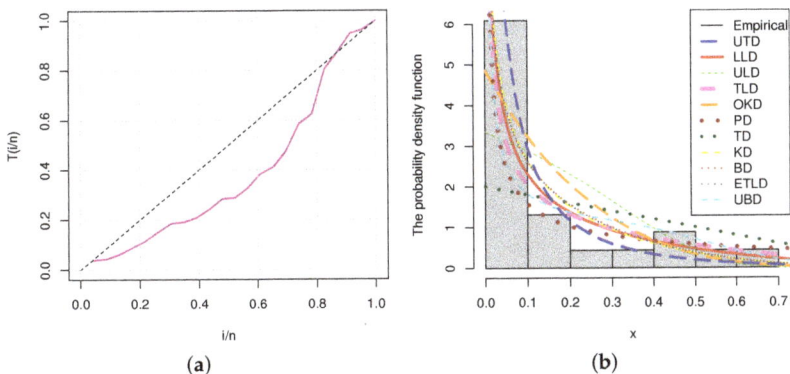

Figure 6. (a) TTT plot and (b) histogram for secondary reactor pumps data.

7. Conclusions

Through this study, the introduction of a bounded form of the TD by the exponential transformation was performed. It is named UTD. Certain statistical properties, such as shape characteristics, moments, incomplete moments, and quantile function were derived. The Shannon entropy and extropy were also obtained. Based on the ML, LS, WLS, and Bayesian methods, estimation of the model parameter was established and examined by simulation studies. The proposed model's dominance was demonstrated using real-world data sets, and it is concluded that the UTD is a good candidate in unit interval distributions. Possible perspectives of this work include the construction of quantile regression models, as in [36,37], as well as bivariate and discrete versions of the UTD. This work requires additional developments and investigations, which we will leave for future research.

Author Contributions: Conceptualization, A.K., R.M., C.C. and M.R.I.; methodology, A.K., R.M., C.C. and M.R.I.; validation, A.K., R.M., C.C. and M.R.I.; formal analysis, A.K., R.M., C.C. and M.R.I.; investigation, A.K., R.M., C.C. and M.R.I.; writing—original draft preparation, A.K., R.M., C.C. and M.R.I.; writing—review and editing, A.K., R.M., C.C. and M.R.I. All authors have read and agreed to the published version of the manuscript.

Funding: This research received no external funding.

Acknowledgments: We would like to thank the four reviewers for the constructive comments on the paper.

Conflicts of Interest: No potential conflict of interest was reported by the authors.

References

1. Lemonte, A.J.; Bazán, J.L. New class of Johnson distributions and its associated regression model for rates and proportions. *Biom. J.* **2016**, *58*, 727–746. [CrossRef]
2. Smithson, M.; Shou, Y. CDF-quantile distributions for modelling random variables on the unit interval. *Br. J. Math. Stat. Psychol.* **2017**, *70*, 412–438. [CrossRef] [PubMed]
3. Topp, C.W.; Leone, F.C. A family of J-shaped frequency functions. *J. Am. Stat. Assoc.* **1955**, *50*, 209–219. [CrossRef]
4. Kumaraswamy, P. A generalized probability density function for double-bounded random processes. *J. Hydrol.* **1980**, *46*, 79–88. [CrossRef]
5. Jorgensen, B. *The Theory of Dispersion Models. CRC Monographs on Statistics and Applied Probability*; Chapman & Hall: London, UK, 1997.
6. Gómez-Déniz, E.; Sordo, M.A.; Calderin-Ojeda, E. The Log–Lindley distribution as an alternative to the beta regression model with applications in insurance. *Insur. Math. Econ.* **2014**, *54*, 49–57. [CrossRef]
7. Pourdarvish, A.; Mirmostafaee, S.M.T.K.; Naderi, K. The exponentiated Topp–Leone distribution: Properties and application. *J. Appl. Environ. Biol. Sci.* **2015**, *5*, 251–256.
8. Ghitany, M.E.; Mazucheli, J.; Menezes, A.F.B.; Alqallaf, F. The unit-inverse Gaussian distribution: A new alternative to two-parameter distributions on the unit interval. *Commun. Stat. Theory Methods* **2019**, *48*, 3423–3438. [CrossRef]

9. Mazucheli, J.; Menezes, A.F.B.; Chakraborty, S. On the one parameter unit-Lindley distribution and its associated regression model for proportion data. *J. Appl. Stat.* **2019**, *46*, 700–714. [CrossRef]
10. Mazucheli, J.; Menezes, A.F.B.; Fernandes, L.B.; de Oliveira, R.P.; Ghitany, M.E. The unit-Weibull distribution as an alternative to the Kumaraswamy distribution for the modeling of quantiles conditional on covariates. *J. Appl. Stat.* **2020**, *47*, 954–974. [CrossRef]
11. Modi, K.; Gill, V. Unit Burr-III distribution with application. *J. Stat. Manag. Syst.* **2020**, *23*, 579–592. [CrossRef]
12. Bakouch, H.S.; Nik, A.S.; Asgharzadeh, A.; Salinas, H.S. A flexible probability model for proportion data: Unit-half-normal distribution. *Commun. Stat. Case Stud. Data Anal. Appl.* **2021**, *7*, 1–18. [CrossRef]
13. Teissier, G. Recherches sur le vieillissement et sur les lois de la mortalité. *Ann. Physiol. Physicochim. Biol.* **1934**, *10*, 237–284.
14. Laurent, A.G. Failure and mortality from wear and ageing. The Teissier model. In *A Modern Course on Statistical Distributions in Scientific Work*; Patil, G.P., Kotz, S., Ord, J.K., Eds.; ASIC; Springer: Berlin/Heidelberg, Germany, 1975; Volume 17, pp. 301–320.
15. Muth, E.J. Reliability models with positive memory derived from the mean residual life function. *Theory Appl. Reliab.* **1977**, *2*, 401–436.
16. Rinne, H. Estimating the lifetime distribution of private motor-cars using prices of used cars: The Teissier model. In *Statistiks Zwischen Theorie und Praxis*; Vandenhoeck & Ruprecht: Göttingen, Germany, 1985; pp. 172–184.
17. Leemis, L.M.; McQueston, J.T. Univariate distribution relationships. *Am. Stat.* **2008**, *62*, 45–53. [CrossRef]
18. Jodrá, P.; Jiménez-Gamero, M.D.; Alba-Fernandez, M.V. On the Muth distribution. *Math. Model. Anal.* **2015**, *20*, 291–310. [CrossRef]
19. Jodrá, P.; Gómez, H.W.; Jiménez-Gamero, M.D.; Alba-Fernández, M.V. The power Muth distribution. *Math. Model. Anal.* **2017**, *22*, 186–201. [CrossRef]
20. Al-Babtain, A.A.; Elbatal, I.; Chesneau, C.; Jamal, F. The transmuted Muth generated class of distributions with applications. *Symmetry* **2020**, *12*, 1677. [CrossRef]
21. Biçer, C.; Bakouch, H.S.; Biçer, H.D. Inference on parameters of a geometric process with scaled Muth distribution. *Fluct. Noise Lett.* **2021**, *20*, 2150006. [CrossRef]
22. Irshad, M.R.; Maya, R.; Arun, S.P. Muth distribution and estimation of a parameter using order statistics. *Statistica* **2021**, *81*, 93–119.
23. Irshad, M.R.; Maya, R.; Krishna, A. Exponentiated power Muth distribution and associated inference. *J. Indian Soc. Probab. Stat.* **2021**, *22*, 265–302. [CrossRef]
24. Abd EL-Baset, A.A.; Ghazal, M.G.M. Exponentiated additive Weibull distribution. *Reliab. Eng. Syst. Saf.* **2020**, *193*, 106663.
25. Alamgir Khalil, M.I.; Ali, K.; Mashwani, W.K.; Shafiq, M.; Kumam, P.; Kumam, W. A novel flexible additive Weibull distribution with real-life applications. *Commun. Stat. Theory Methods* **2021**, *50*, 1557–1572. [CrossRef]
26. Irshad, M.R.; Shibu, D.S.; Maya, R.; D'cruz, V. Binominal mixture Lindley distribution: Properties and Applications. *J. Indian Soc. Probab. Stat.* **2020**, *21*, 437–469. [CrossRef]
27. Korkmaz, M.Ç. A new heavy-tailed distribution defined on the bounded interval: The logit slash distribution and its application. *J. Appl. Stat.* **2020**, *47*, 2097–2119. [CrossRef]
28. Haq, M.A.U.; Hashmi, S.; Aidi, K.; Ramos, P.L.; Louzada, F. Unit modified Burr-III distribution: Estimation, characterizations and validation test. *Ann. Data Sci.* **2020**, 1–26. [CrossRef]
29. Bebbington, M.; Lai, C.D.; Murthy, D.N.P.; Zitikis, R. Modelling N-and W-shaped hazard rate functions without mixing distributions. *Proc. Inst. Mech. Eng. Part O J. Risk Reliab.* **2009**, *223*, 59–69. [CrossRef]
30. Corless, R.M.; Gonnet, G.H.; Hare, D.E.G.; Jeffrey, D.J.; Knuth, D.E. On the Lambert W function. *Adv. Comput. Math.* **1996**, *5*, 329–359. [CrossRef]
31. R Development Core Team. *R: A Language and Environment for Statistical Computing*; R Foundation for Statistical Computing: Vienna, Austria, 2021. Available online: http://www.R-project.org/ (accessed on 28 December 2021).
32. Swain, J.J.; Venkatraman, S.; Wilson, J.R. Least-squares estimation of distribution functions in Johnson's translation system. *J. Stat. Comput. Simul.* **1988**, *29*, 271–297. [CrossRef]
33. Gelman, A.; Hill, J. *Data Analysis Using Regression and Multilevel/Hierarchical Models*; Cambridge University Press: Cambridge, UK, 2006.
34. Dumonceaux, R.; Antle, C.E. Discrimination between the log-normal and the Weibull distributions. *Technometrics* **1973**, *15*, 923–926. [CrossRef]
35. Suprawhardana, M.S.; Prayoto, S. Total time on test plot analysis for mechanical components of the RSG-GAS reactor. *Atom Indones* **1999**, *25*, 81–90.
36. Korkmaz, M.Ç.; Chesneau, C.; Korkmaz, Z.S. On the arcsecant hyperbolic normal distribution. Properties, quantile regression modeling and applications. *Symmetry* **2021**, *13*, 117. [CrossRef]
37. Korkmaz, M.Ç.; Chesneau, C. On the unit Burr-XII distribution with the quantile regression modeling and applications. *Comput. Appl. Math.* **2021**, *40*, 29. [CrossRef]

Article
The Minimum Lindley Lomax Distribution: Properties and Applications

Sadaf Khan [1], Gholamhossein G. Hamedani [2], Hesham Mohamed Reyad [3], Farrukh Jamal [1,*], Shakaiba Shafiq [1] and Soha Othman [4]

1. Department of Statistics, The Islamia University of Bahawalpur, Bahawalpur 63100, Pakistan; smkhan6022@gmail.com (S.K.); shakaiba.hashmi@gmail.com (S.S.)
2. Department of Mathematics, Statistics and Computer Science, Marquette University, Milwaukee, WI 53233, USA; gholamhoss.hamedani@marquette.edu
3. Department of Information Systems and Production Management, Qassim University, Buraydah 52571, Saudi Arabia; hesham_reyad@yahoo.com
4. Department of Applied Statistics, Cairo University, Giza 12613, Egypt; soha_othman@yahoo.com
* Correspondence: farrukh.jamal@iub.edu.pk

Abstract: By fusing the Lindley and Lomax distributions, we present a unique three-parameter continuous model titled the minimum Lindley Lomax distribution. The quantile function, ordinary and incomplete moments, moment generating function, Lorenz and Bonferroni curves, order statistics, Rényi entropy, stress strength model, and stochastic sequencing are all carefully examined as basic statistical aspects of the new distribution. The characterizations of the new model are investigated. The proposed distribution's parameters were evaluated using the maximum likelihood procedures. The stability of the parameter estimations is explored using a Monte Carlo simulation. Two applications are used to objectively assess the new model's extensibility.

Keywords: compounding distributions; Lindley distribution; Lomax distribution; stochastic ordering; stress strength model; characterization

1. Introduction

Appropriate data modeling is believed to provide greater insight into the data, divulging its properties and allowing for tracking its characteristics. Consequently, there is a potential for developing efficient methods for clearer grasp of real-world occurrences. We developed a coherent model to help meet the aspirations of applied practitioners in a wide range of scientific domains, inspired by the application of theoretical probability models in applied research. Tahir and Nadarajah [1] provided a deep review of novel approaches that can be adopted to develop new generalized classes ("G-classes" for short) of distributions. In parallel to G-classes, Tahir and Cordiero [2] presented a review on compounding univariate distributions, their expansions, and classes to detect anomaly scenarios under series and parallel structures. In the current article, we adopted the approach extensively discussed in Section 7 of [2], by integrating two continuous cumulative distribution functions (cdfs) together. Cordeiro et al. [3] initiated this idea and proposed the Exponential-Weibull distribution. In the same vein, we proposed minimum Lindley Lomax (minLLx) distribution by compounding the Lindley and Lomax distributions.

The Lindley (L) and Lomax (Lx) distributions are indispensable models for characterizing data, notably in engineering, for the replacement and maintenance of various goods, systems, and reliability processes. For the stated reason, researchers have found ample evidence of studies that conformed to these distributions, namely, Ghitany et al. [4], Ramos and Louzada [5], Singh et al. [6], Oguntunde et al. [7], Wei et al. [8], and Elgarhy et al. [9], just to mention a few. It is an intriguing fact that both the Lindley and the Lomax distributions emerged from an extension of the exponential model, which is commonly used

to quantify the lifetime of a process or device. Assume that a system comprises of two sub-systems that are operating in tandem at the same time, and that the system will collapse if the first sub-system falters. Let us assume further that the failure times of subsystems follow the Lindley and Lomax distributions with Y and Z independent variables having cdfs, respectively, as follows

$$G(y) = 1 - \left(\frac{1+\theta+\theta y}{1+\theta}\right) e^{-\theta y}, \; y \geq 0, \; \theta > 0$$

$$H(z) = 1 - (1+\lambda z)^{-\beta}, \; z \geq 0, \; \lambda, \beta > 0.$$

Then, the new arbitrary variable (av) $X = \min(Y, Z)$ will be called the min Lindley Lomax (minLLx) to determine the system's failure mechanism. The cdf of the minLLx av is follows as

$$F(x) = 1 - \frac{e^{-\theta x}}{(1+\lambda x)^\beta} \left(\frac{1+\theta+\theta x}{1+\theta}\right), \qquad x \geq 0, \; \theta, \lambda, \beta > 0. \tag{1}$$

The probability density function (pdf), survival function (sf), and hazard rate function (hrf) in harmony with Equation (1) are given, respectively, by

$$f(x) = \frac{e^{-\theta x}}{(1+\theta)(1+\lambda x)^{\beta+1}} \left[\lambda \beta(1+\theta+\theta x) + \theta^2(1+x)(1+\lambda x)\right], \; x > 0, \; \theta, \lambda, \beta > 0, \tag{2}$$

$$S(x) = \frac{e^{-\theta x}}{(1+\lambda x)^\beta} \left(\frac{1+\theta+\theta x}{1+\theta}\right)$$

and

$$h(x) = \frac{\lambda \beta(1+\theta+\theta x) + \theta^2(1+x)(1+\lambda x)}{(1+\lambda x)(1+\theta+\theta x)}, \quad x > 0. \tag{3}$$

From now on, an av $X \sim \text{minLLx}(\theta, \lambda, \beta)$ with a pdf is defined by Equation (2).

The purpose of this research is to present and explore the mathematical configurations of a newly developed three-parameter distribution, the minimum Lindley Lomax model, in the perspective of compounding. The rest of the article is composed of seven main components. The minLLx model's essential mathematical features are examined in Section 2. Specific characterizations of the new distribution are pursued in Section 3. The minLLx model's maximum likelihood estimates and observed information matrix are established in Section 4. In Section 5, a simulation study is carried out. Two applications are provided in Section 6. Eventually, in Section 7, there are some closing remarks.

2. Structural Properties

The standard mathematical characteristics of the newly suggested minLLx distribution, as stipulated by the cdf in Equation (1), are explored in this phase. In each subcategory, we report a few explicit results.

2.1. Quantile Function

Let the pth quantile of the minLLx distribution, say x_p, is demarcated by $F(x_p) = p$, such that $0 < p < 1$. Then the root of

$$x_p = \frac{1}{\lambda} \left\{ \left[\frac{(1+\theta)(1-p)e^{\theta x_p}}{1+\theta+\theta x_p}\right]^{-1/\beta} - 1 \right\}. \tag{4}$$

2.2. The Shape of the minLLx Distribution

Mathematically, the forms of the minLLx distribution's density and hazard functions can be defined. The acute points of the density function are the roots of the following equation:

$$\frac{-\lambda(1+\beta)}{1+\lambda x} + \left\{\frac{\theta[\lambda\beta + 2\theta(1+\lambda x)]}{\lambda\beta(1+\theta+\theta x) + \theta^2(1+x)(1+\lambda x)}\right\} = 0.$$

Furthermore, the acute points of the hazard function are the roots of the following equation:

$$\left\{\frac{\theta[\lambda\beta + 2\theta(1+\lambda x)]}{\lambda\beta(1+\theta+\theta x) + \theta^2(1+x)(1+\lambda x)}\right\} - \frac{\lambda}{1+\lambda x} - \frac{\theta}{1+\theta+\theta x} = 0.$$

The density and hazard functions are visualized in Figures 1 and 2, respectively. The density function has a reverse-J and right-skewed shape with different peeks, while hrf can sometimes be a monotonic (increasing or decreasing), non-monotonic (bathtub), or constant in shape. The standard L and Lx statistical distributions can only create two shapes, whereas the minLLx model can produce a wide number of shapes based on the power parameter beta.

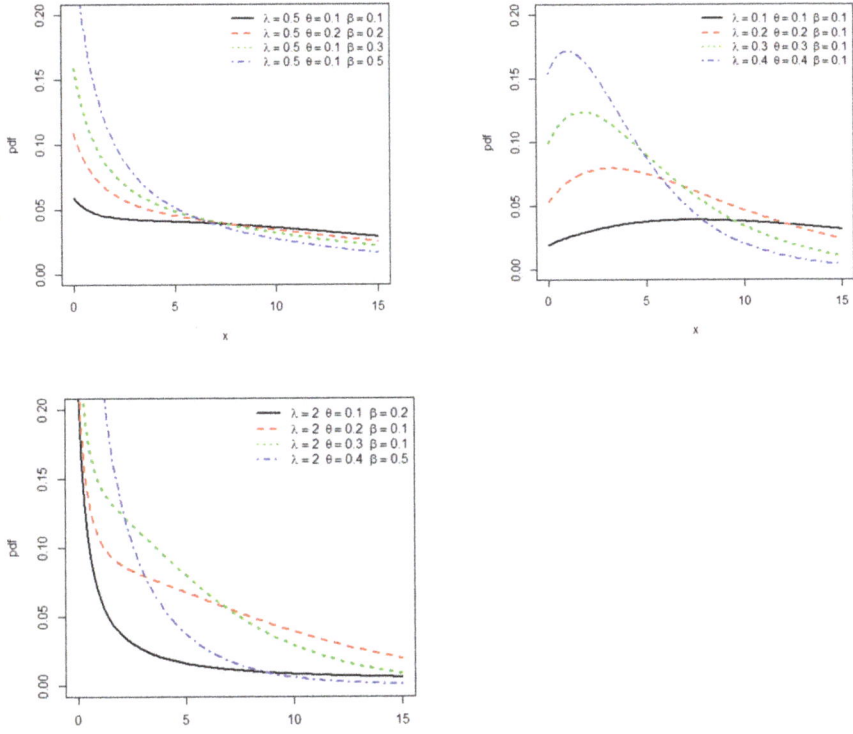

Figure 1. Possible figures of the minLLx pdf for parameter values chosen at random.

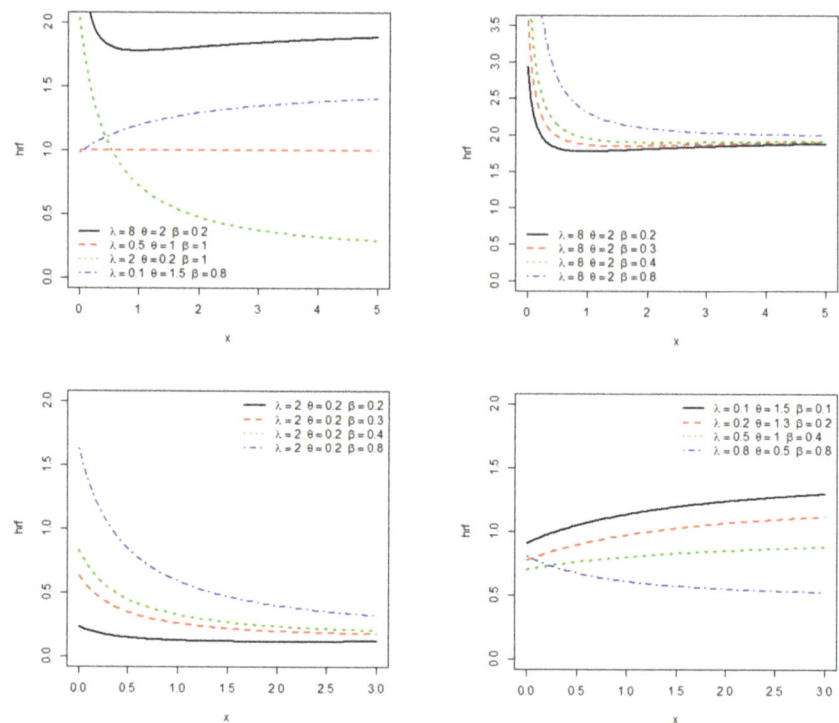

Figure 2. Possible figures of the minLLx hrf for parameter values chosen at random.

2.3. Moments and Moment Generating Function

Let X be an av with the minLLx distribution, then the ordinary moment, say μ'_r, is given by

$$\begin{aligned}
\mu'_r = E(X^r) &= \int_{-\infty}^{\infty} x^r f(x)\,dx \\
&= \frac{\lambda\beta}{1+\theta}\int_0^\infty x^r(1+\theta+\theta x)(1+\lambda x)^{-\beta-1}e^{-\theta x}dx + \frac{\theta^2}{1+\theta}\int_0^\infty x^r(1+x)(1+\lambda x)^{-\beta}e^{-\theta x}dx \\
&= \sum_{j=0}^{\infty}\binom{-\beta-1}{j}\frac{\lambda^{j+1}\beta}{1+\theta}\int_0^\infty x^{r+j}(1+\theta+\theta x)e^{-\theta x}dx + \sum_{j=0}^{\infty}\binom{-\beta}{j}\frac{\theta^2\lambda^j}{1+\theta}\int_0^\infty x^{r+j}(1+x)e^{-\theta x}dx \\
&= \sum_{j=0}^{\infty}\binom{-\beta-1}{j}\frac{\lambda^{j+1}\beta(r+\theta+j+2)\Gamma(r+j+1)}{(1+\theta)\theta^{r+j+1}} + \sum_{j=0}^{\infty}\binom{-\beta}{j}\frac{\lambda^j(r+\theta+j+1)\Gamma(r+j+1)}{(1+\theta)\theta^{r+j}} \\
&= \sum_{j=0}^{\infty}\frac{\lambda^j\Gamma(r+j+1)}{(1+\theta)\theta^{r+j+1}}\left\{\lambda\beta(r+\theta+j+2)\binom{-\beta-1}{j} + \theta(r+\theta+j+1)\binom{-\beta}{j}\right\},
\end{aligned} \quad (5)$$

where $\Gamma(n) = \int_0^\infty x^{n-1}e^{-x}dx$ is the gamma function. Substituting $r = 1, 2, 3, 4$ into (5), we obtain the mean $= \mu'_1$, variance $= \mu'_2 - \mu'^2_1$, skewness $= \{\mu'_3 - 3\mu'_2\mu'_1 + 2\mu'^3_1\}\{\mu_2 - (\mu'_1)^2\}^{-3}$ and kurtosis $= \{\mu'_4 - 4\mu'_3\mu'_1 + 6\mu'_2\mu'^2_1 - 3\mu'^4_1\}\{\mu'_2 - (\mu'_1)^2\}^{-2}$. Table 1 provides the mean, variance, standard deviation, skewness, and kurtosis of X for different combinations of θ, λ, β as $A_1: \theta = 3.5, \lambda = 0.4, \beta = 0.5; A_2: \theta = 0.3, \lambda = 1, \beta = 0.8; A_3: \theta = 1.5, \lambda = 0.1, \beta = 1.5$, and $A_4: \theta = 0.3, \lambda = 0.5, \beta = 0.3$.

Table 1. Moments, variance, standard deviation, skewness and kurtosis of X for randomly selected parameter values of minLLx(θ,λ,β).

μ'_r	A_1	A_2	A_3	A_4
μ'_1	1.126862	0.3661565	0.9344674	0.164529
μ'_2	1.406321	0.2563243	1.76341	0.1045506
μ'_3	1.896405	0.2028995	5.733562	0.07738473
μ'_4	2.72086	0.172917	28.11525	0.06041124
Variance	0.1365036	0.1222537	0.890181	0.07748082
S.D	0.369464	0.349648	0.9434941	0.2783538
Skewness	1.137115	1.563494	2.448468	2.289103
Kurtosis	1.375744	1.731832	0.84139	1.526684

The empirical findings from Table 1 allow us to deduce that the skewness is greater than zero, indicating a lack of symmetry of the tails, specifically an elongated right tail. This signifies that the mean and median are pulled to the right. Moreover, kurtosis values are less than three, demonstrating that the distribution is platykurtic.

The nth principal moment of the minLLx distribution, say μ_n, can be acquired from

$$\mu_n = \sum_{r=0}^{n} \binom{n}{r} (-\mu'_1)^{n-r} E(x^r) \qquad (6)$$
$$= \sum_{r=0}^{n} \sum_{j=0}^{\infty} \binom{n}{r} \frac{(-\mu'_1)^{n-r} \lambda^j \Gamma(r+j+1)}{(1+\theta)\theta^{r+j+1}} \left\{ \lambda\beta(r+\theta+j+2)\binom{-\beta-1}{j} + \theta(r+\theta+j+1)\binom{-\beta}{j} \right\}.$$

The rth incomplete moment of the minLLx distribution, symbolized by $\varphi_s(t)$, is

$$\varphi_s(t) = \int_{-\infty}^{t} x^s f(x)\, dx$$
$$= \sum_{i=0}^{\infty} \frac{\lambda^i}{(1+\theta)\theta^{s+i+1}} \left\{ \begin{array}{l} \lambda\beta[(1+\theta)\gamma(s+i+1,t) + \gamma(s+i+2,t)]\binom{-\beta-1}{i} \\ + \theta[\theta\gamma(s+i+1,t) + \gamma(s+i+2,t)]\binom{-\beta}{ji} \end{array} \right\}, \qquad (7)$$

where $\gamma(a,x) = \int_0^x t^{a-1} e^{-t}\, dt$ is the lower incomplete gamma function.

The moment generating function, signified by $M_x(t)$, of the minLLx distribution can be acquired as

$$M_x(t) = E(e^{tx}) = \sum_{j=0}^{\infty} \frac{\lambda^j \Gamma(j+1)}{(1+\theta)\theta^{j+2}} \left\{ \begin{array}{l} \lambda\beta[\theta(\theta+j-t+2)-t]\binom{-\beta-1}{j} \\ + \theta^2(\theta+j-t+1)\binom{-\beta}{j} \end{array} \right\}. \qquad (8)$$

2.4. Probability Weighted Moments

Ordinary moments of order statistics are generalized by probability weighted moments of a stochastic process, which naturally arise while dealing with ordinary moments. They also play a significant role in several parametric estimate techniques. The formulation for the probability weighted moments of a chance variable with the minLLx distribution is as follows.

The $(r+s)$th probability weighted moments (PWMs) of a chance variable X with the minLLx distribution, about $M_{r,s}$, follows

$$M_{r,s} = E(X^r F(x)^s) = \int_{-\infty}^{\infty} x^r F(x)^s f(x) dx$$

$$= \int_{-\infty}^{\infty} x^r \frac{(1+\lambda x)^{-\beta-1}}{1+\theta} \{\lambda\beta(1+\theta+\theta x) + \theta^2(1+x)(1+\lambda x)\} e^{-\theta x}$$

$$\times \left\{1 - (1+\lambda x)^{-\beta}\left(\frac{1+\theta+\theta x}{1+\theta}\right) e^{-\theta x}\right\}^s dx$$

$$= \sum_{j=0}^{\infty} \frac{(-1)^j}{(1+\theta)^{j+1}} \binom{s}{j} \int_{-\infty}^{\infty} x^r (1+\lambda x)^{-\beta(j+1)-1} (1+\theta+\theta x)^j e^{-\theta(j+1)x}$$

$$\times \{\lambda\beta(1+\theta+\theta x) + \theta^2(1+x)(1+\lambda x)\} dx$$

$$= \underbrace{\sum_{j=0}^{\infty} \frac{(-1)^j \lambda\beta}{(1+\theta)^{j+1}} \binom{s}{j} \int_{-\infty}^{\infty} x^r (1+\lambda x)^{-\beta(j+1)-1} (1+\theta+\theta x)^{j+1} e^{-\theta(j+1)x} dx}_{A}$$

$$+ \underbrace{\sum_{j=0}^{\infty} \frac{(-1)^j \theta^2}{(1+\theta)^{j+1}} \binom{s}{j} \int_{-\infty}^{\infty} x^r (1+\lambda x)^{-\beta(j+1)} (1+x)(1+\theta+\theta x)^j e^{-\theta(j+1)x} dx}_{B},$$

where

$$A = \sum_{i=0}^{\infty} \sum_{w=0}^{j+1} \frac{\lambda^i \theta^w (1+\theta)^{j-w-1} \Gamma(r+i+w+1)}{(\theta(j+1))^{r+i+w+1}} \binom{-\beta(j+1)-1}{i} \binom{j+1}{w}$$

and

$$B = \sum_{i=0}^{\infty} \sum_{w=0}^{j} \frac{\lambda^i \theta^w (1+\theta)^{j-w} [\theta(j+1)+r+i+w+1]\Gamma(r+i+w+1)}{(\theta(j+1))^{r+i+w+2}} \binom{-\beta(j+1)}{i} \binom{j}{w}.$$

Consequently, we arrive at

$$M_{r,s} = \sum_{j,i=0}^{\infty} \frac{(-1)^j \lambda^i}{(1+\theta)^{j+1} \theta^{r+i}(1+j)^{r+i}} \binom{s}{j}$$

$$\times \left\{ \sum_{w=0}^{j+1} \frac{\lambda\beta(1+\theta)^{j-w-1}\Gamma(r+i+w+1)}{(1+j)^{w+1}} \binom{-\beta(j+1)-1}{i} \binom{j+1}{w} \right. \tag{9}$$

$$\left. + \sum_{w=0}^{j} \frac{(1+\theta)^{j-w}[\theta(j+1)+r+i+w+1]\Gamma(r+i+w+1)}{(1+j)^{w+2}} \binom{-\beta(j+1)}{i} \binom{j}{w} \right\}$$

2.5. Order Statistics

The inclusion of sorted random variables, often known as order statistics, is crucial in the modeling of various longevity systems with distinct component structures. David and Nagaraja [10] laid the all-important foundation for this paradigm. The order statistics of the minLLx distribution are linked to having conventional distributional modules; hence their importance is an inarguable fact.

Consider the given scenario as $X_{1:n} \leq X_{2:n}, \ldots \leq X_{n:n}$ be the $X_{k:n}$ th order statistics corresponding to a sample of size n from the minLLx distribution. The pdf of $X_{k:n}$, the kth order statistic, is given by

$$f_{X_{k:n}}(x) = \frac{1}{\beta(k, n-k+1)} \sum_{w=0}^{n-k} (-1)^w \binom{n-k}{w} f(x) F(x)^{k+w-1}, \tag{10}$$

where $\beta(.,.)$ is the exact beta function. From (5) and (6), we have

$$f(x)F(x)^{k+w-1} = \sum_{j=0}^{\infty} \frac{(-1)^j(1+\lambda x)^{-\beta(j+1)-1}e^{-\theta(j+1)x}}{(1+\theta)^{j+1}}$$
$$\times \{\lambda\beta(1+\theta+\theta x) + \theta^2(1+x)(1+\lambda x)\} \binom{k+w-1}{j}. \quad (11)$$

Inserting Equation (11) into Equation (10), we have

$$f_{X_{k:n}}(x) = \sum_{w=0}^{n-k} \sum_{j=0}^{\infty} \frac{(-1)^{w+j}(1+\lambda x)^{-\beta(j+1)-1}e^{-\theta(j+1)x}}{\beta(k,n-k+1)(1+\theta)^{j+1}}$$
$$\times \{\lambda\beta(1+\theta+\theta x) + \theta^2(1+x)(1+\lambda x)\} \binom{n-k}{w}\binom{k+w-1}{j}. \quad (12)$$

Furthermore, the rth moment of kth order statistic for the minLLx distribution is given by

$$E(x_{k:n}^r) = \sum_{w=0}^{n-k} \sum_{j,i=0}^{\infty} \frac{(-1)^{w+j}\lambda^i \Gamma(r+i+1)}{\beta(k,n-k+1)(1+\theta)^{j+1}(\theta(1+j))^{r+i+1}} \binom{n-k}{w}\binom{k+w-1}{j}$$
$$\times \left\{\lambda\beta(r+\theta+i+2)\binom{-\beta(j+1)-1}{i} + \theta(1+j)(\theta(1+j)+r+i+1)\binom{-\beta(j+1)}{i}\right\}. \quad (13)$$

2.6. Rényi Entropy

Entropy is a mathematical concept that encapsulates the logical understanding of quantifying various mechanisms. The entropy technique is adaptable in different fields, including bioenergetics, queuing theory, thermodynamics, colligative properties of solutions, and statistics. There are several mechanisms to quantify the entropy of the minLLx distribution. Rényi entropy is established here by subjecting a feasible expression that may be appraised using any analytical software. In the perspective of the minLLx distribution, the following result incorporates a series expansion of this entropy system of measurement.

Rényi entropy is defined as

$$I_R(X) = (1-\mu)^{-1} \log \int_{-\infty}^{\infty} f(x)^\mu dx, \quad \mu > 0, \mu \neq 0.$$

Using Equation (6) and after some manipulations, we have

$$I_R(X) = (1-\mu)^{-1} \log \left\{ \sum_{i,\ell,w=0}^{\infty} \sum_{j=i}^{\infty} \frac{\lambda^{\mu+w-j}\beta^{\mu-j}\Gamma(i+\ell+w+1)}{\theta^{i+w-2j+1}(1+\theta)^{j+\ell}\mu^{i+\ell+w+1}} \binom{\mu}{j}\binom{j}{i}\binom{\mu-j}{\ell}\binom{j-\mu(\beta+1)}{w} \right\}. \quad (14)$$

2.7. Stochastic Dominance

Across many distinct fields of probability and statistics, stochastic ordering and inequalities are being employed more extensively to examine the comparative behavior. Biometrics, robustness, econometrics, and actuarial sciences are all fields that have developed this presumption. According to Shaked and Shanthikumar [11], an av X_1 is said to be smaller than another av X_2 in the likelihood ratio order $(X_1 \leq_{lr} X_2)$ if $f_1(x)/f_2(x)$ decreases in x. The following theorem shows that the minLLx distribution is ordered in likelihood ratio ordering if the appropriate assumptions exist.

Theorem 1: Let $X_1 \sim \text{minLLx}(\theta_1, \lambda_1, \beta_1)$ and $X_2 \sim \text{minLLx}(\theta_2, \lambda_2, \beta_2)$. If $\theta_1 = \theta_2$, $\lambda_1 = \lambda_2$ and $\beta_1 \geq \beta_2$ (or if $\theta_1 = \theta_2$, $\beta_1 = \beta_2$ and $\lambda_1 \geq \lambda_2$), then $X_1 \leq_{lr} X_2$.

Proof: We have

$$\frac{f_1(x)}{f_2(x)} = \frac{(1+\theta_2)(1+\lambda_2 x)^{1+\beta_2} e^{-(\theta_1-\theta_2)x}}{(1+\theta_1)(1+\lambda_1 x)^{1+\beta_1}} \left\{ \frac{\lambda_1 \beta_1 (1+\theta_1+\theta_1 x) + \theta_1^2 (1+x)(1+\lambda_1 x)}{\lambda_2 \beta_2 (1+\theta_2+\theta_2 x) + \theta_2^2 (1+x)(1+\lambda_2 x)} \right\}.$$

Then

$$\log \frac{f_1(x)}{f_2(x)} = -(\theta_1 - \theta_2) - (1+\beta_1)\log(1+\lambda_1 x) + (1+\beta_2)\log(1+\lambda_2 x) + \log\left(\frac{1+\theta_2}{1+\theta_1}\right)$$
$$+ \log\left[\lambda_1 \beta_1 (1+\theta_1+\theta_1 x) + \theta_1^2 (1+x)(1+\lambda_1 x)\right]$$
$$- \log\left[\lambda_2 \beta_2 (1+\theta_2+\theta_2 x) + \theta_2^2 (1+x)(1+\lambda_2 x)\right].$$

If $\theta_1 = \theta_2$, $\lambda_1 = \lambda_2$ and $\beta_1 \geq \beta_2$ or if $\theta_1 = \theta_2$, $\beta_1 = \beta_2$ and $\lambda_1 \geq \lambda_2$, then we have

$$\frac{d}{dx}\log\frac{f_1(x)}{f_2(x)} = \frac{-\lambda_1(1+\beta_1)}{1+\lambda_1 x} + \frac{\lambda_2(1+\beta_2)}{1+\lambda_2 x} + \frac{\theta_1\{\lambda_1\beta_1 + \theta_1[1+\lambda_1(1+2x)]\}}{\lambda_1 \beta_1(1+\theta_1+\theta_1 x) + \theta_1^2(1+x)(1+\lambda_1 x)}$$
$$- \frac{\theta_2\{\lambda_2\beta_2 + \theta_2[1+\lambda_2(1+2x)]\}}{\lambda_2 \beta_2(1+\theta_2+\theta_2 x) + \theta_2^2(1+x)(1+\lambda_2 x)} < 0.$$

Resultantly, $f_1(x)/f_2(x)$ declines in x and hence $X_1 \leq_{lr} X_2$. □

2.8. Stress Strength Model

Acquired resistance metrics are used in lifetime testing to ascertain a system's durability. The stress-strength parameter, for instance, is based on the likelihood that a framework would work proficiently if the stress concentration will be less than its toughness. In the perspective of the minLLx distribution, the following result exemplifies a primitive outline for this parameter.

Let X_1 and X_2 be two independent chance variables with minLLx$(\theta_1, \lambda_1, \beta_1)$ and minLLx$(\theta_2, \lambda_2, \beta_2)$ distributions. Then, the stress−strength model is given by

$$R = \Pr(X_2 < X_1) = \int_0^\infty f_1(\theta_1, \lambda_1, \beta_1) F_2(\theta_2, \lambda_2, \beta_2) \, dx$$

$$= 1 - \frac{\lambda_1 \beta_1}{(1+\theta_1)(1+\theta_2)} \underbrace{\int_0^\infty (1+\lambda_1 x)^{-\beta_1 - 1}(1+\lambda_2 x)^{-\beta_2}(1+\theta_1+\theta_2)(1+\theta_1+\theta_2 x)e^{-(\theta_1+\theta_2)x} \, dx}_{H}$$

$$- \frac{\theta_1^2}{(1+\theta_1)(1+\theta_2)} \underbrace{\int_0^\infty (1+\lambda_1 x)^{-\beta_1}(1+\lambda_2 x)^{-\beta_2}(1+x)(1+\theta_2+\theta_2 x)e^{-(\theta_1+\theta_2)x} \, dx}_{E},$$

where

$$H = \sum_{j,i=0}^{\infty} \frac{\lambda_1^j \lambda_2^i \Gamma(j+i+1)}{(\theta_1+\theta_2)^{j+i+3}} \left\{ \begin{array}{l} (1+\theta_1)(1+\theta_2)(\theta_1+\theta_2)^2 + (\theta_1+\theta_2)(j+i+1) \\ \times[\theta_2(1+\theta_1) + \theta_1(1+\theta_2)] + \theta_1\theta_2(j+i+1)(j+i+2) \end{array} \right\} \binom{-\beta_1-1}{j} \binom{-\beta_2}{i},$$

and

$$E = \sum_{j,i=0}^{\infty} \frac{\lambda_1^j \lambda_2^i \Gamma(j+i+1)}{(\theta_1+\theta_2)^{j+i+3}} \left\{ \begin{array}{l} (1+\theta_2)(\theta_1+\theta_2)^2 + (\theta_1+\theta_2)(1+2\theta_2)(j+i+1) \\ +\theta_2(j+i+1)(j+i+2) \end{array} \right\} \binom{-\beta_1}{j} \binom{-\beta_2}{i}.$$

Therefore, the stress−strength model for the minLLx distribution is

$$R = 1 - \sum_{j,i=0}^{\infty} \frac{\lambda_1^j \lambda_2^i \Gamma(j+i+1)}{(1+\theta_1)(1+\theta_2)(\theta_1+\theta_2)^{j+i+3}} \binom{-\beta_2}{i}$$

$$\times \left(\lambda_1 \beta_1 \left\{ \begin{array}{c} (1+\theta_1)(1+\theta_2)(\theta_1+\theta_2)^2 + (\theta_1+\theta_2)(j+i+1) \\ \times [\theta_2(1+\theta_1) + \theta_1(1+\theta_2)] + \theta_1\theta_2(j+i+1)(j+i+2) \end{array} \right\} \binom{-\beta_1-1}{j} \right. \\ \left. + \theta_1^2 \left\{ \begin{array}{c} (1+\theta_2)(\theta_1+\theta_2)^2 + (\theta_1+\theta_2)(1+2\theta_2)(j+i+1) \\ + \theta_2(j+i+1)(j+i+2) \end{array} \right\} \binom{-\beta_1}{j} \right). \quad (15)$$

3. Characterization Results

This section outlines how to characterize the minLLx distribution in two ways: (i) on the basis of ratio of two truncated moments and (ii) by using the conditional expectation of certain functions of the av. It is worth emphasizing that for the characterization, (i) the cdf need not have a closed form, but instead relies on the solution of a first order differential equation, which serves as a link between the probability and differential equation. We would also like to highlight that due to the nature of minLLx density function, our characterizations may be the only versions available. Further bear in mind that the characterization (i) is stable in the sense of weak convergence (Glanzel [12]). We present our characterizations (i)–(ii) in the following two subsections.

3.1. Characterizations on the Basis of Two Truncated Moments

This subsection deals with the characterizations of minLLx distribution based on the ratio of two truncated moments. Our initial characterization employs a theorem of Glanzel [13], see Theorem A1 of Appendix A. The result is robust even if interval H is not closed, whereas the Theorem's constraint is on the interior of interval H.

Proposition 1. Let $X : \Omega \to (0, \infty)$ be a continuous av and let $q_1 = [\lambda\beta(1+\theta+\theta x) + \theta^2(1+x)(1+\lambda x)]^{-1} e^{\theta x}$ and $q_2(x) = q_1(x)(1+\lambda x)^{-1}$ for $x > 0$. The av X has pdf (2) iff the function ψ defined in Theorem 1 is of the expression

$$\psi(x) = \frac{\beta(1+\beta)^{-1}}{(1+\lambda x)}, \quad x > 0.$$

Proof. Let us presume that the av X has pdf(2), then

$$(1 - F(x)) E[q_1(X)|X \geq x] = \frac{(1+\theta)^{-1}}{\lambda\beta(1+\lambda x)^\beta}, \quad x > 0,$$

and

$$(1 - F(x)) E[q_2(X)|X \geq x] = \frac{(1+\theta)^{-1}}{\lambda(\beta+1)(1+\lambda x)^{(\beta+1)}}, \quad x > 0.$$

Furthermore,

$$\psi(x) q_1(x) - q_2(x) = -\frac{q_1(x)}{(\beta+1)(1+\lambda x)} < 0, \quad \text{for } x > 0.$$

Conversely, if ξ is of the above form, then

$$s'(x) = \frac{\psi'(x) q_1(x)}{\psi(x) q_1(x) - q_2(x)} = \frac{\lambda\beta}{(1+\lambda x)}, \quad x > 0,$$

and consequently

$$s(x) = -\log\left\{(1+\lambda x)^{-\beta}\right\}, \quad x > 0.$$

Now, according to Theorem 1, X has density (2). $\square$

Corollary 1. Let $X : \Omega \to (0, \infty)$ be a continuous av and let $q_1(x)$ be as in proposition 3.1. The chance variable X has pdf (2) iff there exist functions q_2 and ψ defined in theorem 1 fullfilling the following differential equation

$$\frac{\psi'(x)\, q_1(x)}{\psi(x)\, q_1(x) - q_2(x)} = \frac{\lambda \beta}{(1 + \lambda x)}, \qquad x > 0.$$

Corollary 2. The general solution of the differential equation in Corollary 1 is

$$\psi(x) = (1 + \lambda x)^\beta \left[-\int \lambda \beta (1 + \lambda x)^{-1} \, (1 + \lambda x)^{-1} (q_1(x))^{-(\beta+1)} q_2(x)\, dx + D \right],$$

where D is a constant. It is worth emphasizing that one set of functions satisfying the above differential equation is given in Proposition 1 with $D = 0$. Clearly, there are other triplets (q_1, q_2, ψ) that satisfy constraints of Theorem 1.

3.2. Characterizations on the Basis of Conditional Expectation of Certain Functions of an Arbitrary Variable

In this subsection, we employ a single function Ψ of X and characterize the distribution of X in terms of the truncated moment of $\Psi(X)$. The following proposition has already appeared in Hamedani [14], so we will just state it here that it can be used to characterize the minLLx distribution.

Proposition 2. Let $X : \Omega \to (e, f)$ be a continuous av with cdf F. Let $\Psi(x)$ be a differentiable function on (e, f) with $\lim_{x \to e^+} \Psi(x) = 1$. Then for $\delta \neq 1$,

$$E[\Psi(X)|X \geq x] = \delta \Psi(x), \qquad x \in (e, f)$$

iff

$$\Psi(x) = [1 - F(x)]^{\frac{1}{\delta} - 1}, \qquad x \in (e, f).$$

Remark 1. For $(e, f) = (0, \infty)$, $\Psi(x) = \frac{e^{-\theta x/\beta}}{(1+\lambda x)} \left(\frac{1+\theta+\theta x}{1+\theta} \right)^{1/\beta}$ and $\delta = \frac{\beta}{\beta+1}$, Proposition 2. provides a characterization of the minLLX.

4. Maximum Likelihood Estimation

The maximum likelihood estimates (MLEs) and the observed information matrix for the model parameters of the minLLx distribution will be investigated in this section. Let $x_1, x_2, \ldots, x_n$ be a random sample from the minLLx distribution, then the corresponding log-likelihood function is given by

$$\ell = -n \log(1 + \theta) - \theta \sum_{i=1}^{n} x_i - (1 + \beta) \sum_{i=1}^{n} \log(1 + \lambda x_i)$$
$$+ \sum_{i=1}^{n} \log\{\lambda \beta (1 + \theta + \theta x_i) + \theta^2 (1 + \lambda x_i)(1 + \lambda x_i)\}. \tag{16}$$

The modules of the score vector $\nabla \ell = \left(\frac{\partial \ell}{\partial \theta}, \frac{\partial \ell}{\partial \lambda}, \frac{\partial \ell}{\partial \beta} \right)$ are:

$$\frac{\partial \ell}{\partial \theta} = \frac{-n}{1+\theta} - \sum_{i=1}^{n} x_i + \sum_{i=1}^{n} \left\{ \frac{(1+x_i)[\lambda \beta + 2\theta(1+\lambda x_i)]}{\lambda \beta (1 + \theta + \theta x_i) + \theta^2 (1+\lambda x_i)(1+\lambda x_i)} \right\}, \tag{17}$$

$$\frac{\partial \ell}{\partial \lambda} = -(1+\beta) \sum_{i=1}^{n} \left(\frac{x_i}{1+\lambda x_i} \right) + \sum_{i=1}^{n} \left\{ \frac{\beta(1+\theta+\theta x_i) + \theta^2 x_i (1+x_i)}{\lambda \beta (1+\theta+\theta x_i) + \theta^2 (1+\lambda x_i)(1+\lambda x_i)} \right\}, \tag{18}$$

and
$$\frac{\partial \ell}{\partial \beta} = -\sum_{i=1}^{n} \log(1 + \lambda x_i) + \sum_{i=1}^{n} \left\{ \frac{\lambda(1 + \theta + \theta x_i)}{\lambda \beta(1 + \theta + \theta x_i) + \theta^2(1 + \lambda x_i)(1 + \lambda x_i)} \right\}. \quad (19)$$

The MLEs, say $\hat{\Theta} = (\hat{\theta}, \hat{\lambda}, \hat{\beta})$, of $\Theta = (\theta, \lambda, \beta)^T$, can be obtained by equating the system of nonlinear Equations (17)–(19) to zero and solving them concurrently. The components of the observed information matrix $J(\Theta) = \{J_{wv}\}$ (for $w, v = \theta, \lambda, \beta$(of $\Theta = (\theta, \lambda, \beta)^T$ are given in Appendix B.

5. Simulation Study

It is very difficult to compare the theoretical performances of the different estimators for the minLLx distribution. Therefore, simulation is needed to compare the performances of the different methods of estimation, mainly with respect to their biases, mean square errors, and variances for different sample sizes. A numerical study is performed using Mathematica (v9) software. A portion of the used codes are provided as Supplementary Materials. Different sample sizes are considered through the experiments at size $n = 50, 100, 200, 300$, and 500. For the defined sample size n, the experimental bias and MSE values are the aggregate of values from $N = 2000$ replicated samples of the different values of parameters θ, λ and β, respectively. Traditionally, qf, which is the inverse of cdf, i.e., $Q(u) = F^{-1}(p) = \min\{x : F(x) \geq p\}$, is employed. However, in this case, it is not possible to obtain the qf of the minLLx distribution unequivocally. To obtain the minLLx variates, instead, we can implement the Newton–Raphosn algorithm as follows:

I. Set the values for n, λ, θ, and β, as well as the starting value of x_0.
II. Develop $U \sim Uniform\ (0,1)$.
III. Update x_0 each time via the Newton–Raphson's methodology, as shown below.

$$x_* = x_0 - R(x_0; \lambda, \theta, \beta)$$

where $R(x_0; \lambda, \theta, \beta) = \frac{F(x_0; \lambda, \theta, \beta)}{f(x_0; \lambda, \theta, \beta)}$, and $F(x_0; \lambda, \theta, \beta)$ and $f(x_0; \lambda, \theta, \beta)$ are cdf and pdf (in Equations (1) and (2)) of minLLx distribution, respectively.

I. If $|x_0 - x_*| \leq \varepsilon$, where ε is very small tolerance limit, then store $x_0 = x_*$ as a variate from minLLX (λ, θ, β) distribution.
II. If $|x_0 - x_*| \geq \varepsilon$, fix $x_0 = x_*$ and then proceed to step III.
III. In order to develop $x_1, x_2, x_3, \ldots, x_n$, steps II-V are repeated n times.

The average estimates, biases, MSEs, coverage probabilities (CPs), and confidence intervals (CIs), at 95% and 99%, on the basis of different parameter combinations, are reported in Tables 2–5 respectively.

Table 2. The MLEs, Bias, MSE, and CPs for the model parameters of the minLLx distribution based on some initial (Init) values.

n	Para	Init.	MLE	Bias	MSE	95% CI			99% CI		
						CPs	LB	UB	CPs	LB	UB
50	θ	1.5	2.554	1.054	1.250	0.99	2.451	2.657	1.00	2.448	2.793
	β	0.85	1.763	0.913	0.857	0.96	1.746	1.797	0.99	1.719	1.808
	λ	0.72	1.334	0.614	0.889	0.92	1.309	1.395	0.97	1.288	1.443
100	θ	1.5	2.527	1.027	1.137	0.94	2.471	2.583	0.97	2.454	2.601
	β	0.85	1.667	0.817	0.698	0.97	1.656	1.781	0.98	1.637	1.798
	λ	0.72	1.227	0.507	0.733	0.95	1.215	1.266	0.96	1.202	1.291

Table 2. Cont.

n	Para	Init.	MLE	Bias	MSE	95% CI			99% CI		
						CPs	LB	UB	CPs	LB	UB
200	θ	1.5	2.495	0.995	1.024	0.90	2.469	2.521	0.98	2.520	2.599
	β	0.85	1.601	0.751	0.583	0.97	1.586	1.625	0.95	1.547	1.643
	λ	0.72	1.111	0.391	0.526	0.95	1.084	1.159	0.94	1.005	1.187
300	θ	1.5	1.738	0.238	0.556	0.94	1.721	1.755	1.00	1.727	1.779
	β	0.85	1.229	0.379	0.273	0.96	1.189	1.242	0.97	1.147	1.267
	λ	0.72	0.997	0.277	0.377	0.95	0.979	1.015	0.97	0.958	1.093
500	θ	1.5	1.712	0.212	0.484	0.96	1.701	1.723	0.98	1.694	1.754
	β	0.85	1.003	0.153	0.097	0.94	0.985	1.036	0.98	0.970	1.088
	λ	0.72	0.837	0.117	0.114	0.96	0.826	0.877	0.99	0.811	0.893

Table 3. The MLEs, Bias, MSE, CPs for the model parameters of the minLLx distribution based on some initial (Init) values.

n	Para	Init.	MLE	Bias	MSE	95% CI			99% CI		
						CPs	LB	UB	CPs	LB	UB
50	θ	2.4	3.807	1.407	2.230	0.90	3.648	3.966	0.97	3.466	3.886
	β	0.5	1.128	0.628	0.604	0.98	0.932	1.324	0.94	0.87	1.386
	λ	0.5	0.981	0.481	0.481	0.96	0.785	1.177	0.96	0.723	1.239
100	θ	2.4	3.595	1.195	1.678	0.97	3.719	3.870	0.98	3.454	3.627
	β	0.5	0.967	0.467	0.398	0.94	0.575	1.359	0.99	0.451	1.483
	λ	0.5	0.864	0.364	0.382	0.97	0.472	1.256	0.98	0.348	1.38
200	θ	2.4	2.753	0.353	1.888	0.94	2.721	2.786	0.99	2.503	2.597
	β	0.5	0.881	0.381	0.395	0.96	0.691	1.071	0.96	0.631	1.131
	λ	0.5	0.722	0.222	0.199	0.97	0.532	0.912	0.97	0.472	0.972
300	θ	2.4	2.532	0.132	0.833	0.95	2.705	2.762	1.00	2.499	2.569
	β	0.5	0.646	0.146	0.271	0.96	0.42452	0.867	0.98	0.354	0.938
	λ	0.5	0.637	0.137	0.269	0.97	0.415	0.858	0.99	0.345	0.929
500	θ	2.4	2.518	0.118	0.270	0.96	2.506	2.531	1.00	2.537	2.577
	β	0.5	0.557	0.057	0.253	0.95	0.5276	0.586	0.99	0.518	0.596
	λ	0.5	0.597	0.097	0.259	0.96	0.5676	0.626	1.00	0.558	0.636

From Tables 2 and 3, we deduced that when the postulated model differs significantly from the genuine model, as anticipated, the MSE of the estimators rises. The MSE drops as the sample size is increased and the homogeneity disintegrates. In general, when the kurtosis increases the MSE declines. Likewise, if the asymmetry widens, so does the bias, and vice versa. The bias lessens as the kurtosis increases. Therefore, it is evident that as sample size n gets larger, the MSEs and biases reduce. Similarly, the CPs of the confidence interval seems to be quite near to the conventional levels of certainty (95% and 99%), which endorses the already established empirical findings. In a nutshell, we may infer that MLEs perform impressively in estimating the parameters of the minLLx distribution.

Table 4. The MLEs, Bias, MSE, and CPs for the model parameters of the minLLx distribution based on some initial (Init) values.

n	Para	Init.	MLE	Bias	MSE	95% CI			99% CI		
						CPs	LB	UB	CPs	LB	UB
50	θ	2.4	3.551	1.151	1.575	0.90	3.648	3.966	1.00	3.466	3.886
	β	0.15	0.667	0.517	0.477	0.99	0.471	0.863	0.94	0.409	0.925
	λ	1.5	2.778	1.278	1.883	0.92	2.582	2.974	0.97	2.52	3.036
100	θ	2.4	3.295	0.895	1.051	0.98	3.719	3.870	0.96	3.454	3.627
	β	0.15	0.546	0.396	0.337	0.97	0.154	0.938	0.98	0.03	1.062
	λ	1.5	2.337	0.837	0.951	0.94	1.945	2.729	0.99	1.821	2.853
200	θ	2.4	3.016	0.616	0.629	0.96	2.721	2.786	0.95	2.503	2.597
	β	0.15	0.881	0.731	0.784	0.96	0.691	1.071	0.97	0.631	1.131
	λ	1.5	1.836	0.336	0.263	0.95	1.646	2.026	0.97	1.586	2.086
300	θ	2.4	2.842	0.442	0.345	0.97	2.705	2.762	0.98	2.499	2.569
	β	0.15	0.646	0.496	0.496	0.96	0.425	0.867	0.99	0.354	0.938
	λ	1.5	1.772	0.272	0.324	0.95	1.551	1.993	0.97	1.480	2.064
500	θ	2.4	2.537	0.137	0.27	0.95	2.506	2.531	0.98	2.537	2.577
	β	0.15	0.557	0.407	0.416	0.96	0.5276	0.5864	0.99	0.5183	0.5957
	λ	1.5	1.606	0.106	0.261	0.95	1.5766	1.6354	0.98	1.5673	1.6447

Table 5. The MLEs, Bias, MSE, and CPs for the model parameters of the minLLx distribution based on some initial (Init) values.

n	Para	Init.	MLE	Bias	MSE	95% CI			99% CI		
						CPs	LB	UB	CPs	LB	UB
50	θ	2.4	3.851	1.451	2.355	0.99	3.648	3.966	1.00	3.466	3.886
	β	0.15	0.767	0.617	0.631	0.93	0.571	0.963	0.94	0.509	1.025
	λ	3.5	4.708	1.208	1.709	0.98	4.512	4.904	0.92	4.45	4.966
100	θ	2.4	3.529	1.129	1.525	0.98	3.719	3.870	0.98	3.454	3.627
	β	0.15	0.665	0.515	0.515	0.97	0.273	1.057	0.95	0.149	1.181
	λ	3.5	4.553	1.053	1.359	0.96	4.161	4.945	0.93	4.037	5.069
200	θ	2.4	3.119	0.719	0.767	0.98	2.721	2.786	0.94	2.503	2.597
	β	0.15	0.498	0.348	0.371	0.97	0.308	0.688	0.98	0.248	0.748
	λ	3.5	4.078	0.578	0.584	0.96	3.888	4.268	0.99	3.828	4.328
300	θ	2.4	2.728	0.328	0.358	0.96	2.705	2.762	0.98	2.499	2.569
	β	0.15	0.367	0.217	0.297	0.97	0.146	0.588	0.99	0.075	0.659
	λ	3.5	3.876	0.376	0.391	0.94	3.655	4.097	0.98	3.584	4.168
500	θ	2.4	2.643	0.243	0.209	0.96	2.506	2.531	0.99	2.537	2.577
	β	0.15	0.268	0.118	0.164	0.95	0.2386	0.2974	0.98	0.2293	0.3067
	λ	3.5	3.711	0.211	0.195	0.95	3.6816	3.7404	1.00	3.6723	3.7497

6. Applications

In this portion, we consider two actual cases of the minLLx distribution to showcase its effectiveness. When the pressure is at % anxiety levels, the first data set reflects the failure times of the Kevlar 49/epoxy strands. This data are leptokurtic, unimodal, and substantially right skewed, with a likely outlier (skewness = 3.05 and kurtosis = 14.47). This data set is taken from Andrews and Herzberg [15] and the original source is Barlow et al. [16].The data are: 0.01, 0.01,0.02, 0.02, 0.02, 0.03, 0.03, 0.04, 0.05, 0.06, 0.07, 0.07, 0.08, 0.09, 0.09, 0.10, 0.10, 0.11, 0.11, 0.12, 0.13, 0.18, 0.19, 0.20, 0.23, 0.24, 0.24, 0.29, 0.34, 0.35, 0.36, 0.38, 0.40, 0.42, 0.43, 0.52, 0.54, 0.56, 0.60, 0.60, 0.63, 0.65, 0.67, 0.68, 0.72, 0.72, 0.72, 0.73, 0.79, 0.79, 0.80, 0.80, 0.83, 0.85, 0.90, 0.92, 0.95, 0.99, 1.00, 1.01, 1.02, 1.03, 1.05, 1.10, 1.10, 1.11, 1.15, 1.18, 1.20, 1.29, 1.31, 1.33, 1.34, 1.40, 1.43, 1.45, 1.50, 1.51, 1.52, 1.53, 1.54, 1.54, 1.55, 1.58, 1.60, 1.63, 1.64, 1.80, 1.80, 1.81, 2.02, 2.05, 2.14, 2.17, 2.33, 3.03, 3.03, 3.34, 4.20, 4.69, and 7.89. These data are also used by Cooray and Ananda [17] and Al-Aqtash et al. [18].

The second data set signifies the failure time of 20 components from Murthy et al. [19]. The data are: 0.072, 4.763, 8.663, 12.089, 0.477, 5.284, 9.511, 13.036, 1.592, 7.709, 10.636, 13.949, 2.475, 7.867, 10.729, 16.169, 3.597, 8.661, 11.501, and 19.809.

We obtained the MLEs for the unknown parameters of all competitive models and then compared the results via goodness-of-fit statistics: Anderson-Darling (A*), Cramér-von Mises (W*), AIC (Akaike information criterion), and BIC (Bayesian information criterion). The better model corresponds to the smaller of these criteria. The values for the Kolmogorov Smirnov (KS) statistic and its p-value are also presented.

We compared the minLLx distribution with those of Weibull Lindley (WL) (Asgharzadeh et al. [20]), Lomax (Lx), Lindley (L), quasi Lindley (QL) (Shanker and Mishra [21]), and power Lomax (PLx) (Rady et al. [22]). The MLEs, their standard errors (SEs), and some goodness of fit statistics of the models for the respective data sets are introduced in Tables 6–9. The estimated pdf and cdf plots of all competitive distributions for the two data sets are displayed in Figures 3 and 4, respectively.

Table 6. The MLEs alongside their accompanying SEs (in parenthesis) for the first data set.

Distribution	ML Estimates with SEs					
	$\hat{\lambda}$	$\hat{\beta}$	$\hat{\theta}$	$\hat{\alpha}$	$\hat{a}$	$\hat{b}$
minLLx	29.1543 (24.5461)	1.1967 (0.1353)	0.0565 (0.0444)	-	-	-
WL	-	-	-	54.8909 (46.5022)	0.1262 (0.0029)	1.3776 (0.1066)
Lx	-	0.0649 (0.0730)	-	16.0324 (11.8945)	-	-
L	-	—	-	1.3848 (0.1068)	-	-
QL	-	16.2215 (18.4297)	-	1.0312 (0.1876)	-	-
PLx	-	-	49.8009 (55.9286)	-	0.9381 (0.0842)	48.6282 (64.3737)

Table 7. Some goodness of fit statistics for the fitted models to the first data set.

Distribution	Goodness-of-Fit Statistics						
	−LL	A*	W*	KS	p-Value	AIC	BIC
minLLx	101.7467	0.73166	0.1174	0.0751	0.6188	209.4934	217.3388
WL	103.7773	0.8412	0.1372	0.1069	0.1985	213.5547	221.4001

Table 7. Cont.

Distribution	Goodness-of-Fit Statistics						
	−LL	A*	W*	KS	p-Value	AIC	BIC
Lx	103.2335	1.1543	0.2082	0.0836	0.4803	210.4669	215.6972
L	104.6558	0.8349	0.1377	0.1062	0.2046	211.3115	213.9267
QL	103.5036	1.0226	0.1796	0.0892	0.3968	211.0071	216.2374
PLx	102.9973	1.1376	0.2044	0.0912	0.3694	211.9947	219.8400

Table 8. The MLEs alongside their accompanying SEs (in parenthesis) for the second data set.

Distribution	ML Estimates with SEs					
	$\hat{\lambda}$	$\hat{\beta}$	$\hat{\theta}$	$\hat{\alpha}$	$\hat{a}$	$\hat{b}$
minLLx	23.2537 (6.2332)	0.2000 (0.0357)	0.0176 (0.0242)	-	-	-
WL	-	-	-	0.5063 (0.2646)	0.0022 (0.0049)	0.1936 (0.0376)
Lx	-	0.0063 (0.0050)	-	19.2257 (15.1770)	-	-
L	-	-	-	0.2161 (0.0344)	-	-
QL	-	12.7561 (8.1217)	-	0.1276 (0.0188)	-	-
PLx	-	-	5.1542 (4.2880)	—	1.2999 (0.2549)	77.2599 (64.2934)

Table 9. Some goodness of fit statistics for the models fitted to the second data set.

Distribution	Goodness-of-Fit Statistics						
	−LL	A*	W*	KS	p-Value	AIC	BIC
minLLx	60.4860	0.4993	0.0891	0.2013	0.3319	126.1758	129.0630
WL	60.8537	0.5622	0.0992	0.2051	0.3237	127.7075	128.2906
Lx	62.9558	0.9314	0.1602	0.2484	0.1422	129.9117	131.9032
L	61.3791	0.6909	0.1203	0.2022	0.3298	126.9583	129.7541
QL	62.6023	0.8804	0.1514	0.2493	0.1396	129.2046	131.1960
PLx	62.5202	0.9067	0.1561	0.2315	2000	131.0405	134.0277

The values in Tables 7 and 9 clearly show that the minLLx distribution has the smallest values for A*, W*, AIC, BIC, and KS, and the largest p-values among all competitive models, compelling it to be chosen as the best model. It is clear from Figures 3 and 4, that the new minLLx distribution provides the best fits for the two data sets.

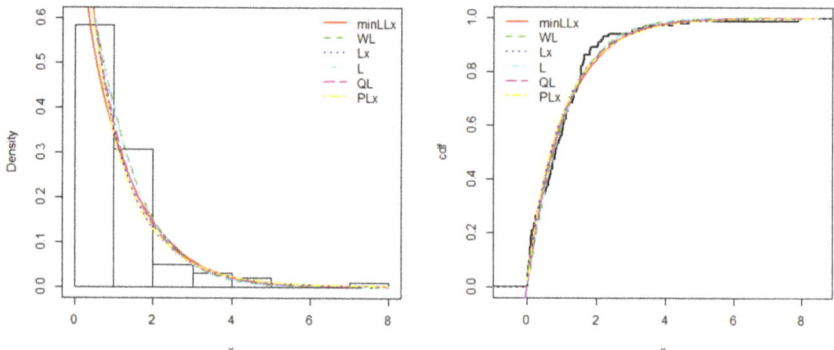

Figure 3. Estimated pdf and cdf plots of the minLLx distribution for the first data set.

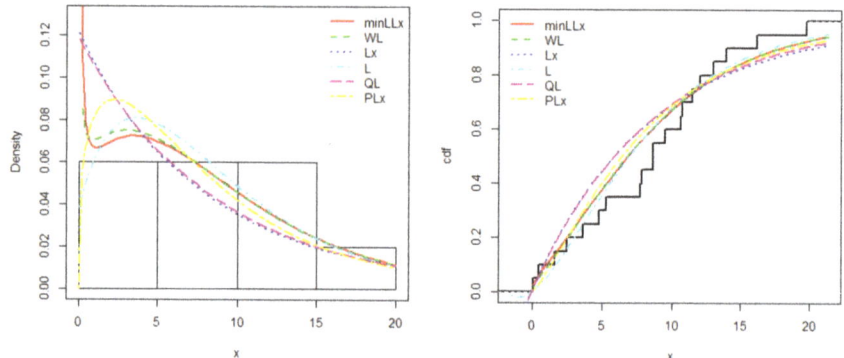

Figure 4. Estimated pdf and cdf plots of the minLLx distribution for the second data set.

7. Conclusions

By unifying the Lindley and Lomax distributions, we establish a three-parameter distribution called the minimum Lindley Lomax (minLLx). The quantile function, ordinary and incomplete moments, moment generating function, Lorenz and Bonferroni curves, order statistics, Rényi entropy, stress–strength model, and stochastic ordering are all considered as defining attributes of the new model. The envisaged model's characterizations are evaluated. The model parameters are determined using the optimum likelihood criterion, and these projections are assessed using numerical simulations. Two real-world applications exemplify the utility of the new model.

Supplementary Materials: Partial codes used in Section 5 are available online at https://www.mdpi.com/article/10.3390/mca27010016/s1.

Author Contributions: Conceptualization, S.K. and G.G.H.; methodology, H.M.R.; software, S.O.; validation, F.J., S.K. and H.M.R; formal analysis, F.J. and H.M.R.; investigation, S.S.; resources, S.O.; data curation, S.S.; writing—original draft preparation, F.J. and G.G.H.; writing—review and editing S.K. and F.J.; visualization, S.K. and S.S.; supervision, S.O.; project administration, F.J. All authors have read and agreed to the published version of the manuscript.

Funding: This research received no external funding.

Conflicts of Interest: The authors declare no conflict of interest.

Appendix A

Theorem A1. Let (Ω, F, P) be a given probability space and let $H = [a, b]$ be an interval for some $d < b$ ($a = -\infty$, $b = \infty$ might as well be allowed). Let $X : \Omega \to H$ be a continuous av with the distribution function F and let q_1 and q_2 be two real functions defined on H, such that

$$E[q_2(X)|X \geq x] = E[q_1(X)|X \geq x]\psi(x), \quad x \in H,$$

is defined with some real function η. Assume that $q_1, q_2 \in C^{-1}(H)$, $\psi \in C^2(H)$ and F is a twice continuously differentiable and strictly monotone function on the set H. Finally, assume that the equation $\psi q_1 = q_2$ has no real solution in the interior of H. Then F is uniquely determined by the functions q_1, q_2, and ψ, particularly

$$F(x) = \int_a^x C \left| \frac{\psi'(u)}{\psi(u) q_1(u) - q_2(u)} \right| \exp(-s(u))\, du,$$

where function s is a solution of the differential equation $s' = \frac{\psi' q_1}{\psi q_1 - q_2}$ and C is the normalization constant, such that $\int_H dF = 1$.

We like to mention that this kind of characterization based on the ratio of truncated moments is stable in the sense of weak convergence (see Glanzel [12]), in particular, let us assume that there is a sequence $\{X_n\}$ of avs with a distribution function $\{F_n\}$, such that the functions q_{1n}, q_{2n}, and ψ_n ($n \in N$) satisfy the conditions of Theorem 1, and let $q_{1n} \to q_1$, $q_{2n} \to q_2$ for some continuously differentiable real functions q_1 and q_2. Finally, let X be a chance variable with distribution F. Under the condition that $q_{1n}(X)$ and $q_{2n}(X)$ are uniformly integrable and the family $\{F_n\}$ is relatively compact, the sequence X_n converges to X in distribution if and only if ψ_n converges to ψ, where

$$\psi(x) = \frac{E[q_2(X)|X \geq x]}{E[q_1(X)|X \geq x]}.$$

This stabilization theorem ensures that the precision of the distribution function is duplicated in the subsequent convergence of functions q_1, q_2, and ψ_n. It ensures, e.g., that the characterization on the Wald distribution coincides with that on the Levy-Smirnov distribution if $\alpha \to \infty$. The application of this theorem over certain challenges in analytical techniques, such as the estimation of the parameters of discrete distributions, is yet another corollary of Theorem 1's stability condition. The functions q_1, q_2, and in particular, ψ should be as straightforward and feasible for this reason. Although the function quartet is not distinctive, it is frequently possible to choose ψ as a linear combination. As a direct consequence, it is worth considering a few specific instances in order to develop innovative characterizations that capture the link between individual continuous univariate distributions and are relevant in other disciplines of science.

Appendix B

The components of the observed information matrix are the following

$$\frac{\partial^2 \ell}{\partial \theta^2} = \frac{-n}{(1+\theta)^2} + \sum_{i=1}^{n} \left\{ \frac{2(1+x_i)(1+\lambda x_i)[\lambda\beta(1+\theta+\theta x_i) + \theta^2(1+\lambda x_i)(1+\lambda x_i)] - [\lambda\beta(1+x_i) + 2\theta(1+x_i)(1+\lambda x_i)]^2}{[\lambda\beta(1+\theta+\theta x_i) + \theta^2(1+\lambda x_i)(1+\lambda x_i)]^2} \right\},$$

$$\frac{\partial^2 \ell}{\partial \theta \, \partial \lambda} = \sum_{i=1}^{n} \left\{ \frac{(1+x_i)\left\{ \begin{array}{c} (\beta + 2\theta x_i)[\lambda \beta(1+\theta+\theta x_i) + \theta^2(1+x_i)(1+\lambda x_i)] \\ -[\lambda \beta + 2\theta(1+\lambda x_i)][\beta(1+\theta+\theta x_i) + \theta^2 x_i(1+x_i)] \end{array} \right\}}{[\lambda \beta(1+\theta+\theta x_i) + \theta^2(1+\lambda x_i)(1+\lambda x_i)]^2} \right\},$$

$$\frac{\partial^2 \ell}{\partial \theta \, \partial \beta} = \lambda \sum_{i=1}^{n} \left\{ \frac{\begin{array}{c}(1+x_i)[\lambda \beta(1+\theta+\theta x_i) + \theta^2(1+x_i)(1+\lambda x_i)] \\ -(1+\theta+\theta x_i)[\lambda \beta(1+x_i) + 2\theta(1+x_i)(1+\lambda x_i)]\end{array}}{[\lambda \beta(1+\theta+\theta x_i) + \theta^2(1+\lambda x_i)(1+\lambda x_i)]^2} \right\},$$

$$\frac{\partial^2 \ell}{\partial \lambda^2} = -\sum_{i=1}^{n} \left\{ \frac{[\beta(1+\theta+\theta x_i) + \theta^2 x_i(1+x_i)]^2}{[\lambda \beta(1+\theta+\theta x_i) + \theta^2(1+\lambda x_i)(1+\lambda x_i)]^2} \right\},$$

$$\frac{\partial^2 \ell}{\partial \lambda \, \partial \beta} = \sum_{i=1}^{n} \left(\frac{x_i}{1+\lambda x_i} \right) + \sum_{i=1}^{n} \left\{ \frac{\theta^2(1+x_i)(1+\theta+\theta x_i)}{[\lambda \beta(1+\theta+\theta x_i) + \theta^2(1+\lambda x_i)(1+\lambda x_i)]^2} \right\},$$

$$\frac{\partial^2 \ell}{\partial \beta^2} = -\lambda^2 \sum_{i=1}^{n} \left\{ \frac{(1+\theta+\theta x_i)^2}{[\lambda \beta(1+\theta+\theta x_i) + \theta^2(1+\lambda x_i)(1+\lambda x_i)]^2} \right\}.$$

References

1. Tahir, M.H.; Nadarajah, S. Parameter induction in continuous univariate distributions: Well-established G families. *Ann. Braz. Acad. Sci.* **2015**, *87*, 539–568. [CrossRef] [PubMed]
2. Tahir, M.H.; Cordeiro, G.M. Compounding of distributions: A survey and new generalized classes. *J. Stat. Distrib. Appl.* **2016**, *3*, 13. [CrossRef]
3. Cordeiro, G.M.; Ortega, E.; Lemonte, A.J. The exponential–Weibull lifetime distribution. *J. Stat. Comput. Simul.* **2013**, *84*, 2592–2606. [CrossRef]
4. Ghitany, M.E.; Atieh, B.; Nadarajah, S. Lindley distribution and its application. *Math. Comput. Simul.* **2008**, *78*, 493–506. [CrossRef]
5. Ramos, P.; Louzada, F. The generalized weighted Lindley distribution: Properties, estimation, and applications. *Cogent Math.* **2016**, *3*, 1256022. [CrossRef]
6. Singh, S.K.; Singh, U.; Sharma, V.K. Estimation and prediction for Type-I hybrid censored data from generalized Lindley distribution. *J. Stat. Manag. Syst.* **2016**, *19*, 367–396. [CrossRef]
7. Oguntunde, P.E.; Khaleel, M.A.; Ahmed, M.T.; Adejumo, A.O.; Odetunmibi, O. A New Generalization of the Lomax Distribution with Increasing, Decreasing, and Constant Failure Rate. *Model. Simul. Eng.* **2017**, *2017*, 6043169. [CrossRef]
8. Wei, S.; Wang, C.; Li, Z. Bayes estimation of Lomax distribution parameter in the composite LINEX loss of symmetry. *J. Interdiscip. Math.* **2017**, *20*, 1277–1287. [CrossRef]
9. Elgarhy, M.; Sharma, V.K.; ElBatal, I. Transmuted Kumaraswamy Lindley distribution with application. *J. Stat. Manag. Syst.* **2018**, *21*, 1083–1104. [CrossRef]
10. David, H.A.; Nagaraja, H. *Order Statistics*, 3rd ed.; Wiley: New York, NY, USA, 2003.
11. Shaked, M.; Shanthikumar, J.G. *Stochastic Orders*; Wiley: New York, NY, USA, 2007.
12. Glanzel, W. Some consequences of a characterization theorem based on truncated moments. *J. Theor. Appl. Stat.* **1990**, *21*, 613–618. [CrossRef]
13. Glanzel, W. A Characterization Theorem Based on Truncated Moments and its Application to Some Distribution Families. In *Mathematical Statistics and Probability Theory*; Springer: Dordrecht, The Netherlands, 1987; pp. 75–84. [CrossRef]
14. Hamedani, G.G. *On Certain Generalized Gamma Convolution Distribution II*; Technical Report No. 484, MSCS; Marquette University: Milwaukee, WI, USA, 2013.
15. Andrews, D.F.; Herzberg, A.M. *Data: A Collection of Problems from Many Fields for the Student and Research Worker (Springer Series in Statistics)*; Springer: New York, NY, USA, 1985.
16. Barlow, R.E.; Toland, R.H.; Freeman, T. A Bayesian analysis of stress-rupture life of Kevlar 49/epoxy spherical pressure vessels. In Proceedings of the Canadian Conference in Applied Statistics, Dwivedi, T.D., Ed. Marcel Dekker: New York, NY, USA, 1984.
17. Cooray, K.; Ananda, M.M.A. A Generalization of the Half-Normal Distribution with Applications to Lifetime Data. *Commun. Stat. Theory Methods* **2008**, *37*, 1323–1337. [CrossRef]
18. Al-Aqtash, R.; Lee, C.; Famoye, F. Gumbel-Weibull distribution: Properties and applications. *J. Mod. Appl. Stat. Methods* **2014**, *13*, 201–225. [CrossRef]
19. Murthy, D.N.P.; Xie, M.; Jiang, R. *Weibull Models*; Wiley: New York, NY, USA, 2004.
20. Asgharzadeh, A.; Nadarajah, S.; Sharafi, F. Weibull Lindley distribution. *REVSTAT Stat. J.* **2018**, *16*, 87–113.

21. Shanker, R.; Mishra, A. A quasi Lindley distribution. *Afr. J. Math. Comput. Sci. Res.* **2013**, *6*, 64–71.
22. Rady, E.-H.A.; Hassanein, W.A.; Elhaddad, T.A. The power Lomax distribution with an application to bladder cancer data. *SpringerPlus* **2016**, *5*, 1838. [CrossRef] [PubMed]

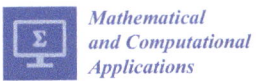

Article

Nadarajah–Haghighi Lomax Distribution and Its Applications

Vasili B. V. Nagarjuna [1], Rudravaram Vishnu Vardhan [1] and Christophe Chesneau [2,*]

[1] Department of Statistics, Pondicherry University, Pondicherry 605 014, India; arjun.vasili@gmail.com (V.B.V.N.); vrstatsguru@gmail.com (R.V.V.)
[2] Department of Mathematics, LMNO, Université de Caen-Normandie, Campus II, Science 3, 14032 Caen, France
* Correspondence: christophe.chesneau@unicaen.fr

Abstract: Over the years, several researchers have worked to model phenomena in which the distribution of data presents more or less heavy tails. With this aim, several generalizations or extensions of the Lomax distribution have been proposed. In this paper, an attempt is made to create a hybrid distribution mixing the functionalities of the Nadarajah–Haghighi and Lomax distributions, namely the Nadarajah–Haghighi Lomax (NHLx) distribution. It can also be thought of as an extension of the exponential Lomax distribution. The NHLx distribution has the features of having four parameters, a lower bounded support, and very flexible distributional functions, including a decreasing or unimodal probability density function and an increasing, decreasing, or upside-down bathtub hazard rate function. In addition, it benefits from the treatable statistical properties of moments and quantiles. The statistical applicability of the NHLx model is highlighted, with simulations carried out. Four real data sets are also used to illustrate the practical applications. In particular, results are compared with Lomax-based models of importance, such as the Lomax, Weibull Lomax, and exponential Lomax models, and it is observed that the NHLx model fits better.

Keywords: Nadarajah–Haghighi distribution; moments; Lomax distribution; data analysis

Citation: Nagarjuna, V.B.V.; Vardhan, R.V.; Chesneau, C. Nadarajah–Haghighi Lomax Distribution and Its Applications. *Math. Comput. Appl.* **2022**, *27*, 30. https://doi.org/10.3390/mca27020030

Received: 1 December 2021
Accepted: 18 March 2022
Published: 1 April 2022

Publisher's Note: MDPI stays neutral with regard to jurisdictional claims in published maps and institutional affiliations.

Copyright: © 2022 by the authors. Licensee MDPI, Basel, Switzerland. This article is an open access article distributed under the terms and conditions of the Creative Commons Attribution (CC BY) license (https://creativecommons.org/licenses/by/4.0/).

1. Introduction

Modeling heavy-tailed data is one of the important aspects in many engineering and medical domains. Initial work on this topic was carried out by Pareto [1] to model income data. In later years, the applications of Pareto, particularly the type II Lomax distribution (see [2]), usually referred to as Lomax (Lx) distribution, branched into scientific fields such as engineering sciences, actuarial sciences, medicine, income, and many more. The distribution function (cdf) and probability density function (pdf) of the Lx distribution are given by

$$F_{Lx}(x;\eta) = 1 - \left(\frac{\beta}{x+\beta}\right)^{\alpha}$$

and

$$f_{Lx}(x;\eta) = \frac{\alpha}{\beta}\left(\frac{\beta}{x+\beta}\right)^{\alpha+1}, \quad x \geq 0, \eta = (\alpha,\beta) > 0,$$

respectively, where α is a shape parameter, and β is a scale parameter. We have $F_{Lx}(x;\eta) = f_{Lx}(x;\eta) = 0$ for $x < 0$. References [3,4] considered the Lx distribution to model income and wealth data. Reference [5] used the Lx distribution as an alternative to the exponential, gamma, and Weibull distributions for heavy-tailed data. Reference [6] derived various estimation techniques based on the Lx distribution. References [7,8] examined the various structural properties and record value moments of the Lx distribution. Reference [9] extensively studied and extended the family of distributions that were used in the Lx distribution. Reference [10] considered the Lx distribution as an important distribution to model lifetime data, since it belongs to the family of decreasing hazard rate.

In continuation of this, many researchers have proposed several distributions that deal with heavy-tailed data by generalizing the functional forms of the Lx distribution. It mainly consists of adding scale/shape parameters accordingly. A few to mention are the exponentiated Lx (EL) distribution in [11], beta Lx (BL) distribution in [12], Poisson Lx distribution in [13], exponential Lx (EXL) distribution in [14], gamma Lx (GL) distribution in [15], Weibull Lx (WL) distribution in [16], beta exponentiated Lx distribution in [17], power Lx distribution in [18], exponentiated Weibull Lx distribution in [19], Marshall–Olkin exponential Lx distribution in [20], type II Topp–Leone power Lx distribution in [21], Marshall–Olkin length biased Lomax distribution in [22], Kumaraswamy generalized power Lx distribution in [23] and sine power Lx distribution in [24]. For the purpose of this study, a retrospective on the EXL distribution is required. To begin, it is defined by the following cdf and pdf:

$$F_{EXL}(x;\xi) = 1 - e^{-\lambda\left(\frac{\beta}{x+\beta}\right)^{-\alpha}}$$

and

$$f_{EXL}(x;\xi) = \frac{\lambda\alpha}{\beta}\left(\frac{\beta}{x+\beta}\right)^{-\alpha+1} e^{-\lambda\left(\frac{\beta}{x+\beta}\right)^{-\alpha}}, \quad x \geq -\beta, \xi = (\alpha, \beta, \lambda) > 0,$$

respectively, where α is a shape parameter, and β and λ are scale parameters. We have $F_{EXL}(x;\xi) = f_{EXL}(x;\xi) = 0$ for $x < -\beta$. Thus, the EXL distribution combines the functionalities of the exponential and Lx distributions through a specific composition scheme. This scheme may be called the extended Lx scheme (it will be discussed mathematically later). As immediate remarks, the EXL distribution has three parameters and is with a lower bounded support. It is shown in [14] that the pdf of the EXL distribution is unimodal and has an increasing hazard rate function (hrf). Moreover, its quantile and moment properties are manageable. On the statistical side, by considering the aircraft windshield data collected in [25], it is proven in [14] that the EXL model outperforms several three- or four-parameter extensions of the Lx model, including the EL, BL, and GL models. Thus, strong evidence is for the use of the extended Lx scheme for the construction of efficient distributions and models.

On the other hand, recently, a generalized version of the exponential distribution was given by Nadarajah and Haghighi [26]. It can be presented as an alternative to the Weibull, gamma, and exponentiated exponential (EE) distributions. It is called the Nadarajah–Haghighi (NH) distribution. The cdf and pdf of the NH distribution are

$$F_{NH}(x;\tau) = 1 - e^{1-(1+bx)^a}$$

and

$$f_{NH}(x;\tau) = ab(1+bx)^{a-1} e^{1-(1+bx)^a}, \quad x \geq 0, \tau = (a,b) > 0,$$

respectively. We have $F_{NH}(x;\tau) = f_{NH}(x;\tau) = 0$ for $x < 0$. Among its main features, the pdf can have decreasing and uni-modal shapes, and the hrf exhibits increasing, decreasing, and constant shapes. According to [26], if the pdfs of the gamma, Weibull, and exponentiated exponential are monotonically decreasing, then it is not possible to allow increasing hrf. However, such a hrf property can be achieved by the NH distribution.

In light of the above research work, we present a new distribution based on the extended Lx scheme with the use of the NH distribution as the main generator. It is called the NH Lx (NHLx) distribution. In this sense, the NHLx distribution is to the NH distribution what the EXL distribution is to the exponential distribution. The NHLx distribution can also be presented as a generalization of the EXL distribution through the introduction of an additional shape parameter. We investigate the theoretical and practical facets of the NHLx distribution. Among its functional features, it has four parameters, it is lower-bounded (as with the EXL distribution, with a bound governed by a scale parameter), its pdf exhibits non-increasing and inverted J-shaped curves, and its hrf possesses increasing, decreasing, and upside-down bathtub shapes. This combination of qualitative characteristics is rare for a lower-bounded distribution and, in this way, it has better functionality to model lifetime data than the EXL and Lx distributions, among others. We illustrate this aspect by considering four different data sets referenced in the literature.

The rest of the article covers the following aspects: Section 2 presents the most important functions of the NHLx distribution, namely the cdf, pdf, hrf, and quantile function (qf), along with a graphical analysis when necessary. Section 3 is devoted to moment analysis and related functions. Section 4 concerns the maximum likelihood estimates of the NHLx model parameters. The above section is completed by a simulation study in Section 5. Concrete applications of the NHLx model are developed in Section 6. A conclusion is formulated in Section 7.

2. NHLx Distribution

In order to understand the essence of the NHLx distribution, let us describe more precisely the extended Lx scheme on the basis of the EXL distribution. One can remark that $F_{EXL}(x;\xi) = F_E\left(\frac{1}{1-F_{Lx}^*(x;\eta)};\lambda\right)$, where $F_E(x;\lambda)$ denotes the cdf of the exponential distribution with parameter λ, and $F_{Lx}^*(x;\eta) = 1 - \left(\frac{\beta}{x+\beta}\right)^\alpha$ for $x \geq -\beta$, and $F_{Lx}^*(x;\eta) = 0$ otherwise. Thus, $F_{Lx}^*(x;\eta)$ can be thought of as a support-extended version of $F_{Lx}(x;\eta)$ over the semi-finite interval $[-\beta,\infty)$. It is worth noting that $F_{Lx}^*(x;\eta)$ is not a cdf anymore, but it is increasing and satisfies $\lim_{x\to-\beta}\frac{1}{1-F_{Lx}^*(x;\eta)} = 0$ and $\lim_{x\to\infty}\frac{1}{1-F_{Lx}^*(x;\eta)} = \infty$, which ensure that $F_{EXL}(x;\xi)$ as a cdf is mathematically correct. It is worth noting that it can be applied to any lifetime distribution in place of the generator exponential distribution.

Based on the extended Lx scheme with the NH distribution as a generator, the cdf and pdf of the NHLx distribution are specified by

$$F_{NHLx}(x;\zeta) = 1 - e^{1-\left(1+b\left(\frac{\beta}{x+\beta}\right)^{-\alpha}\right)^a}$$

and

$$f_{NHLx}(x;\zeta) = \frac{ab\alpha}{\beta}\left(\frac{\beta}{x+\beta}\right)^{-\alpha+1}\left(1+b\left(\frac{\beta}{x+\beta}\right)^{-\alpha}\right)^{a-1} e^{1-\left(1+b\left(\frac{\beta}{x+\beta}\right)^{-\alpha}\right)^a},$$

$$x \geq -\beta \ \zeta = (a,b,\alpha,\beta) > 0,$$

respectively, where a and α are shape parameters, and b and β are scale parameters. We have $F_{NHLx}(x;\zeta) = f_{NHLx}(x;\zeta) = 0$ for $x < -\beta$. Thus, the cdf has been derived from the following formula: $F_{EXL}(x;\xi) = F_{NH}\left(\frac{1}{1-F_{Lx}^*(x;\eta)};\tau\right)$, $x \in \mathbb{R}$. By taking $a = 1$, we remark that $F_{NHLx}(x;\zeta) = F_{EXL}(x;\xi)$; the NHLx distribution is reduced to the EXL distribution with $\lambda = b$. The asymptotic properties of the pdf depend on the values of α mainly; with the use of standard asymptotic techniques, we establish that

$$\lim_{x\to-\beta} f_{NHLx}(x;\zeta) = \begin{cases} \infty & \text{if } \alpha < 1 \\ \frac{ba}{\beta} & \text{if } \alpha = 1 \\ 0 & \text{if } \alpha > 1 \end{cases}, \quad \lim_{x\to\infty} f_{NHLx}(x;\zeta) = 0.$$

Figure 1 completes these asymptotic results by showing some curves of the pdf for several parameter values.

In Figure 1, we see that the pdf can be inverted J decreasing or have uni-modal shapes. It is very flexible to skewness, peakedness, and platness curves at a small value of β (at least), and different selected parameter values of a, b, and α. Such flexibility is not observed for the pdf of the EXL distribution, as visually shown in the figures in [14].

The analysis of the corresponding hrf is now examined. By applying the definition $h_{NHLx}(x;\zeta) = f_{NHLx}(x;\zeta)/[1 - F_{NHLx}(x;\zeta)]$, it is given by

$$h_{NHLx}(x;\zeta) = \frac{ab\alpha}{\beta}\left(\frac{\beta}{x+\beta}\right)^{-\alpha+1}\left(1+b\left(\frac{\beta}{x+\beta}\right)^{-\alpha}\right)^{a-1}, \quad x \geq -\beta,$$

and $h_{NHLx}(x;\zeta) = 0$ for $x < -\beta$. Contrary to the pdf, the asymptotic properties of the hrf mainly depend on the values of a and α; we have

$$\lim_{x \to -\beta} h_{NHLx}(x;\zeta) = \begin{cases} \infty & \text{if } \alpha < 1 \\ \frac{ba}{\beta} & \text{if } \alpha = 1 \\ 0 & \text{if } \alpha > 1 \end{cases}, \quad \lim_{x \to \infty} h_{NHLx}(x;\zeta) = \begin{cases} \infty & \text{if } a\alpha > 1 \\ \frac{b^a}{\beta} & \text{if } a\alpha = 1 \\ 0 & \text{if } a\alpha < 1 \end{cases}.$$

In full generality, the possible shapes of the hrf are determinant for modeling purposes: the more different shapes it has, the more the associated model is applicable to a wide panel of data sets.

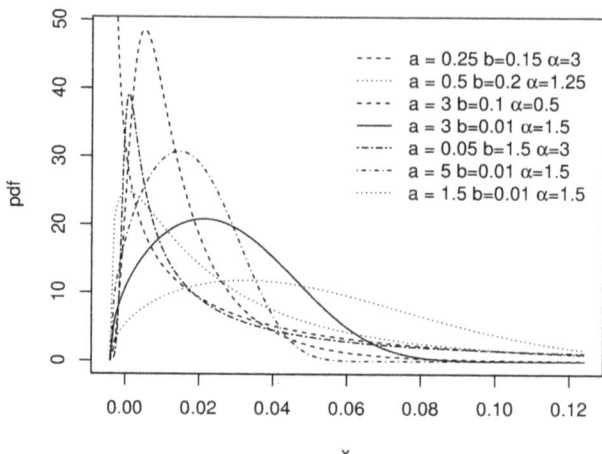

Figure 1. Curves of the pdf of the NHLx distribution for various parameter values, but with the fixed value: $\beta = 0.005$.

Figure 2 presents the identified shapes for the hrf of the NHLx distribution. From Figure 2, we see that the hrf can be increasing, decreasing, or upside-down bathtub-shaped, with flexible convex–concave properties. In particular, these curve modulations are possible thanks to the variation of the new additional parameters a. We are far beyond the curve possibilities of the hrf of the EXL distribution, which is only increasing according to [14]. Thus, from one perspective, the NHLx distribution adds a new shape parameter a to the EXL distribution in a thorough fashion, considerably improving its modeling properties.

The qf of the NHLx distribution is now studied. To begin, it is defined in function of $F_{NHLx}(u;\zeta)$ by $Q_{NHLx}(u;\zeta) = F_{NHLx}^{-1}(u;\zeta)$, $u \in (0,1)$. After some mathematical development, we establish that

$$Q_{NHLx}(u;\zeta) = \beta\left\{\left[\frac{1}{b}\left((1-\log(1-u))^{\frac{1}{a}}-1\right)\right]^{\frac{1}{\alpha}}-1\right\}, \quad u \in (0,1).$$

Based on this qf, the main quartiles of the NHLx distribution can be explicated: by taking $u = 1/4$, $u = 1/2$, and $u = 3/4$ into $Q_{NHLx}(u;\zeta)$, we get the first, second, and third quartiles. In addition, several quantile-based functions, and skewness and kurtosis measures, can be listed and analyzed (see [27]). In addition, various quantile regression models can be constructed (see [28]).

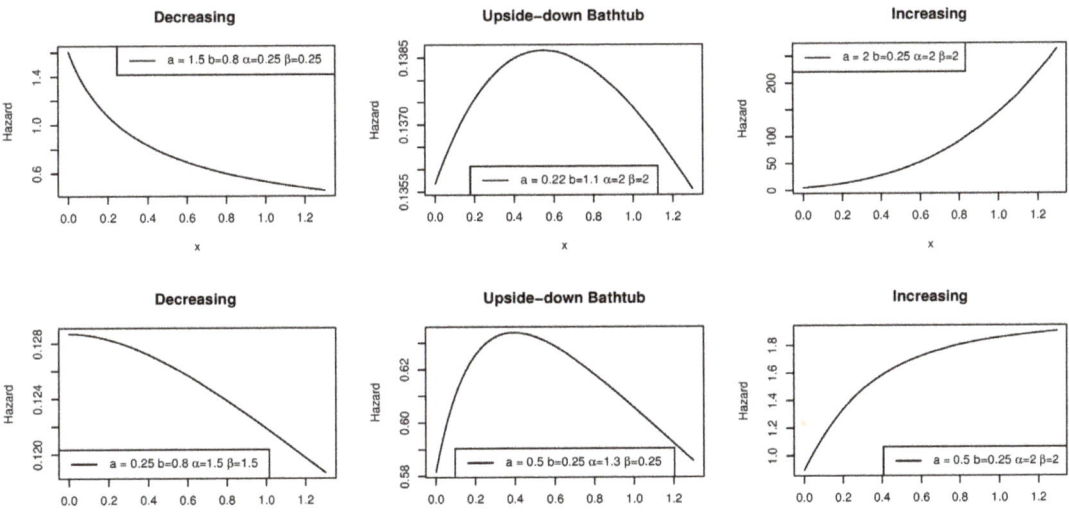

Figure 2. Curves of the hrf of the NHLx distribution for various parameter values.

3. Moment Properties of the NHLx Distribution

The moment properties of the NHLx distribution are now under investigation. First, for a random variable X with the NHLx distribution and any integer r, the rth moment of X is defined by

$$\mu_r^* = E(X^r) = \int_{-\infty}^{\infty} x^r f_{NHLx}(x;\zeta)dx,$$

which can be explicated as

$$\mu_r^* = \int_{-\beta}^{\infty} x^r \frac{ab\alpha}{\beta}\left(\frac{\beta}{x+\beta}\right)^{-\alpha+1}\left(1+b\left(\frac{\beta}{x+\beta}\right)^{-\alpha}\right)^{a-1} e^{1-\left(1+b\left(\frac{\beta}{x+\beta}\right)^{-\alpha}\right)^a} dx.$$

For given distribution parameters, this integral can be computed numerically with the help of scientific software. An analytical expression involving sums is given in the next proposition.

Proposition 1. *Let X be a random variable with the NHLx distribution. Then, its rth moment can be expressed as*

$$\mu_r^* = \beta^r e \sum_{j=0}^{r} \sum_{k=0}^{\infty} \binom{r}{j}\binom{\frac{j}{\alpha}}{k}\frac{(-1)^{r-j+k}}{b^{\frac{j}{\alpha}}}\Gamma\left(\frac{1}{a}\left(\frac{j}{\alpha}-k\right)+1,1\right),$$

where $e = \exp(1)$ and $\Gamma(x,y) = \int_x^{\infty} t^{y-1}e^{-t}dt$ with $y \in \mathbb{R}$ and $x > 0$, which defines the incomplete gamma function.

Proof. Let us apply the following change of variable:

$$u = \left(1+b\left(\frac{\beta}{x+\beta}\right)^{-\alpha}\right)^a, \quad du = \frac{ab\alpha}{\beta}\left(\frac{\beta}{x+\beta}\right)^{-\alpha+1}\left(1+b\left(\frac{\beta}{x+\beta}\right)^{-\alpha}\right)^{a-1}dx,$$

which satisfies $\lim_{x\to-\beta} u = 1$ and $\lim_{x\to\infty} u = \infty$. Then, we have

$$\mu_r^* = \beta^r e \int_1^\infty \left\{ \left[\frac{1}{b}\left(u^{\frac{1}{a}}-1\right)\right]^{\frac{1}{\alpha}} - 1 \right\}^r e^{-u} du.$$

By applying the standard and generalized binomial formulas, we get

$$\mu_r^* = \beta^r e \sum_{j=0}^r \binom{r}{j} \frac{(-1)^{r-j}}{b^{\frac{j}{\alpha}}} \int_1^\infty \left(u^{\frac{1}{a}}-1\right)^{\frac{j}{\alpha}} e^{-u} du$$

$$= \beta^r e \sum_{j=0}^r \binom{r}{j} \frac{(-1)^{r-j}}{b^{\frac{j}{\alpha}}} \sum_{k=0}^\infty \binom{\frac{j}{\alpha}}{k} (-1)^k \int_1^\infty u^{\frac{1}{a}\left(\frac{j}{\alpha}-k\right)} e^{-u} du$$

$$= \beta^r e \sum_{j=0}^r \sum_{k=0}^\infty \binom{r}{j}\binom{\frac{j}{\alpha}}{k} \frac{(-1)^{r-j+k}}{b^{\frac{j}{\alpha}}} \Gamma\left(\frac{1}{a}\left(\frac{j}{\alpha}-k\right)+1,1\right).$$

This ends the proof of Proposition 1. □

Based on Proposition 1, the mean of X can be expanded as

$$\mu_1^* = \beta e \sum_{j=0}^1 \sum_{k=0}^\infty \binom{1}{j}\binom{\frac{j}{\alpha}}{k} \frac{(-1)^{1-j+k}}{b^{\frac{j}{\alpha}}} \Gamma\left(\frac{1}{a}\left(\frac{j}{\alpha}-k\right)+1,1\right)$$

and the moment of order 2 of X can be expressed as

$$\mu_2^* = \beta^2 e \sum_{j=0}^2 \sum_{k=0}^\infty \binom{2}{j}\binom{\frac{j}{\alpha}}{k} \frac{(-1)^{k-j}}{b^{\frac{j}{\alpha}}} \Gamma\left(\frac{1}{a}\left(\frac{j}{\alpha}-k\right)+1,1\right).$$

From the above moments, we derive the variance of X by $V = \mu_2^* - (\mu_1^*)^2$. Several other moment measures can be expressed in a similar manner, including the dispersion index, coefficient of variation, moment skewness, and moment kurtosis. More details on the moment skewness and moment kurtosis will be provided later.

The two following points can be proven by following the lines of the proof of Proposition 1.

- The rth moment of X about the mean can be expressed as

$$\mu_r = E[(X - \mu_1^*)^r]$$

$$= \beta^r e \sum_{j=0}^r \sum_{k=0}^\infty \binom{r}{j}\binom{\frac{j}{\alpha}}{k} \frac{(-1)^{r-j+k}}{b^{\frac{j}{\alpha}}} \left(1+\frac{\mu_1^*}{\beta}\right)^{r-j} \Gamma\left(\frac{1}{a}\left(\frac{j}{\alpha}-k\right)+1,1\right).$$

Based on it, the standard moment skewness measure is defined by $SK = \mu_3/V^{\frac{3}{2}}$, and the standard moment kurtosis measure is defined by $KU = \mu_4/V^2$, among other moment measures.

- The rth unconditional moment of X at a certain $t > 0$ can be expanded as

$$\mu_r(t) = E[X^r \mid X \leq t] = \frac{\beta^r e}{1 - e^{1-\left(1+b\left(\frac{\beta}{t+\beta}\right)^{-\alpha}\right)^a}} \sum_{j=0}^r \sum_{k=0}^\infty \binom{r}{j}\binom{\frac{j}{\alpha}}{k} \frac{(-1)^{r-j+k}}{b^{\frac{j}{\alpha}}} \times$$

$$\left[\Gamma\left(\frac{1}{a}\left(\frac{j}{\alpha}-k\right)+1,1\right) - \Gamma\left(\frac{1}{a}\left(\frac{j}{\alpha}-k\right)+1, \left(1+b\left(\frac{\beta}{t+\beta}\right)^{-\alpha}\right)^a\right)\right].$$

It is immediate that $\lim_{t\to\infty} \mu_r(t) = \mu_r$. The unconditional moments are useful in the expression of various important functions, such as the mean residual life and reversed mean residual life functions. For more information on these functions, see [29].

4. Maximum Likelihood Estimates of the Parameters

We now consder the NHLx distribution as a statistical model, and we assume that the parameters a, b, α, and β are unknown. We aim to give some details on the maximum likelihood estimates (MLEs) of the parameters. First, let n be a positive integer, $X_1, X_2, \ldots X_n$ be independent and identically distributed random variables drawn from the NHLx distribution, and $x_1, x_2, \ldots, x_n$ be corresponding observations. Then, provided that $\inf(x_1, x_2, \ldots, x_n) \geq -\beta$, the likelihood function and log-likelihood functions are defined by

$$L(x_1, x_2, \ldots, x_n; \zeta) = \prod_{i=1}^{n} f_{NHLx}(x_i; \zeta)$$

$$= \left(\frac{ab\alpha}{\beta}\right)^n \prod_{i=1}^{n} \left(\frac{\beta}{x_i + \beta}\right)^{-\alpha+1} \prod_{i=1}^{n} \left(1 + b\left(\frac{\beta}{x_i + \beta}\right)^{-\alpha}\right)^{a-1} e^{n - \sum_{i=1}^{n}\left(1 + b\left(\frac{\beta}{x_i+\beta}\right)^{-\alpha}\right)^a}$$

and

$$\ell(x_1, x_2, \ldots, x_n; \zeta) = \log L(x_1, x_2, \ldots, x_n; \zeta)$$

$$= n \log\left(\frac{ab\alpha}{\beta}\right) + (1-\alpha) \sum_{i=1}^{n} \log\left(\frac{\beta}{x_i + \beta}\right) + (a-1) \sum_{i=1}^{n} \log\left(1 + b\left(\frac{\beta}{x_i + \beta}\right)^{-\alpha}\right)$$

$$+ n - \sum_{i=1}^{n} \left(1 + b\left(\frac{\beta}{x_i + \beta}\right)^{-\alpha}\right)^a,$$

respectively. Then, the MLEs of the parameters a, b, α, and β, say $\hat{a}$, $\hat{b}$, $\hat{\alpha}$, and $\hat{\beta}$, respectively, are defined by

$$\hat{\zeta} = (\hat{a}, \hat{b}, \hat{\alpha}, \hat{\beta}) = \operatorname{argmax}_{\zeta} \ell(x_1, x_2, \ldots, x_n; \zeta).$$

In the case where β is known and we have surely $\inf(x_1, x_2, \ldots, x_n) \geq -\beta$, the MLEs of a, b, and α are the solution of the following equations: $\frac{\partial}{\partial a}\ell(x_1, x_2, \ldots, x_n; \zeta) = 0$, $\frac{\partial}{\partial b}\ell(x_1, x_2, \ldots, x_n; \zeta) = 0$ and $\frac{\partial}{\partial \alpha}\ell(x_1, x_2, \ldots, x_n; \zeta) = 0$, where

$$\frac{\partial}{\partial a}\ell(x_1, x_2, \ldots, x_n; \zeta) = \frac{n}{a} + \sum_{i=1}^{n} \log\left(1 + b\left(\frac{\beta}{x_i + \beta}\right)^{-\alpha}\right)$$

$$- \sum_{i=1}^{n} \left(1 + b\left(\frac{\beta}{x_i + \beta}\right)^{-\alpha}\right)^a \log\left(1 + b\left(\frac{\beta}{x_i + \beta}\right)^{-\alpha}\right),$$

$$\frac{\partial}{\partial b}\ell(x_1, x_2, \ldots, x_n; \zeta) = \frac{n}{b} + (a-1) \sum_{i=0}^{n} \left(1 + b\left(\frac{\beta}{x_i + \beta}\right)^{-\alpha}\right)^{-1} \left(\frac{\beta}{x_i + \beta}\right)^{-\alpha}$$

$$- a \sum_{i=1}^{n} \left(1 + b\left(\frac{\beta}{x_i + \beta}\right)^{-\alpha}\right)^{a-1} \left(\frac{\beta}{x_i + \beta}\right)^{-\alpha}$$

and

$$\frac{\partial}{\partial \alpha}\ell(x_1, x_2, \ldots, x_n; \zeta) = \frac{n}{\alpha} - \sum_{i=1}^{n} \log\left(\frac{\beta}{x_i + \beta}\right)$$

$$+ (1-a)b \sum_{i=1}^{n} \left(1 + b\left(\frac{\beta}{x_i + \beta}\right)^{-\alpha}\right)^{-1} \left(\frac{\beta}{x_i + \beta}\right)^{-\alpha} \log\left(\frac{\beta}{x_i + \beta}\right)$$

$$+ ab \sum_{i=1}^{n} \left(1 + b\left(\frac{\beta}{x_i + \beta}\right)^{-\alpha}\right)^{a-1} \left(\frac{\beta}{x_i + \beta}\right)^{-\alpha} \log\left(\frac{\beta}{x_i + \beta}\right).$$

The above expressions do not have closed-form solutions; hence, they are to be solved numerically by iterative methods. These numerical values can be easily obtained using specific tools in statistical software such as the R software, and the MLE of β is obtained by taking its first-order statistics, as in [14]. It is also possible to determine the values of the standard errors (SEs) of the MLEs. For more information, see [30].

Based on the MLEs, we define the estimated pdf of the NHLx distribution by $f_{NHLx}(x;\hat{\zeta})$. Conceptually, the curve of this estimated function must be close to the shape of the histogram of the data, among other visual criteria.

5. Simulation Study

In this section, we perform 1000 Monte Carlo simulation studies for three different sets of parameters and each of the sample sizes of $n \in \{50, 100, 250, 500, 750, 1000\}$. By considering the order (a, b, α, β), these sets of parameters are Set I = $(0.5, 1.5, 5, 0.5)$, Set II = $(0.5, 1.5, 4, 0.75)$, and Set III = $(1.5, 0.5, 4, 0.5)$. Table 1 shows the mean MLEs (MMLEs), biases and mean squared errors (MSEs) of the studies.

From Table 1, it can be observed that as the sample size increases, the biases and MSEs of the MLEs decrease, and with the increase in the sample sizes, the MMLEs are closer to the true parameter values. These results prove the accuracy of the considered parameter strategy estimation.

Table 1. Simulation results related to the MLEs of the NHLx model parameters.

n	$\hat{a}$			$\hat{b}$			$\hat{\alpha}$			$\hat{\beta}$		
	MMLE	Bias	MSE	MMLE	Bias	MSE	MMLE	Bias	MSE	MMLE	Bias	MSE
						Set I						
50	2.4691	1.9691	13.3180	2.9450	1.4450	54.9364	9.2113	4.2113	310.9836	0.7425	0.2425	1.2292
100	1.9081	1.4081	3.6890	2.4782	0.9782	40.3076	6.9163	1.9163	128.6739	0.5886	0.0886	0.4186
250	1.5403	1.0403	2.0234	1.8814	0.3814	7.3395	6.0152	1.0152	58.5405	0.5465	0.0465	0.1517
500	1.2723	0.7723	0.9669	1.6510	0.1510	2.3173	5.3303	0.3303	11.0218	0.5143	0.0143	0.0322
750	1.1679	0.6679	0.6544	1.5840	0.0840	0.8996	5.1302	0.1302	2.9589	0.5053	0.0053	0.0090
1000	1.1355	0.6355	0.5534	1.5507	0.0507	0.6974	5.1039	0.1039	2.2337	0.5046	0.0046	0.0067
						Set II						
50	2.4179	1.9179	6.9777	3.2423	1.7423	170.7508	6.0725	2.0725	268.0909	0.8891	0.1391	1.7534
100	2.0139	1.5139	4.3307	2.3196	0.8196	34.8643	4.3855	0.3855	28.9119	0.7571	0.0071	0.1972
250	1.5278	1.0278	1.8965	1.7518	0.2518	6.5711	4.1402	0.1402	6.6622	0.7548	0.0048	0.0465
500	1.2650	0.7650	0.9185	1.6067	0.1067	1.7551	4.0561	0.0561	2.5538	0.7509	0.0009	0.0198
750	1.1526	0.6526	0.5826	1.5615	0.0615	1.0451	3.9887	−0.0113	1.1312	0.7459	−0.0041	0.0083
1000	1.1297	0.6297	0.5391	1.5262	0.0262	0.5502	3.9882	−0.0118	0.6922	0.7470	−0.0030	0.0054
						Set III						
50	2.6338	1.1338	6.5559	0.6993	0.1993	18.5802	5.2280	1.2280	246.0191	0.5697	0.0697	1.6784
100	2.0040	0.5040	2.9493	0.5758	0.0758	1.3542	4.7413	0.7413	46.4162	0.5341	0.0341	0.1676
250	1.5413	0.0413	1.1911	0.4969	−0.0031	0.1653	4.1298	0.1298	8.1507	0.5022	0.0022	0.0432
500	1.2694	−0.2306	0.4318	0.4946	−0.0054	0.0944	4.0379	0.0379	2.6806	0.4975	−0.0025	0.0139
750	1.1916	−0.3084	0.2963	0.4779	−0.0221	0.0413	3.9693	−0.0307	0.9247	0.4957	−0.0043	0.0055
1000	1.1182	−0.3818	0.2738	0.4934	−0.0066	0.0286	3.9998	−0.0002	0.6297	0.4989	−0.0011	0.0038

6. Applications of the NHLx Model

6.1. Heavy-Tailed Data Applications

Two real data sets taken from [31], namely the theft and claim data, are considered to illustrate the proposed methodology. These data sets are known to have heavy tail features. Table 2 presents the estimation of the tails of several standard distributions, namely the lognormal, Weibull, gamma, and exponential distributions, and the proposed NHLx distribution, taken at several values. The survival function, denoted by $S(x)$ for all distributions in full generality, determines the tail probabilities at the point x.

Table 2. Estimation of the tail probabilities of various distributions for the considered data sets.

Models (Theft Data)	S(8000)	S(10,000)	S(20,000)
NHLx	0.12090	0.10840	0.07531
lognormal	0.05972	0.04418	0.01536
Weibull	0.03970	0.02270	0.00200
gamma	0.03755	0.01897	0.00070
exponential	1.9067×10^{-2}	7.0851×10^{-3}	5.0197×10^{-5}
(Claim data)	S(330)	S(430)	S(530)
NHLx	0.94477	0.86505	0.73682
Weibull	0.370878	0.19394	0.08713
gamma	0.34625	0.19193	0.10045
lognormal	0.32961	0.20544	0.13060
exponential	0.32557	0.23171	0.16491

It is obvious from Table 2 that the NHLx model has a better fit in both data sets, and its corresponding tail probabilities are also fairly high. This means that the proposed distribution is also a heavy-tailed distribution, which was compared to other heavy-tailed distributions and contains more mass at the tail ends than the other distributions considered for comparison.

The rest of the study is devoted to the in-depth analysis of two famous data sets in the literature, highlighting the efficiency of the estimated NHLx model under real-life scenarios.

6.2. Practical Applications

The first data set contains 65 successive eruptions of the waiting times (in seconds) of the Kiama Blowhole data. It was studied in [32,33]. The second data set is about intensive care unit (ICU) patients for varying time periods of 37 patients. It was analyzed in [34] and, more recently, in [35].

The descriptive measures such as mean, median, skewness, and kurtosis have been computed for both the eruption data and ICU data sets. The results are presented in Table 3.

Table 3. Descriptive measures for the two data sets.

Data Sets	Mean	Median	Skewness	Kurtosis
Eruption data	46.5486	29.3811	47.9415	43.2141
ICU data	19.9494	12.5919	47.9426	43.2148

From the measures of skewness and kurtosis, it is clear that the data are highly skewed and heavy-tailed. Furthermore, the mean value is larger than the median.

For comparison purposes, we consider some of the most accurate extended Lx models: the WL, EXL, and Lx models.

The MLEs and the corresponding SEs of these models are listed in Table 4.

Table 4. MLEs with SEs in parentheses of the considered models for the two data sets.

Data Sets	Models	a	b	α	β
Eruption data	NHLx	0.1052 (0.0246)	0.0095 (0.0080)	6.0546 (1.2023)	7 (-)
	WL	1.9842 (3.8721)	2.9883 (2.8437)	0.1915 (0.1422)	2.0231 (8.2844)
	EXL	-	1.5369 (0.1418)	0.0452 (0.0016)	7 (-)
	Lx	-	-	1.8007 (0.4426)	46.7964 (13.9539)
ICU data	NHLx	0.0886 (0.0443)	0.0132 (0.0381)	8.0704 (4.2954)	3 (-)
	WL	4.3066 (5.4222)	1.6119 (0.4737)	0.2584 (0.1369)	3.7699 (4.7682)
	EXL	-	1.4432 (0.1743)	0.0951 (0.0366)	3 (-)
	Lx	-	-	2.7744 (1.2094)	21.0499 (10.8889)

The measures of goodness of fit are used to verify whether a data set is distributionally compatible with a given model. To judge the accuracy of a model, we use the Cramér–von Mises (W^*), Anderson–Darling (A^*), and Kolmogorov–Smirnov (K-S) statistics (D), along with the K-S p-Value related to D. Adequacy measures are widely used to determine which model is best. Here, we traditionally consider the Akaike information criterion (AIC), consistent AIC (CAIC), Bayesian information criterion (BIC), and Hannan–Quinn information criterion (HQIC), which are based on the MLEs of the models. The model with the minimum W^*, A^*, D, AIC, CAIC, BIC, and HQIC value and maximum p-Value is chosen as the best one that fits the data. We may refer to [36] for the precise definitions of these measures. Their values for the considered models and the two data sets are collected in Table 5.

From Table 5, it is witnessed that the two data sets have a better fit for the proposed NHLx model than the other three models.

Table 5. Values of the statistical measures for the considered models.

Models	W*	A*	D	p-Value	AIC	CAIC	BIC	HQIC
Eruption data								
NHLx	0.0998	0.7471	0.0761	0.8520	592.9289	593.3289	599.4056	595.4804
WL	0.1119	0.8036	0.1062	0.4656	597.1462	597.8242	605.7818	600.5482
EXL	0.2213	1.4388	0.1230	0.2873	607.4000	607.5968	611.7178	609.101
Lx	0.1182	0.8570	0.2317	0.0021	619.4907	619.6874	623.8085	621.1917
ICU data								
NHLx	0.2666	1.7021	0.2047	0.0905	242.5222	243.2495	247.355	244.226
WL	0.3459	2.1002	0.2071	0.0838	252.6803	253.9303	259.124	254.952
EXL	0.547	3.044	0.2266	0.0448	264.4479	264.8008	267.6697	265.5837
Lx	0.3404	2.0824	0.3091	0.0017	256.5615	256.9144	259.7833	257.6973

The histogram plots and estimated pdfs of the considered models are reported in Figure 3.

From Figure 3, we see that both histograms exhibit the skewed nature of the two data sets, and the estimated pdf curves depict that the NHLx model is observed to have a better pattern of closeness to the histogram plot when compared to the other three models.

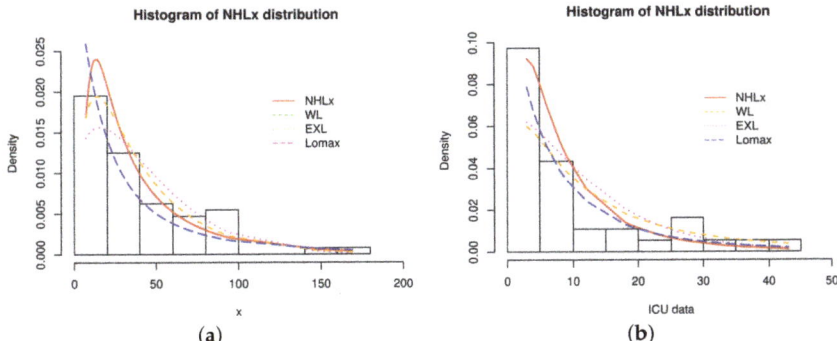

Figure 3. Curves of the estimated pdfs of the considered models for the two data sets. (**a**) Eruption data. (**b**) ICU data.

7. Conclusions

In this paper, we propose a new four-parameter Lomax distribution called the Nadarajah–Haghighi Lomax distribution. It aims to provide a new lower-bounded distribution that combines the functionalities of the Nadarajah–Haghighi and Lomax distributions, and extends the modeling scope of the so-called exponential Lomax distribution. We have derived various properties, including the expression of the probability density, hazard and quantile functions, and diverse kinds of moments. The maximum likelihood method is used for estimating the model parameters. Simulation studies show its effectiveness by considering different sets of parameters. Furthermore, the support of two real data sets is taken to illustrate the applications of the Nadarajah–Haghighi Lomax distribution and it is compared with other Lomax-based distributions. From the obtained results, it is very easy to understand that the Nadarajah–Haghighi Lomax distribution has a better fit than the other Lomax models. The perspectives of new work based on the Nadarajah–Haghighi Lomax distribution are numerous, including:

- the development of various extensions, such as parametric-functional, multivariate, and discrete versions;
- the creation of new families of distributions;
- the construction of diverse regression models;
- by viewing the related cdf as a sigmoidal function, one can think of studying the "confidential intervals" (or "confidential bounds") and "supersaturation" to the horizontal asymptote (at the median level) in the Hausdorff sense (see [37]). These two characteristics are important for researchers in choosing an appropriate model for approximating specific data from very different branches of scientific knowledge, such as computer virus propagation (see [38]).

Author Contributions: Conceptualization, V.B.V.N., R.V.V. and C.C.; methodology, V.B.V.N., R.V.V. and C.C.; software, V.B.V.N., R.V.V. and C.C.; validation, V.B.V.N., R.V.V. and C.C.; formal analysis, V.B.V.N., R.V.V. and C.C.; investigation, V.B.V.N., R.V.V. and C.C.; data curation, V.B.V.N., R.V.V. and C.C.; writing—original draft preparation, V.B.V.N., R.V.V. and C.C.; writing—review and editing, V.B.V.N., R.V.V. and C.C.; visualization, V.B.V.N., R.V.V. and C.C. All authors have read and agreed to the published version of the manuscript.

Funding: This research received no external funding.

Acknowledgments: We thank the three reviewers and the associate editor for their in-depth comments on the first version of the article.

Conflicts of Interest: The authors declare no conflict of interest.

References

1. Pareto, V. *Cours d'Économie Politique*; Rouge: Lausanne, Switzerland, 1897; Volume II.
2. Lomax, K. Business failures: Another example of the analysis of failure data. *J. Am. Stat. Assoc.* **1954**, *49*, 847–852. [CrossRef]
3. Harris, C.M. The Pareto distribution as a queue service discipline. *Oper. Res.* **1968**, *16*, 307–313. [CrossRef]
4. Atkinson, A.B.; Harrison, A.J. *Distribution of Personal Wealth in Britain*; Cambridge University Press: Cambridge, UK, 1978.
5. Bryson, M.C. Heavy-tailed distributions: Properties and tests. *Technometrics* **1974**, *16*, 61–68. [CrossRef]
6. Lingappaiah, G. Bayes prediction in exponential life-testing when sample size is a random variable. *IEEE Trans. Reliab.* **1986**, *35*, 106–110. [CrossRef]
7. Ahsanullah, M. Record values of the Lomax distribution. *Stat. Neerl.* **1991**, *45*, 21–29. [CrossRef]
8. Balakrishnan, N.; Ahsanullah, M. Relations for single and product moments of record values from Lomax distribution. *Sankhya Indian J. Stat. Ser. B* **1994**, *56*, 140–146.
9. Marshall, A.W.; Olkin, I. A new method for adding a parameter to a family of distributions with application to the exponential and Weibull families. *Biometrika* **1997**, *84*, 641–652. [CrossRef]
10. Chahkandi, M.; Ganjali, M. On some lifetime distributions with decreasing failure rate. *Comput. Stat. Data Anal.* **2009**, *53*, 4433–4440. [CrossRef]
11. Abdul-Moniem, I.B. Recurrence relations for moments of lower generalized order statistics from Exponentiated Lomax Distribution and its characterization. *J. Math. Comput. Sci.* **2012**, *2*, 999–1011.
12. Rajab, M.; Aleem, M.; Nawaz, T.; Daniyal, M. On Five Parameter Beta Lomax Distribution. *J. Stat.* **2013**, *20*, 102–118.
13. Al-Jarallah, R.A.; Ghitany, M.E.; Gupta, R.C. A proportional hazard Marshall–Olkin extended family of distributions and its application to Gompertz distribution. *Commun. Stat. Theory Methods* **2014**, *43*, 4428–4443. [CrossRef]
14. El-Bassiouny, A.H.; Abdo, N.F.; Shahen, H.S. Exponential Lomax Distribution. *Int. J. Comput. Appl.* **2015**, *121*, 24–29.
15. Cordeiro, G.M.; Ortega, E.M.; Popovic, B.V. The Gamma–Lomax Distribution. *J. Stat. Comput. Simul.* **2015**, *85*, 305–319. [CrossRef]
16. Tahir, M.H.; Cordeiro, G.M.; Mansoor, M.; Zubair, M. The Weibull–Lomax distribution: Properties and applications. *Hacet. J. Math. Stat.* **2015**, *44*, 455–474. [CrossRef]
17. Mead, M.E. On five-parameter Lomax distribution: Properties and applications. *Pak. J. Stat. Oper. Res.* **2016**, *12*, 185–200.
18. Rady, E.H.A.; Hassanein, W.A.; Elhaddad, T.A. The power Lomax distribution with an application to bladder cancer data. *SpringerPlus* **2016**, *5*, 1–22. [CrossRef]
19. Hassan, A.S.; Abd-Allah, M. Exponentiated Weibull–Lomax distribution: Properties and estimation. *J. Data Sci.* **2018**, *16*, 277–298. [CrossRef]
20. Nagarjuna, B.V.; Vishnu Vardhan, R. Marshall–Olkin exponential Lomax distribution: Properties and its application. *Stoch. Model. Appl.* **2020**, *24*, 161–177.
21. Al-Marzouki, S.; Jamal, F.; Chesneau, C.; Elgarhy, M. Type II Topp–Leone power Lomax distribution with applications. *Mathematics* **2020**, *8*, 4. [CrossRef]
22. Mathew, J.; Chesneau, C. Some new contributions on the Marshall–Olkin length biased Lomax distribution: Theory, modelling and data analysis. *Math. Comput. Appl.* **2020**, *25*, 79. [CrossRef]
23. Nagarjuna, B.V.; Vishnu Vardhan, R.; Chesneau, C. Kumaraswamy Generalized Power Lomax Distribution and Its Applications. *Stats* **2021**, *4*, 28–45. [CrossRef]
24. Nagarjuna, B.V.; Vishnu Vardhan, R.; Chesneau, C. On the Accuracy of the Sine Power Lomax Model for Data Fitting. *Modelling* **2021**, *2*, 78–104. [CrossRef]
25. Murthy, D.N.P.; Xie, M.; Jiang, R. *Weibull Models*; John Wiley & Sons: New York, NY, USA, 2004.
26. Nadarajah, S.; Haghighi, F. An extension of the exponential distribution. *Statistics* **2011**, *45*, 543–558. [CrossRef]
27. Gilchrist, W.G. *Statistical Modelling with Quantile Functions*; Chapman & Hall/CRC: London, UK, 2000.
28. Koenker, R. *Quantile Regression*; Cambridge University Press: Cambridge, UK, 2005.
29. Cordeiro, G.M.; Silva, R.B.; Nascimento, A.D.C. *Recent Advances in Lifetime and Reliability Models*; Bentham Books: Sharjah, United Arab Emirates, 2020. [CrossRef]
30. Casella, G.; Berger, R.L. *Statistical Inference*; Brooks/Cole Publishing Company: Pacific Grove, CA, USA, 1990.
31. Boland, P.J. *Statistical and Probabilistic Methods in Actuarial Science*; CRC Press: Boca Raton, FL, USA, 2007.
32. da Silva, R.V.; de Andrade, T.A.; Maciel, D.B.; Campos, R.P.; Cordeiro, G.M. A New Lifetime Model: The Gamma Extended Fréchet Distribution. *J. Stat. Theory Appl.* **2013**, *12*, 39–54. [CrossRef]
33. Pinho, L.G.B.; Cordeiro, G.M.; Nobre, J.S. The Harris extended exponential distribution. *Commun. Stat. Theory Methods* **2015**, *44*, 3486–3502. [CrossRef]
34. Kang, I.; Hudson, I.; Rudge, A.; Chase, J.G. Density estimation and wavelet thresholding via Bayesian methods: A wavelet probability band and related metrics approach to assess agitation and sedation in ICU patients. In *Discrete Wavelet Transforms—A Compendium of New Approaches and Recent Applications*; IntechOpen: London, UK, 2013.
35. Khan, M.S.; King, R.; Hudson, I.L. Transmuted generalized exponential distribution: A generalization of the exponential distribution with applications to survival data. *Commun. Stat. Simul. Comput.* **2017**, *46*, 4377–4398. [CrossRef]
36. Chen, G.; Balakrishnan, N. A general purpose approximate goodness-of-fit test. *J. Qual. Technol.* **1995**, *27*, 154–161. [CrossRef]

37. Sendov, B. *Hausdorff Approximations*; Kluwer: Boston, MA, USA, 1990.
38. Iliev, A.; Kyurkchiev, N.; Rahnev, A.; Terzieva, T. *Some Models in the Theory of Computer Viruses Propagation*; LAP LAMBERT Academic Publishing: Saarbrucken, Germany, 2019.

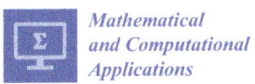

Article

On the Prediction of Evaporation in Arid Climate Using Machine Learning Model

Mansura Jasmine, Abdolmajid Mohammadian * and Hossein Bonakdari

Department of Civil Engineering, University of Ottawa, 161 Louis Pasteur Private, Ottawa, ON K1N 6N5, Canada; mjasm083@uottawa.ca (M.J.); hossein.bonakdari@uottawa.ca (H.B.)
* Correspondence: amohamma@uottawa.ca

Abstract: Evaporation calculations are important for the proper management of hydrological resources, such as reservoirs, lakes, and rivers. Data-driven approaches, such as adaptive neuro fuzzy inference, are getting popular in many hydrological fields. This paper investigates the effective implementation of artificial intelligence on the prediction of evaporation for agricultural area. In particular, it presents the adaptive neuro fuzzy inference system (ANFIS) and hybridization of ANFIS with three optimizers, which include the genetic algorithm (GA), firefly algorithm (FFA), and particle swarm optimizer (PSO). Six different measured weather variables are taken for the proposed modelling approach, including the maximum, minimum, and average air temperature, sunshine hours, wind speed, and relative humidity of a given location. Models are separately calibrated with a total of 86 data points over an eight-year period, from 2010 to 2017, at the specified station, located in Arizona, United States of America. Farming lands and humid climates are the reason for choosing this location. Ten statistical indices are calculated to find the best fit model. Comparisons shows that ANFIS and ANFIS–PSO are slightly better than ANFIS–FFA and ANFIS–GA. Though the hybrid ANFIS–PSO (R^2 = 0.99, VAF = 98.85, RMSE = 9.73, SI = 0.05) is very close to the ANFIS (R^2 = 0.99, VAF = 99.04, RMSE = 8.92, SI = 0.05) model, preference can be given to ANFIS, due to its simplicity and easy operation.

Keywords: evaporation; adaptive neuro fuzzy system; firefly algorithm; particle swarm optimization; genetic algorithm; statistical indices

Citation: Jasmine, M.; Mohammadian, A.; Bonakdari, H. On the Prediction of Evaporation in Arid Climate Using Machine Learning Model. *Math. Comput. Appl.* **2022**, *27*, 32. https://doi.org/10.3390/mca27020032

Academic Editor: Nicholas Fantuzzi

Received: 27 March 2022
Accepted: 31 March 2022
Published: 5 April 2022

Publisher's Note: MDPI stays neutral with regard to jurisdictional claims in published maps and institutional affiliations.

Copyright: © 2022 by the authors. Licensee MDPI, Basel, Switzerland. This article is an open access article distributed under the terms and conditions of the Creative Commons Attribution (CC BY) license (https://creativecommons.org/licenses/by/4.0/).

1. Introduction

Currently, water deficiency is increasing and becoming a challenge for human society. It is increasingly becoming the most important environmental limitation, which is limiting plant growth. According to the statistics, over 30 arid and semi-arid countries are expected to experience water deficiency in 2025 [1]. This will limit agricultural development, threaten food supplies, and inflame rural poverty. Evaporation estimations are essential for controlling and modelling the integrated hydrological resources connected to hydrology, agricultural business, arboriculture, irrigation, flooding, and lake ecosystems. Evaporation is described as the reduction of deposited water due to the conversion of liquid phase to steam phase, which is influenced by the climate situation, such as weather, wind velocities, relative humidity, and sunshine. According to the World Meteorological Organization (WMO), more than half of the total inflow (rainfall or any other sources) to Lake Victoria in the U.S. is lost due to evaporation, which results in relatively humid conditions [1].

The evaluation of evaporation from reservoirs in arid and semi-arid areas is also important. For example, Libya has built one of the largest civil engineering groundwater pumping and transferring systems to overcome water limitations and climate hindrance (high temperature and low rainfall). This project is known as the Manmade River Project (MRP) [1]. The purpose of this project was to supply the water demand of Libya by pumping underground water underneath the Sahara Desert and transfer it using a network of huge underground pipes, especially for irrigation. The high cost of water pumping and lack of appropriate planning are the main concerns. In Egypt's Lake Nasser (located in an arid area), where the Nile's water is stored, downstream water loss, due to evaporation, is estimated to be 3 m in depth, or double that of Lake Victoria [1]. In Australia, it is calculated that around 95% of the precipitation evaporates and has no contribution to runoff [1].

Artificial intelligence models are becoming increasingly popular for forecasting data, instead of traditional models [2]. ANFIS model is one of them, which is also called a data-driven model [3,4], and can be used for different measurements, such as rainfall, streamflow, evaporation, water quality, and many others. A comparison has been made by Moghaddamnia et al. [5] on evaporation evaluation using an artificial neural network (ANN) and adaptive neuro fuzzy inference system (ANFIS). The ANFIS model was compared with the regression-based method by Dogana et al. [2], and ANFIS was declared to be the finest. A group of researchers [6] has published their work on ANN, LS-SVR, fuzzy logic, and ANFIS on daily pan evaporation, with the conclusion of fuzzy logic as being the best performer.

More recent research on evaporation is also conducted by AI methods. A new artificial technique, support vector regression (SVR), and a few nature-inspired algorithms (whale optimization algorithm, particle swarm optimization, and salp swarm algorithm) were investigated by a bunch of researchers in 2021 [7]. A unique contribution to evaporation estimation, based on maximum air temperature, was published earlier this year by scientists [8]. They became successful in the application of deep learning-based model to predict evaporation. However, a group of scholars found effective results of the application of the multiple learning artificial intelligence model in 2020 [9]. They analyzed multiple model-artificial neural networks (MM-ANN), multivariate adaptive regression spline (MARS), support vector machine (SVM), multi-gene genetic programming (MGGP), and 'M5Tree' to simulate the evaporation on a monthly scale basis (EP_m) at two stations in India. Artificial neural network (MM-ANN) and multi-gene genetic programming (MGGP) posed the best results.

Some analysis was performed based on four climate variables, whereas some depended only on maximum temperature. Additionally, different researchers worked on different models for different locations. This is the first time ANFIS model, and few optimizers were adopted for this data set of Arizona, United States, along with six weather variables inputs. In this study, adaptive neuro-fuzzy inference system (ANFIS), ANFIS with firefly algorithm [10,11], ANFIS with genetic algorithm [5], and PSO [12] were analyzed and compared, for the first time, in order to investigate the best modeling approach for evaporation. The main objectives of this study are:

1. To evaluate the performance of all four models, using the climate information of Arizona, United States, and compare the results by using statistical analysis.
2. To explore the ability of the ANFIS model to improve the accuracy of daily evaporation estimation for the data set.
3. To obtain the best model, in terms of accuracy and efficiency, for the arid environments in the United States.

2. Methodology

2.1. Adaptive Neuro Fuzzy Inference System (ANFIS)

The ANFIS model is a mixture of fuzzy inference system (FIS) and artificial neural network (ANN). The fuzzy inference system (FIS) is a very successful and popular model, based on fuzzy logic, which was first proposed by Chang in his study [13]. For the modeling of reservoir performance and problems regarding data uncertainty, fuzzy logic is a highly recommended system [13]. This model is adopted mainly due to its good capacity of extraction of data from input to fuzzy values, in a range of 0 to 1. The ANN model was combined to overcome the limitation of the FIS model. ANN is adopted due to its ability to arrange input and output in pairs and make the structure ready to calibrate. ANN also has the following characteristics:

(a) Can identify the relation between input and output without direct physical consideration.
(b) It can work even when the training sets carry noise and/or measurement errors.
(c) It can adapt situations in changing environments. Therefore, an adaptive neuro-fuzzy inference system (ANFIS) is preferred to maximize the benefit from the combination of both FIS and ANN model in one structure. ANFIS can be well-understood by the following diagram, as shown in Figure 1.

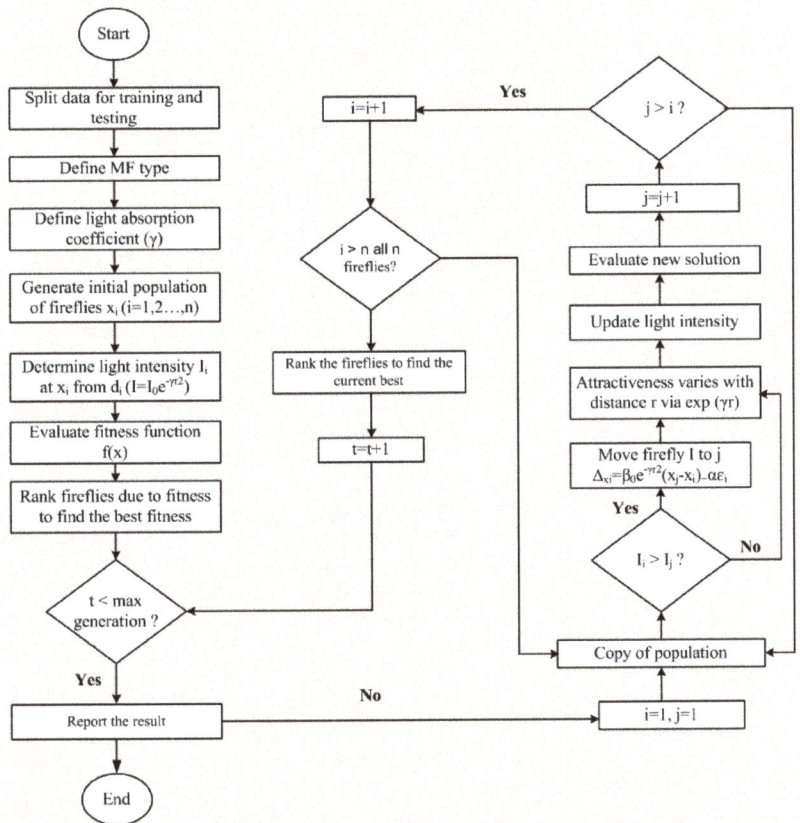

Figure 1. Flow chart of the ANFIS model.

2.2. Firefly Algorithm (FFA)

The mechanism of FFA is based on the nature of the firefly (flashing behavior). This algorithm is applied during the training phase to select the best set of data. This model depends on three basic principles:

1. Each firefly can engage another firefly.
2. The attractiveness between two fireflies is calculated by the light intensity of each firefly.
3. The brightness is correspondingly related to the light released by fireflies [14].

Thus, the objective function of the FFA model is introduced by the intensity of the light produced by, and brightness of, the firefly. The following equations present the intensity (I) and attractiveness, respectively [14,15]:

$$I = I_0 e^{-\gamma r^2} \quad (1)$$

$$w(r) = w_0 e^{-\gamma r^2} \tag{2}$$

where r is the distance between fireflies, I_0 is light intensity, w_0 is attractiveness at $r = 0$ distance, and γ is the light absorption coefficient. β and α are the attraction and movement co-efficient. α, β, and γ are required to be adjusted by trial and error, in order to integrate the ANFIS model with the FFA [14].

2.3. Genetic Algorithm (ANFIS–GA)

This model is highly useful for evapotranspiration calculations. The genetic algorithm (GA) is based on the characters of natural genetics and its selection system. GA includes three major stages: (1) population initialization, (2) GA operators, and (3) evaluation [11]. This system can solve large space problems efficiently and optimize complicated functions. Any hybrid model (hybrid ANFIS) can optimize the MF by using GA. This fuzzy-genetic algorithm has a potential to minimize model errors [11]. The development begins from the population of random chromosomes, thus generating form. In each generation, the fitness of the whole population is estimated. Then, based on the fitness, multiple chromosomes are stochastically adopted from the current population and adjusted by utilizing genetic operators, such as crossover and mutation, to create a new population. The current population is applied in the following iteration of the algorithm [10].

2.4. Particle Swarm Optimization (PSO)

The PSO technique was invented by Kennedy and Eberhart [16], based on the characteristics of bird and fish swarms in a multi-dimensional area, for example, looking for food and running away from hazards [16]. Every element in this algorithm is identified as a "particle"; particles create the density (population), and each density is identified as a "swarm". Every particle is considered a candidate for the answer to the question in this algorithm. The swarm and particle values of this technique depend on the chromosome and density (population) items, which are similar to the genetic algorithm [12]. PSO is a trial-and-error solution procedure that explores the characteristics of swarm particles in a multi-dimensional exploration zone. The computation process is different in the case of large data sets because of the higher expenses of developing a significant number of models. PSO can optimize with a large possibility and high meeting (convergence) rate. This optimizer works through the following mathematical expression [12].

$$V_d^{Max} = \left(x_d^{Max} - x_d^{Min}\right)/2 \tag{3}$$

$$V_d^{Min} = -V_d^{Max} \tag{4}$$

The values of x_d^{Max} and x_d^{Min} are selected according to the limit of the variables. The starting position and velocities of the individuals are irregularly calculated, based on the following equations:

$$x_{prd}^k = x_d^{Min} + r\left(x_d^{Max} - x_d^{Min}\right) \tag{5}$$

$$v_{prd}^k = V_d^{Max}(2r - 1) \tag{6}$$

where p, d, v, x, and r denote particle number, exploration direction, particle velocity, position of particle, and irregularly created number close to unvaried distribution with the limit (0, 1), respectively. Each particle upgrades its own position, until the position and velocity values face the stopping condition, based on the earlier steps and position of the finest particle in the entire swarm.

$$v_{prd}^{k+1} = \omega v_{prd}^k + c_1 r_1 \left(x_{prd}^{ind} - x_{prd}^k\right) + c_2 r_2 \left(x_d^{glo} - x_{prd}^k\right) \tag{7}$$

$$x_{prd}^{k+1} = x_{prd}^k + v_{prd}^{k+1} \tag{8}$$

where k indicates the number of repetitions needed for the trial-and-error process. ω, c_1, and c_2 are explore variables; r_1 and r_2 are two irregular numbers with an unvaried distribution with the limit (0, 1). x_{prd}^{ind} is the finest location defined by a particle, while x_d^{glo} is the finest location defined by the entire swarm. Variables c_1 and c_2 are the cognition and social variables, respectively [16]. Kennedy and Eberhart introduced ω as a coefficient, which is 1 in the PSO algorithm.

$$\omega = \omega_{max} - (\omega_{max} - \omega_{min})\frac{k}{k_{max}} \quad (9)$$

k_{max} and k are the highest and current number of repetitions for the trial-and-error process, respectively. Regeneration is chosen with the utilization of linear fitness scaling (LFS) to increase diversity of the iteration process.

$$f_{best} - f_{worst} <_{div} \quad (10)$$

where f_{best} and f_{worst} represent the finest and the least objective functions in the entire swarm and $_{div}$ is the expression for diversity. The following equation presents the objective function.

$$MinF(x) = \frac{1}{N}\sum_{i=1}^{N}\left(I_i^{obs} - I_i^{est}\right)^2 \quad (11)$$

where I_i^{obs} is observed, and I_i^{est} is the estimated evaporation intensity; N is the observation number. This optimization process with the particle swarm technique extends up to a required concluding situation. In this analysis, the aim of the PSO algorithm is to minimize the objective function. The levels of computation of this process, using PSO, can be found in reference [12].

3. Results and Discussion

3.1. Data Description

Arizona is the sixth biggest state of USA, which is situated next to the state of California. The area of this state is 113,000 square miles and partly surrounded by the Pacific Ocean. The weather conditions in Arizona are quite caustic, with tropical summers and muggy winters. Phoenix is the capital of Arizona state, located in the Northeastern part of the Sonoran Desert; therefore, it has a hot desert climate condition. This city has an agricultural neighborhood, which is close to the confluence of the Salt and Gila River. The study area was chosen due to the hot climate condition and proximity to an agricultural neighborhood. Figure 2 shows the study area, which is 355.7 m higher from sea level, with 33.4258 latitude and −111.9217 longitude.

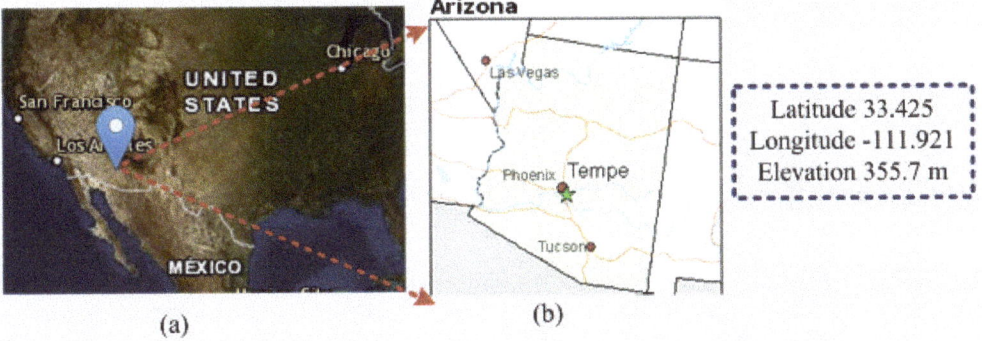

Figure 2. Location of the study area under consideration in this manuscript. (**a**) zoom-out view; (**b**) zoom-in view. Source: Internet.

To assess efficiency, all models are separately calibrated, with a total of 86 data points for an eight-year period of 2010–2017 at each selected station within the United States of America and a one-month lead time. Data were collected from the government database of Arizona state in the US. Study area is humid and has an agricultural neighborhood. Two combinations of data sets were studied to check the results and verify whether they are similar in pattern or not. The data set is initially divided into two parts: the training and test portions. About one-third (~27) of the data points of the total data set was selected as the training data set, whereas the remaining two-thirds (~59) of the data points was considered a testing data set.

Table 1 summarizes the statistical indices of the test, training and all data used in this study. The table contains the skewness, kurtosis, coefficient of variation (CV), standard deviation (SD), and first (1st) and third (3rd) quarters (Q) for all the data points (N). The table also reports the minimum (Min) and maximum (Max), along with the average (Avg), of all data point. It further reports the similar statistical indices for both the training and testing cases, as well.

Table 1. Statistical indices of the evaporation data set used to verify the modelling.

Statistics	N	Min	1st Q	X50	3rd Q	Max	Avg	SD	CV (%)	Skewness	Kurtosis
All	86	44	82.50	158	254.50	331	172.30	89.48	51.93	0.066	−1.45
Train	59	44	83	183	273	331	178.28	91.79	51.48	0.015	−1.48
Test	27	49	74.75	154	247.25	298	158.73	82.40	51.91	0.117	−1.50

Standard deviation shows the distribution nature of data set. For example, the standard deviation of the test data set is 82.40, and the average value of the data is 158.73. This means that most of the test data lies between 78.33 (158.73 − 82.40 = 78.33) to 241.13 (158.73 + 82.40 = 241.13). On the other hand, the coefficient of variation shows the precision of the data set in this table. It is defined as the ratio of standard deviation and mean value (in percentages). Two combinations of the testing and training data were chosen arbitrarily, in order to verify the robustness and repeatability of the proposed modeling techniques.

Combination 1: Training data (September 2010 to September 2015); testing data (October 2015 to December 2017).

Combination 2: Training data (September 2010 to December 2012 & June 2015 to December 2017); testing data (January 2013 to May 2015).

3.2. Model Accuracy Indicator

The performances of all four models were individually evaluated using statistical analysis to monitor accuracy, with respect to the evaporation forecasting data. The accuracy indicators for the ANFIS, FFA, PSO, and GA models were calculated, in terms of the coefficient of determination (R^2) [17], Nash–Sutcliffe coefficient (NSE) [17], root mean square error (RMSE) [10], mean absolute error (MAE) [17], variance account for (VAF) [18], absolute relative error (MARE), scatter index (SI) [17], bias [13], and root mean square relative error (RMSRE) [17]. The root mean squared error (RMSE) represents a good measure of the goodness of fit at high parameter values. The standard RMSE value should be 0, according to the theory. The relative error (MARE) provides a more balanced idea of the goodness of fit at moderate and low values. The standard value of MARE is also 0. The coefficient of determination R^2 should be 1 for a perfect fit model. This coefficient measures the correlation of the predicted values with the observational data—the closer the coefficient is to one, the greater the correlation. The value of this coefficient does not interfere with the data unit considered. The SI index is the relative form of RMSE. The performance factor of the model, expressed as the Nash–Sutcliffe error criterion (E_{NSC}), was used to evaluate the predictive power of the model. A value of unity for the E_{NSC} indicates optimum conformity between predicted and observed data. In this work, both R^2 and E_{NSC}

are expressed in percentages. The closer their magnitude to 100, the better the performance of the model. The ideal value for VAF is 100. All of them can be calculated from designed formulations, which are presented in the appendices section (see Appendix A for details).

3.3. Simulation Results

Computer-based software simulation is performed to validate the proposed model; in particular, MATLAB is used to validate the model. Figure 3 shows the step-by-step work, performed in this manuscript, to validate the proposed model. First of all, the data is collected from the location under interest. Secondly, the models are chosen based on the collected data. In this case, ANFIS is chosen due to the nonlinear nature of the data set. Model accuracy indicators are then selected. On the other hand, the data is partitioned in to two groups: one group (almost two third of the data) is for testing the network, while the other group (remaining data set) is for training the proposed model. In the beginning the network is trained using nearly two-thirds of the total data set. This calibrates the network. Later, the remaining data set (approximately one third) are used to test the network. To verify the overall performance of the observed models, the observed and predicted evaporation values were plotted together for both combinations (combinations 1 and 2, as depicted in Section 3.1). Graphical representation is made in terms of the observed and predicted data. Figures 4–7 show the pattern of the observed and predicted data for all four models. Figures 4 and 5 show the training data pattern for the first and second combination of data sets as mentioned in Section 3.1. These figures show the comparison of target and output sample index of trained data for the (a) ANFIS, (b) FFA, (c) GA, and (d) PSO models. Similarly, Figures 6 and 7 show the test data pattern of all models, and these present the comparison of the target and obtained output sample index of test data for (a) ANFIS, (b) FFA, (c) GA, and (d) PSO, respectively.

According to the graphs, both data sets lie between −15% to +15% of a perfect line. Graphical presentation also demonstrates that the data set are well-trained. According to the analysis, all the models are suitable for the evaporation estimation. However, the pattern for Figure 6a ANFIS (first combination) and Figure 7a ANFIS (second combination) were the best fits, and the pattern for Figure 6b ANFIS–FFA (first combination) and Figure 7b ANFIS–FFA (second combination) show less fitness among the four models. The figures for ANFIS–PSO and ANFIS–GA, for both combinations, were close to each other. Additionally, a few accuracy tests were performed to obtain a better understanding for both training and testing. Few statistical indices tests have been performed and summarized in Tables 2–6.

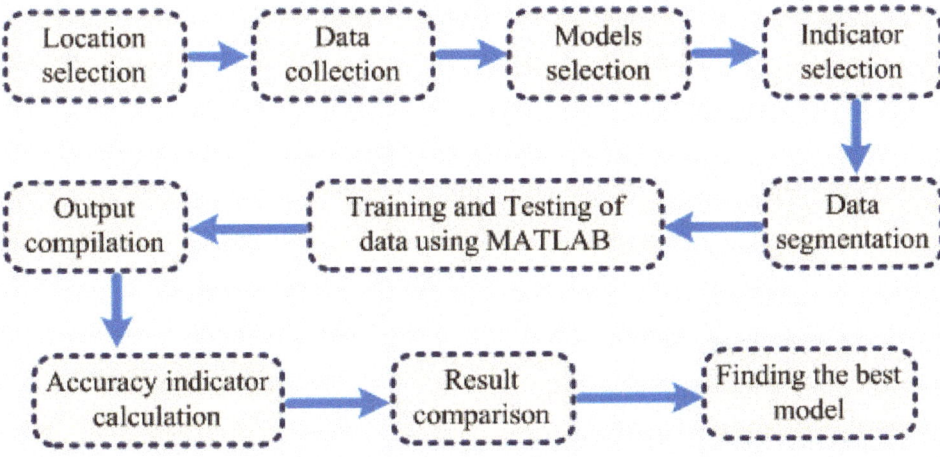

Figure 3. The step-by-step structural outline of the work performed in this manuscript.

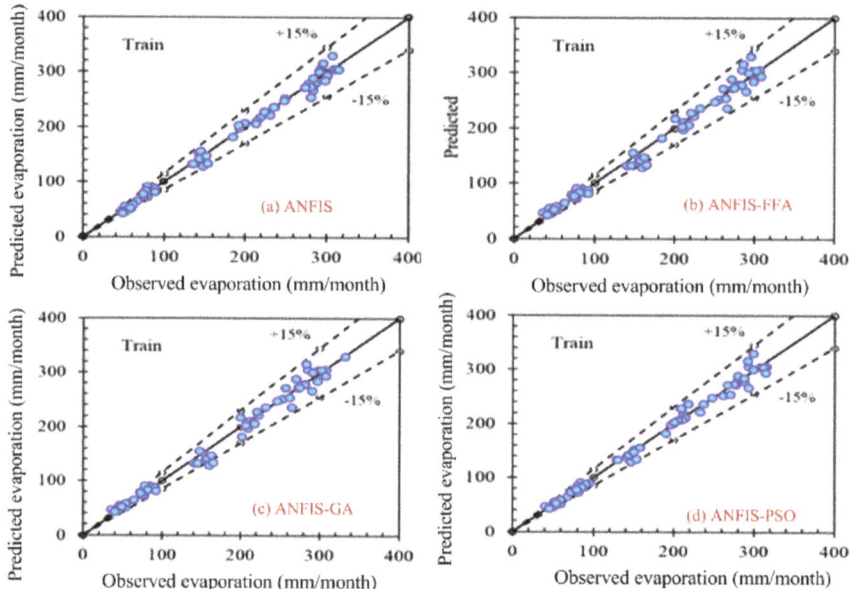

Figure 4. Comparison of the target (predicted) and obtained output sample index of training data set for (**a**) ANFIS, (**b**) ANFIS–FFA, (**c**) ANFIS–GA, and (**d**) ANFIS–PSO, respectively, using the first combination of the data set.

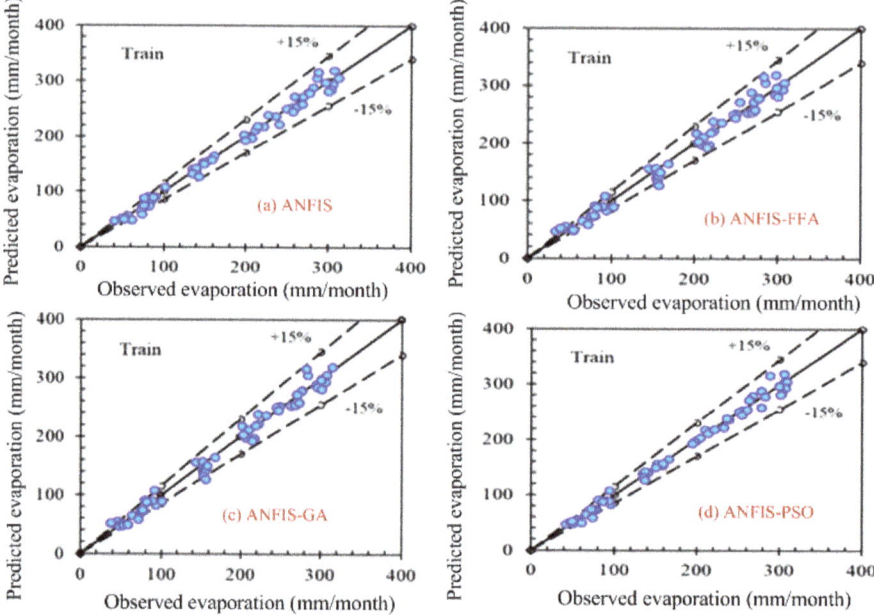

Figure 5. Comparison of the target (predicted) and obtained output sample index of training data set for (**a**) ANFIS, (**b**) ANFIS–FFA, (**c**) ANFIS–GA, and (**d**) ANFIS–PSO, respectively, using the second combination of the data set.

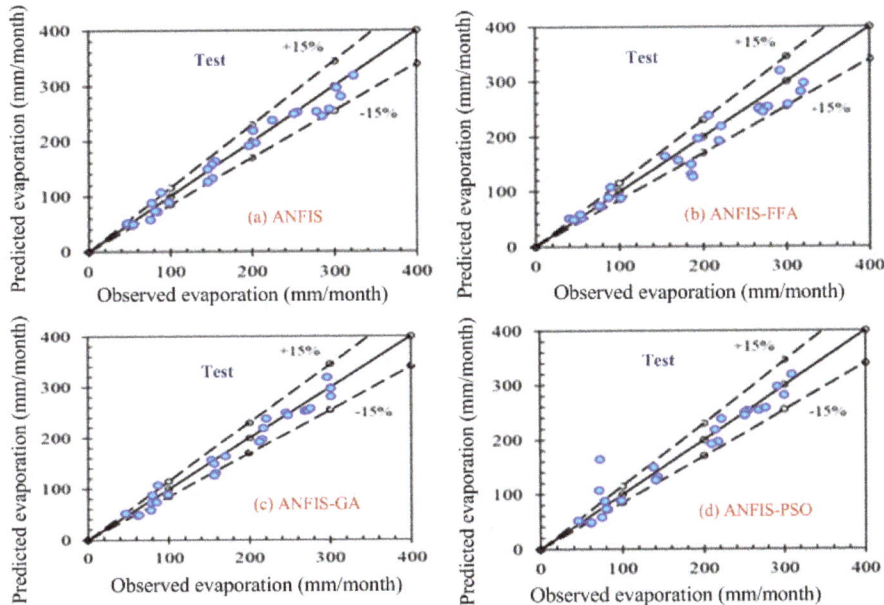

Figure 6. Comparison of the target and obtained output sample index of the test data for (**a**) ANFIS, (**b**) ANFIS–FFA, (**c**) ANFIS–GA, and (**d**) ANFIS–PSO, respectively (first combination of the data set).

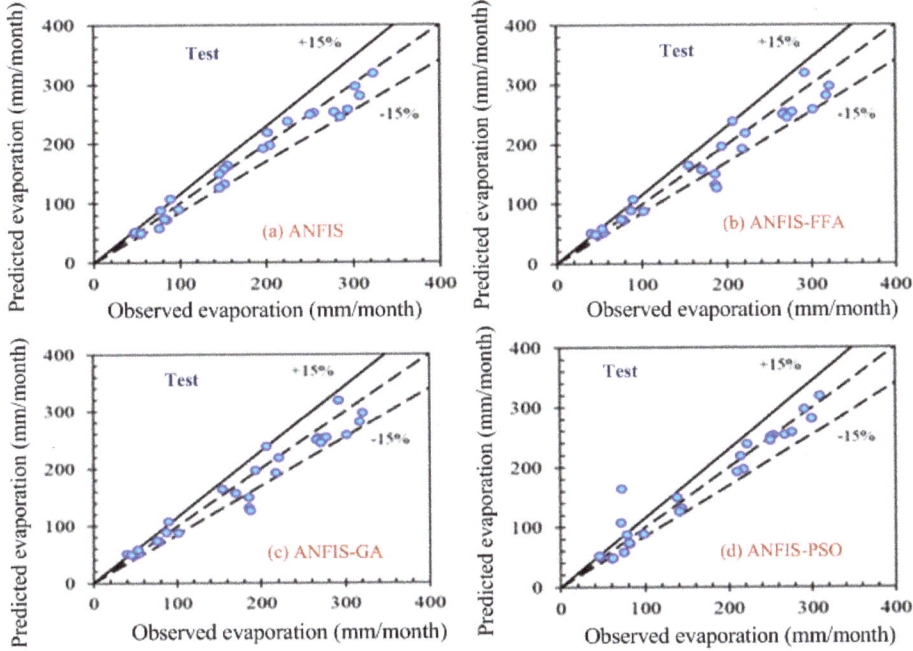

Figure 7. Comparison of the target and obtained output sample index of the test data for (**a**) ANFIS, (**b**) ANFIS–FFA, (**c**) ANFIS–GA, and (**d**) ANFIS–PSO, respectively (second combination of the data set).

Table 2. Summary of model accuracy indicator test for the training data set (for the first combination data set), which was calculated in Excel.

	R^2	VAF	RMSE	SI	MAE	MARE	RMSRE	MRE	BIAS	NASH
ANFIS	0.99	99.04	8.93	0.050	−0.0008	0.044	0.001	0.001	0.001	0.99
FFA	0.97	94.08	24.38	0.140	8.976	0.110	0.079	0.018	8.98	0.92
GA	0.98	97.50	14.38	0.084	4.569	0.095	0.024	0.027	4.57	0.97
PSO	0.99	98.85	9.73	0.054	−0.167	0.040	0.001	−0.001	−1.69	0.98

Table 3. Summary of model accuracy indicator test for the testing data set (for the first combination data set), which was calculated in Excel.

	R^2	VAF	RMSE	SI	MAE	MARE	RMSRE	MRE	BIAS	NASH
ANFIS	0.98	97.04	15.55	0.094	−4.56	0.087	0.018	−0.027	−4.56	0.97
FFA	0.97	93.11	24.39	0.148	−8.98	0.118	0.154	−0.400	−8.98	0.93
GA	0.98	97.51	14.38	0.087	−4.57	0.101	0.033	−0.421	−4.57	0.97
PSO	0.98	97.18	14.60	0.088	1.68	0.101	0.014	0.003	1.68	0.97

Table 4. Summary of model accuracy indicator test for the training data set (for the second combination data set), which was calculated in Excel.

	R^2	VAF	RMSE	SI	MAE	MARE	RMSRE	MRE	BIAS	NASH
ANFIS	0.99	98.99	8.99	0.05	0.0002	0.046	0.0003	0.019	2.733	0.99
FFA	0.98	97.93	12.80	0.07	0.001	0.082	0.0066	−0.015	0.001	0.98
GA	0.99	98.32	11.66	0.07	0.403	0.072	0.0073	−0.011	0.403	0.98
PSO	0.99	99.11	8.44	0.05	−0.040	0.042	0.0016	−0.001	−0.040	0.99

Table 5. Summary of model accuracy indicator test for test data set (for the second combination data set), which was calculated in Excel.

	R^2	VAF	RMSE	SI	MAE	MARE	RMSRE	MRE	BIAS	NASH
ANFIS	0.99	98.42	11.94	0.07	3.73	0.062	0.007	0.025	3.73	0.98
FFA	0.98	97.45	15.03	0.09	4.25	0.076	0.018	0.014	4.25	0.97
GA	0.98	97.52	14.63	0.08	3.50	0.073	0.024	0.010	3.50	0.97
PSO	0.98	97.50	15.08	0.09	4.61	0.081	0.0004	0.024	4.61	0.97

Table 6. Summary of model accuracy indicator test during the testing period, provided by 'MATLAB'.

Type of Model	Training Data				Test Data			
	MSE	RMSE	MEAN	STD	MSE	RMSE	MEAN	STD
GA	146.92	12.12	−2.82	11.89	206.79	14.38	−4.57	13.89
ANFIS	58.23	7.63	−8.16	7.69	241.72	15.54	−4.56	15.14
PSO	58.75	7.66	0.11	7.73	213.05	14.59	1.68	14.77
FFA	507.20	22.52	−4.87	22.17	594.80	24.38	−8.97	23.10

The overall summary of the findings is presented in Table 6. Table 6 presents the results provided by the MATLAB tool. It shows that the MSE values, for all the test models, were very high (MSE for ANFIS 241.72, for FFA 594.80, for GA it is 206.79, and for PSO it is 213.05) for the testing data, and higher for the training data. To ensure a rigorous comparison of the models, an extended analysis was performed using RMSE, R^2, MAE, MARE, RMSRE, SI, MRE, Bias, NASH, and VAF as statistical indices for the estimated values. Tables 2–5 present values of all statistical indices for training and testing data set of all models. According to all statistical indices, especially the R^2, RMSE, VAF, and NASH values, the second combination of the data set presented better results than the first combination of the data set, which is presented in Table 3. The results of the ANFIS and ANFIS–PSO models were almost identical in both combinations. RMSE was lower for ANFIS and ANFIS–GA. ANFIS–FFA posed worse results, among all model, in all the cases. Biasness is less for ANFIS model. According to the test results from Tables 3 and 5, the R^2 for ANFIS, GA, and PSO were almost identical, 0.99, whereas R^2 for FFA was 0.97. This is found to be aligned with the training result. A commonly used correlation measure, i.e., (R^2), in the testing of statistical indices cannot always be accurate, or sometimes it could be misleading, when used to compare the predicted and observed models [1]. The two most widely used statistical indicators, i.e., root mean square error (RMSE) and bias error, were used in this analysis. The model performance is inversely proportional to the RMSE value; lower RMSE values present higher accuracy and vice versa. RMSE is the minimum for PSO and GA, which were 14.59, 14.63, and 14.38, 15.07, respectively, whereas ANFIS was 15.54, and FFA presents the worst value: 24.38. Negative biasness was noticed for all the models, where ANFIS and GA possessed minimum biasness.

Hence, the MSE values are higher, and the relative statistical indices are compared to find better results. The MARE and RMSRE results should also be minimal for the best fit model. Again, ANFIS shows the minimum MARE value (0.087), and PSO gives similar result to ANFIS. However, according to the RMSRE results, PSO shows the best result. For more clarity, NASH has been considered another accuracy indicator, and the value should be close to 1 for the best fit. The table presents the highest NASH value for ANFIS (0.97), GA (0.97), and PSO (0.97). FFA was also close to 1 (0.93). To avoid confusion, VAF was calculated. Here, ANFIS, GA, and PSO showed higher results (all three results were close to 97.11), and FFA indicates 93.11.

Time is an important factor of these calculations. The time frame is given below in Table 7 for all four models. It shows that the ANFIS model took less time than the others, and FFA is the complicated one. After analyzing all the results, the FFA model is considered the least acceptable model among the four. ANFIS, with GA and PSO models, were showing better fit in some situations. Although GA and PSO were showing similar results and took same time to run, ANFIS can be considered more acceptable because of its simplicity.

Table 7. Time taken by four models (approximate).

Model Name	Run Time
ANFIS	5 to 10 min
ANFIS–FFA	30 min to 3 h
ANFIS–PSO	10 to 30 min
ANFIS–GA	10 to 30 min

3.4. Discussion

In this study, evaporation was estimated from six climate variables, i.e, minimum temperature, maximum temperature, average temperature, sunshine hour, wind speed, and relative humidity. Evaporation depends on the combined effect of humidity, temperature variation, sunshine, and wind [11]. Sunshine is an important factor that helps evaporate the water body [7]. Similarly, temperature and humidity also play an important role in evaporation. When they decrease, evaporation increases. Wind takes water away to the atmosphere [7]. Therefore, all of them were considered, as they affect evaporation. Key parameters were selected by trial-and-error method. Only one set of parameters was experimented with.

The findings of this research demonstrated that the FFA model is considered the least acceptable model among the four. ANFIS with GA and PSO models were showing a better fit in some situations. Although GA and PSO were showing similar results, based on all accuracy indicator tests (especially, on maximum R^2 value, minimum RMSE, less Biasness, maximum VAF, minimum RMSRE value, and maximum value of Nash coefficient), and took the same time to run the model, ANFIS can be considered more acceptable because of its simplicity. This model can be used as a role model for any dataset of an arid climate. It can be helpful for the local stakeholder, in terms of the hydrological resource management system. The main advantage of adopting ANFIS for this location is the pattern of the dataset. As the datasets are inherently nonlinear, the ANFIS model was able to achieve high accuracy in the prediction of evaporation. The ANFIS model and this model, with the optimizers (FFA, GA, and PSO), can be widely used for arid climates, with the same weather variables, in any part of the world.

More investigation is needed for this location. Lack of data was a limitation of this study. More climate variables can be added for more accuracy of the model. Other modern machine learning technique should be implemented in the future, in order to use the available resources to enhance the water resource management system. That would be beneficial for the local agri-economical prospect, as well.

4. Conclusions

The comparison among the adaptive neuro fuzzy inference system (ANFIS) and its hybridization, using three different algorithms (FFA, GA, and PSO), has been illustrated in this study, in the context of evaporation estimation, using different climate variables, namely sunshine, relative humidity, average temperature, maximum temperature, minimum temperature, and wind speed. Two combinations of data sets were trained and tested, in order to verify the correlation among the different models. The study illustrated the accuracy of all four models. However, the performance of the models was evaluated based on the various statistical measures (RMSE, RMSRE, MBE, VAF, NASH, biasness, MBE, MARE, SI, and R^2). Result shows that the second combination of the testing and training data set posed slightly better results than the first combination. Overall, all four models are suitable for the estimation of evaporation, but ANFIS and ANFIS, with optimizer PSO, is superior for all accuracy indicator values. Relative and absolute accuracy tests were performed to find the best model in this study. Though all the results of the two models (ANFIS and ANFIS–PSO) were merely identical, ANFIS is recommended, due to its simple formulation and easy development, compared to the ANFIS–PSO model. The computational time of ANFIS model is less, in comparison to the other models with optimizers. The main objective of the adoption of different optimizer techniques is to verify the accuracy of the outcome prediction by ANFIS model. Since the prediction was almost identical in all cases, the ANFIS model is recommended, due to its simplicity. The major challenge of this project was the limitation of data. These models can be applied for different data sets to investigate the results, if they were available. This analysis is limited to a particular location. However, in future work, other locations can be explored, and their performance can be compared with modern machine learning methods. Another optimizer, for example, the ant colony optimizer (ACO), can be investigated in future work. Multi gene-genetic programming (MGGP) can also be explored in the future. Another climate variable, such as, atmospheric pressure, can be considered as an input in the future. However, the evaporation of a given location can easily be modelled from the available data using the ANFIS model. Additionally, this model can be applied as a module for calculating evaporation data in hydrological modeling studies.

Author Contributions: Conceptualization, M.J., H.B., and A.M.; methodology, M.J. and H.B.; software, A.M.; validation, M.J.; formal analysis, M.J. and A.M.; investigation, M.J.; resources, H.B.; data curation, M.J. and H.B.; writing—original draft preparation, M.J.; writing—review and editing, H.B. and A.M.; visualization, H.B.; supervision, A.M.; project administration, H.B. All authors have read and agreed to the published version of the manuscript.

Funding: This research received no external funding.

Data Availability Statement: Part of the data used in this manuscript are available through the corresponding author upon reasonable request.

Acknowledgments: Mansura Jasmine acknowledges the support of Mehedi Hasan during the preparation of this manuscript.

Conflicts of Interest: The authors declare no conflict of interest.

Appendix A

The relationships for statistical indices and error measures used in this paper are provided in the following.

R^2: coefficient of determination, which can be expressed in the following form:

$$r = \frac{n(\sum xy) - (\sum x)(\sum y)}{\sqrt{\left[n\sum x^2 - (\sum x)^2\right]\left[n\sum y^2 - (\sum y)^2\right]}} \tag{A1}$$

RMSE: root mean square error, which can be formulated as follows:

$$RMSE = \left[\frac{\sum_{i=1}^{M}\left(Y_{i(model)} - Y_{i(actual)}\right)}{M}\right]^{\frac{1}{2}} \tag{A2}$$

MARE: absolute relative error. The formula is given below:

$$MARE = \frac{1}{M}\sum_{i=1}^{M}\left(\frac{\left|Y_{i(model)} - Y_{i(actual)}\right|}{Y_{i(actual)}}\right) \tag{A3}$$

$$Bias = \frac{\sum_{i=1}^{M}\left(Y_{i(model)} - Y_{i(actual)}\right)}{M} \tag{A4}$$

SI: scatter index, which can be expressed as follows:

$$SI = \frac{RMSE}{\frac{1}{M}\sum_{i=1}^{M}\left(Y_{i(actual)}\right)} \tag{A5}$$

RMSRE: root mean square relative error. This error can be calculated from the following equation:

$$RMSRE = \frac{1}{N}\sqrt{\sum\left(\frac{y_t - \hat{y}_t}{y_t}\right)^2} \tag{A6}$$

MAE: mean absolute error. This error can be calculated from the following equation:

$$MAE = \frac{1}{n}\sum_{i=1}^{n}|T_{i.Actual} - T_{i.Predicted}| \tag{A7}$$

VAF: variance account for. This term can be presented by the following equation:

$$VAF = \left(\frac{1 - var(T_{i.Actual} - T_{i.Predicted})}{var(T_{i.Actual})}\right) * 100 \tag{A8}$$

NSE: Nash–Sutcliffe coefficient. This coefficient can be formulated as follows:

$$E_{NSC} = 1 - \left(\frac{\sum (y_t - \widehat{y_t})^2}{\sum (y_t - \overline{y_t})^2} \right) \tag{A9}$$

where,

$Y_{i(actual)}$: the output observational parameter;
$Y_{i(model)}$: the y parameter predicted by the models;
$Y_{i(model)}$: the mean predicted y parameter;
M: the number of parameters;
n: number of samples;
E_{NSC}: the Nash–Sutcliffe test statistic;
$T_{i.Actual}$: the ith value of actual data;
$T_{i.Predicted}$: the ith value of predicted data.

References

1. Benzagtha, M.A. Estimation of Evaporation from a Reservoir in Semi-arid Environments Using Artificial Neural Network and Climate Based Models. *Br. J. Appl. Sci. Technol.* **2014**, *4*, 3501–3518. [CrossRef]
2. Dogana, E.; Gumrukcuoglu, M.; Sandalci, M.; Opan, M. Modelling of evaporation from the reservoir of Yuvacik dam using adaptive neuro-fuzzy inference systems. *Eng. Appl. Artif. Intell.* **2010**, *23*, 961–967. [CrossRef]
3. Kisi, O.; Sanikhani, H.; Zounemat-Kermani, M. Comparison of two different adaptive Neuro-Fuzzy inference system in modelling daily reference evapotranspiration. *Water Resour. Manag.* **2014**, *28*, 2655–2675. [CrossRef]
4. Kisi, O.; Sanikhani, H.; Zounemat-Kermani, M.; Niazi, F. Long-term monthly evaporation modeling by several data-driven methods without climate data. *Comput. Electron. Agric.* **2015**, *115*, 66–77. [CrossRef]
5. Moghaddamnia, A.; Gousheh, M.G.; Pirli, J.; Amin, S.; Han, D. Evaporation estimation using artificial neural network and adaptive neuro-fuzzy inference system techniques. *Adv. Water Resour.* **2009**, *32*, 88–97. [CrossRef]
6. Goyal, M.K.; Bharti, B.; Quilty, J.; Adamowsky, J.; Pandey, A. Modeling of daily pan evaporation in subtropical climates using ANN, LS-SVR, Fuzzy Logic, and ANFIS. *Expert Syst. Appl.* **2014**, *41*, 5267–5276. [CrossRef]
7. Malik, A.; Tikhamarine, Y.; Al-Ansari, N.; Shahid, S.; Sekhon, H.S.; Pal, R.K.; Rai, P.; Panday, K.; Singh, P.; Elbeltagi, A.; et al. Daily pan evaporation estimation in different agro-climatic zones using hybrid support vector regression optimized by Salp swarm algorithm in conjunction with gamma test. *Eng. Appl. Comput. Fluid Mech.* **2021**, *15*, 1075–1094. [CrossRef]
8. Malik, A.; Saggi, M.K.; Rehman, S.; Sajjad, H.; Inyurt, S.; Bhatia, A.S.; Farooque, A.A.; Oudah, A.Y.; Yaseen, Z.M. Deep learning versus gradient boosting machine for pan evaporation prediction. *Eng. Appl. Comput. Fluid Mech.* **2022**, *16*, 570–587. [CrossRef]
9. Malik, A.; Kumar, A.; Kimb, S.; Kashani, H.M.; Karimi, V.; Sharafati, A.; Ghorbani, A.M.; Al-Ansari, N.; Salih, S.Q.; Yaseen, Z.M.; et al. Modeling monthly pan evaporation process over the Indian central Himalayas: Application of multiple learning artificial intelligence model. *Eng. Appl. Comput. Fluid Mech.* **2020**, *14*, 323–338. [CrossRef]
10. Ayvaz, M.T.; Elci, A. Identification of the optimum groundwater quality monitoring network using a genetic algorithm-based optimization approach. *J. Hydrol.* **2018**, *563*, 1078–1091. [CrossRef]
11. Wang, L.; Kisi, O.; Sanikhani, H.; Zounemat-Kermani, M.; Li, H. Pan evaporation modelling six different heuristic computing methods in different climates in China. *J. Hydrol.* **2017**, *544*, 407–427. [CrossRef]
12. Karahan, H. Determining rainfall-intensity-duration-frequency relationship using particle swarm optimization. *J. Civil Eng.* **2012**, *16*, 667–675. [CrossRef]
13. Chang, F.J.; Chang, Y.T. Adaptive neuro-fuzzy inference system for prediction of water level in reservoir. *Adv. Water Resour.* **2006**, *29*, 1–10. [CrossRef]
14. Yaseen, Z.M.; Ebtehaj, I.; Bonakdari, H.; Deo, R.C.; Mehr, A.D.; Mohtar, W.H.M.W.; Diop, L.; El-Shafie, A.; Singh, V.P. Novel approach for streamflow forecasting using a hybrid ANFIS-FFA model. *J. Hydrol.* **2017**, *554*, 263–276. [CrossRef]
15. Yaseen, Z.M.; Ghareb, M.I.; Ebtehaj, I.; Bonakdari, H.; Siddique, R.; Heddam, S.; Yusif, A.A.; Deo, R. Rainfall pattern forecasting using novel hybrid intelligent model based ANFIS-FFA. *Water Resour. Manag.* **2018**, *32*, 105–122. [CrossRef]
16. Kennedy, J.; Eberhart, R.C. Particle swarm optimization. In Proceedings of the IEEE international Conference on Neural Networks, Perth, Australia, 27 November–1 December 1995; pp. 1942–1948.
17. Lotfi, K.; Bonakdari, H.; Ebtehaj, I.; Mjalli, S.F.; Zeynoddin, M.; Delatolla, R.; Gharabaghi, B. Predicting wastewater treatment plant quality parameters using a novel hybrid linear-nonlinear methodology. *J. Environ. Manag.* **2019**, *240*, 464–474. [CrossRef] [PubMed]
18. Zeynoddin, M.; Bonakdari, H.; Ebtehaj, I.; Esmaeilbeiki, F.; Gharabaghi, B.; Zare Haghi, D. A reliable schochastic daily soil temperature forecast model. *Soil Tillage Res.* **2019**, *189*, 73–87. [CrossRef]

Article

Multivariable Panel Data Cluster Analysis of Meteorological Stations in Thailand for ENSO Phenomenon

Porntip Dechpichai, Nuttawadee Jinapang, Pariyakorn Yamphli, Sakulrat Polamnuay, Sittisak Injan and Usa Humphries *

Department of Mathematics, Faculty of Science, King Mongkut's University of Technology Thonburi, 126 Pracha Uthit Rd., Bang Mot, Thung Khru, Bangkok 10140, Thailand; porntip.dec@kmutt.ac.th (P.D.); nuttawadee.jina@mail.kmutt.ac.th (N.J.); pariyakorn.tuan@mail.kmutt.ac.th (P.Y.); sakunrat.pol@mail.kmutt.ac.th (S.P.); sittisak_injan@hotmail.com (S.I.)
* Correspondence: usa.wan@kmutt.ac.th; Tel.: +66-2470-8822

Abstract: The purpose of this research is to study the spatial and temporal groupings of 124 meteorological stations in Thailand under ENSO. The multivariate climate variables are rainfall, relative humidity, temperature, max temperature, min temperature, solar downwelling, and horizontal wind from the conformal cubic atmospheric model (CCAM) in years of El Niño (1987, 2004, and 2015) and La Niña (1999, 2000, and 2011). Euclidean distance timed and spaced with average linkage for clustering and silhouette width for cluster validation were employed. Five spatial clusters (SCs) and three temporal clusters (TCs) in each SC with different average precipitation were compared by El Niño and La Niña. The pattern of SCs and TCs was similar for both events except in the case when severe El Niño occurred. This method could be applied using variables forecasted in the future to be used for planning and managing crop cultivation with the climate change in each area.

Keywords: Euclidean distance timed and spaced; meteorological station; multivariable panel data cluster analysis

1. Introduction

In the past, the climate in Thailand was largely influenced by monsoon winds, such as southwest moonsoon and northeast moonsoon, resulting in Thailand having a predominantly rainy season and dry season (summer and winter) taking place at a relatively certain time. Currently, however, there has been an El Niño–La Niña phenomenon known as the ENSO phenomenon (ENSO) that affects the climate. The ENSO phenomenon is caused by variations in the Southern Hemisphere's climate system. It is a phenomenon that has a connection between ocean phenomena and ocean winds. It brings about climatic variations, causing unusually high rainfall and unusual drought [1]. There are three types of weather variability: drought, rain and cold disasters, and tropical cyclones. Thailand's proximity to the Western Pacific makes it directly affected by El Niño during 1997–1998, which resulted in drought, lower than normal rainfall, and higher than normal air temperatures across the country [2]. In 1999–2000, during the La Niña period, Thailand experienced more rainfall than usual and cold weather, breaking records in many provinces [2]. Thailand is in the humid tropics, which is suitable for agriculture. Most of its population is engaged in agriculture, so agricultural products are the main source of the country's income and, therefore, vital to its economy. The 12th Agricultural Development Plan (2017–2021) summarizes the agricultural situation in terms of climate change and seasonal variability, resulting in decreased agricultural productivity. Existing plant species are unable to adapt to changing climate conditions, especially the ongoing drought from 2012 to 2015, damaging important crops. This may be due to insufficient observation or experience by farmers to cope with unprecedented situations in time, posing a risk of loss of productivity and increased production costs [3]. ENSO-related climate variability exerts strong influences on agricultural production in different regions, including in Thailand [4–9].

Cluster analysis, unsupervised learning, have been applied in many studies to define spatial and temporal variability from climate variables. In previous studies, only one variable, mostly focusing on rainfall in a time series format, has been used for spatial and temporal cluster [10–12]. However, there are other climate factors that affect agricultural production such as relative humidity and temperature, which statistically significantly affected sugarcane production, which was likely to decrease in the year of El Niño and to increase in the year of La Niña [13]. Although there are some studies which employed longitudinal meteorological factors such as rainfall, air temperature, humidity, pressure, wind, evaporation, etc., they firstly average data over the time into the general cross-sectional data and then the distance between samples is calculated for clustering [14]. Averaging over the time will result in a high amount of data loss because the mean shows the average change in the data, yet it does not show the distribution of the data [15–18].

It would be beneficial to study variation across different geographic scales using multivariable panel hierarchical clustering from ENSO-effected climate variables in Thailand, obtained from the conformal cubic atmospheric model (CCAM). There are seven weather variables, including rainfall, average temperature, highest temperature, lowest temperature, temperature difference from highest temperature, temperature difference from lowest temperature, relative humidity, and solar radiation according to the locations of the weather stations of the Thailand Meteorological Department. These monthly data have been characterized by a combination of panel data, cross-sectional data, and time-series data representing behavioral units and periods.

Therefore, this research will employ the distance measurement that does not need to average the data, which is Euclidean distance timed and spaced, to cluster meteorological weather stations in Thailand and discover the seasonal pattern for each cluster using climate factors associated with precipitation when ENSO phenomena occur, since changes in rainfall are important variables affecting agricultural productivity. The studied method, cluster analysis on multivariable panel data with climate change application, therefore, could be applied to the future data from weather models to group area and season. The clustering framework applied in this study is shown in Figure 1. The results could be used as a guideline to benefit the agricultural sector or the relevant agencies to prepare for the upcoming changes resulting from climate change. In addition, spatial and timely management plans can also be appropriately executed, including drought monitoring, water management of both agricultural areas, as well as crop management.

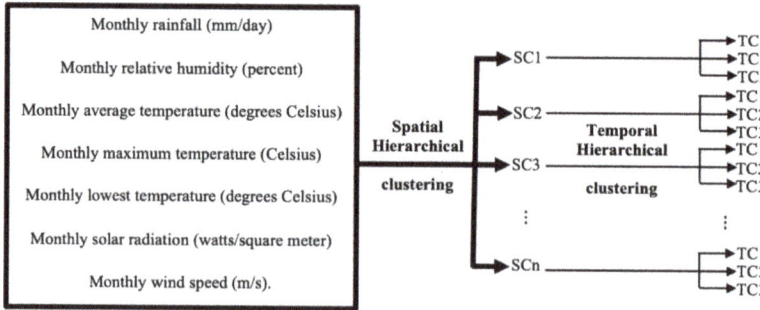

Figure 1. The multivariable panel data clustering framework.

2. Materials and Methods

2.1. Study Area

Thailand is located between latitudes 5°37′ N and 20°27′ N and longitudes 97°22′ E and 105°37′ E. A total of 124 stations of the Thai Meteorological Department (Figure 2) were selected for the cluster analysis.

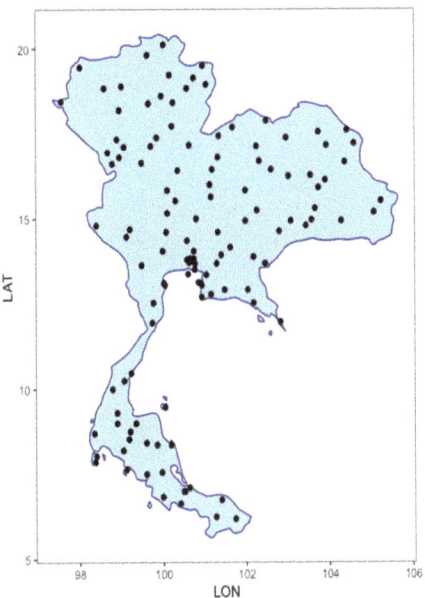

Figure 2. Spatial distribution of 124 meteorological stations in Thailand from the Thai Meteorological Department (TMD).

2.2. El Niño–Southern Oscillation (ENSO)

El Niño–southern oscillation (ENSO) is a periodic change in the oceanic atmosphere system in the tropical Pacific Ocean that affects climate around the world. It occurs every three to seven years (average five years) and typically lasts nine months to two years, associated with floods, droughts, and other global disturbances. During normal or non El Niño conditions, trade winds blow west across the Pacific Ocean. The western part of the equatorial Pacific is characterized by warm, wet, and low-pressure weather conditions due to the accumulation of moisture in the form of typhoons and thunderstorms.

During the ENSO event, there was an increase in air pressure across the Indian Ocean, Indonesia, and Australia, and a decrease in air pressure over Tahiti and the rest of the central and eastern Pacific Ocean. The trade winds in the South Pacific weaken or head east, and warm water spreads eastward from the western Pacific and Indian Ocean to the eastern Pacific. This has led to widespread droughts in the western Pacific and dry eastern Pacific rainfall. While El Niño is characterized by unusually warm ocean temperatures in the central to eastern Pacific Ocean, La Niña is characterized by unusually cold ocean temperatures in the region, but warmer waters in the western Pacific Ocean, as shown in Figure 3. However, as El Niño conditions lasted for several months, more global warming occurred in the oceans.

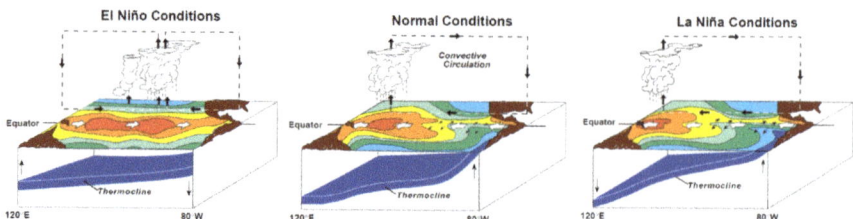

Figure 3. Model of surface temperature, wind, area of rising air and thermoline (blue surface) in the tropical Pacific during El Niño, Normal, and La Niña (https://reefresilience.org/th/stressors/climate-and-ocean-change/el-nino-southern-oscillation/, accessed on 24 March 2022).

In this study, the Oceanic Niño Index (ONI) from the National Oceanic and Atmospheric Administration (2020) was used to identify the El Niño–southern oscillation. The ONI is the 3-month running mean of the sea surface temperature anomaly in the Niño 3.4 region (5° N–5° S, 120°–170° W). The ONI index exceeding +0.5 °C or −0.5 °C for at least five consecutive months was considered as a full-fledged El Niño (E) or La Niña (L). According to Null report, the three latest very strong El Niño events (ONI $\geq$ 2 °C) in 1982, 1997, and 2015 and three latest strong La Niña events (−1.5 to −1.9 °C) in 1999, 2007, and 2011 were selected to study the climate variations [19].

2.3. Conformal Cubic Atmospheric Model (CCAM)

The CCAM is a dynamic global climate model developed by the Commonwealth Scientific and Industrial Research Organization (CSIRO), Division of Atmospheric Research, Australia. It is used to forecast global climate through dynamic scale reduction by generating a grid covering the region's forecast area [20]. The model has also been developed by adding physical parameterization schemes that include longwave radiation, shortwave radiation, aerosol, cumulus convection, cloud distribution, soil temperature, etc., to reduce the climate forecast error. The CCAM dataset was downscaled to 10 km grid resolution, which is sufficient for the analysis of both spatial and temporal forecasts at the regional level [21,22]. Data were changed from grid data to station format, which covers 124 meteorological measurement stations across Thailand (Figure 1).

Climate variables, focusing mainly on agricultural-related variables for cluster analysis, were used in this study. They consist of a total of 7 variables: rainfall (mm/day), relative humidity (percent), average temperature (degrees Celsius), maximum temperature (degree Celsius), minimum temperature (degrees Celsius), solar radiation (watts/square meter), and wind speed (m/s). Monthly data of those variables were collected for the years 1987, 1999, 2000, 2004, 2011, and 2015, of which the ENSO phenomenon occurred.

2.4. Multivariate Panel Data

Panel data is the combination of cross-sectional data and time-series data representing behavioral units over the time $(x_{ij}(t))$. Data were collected from cross-section data, which collects the value of the variables in each unit at a given point in time. Then, the data were repeatedly collected from the same unit at a subsequent time, either yearly, quarterly, monthly, weekly, daily, or hourly. If each panel unit is observed at the same time point, a data set is called balanced panel data. Consequently, if a balanced panel contains n panel units and T periods, the number of observations in the dataset is necessarily $N = n \times T$. However, if at least one panel unit is not observed every period, a data set is called unbalanced panel data. Therefore, the number of observations in the unbalanced panel dataset is $N < n \times T$.

Multivariate panel data has a very complex structure and cannot be represented by a simple two-dimensional table. Table 1 shows the multivariate combination of data in a two-dimensional table format, where n represents the number of samples collected, p represents the number of variables $(x_1, x_2, \ldots, x_p)$, T represents the length of time and represents the data value of the ith sample and jth variable at time t, where $i \in [1,n]; j \in [1,p]; t \in [1,T]$. Descriptive statistics, such as mean and variance of jth variable, is calculated as Equations (1) and (2), respectively [15–18].

$$\overline{x}_j = \frac{1}{T}\frac{1}{n}\sum_{t=1}^{T}\sum_{i=1}^{n} x_{ij}(t), \qquad (1)$$

$$s_j^2 = \frac{1}{T}\sum_{t=1}^{T}\frac{1}{n-1}\sum_{i=1}^{n}\left[x_{ij}(t) - \overline{x}_j(t)\right]^2, \qquad (2)$$

Table 1. Multivariate panel data format for spatail cluster.

Sample (i)	Time Index (t)			
	1	2	...	T
	$x_1, x_2, ..., x_p$	$x_1, x_2, ..., x_p$	...	$x_1, x_2, ..., x_p$
1	$x_{11}(1), x_{12}(1), ..., x_{1p}(1)$	$x_{11}(2), x_{12}(2), ..., x_{1p}(2)$	...	$x_{11}(T), x_{12}(T), ..., x_{1p}(T)$
2	$x_{21}(1), x_{22}(1), ..., x_{2p}(1)$	$x_{21}(2), x_{22}(2), ..., x_{2p}(2)$	...	$x_{21}(T), x_{22}(T), ..., x_{2p}(T)$
⋮	⋮	⋮	...	⋮
n	$x_{n1}(1), x_{n2}(1), ..., x_{np}(1)$	$x_{n1}(2), x_{n2}(2), ..., x_{np}(2)$	...	$x_{n1}(T), x_{n2}(T), ..., x_{np}(T)$

The values for monthly climate variables were organized in two configuration matrices. Matrix $N \times p$ had monthly data (T) for stations (n) in its rows ($N = n \times T$) and the variables (p) in the columns. It was used to identify clusters of similar stations. Furthermore, monthly climate variables within these clusters (N_c) were analyzed to discover seasonality within the spatial cluster. For the second step, monthly climate variables were arranged in $T \times N_c$ rows, and the variables (p) were set up in columns (Table 2).

Table 2. Multivariate panel data format for temporal cluster.

Month (t)	Station Index (i)			
	1	2	...	N_c
	$x_1, x_2, ..., x_p$	$x_1, x_2, ..., x_p$	...	$x_1, x_2, ..., x_p$
Jan	$x_{11}(1), x_{12}(1), ..., x_{1p}(1)$	$x_{11}(2), x_{12}(2), ..., x_{1p}(2)$	...	$x_{11}(N_c), x_{12}(N_c), ..., x_{1p}(N_c)$
Feb	$x_{21}(1), x_{22}(1), ..., x_{2p}(1)$	$x_{21}(2), x_{22}(2), ..., x_{2p}(2)$	...	$x_{21}(N_c), x_{22}(N_c), ..., x_{2p}(N_c)$
⋮	⋮	⋮	...	⋮
Dec	$x_{121}(1), x_{122}(1), ..., x_{12p}(1)$	$x_{121}(2), x_{122}(2), ..., x_{12p}(2)$	...	$x_{121}(N_c), x_{122}(N_c), ..., x_{12p}(N_c)$

2.4.1. Multivariate Cluster Analysis

Cluster analysis is an unsupervised learning technique to identify groups with similar characteristics in the same group [23]. Agglomerative hierarchical clustering was used in this research. The bottom-up hierarchical algorithm treats each sample as a single cluster and then combines pair of clusters that are most similar until every cluster is grouped into one single cluster. In the case of general cross-section data, block distance, Euclidean distance, Minkowski distance, Chebychev distance, or Mahalanobis distance are used to measure the distance between two vectors ($\underline{x}_i' = [x_{i1}, x_{i2}, ..., x_{ip}]$) and $\underline{x}_j' = [x_{j1}, x_{j2}, ..., x_{jp}]$).

Cluster analysis of samples collected from multivariate panel data is often averaged over time data into general cross-section data. Typical Euclidean distance is then calculated for further grouping. However, this will result in information loss because the mean shows the average change in the data but does not show the distributing characteristics of the data, such as the standard deviation. Therefore, in this study, a Euclidean distance timed and spaced (d_{rk}) is used to calculate the distance between sample r and sample k [15–18], as in Equation (3).

$$d_{rk} = \sqrt{\sum_{t=1}^{T} \sum_{j=1}^{p} \left(x_{rj}(t) - x_{kj}(t)\right)^2}, \tag{3}$$

The distance should satisfy some conditions as follows:
1. $d_{rk} \geq 0$, if $x_{rj}(t) = x_{kj}(t)$ then $d_{rk} = 0$
2. $d_{rk} = d_{kr}$, to all $x_{rj}(t), x_{rj}(t)$
3. $d_{rk} \leq d_{rl} + d_{kl}$, to all $x_{rj}(t), x_{kj}(t), x_{lj}(t)$

A distance matrix for spatial grouping analysis contains a distance value between every pair of samples as in Equation (4a), which is the symmetric matrix ($n \times n$) with all diagonal values of zero. At the same time, a distance matrix for temporal grouping analysis within the spatial cluster contains a distance value between every pair of months as in Equation (4b), which is the asymmetric matrix (12×12) with all diagonal values of zero.

$$\begin{bmatrix} 0 & & & & \\ d_{21} & 0 & & & \\ d_{31} & d_{32} & 0 & & \\ \vdots & \vdots & \vdots & \ddots & \\ d_{n1} & d_{n2} & \cdots & d_{n(n-1)} & 0 \end{bmatrix}, \tag{4a}$$

$$\begin{bmatrix} 0 & & & & \\ d_{21} & 0 & & & \\ d_{31} & d_{32} & 0 & & \\ \vdots & \vdots & \vdots & \ddots & \\ d_{121} & d_{T2} & \cdots & d_{12(11)} & 0 \end{bmatrix}, \tag{4b}$$

Average linkage, which is the unweighted pair group method using arithmetic averages (UPGMA), was used to average the distance values between pairs of clusters [24]. It is widely used because it compromises the extreme cases [25].

The multivariate cluster analysis used in this paper was implemented directly using the "philanthropy", "cluster", "factoextra" and "FactoMineR" package in R programming language and RStudio [26].

2.4.2. Cluster Validation

This paper employed silhouette width (S_i) [27] to determine the optimal number of clusters, and it also could be used to validate consistency within clusters of data. The silhouette measures the similarity of i-th observation to its own cluster and the similarity of observation to other clusters as Equation (5).

$$S_i = \frac{b_i - a_i}{\max(b_i, a_i)} \tag{5}$$

where a_i is the average distance between i and all other observations in the same cluster, and b_i is the average distance between i and the observations in the "nearest neighboring cluster" as Equation (6).

$$b_i = \min_{C_k \in \mathcal{C} \setminus C(i)} \sum_{j \in C_k} \frac{d(i,j)}{n(C_k)} \tag{6}$$

where $C(i)$ is the cluster containing observation i, $d(i,j)$ is the Euclidean distance timed and spaced between observations i and j, and $n(C)$ is the cardinality of cluster C.

S_i ranges from -1 to $+1$, where a high value indicates that the observation is well matched to its own cluster, while a low or negative value indicates that observation is poorly matched to its own cluster. The average of observation's silhouette in a cluster was obtained to determine whether the clustering configuration is appropriate. The advantage of using silhouette only depends on the actual partition of the observations, not on the clustering algorithm that was used, and no need to access the original data. This paper implemented this function using the silhouette function in package cluster [28].

3. Results

This section may be divided into subheadings. It should provide a concise and precise description of the experimental results, their interpretation, as well as the experimental conclusions that can be drawn.

3.1. Variable Characteristics

Figure 4 shows boxplots of seven variables; they are varied by month but have the same pattern each year.

Rainfalls were more varied than others in 1997 and 2007 for the El Niño and La Niña phenomenon, respectively. The average rainfall in La Niña phenomenon was higher than that in El Niño phenomenon and the normal average, except for 1999, which was affected by the 1997–1998 very strong El Niño. Furthermore, all factors in each year had a pattern in relation to the season. For example, rainfall was very high and more fluctuated from August to September. It can be concluded that climate factors were different from month to month and year to year. Obviously, the rainfall between El Niño and La Niña differed significantly, while other climate factors were similar. This suggested the rainfall should be more focused to analyse the impact of the ENSO phenomenon on spatial clustering.

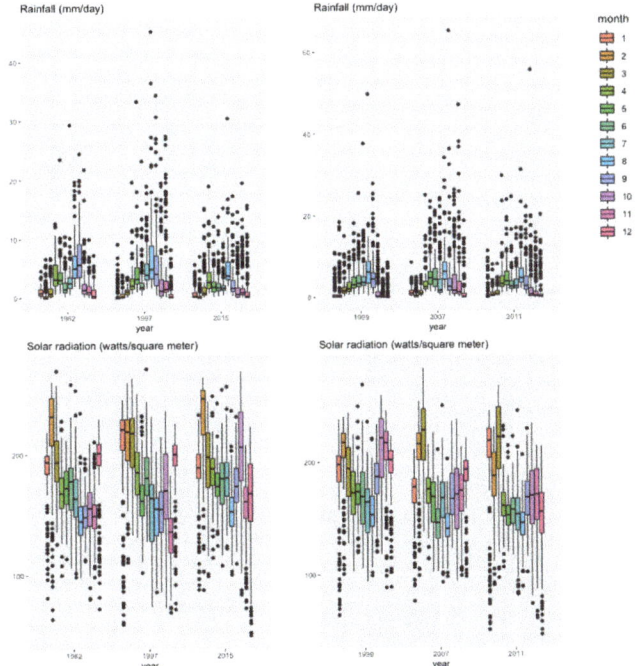

Figure 4. *Cont.*

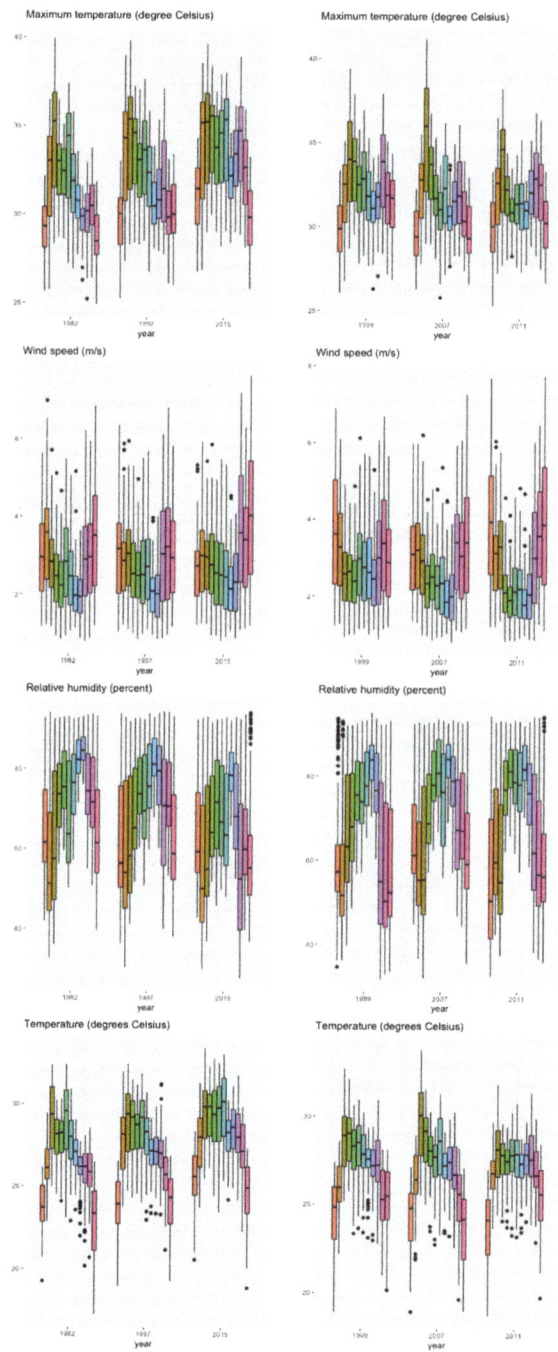

Figure 4. *Cont.*

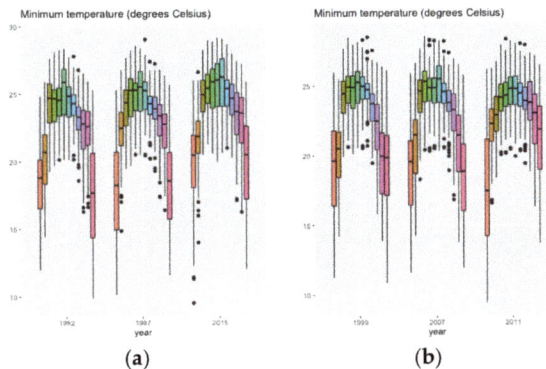

Figure 4. Boxplots of climate factors of 124 Thai meteorological stations by ENSO, years and months: (**a**) El Niño; (**b**) La Niña.

3.2. Spatial Clustering

The average silhouette width was used to determine a suitable number of clusters (k). It suggested the value 4 or 5 for k, due to their maximum width (Figure 5). So, a fair comparison between the ENSO events was achieved for choosing five spatial clusters (SCs) close to height 12.5 (distance between clusters) for all datasets in this study.

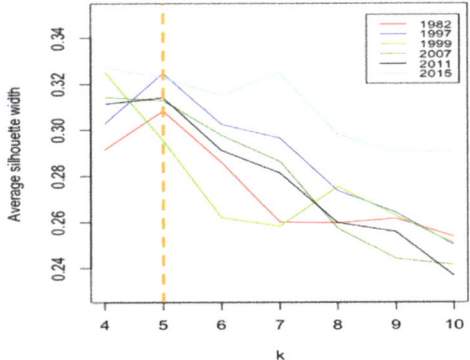

Figure 5. The average silhouette width for spatial cluster analysis by number of clusters and years.

Five spatial clusters, SC1, SC2, SC3, SC4, and SC5, which were sorted according to the amount of precipitation from ascending to high, were formed and displayed on a spatial map in Figure 6. It was obvious that precipitation was the only meteorological data to noticeably differ between clusters. Spatial clustering in El Niño events was mostly grouped in SC2 (yellow) with 62–66 members except in 1982, which mostly in SC1 (red) with 59 members; however, its average rainfalls were nearly the same to SC2, whereas spatial clustering in La Niña events was mostly grouped in SC1 (red) with 61–83 members. While SC5 (pink) was the least populated member with one member, which was the station in the east for both events (Table 3). These showed most areas in Thailand had low precipitation rate.

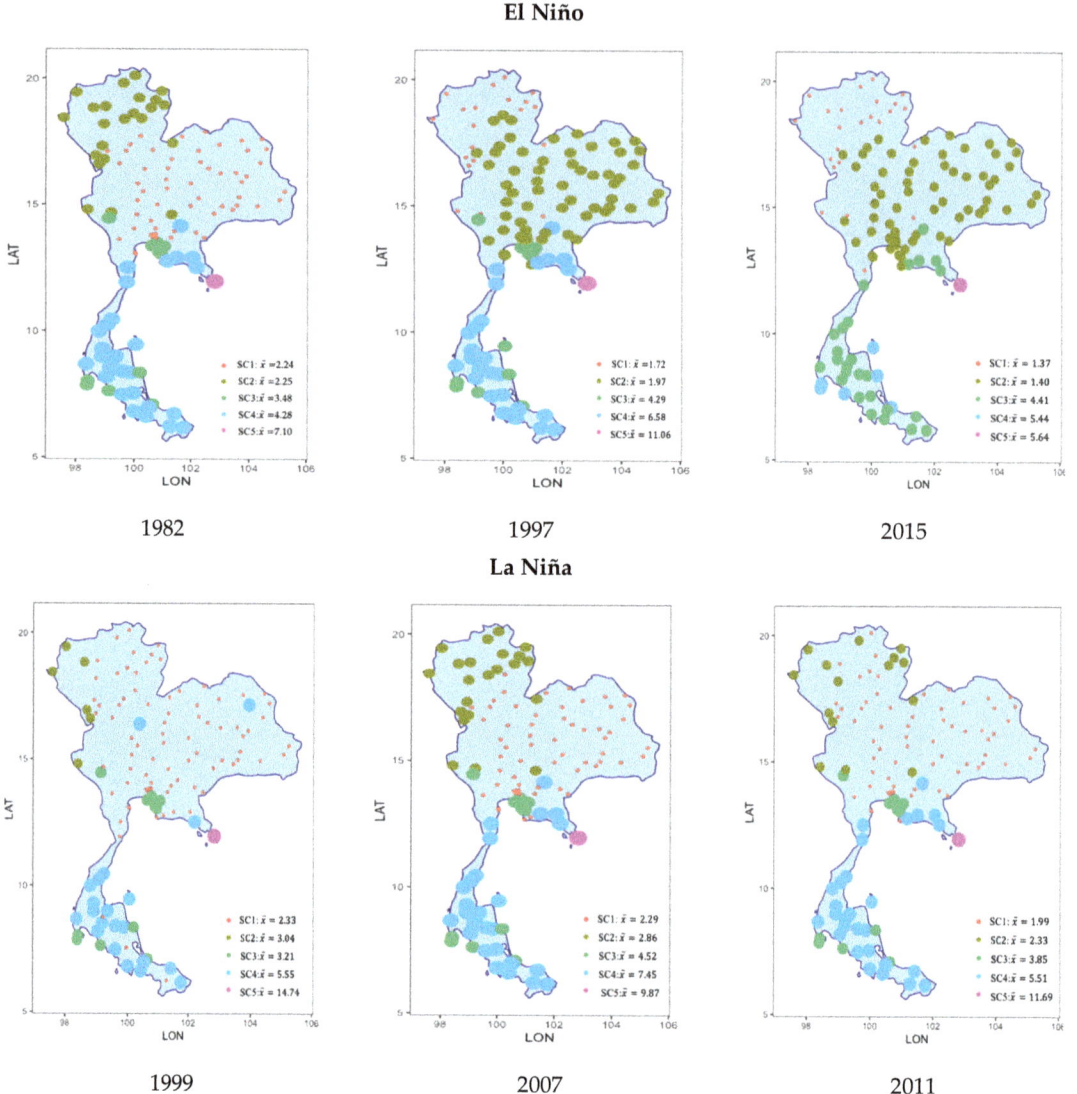

Figure 6. Spatial cluster analysis (SC1–SC5) for the 124 Thai meteorological stations on a map by ENSO events and years ($\bar{x}$ is monthly rainfall average).

Table 3. Number of members and mean of monthly climate factors by ENSO events, years and SCs.

Year	SC	n	C (n = 26)	E (n = 15)	N (n = 16)	NE (n = 28)	S (n = 27)	W (n = 12)	Rainfall	Relative Humidity	Solar Down Welling	Temperature	Max Temperature	Min Temperature	Horizontal Wind
								El Niño							
1982	1	59	23	5	1	26	-	4	2.24	64.56	178.31	27.47	32.46	23.01	3.08
	2	23	1	1	15	2	-	5	2.25	69.52	182.24	25.11	31.42	19.82	1.97
	3	12	2	4	-	-	5	1	3.48	75.41	188.43	27.26	29.32	25.38	4.14
	4	29	-	5	-	-	22	2	4.28	87.29	142.18	25.78	29.74	22.79	1.99
	5	1	-	1	-	-	-	-	7.10	78.55	163.36	25.90	28.25	24.00	3.18
	Total							$\bar{x}$	2.88	71.96	171.45	26.60	31.30	22.60	2.72
								s	3.23	14.26	34.31	2.44	2.56	3.24	1.11
1997	1	20	1	-	12	2	-	5	1.72	66.95	191.63	25.64	32.24	20.07	2.07
	2	62	23	5	4	26	-	4	1.97	62.51	184.51	28.06	33.29	23.30	3.07
	3	13	2	4	-	-	6	1	4.29	76.00	183.52	27.54	29.60	25.67	4.00
	4	28	-	5	-	-	21	2	6.58	87.91	136.45	25.96	29.87	23.00	1.85
	5	1	-	1	-	-	-	-	11.06	79.55	149.37	26.18	28.38	24.37	3.32
	Total							$\bar{x}$	3.29	70.51	174.42	27.13	31.92	22.97	2.73
								s	4.60	14.76	39.19	2.31	2.65	3.15	1.10
2015	1	24	1	-	15	2	-	6	1.37	60.61	195.98	26.79	33.67	20.81	2.12
	2	66	25	9	1	26	-	5	1.40	57.90	191.31	29.24	34.31	24.50	3.48
	3	27	-	5	-	-	21	1	4.41	86.58	140.25	26.66	30.76	23.55	1.92
	4	6	-	5	-	-	6	-	5.44	80.10	163.10	27.84	29.45	26.47	3.60
	5	1	-	1	-	-	-	-	5.64	76.31	167.87	26.99	29.60	24.91	3.21
	Total							$\bar{x}$	2.28	65.89	179.54	28.12	33.14	23.68	2.88
								s	2.85	15.93	36.99	2.34	2.73	3.02	1.22
								La Niña							
1999	1	83	23	9	13	27	3	8	2.33	62.44	192.34	27.38	33.02	22.31	2.99
	2	6	-	-	3	-	-	3	3.04	73.09	184.80	24.29	31.00	19.03	1.83
	3	12	2	4	-	-	5	1	3.21	72.95	197.32	27.81	30.03	25.71	4.41
	4	22	1	4	-	1	19	1	5.55	88.22	142.53	26.12	30.09	23.13	1.99
	5	1	-	1	-	-	-	-	14.74	76.97	161.99	26.40	28.80	24.42	3.79
	Total							$\bar{x}$	3.12	68.66	183.38	27.04	32.08	22.64	2.90
								s	3.89	16.07	34.45	2.09	2.28	3.16	1.16
2007	1	61	23	6	2	26	-	4	2.29	64.40	181.10	27.71	32.82	23.08	3.05
	2	22	1	-	14	2	-	5	2.86	70.56	187.98	25.10	31.40	19.78	1.96
	3	12	2	4	-	-	5	1	4.52	74.00	184.46	27.87	29.99	25.89	4.16
	4	28	-	4	-	-	22	2	7.45	87.43	139.07	26.22	30.18	23.21	1.91
	5	1	-	1	-	-	-	-	9.87	77.86	162.88	26.42	28.75	24.43	3.33
	Total							$\bar{x}$	3.83	71.73	173.01	26.92	31.67	22.81	2.71
								s	5.38	14.20	35.00	2.33	2.46	3.15	1.12
2011	1	67	23	5	7	26	-	6	1.99	62.83	177.59	27.60	32.78	22.93	3.00
	2	15	1	-	9	2	-	3	2.33	68.55	181.98	24.91	31.09	19.76	1.95
	3	12	2	4	-	-	5	1	3.85	74.46	178.97	27.75	29.76	25.84	4.22
	4	29	-	5	-	-	22	2	5.51	87.37	132.16	26.20	29.91	23.45	2.03
	5	1	-	1	-	-	-	-	11.69	77.86	155.50	26.31	28.65	24.33	3.21
	Total							$\bar{x}$	3.11	70.51	167.45	26.95	31.58	22.96	2.76
								s	3.96	15.88	37.61	1.88	2.16	2.92	1.34

n—number of members, C—Central, E—East, N—North, NE—Northeast, S—South, W—West.

In La Niña event, SC1 (red) was found mostly in Northeast and Central areas, which had the least amount of rainfall, and SC2 (yellow) was widely distributed in the North, which had low rainfall. SC3 (green) with moderate rainfall were distributed among all regions, except North and Northeast, while SC4 (blue) are in the south which had quite a lot of rainfall. Lastly, SC5 (pink) with the highest rainfall had one station in the East (Table 3).

While, spatial clustering in El Niño was differently distributed by years. In 1997 and 2015, SC1 (red) was found mostly in the North and SC2 (yellow) was widely distributed in the Northeast, and vice versa in 1982. In 1982 and 1997, SC3 (green) with moderate rainfalls were distributed among all regions, except North and Northeast, and SC4 (blue) were in the South which had quite a lot of rainfall, and vice versa in 2015. In every year, SC5 (pink) with the highest rainfall had one station in the East (Table 3).

The spatial clistering extracted the drought areas in the North region, classified as SC1 with less rainfall than SC2 and in the South region classified as SC3 with less rainfall than SC4 for the El Niño event. These areas would be at risk to be the most drought-prone areas. This suggested the effect of ENSO on spatial clustering.

The distribution of SGs over six regions, showing a clear trend in the redistribution of SGs observed in this study, is shown in Table 4. More diverse climate was found in the East and West than other regions. All regions had a heterogenous meteorological distribution. Every year for both El Niño and La Niña events had 2–4 SCs. However, less distribution for El Niño (2015) in Central, East, and West regions and for La Niña (1999) in the West, and more distribution for La Niña (1999) in the South were noted. These would be due to changes in TGs and intensity of climate factors.

Table 4. The distribution of SGs over six regions for ENSO.

Region	Number of Members	El Niño			La Niña		
		1982	1997	2015	1999	2007	2011
Central	26	3	3	2	3	3	3
East	15	4	4	3	4	4	4
North	16	2	2	2	2	2	2
Northeast	28	2	2	2	2	2	2
South	27	2	2	2	3	2	2
West	12	4	4	3	3	4	4

3.3. Temporal Clustering

After spatial cluster analysis had been obtained, a Euclidean distance timed and spaced with average linkage was next applied to the monthly climate factors for each SC to find temporal clusters (TCs) within each SG. Normally, Thailand has three seasons, summer (February–May), rainy (May–October), and winter (October–February). To compare temporal clusters of the ENSO phenomenon, three TCs within each SC were compared in this study. TC1, TC2, and TC3, which were sorted according to the amount of ascending precipitation, were represented by orange, blue, and green, respectively. TCs corresponding to each SG is shown in the dendrogram to depict the groups of clusters and their combination, indicating dissimilarity in the vertical scale and the samples (months) in clustering order on the horizontal axis. They help to see how long each season lasts and the different period of seasons in each spatial grouping (Figure 7).

Figure 7. Dendrograms of TGs for the five different SCs discovered by ENSO events, years and SCs: (**a**) SC1; (**b**) SC2; (**c**) SC3; (**d**) SC4; (**e**) SC5.

For example, in 1982, TC1 and TC2 in SC1 depicted a very dry season with average precipitation intensity of less than 2 mm/day (Table 5). They were composed of three months. Months of TC1 were December and of TC2 were January and February. TC3, on the other hand, was a slightly wet season with an average precipitation of 2 mm/day or more for 9 months, March–November.

Table 5. Number of members (months) and mean of monthly climate factors by ENSO events, years, SCs and TCs.

Year	SCs			TC1						TC2						TC3					
	SC	n	n	RF	RH	SDW	Temp	Wind	n	RF	RH	SDW	Temp	Wind	n	RF	RH	SDW	Temp	Wind
									El Niño											
1982	1	59	2	0.45	51.90	217.32	25.64	3.53	1	0.64	56.18	203.21	22.77	3.91	9	2.82	68.30	166.87	28.39	2.89
	2	23	3	0.33	47.45	233.93	27.23	2.18	2	0.41	60.00	199.95	20.75	2.03	8	3.19	77.42	164.89	25.67	1.90
	3	12	1	0.78	67.94	199.38	25.38	5.82	2	1.08	70.43	202.37	26.12	5.22	9	4.31	77.35	184.12	27.72	3.71
	4	29	2	1.91	85.07	145.13	24.52	2.63	2	2.24	84.68	139.28	24.04	2.92	9	5.03	88.07	141.84	26.26	1.75
	5	1	2	1.21	68.36	216.51	24.63	4.14	1	4.50	71.78	206.18	23.65	5.33	9	8.70	81.57	146.79	26.43	2.72
1997	1	20	4	1.53	66.75	193.05	25.77	2.02	7	1.78	67.14	190.66	25.57	2.09	1	2.07	66.46	192.71	25.56	2.24
	2	62	4	1.91	62.18	185.43	28.02	3.03	4	1.97	62.49	184.67	28.11	3.11	4	2.02	62.85	183.43	28.04	3.06
	3	13	5	3.38	74.61	188.82	27.43	3.96	2	3.82	76.89	186.10	27.56	4.19	5	5.39	77.02	177.19	27.64	3.95
	4	28	3	6.42	87.73	136.62	25.90	1.83	4	6.49	87.92	134.76	25.97	1.84	5	6.75	88.00	137.71	25.99	1.87
	5	1	2	0.01	70.45	221.12	24.34	5.25	4	4.12	77.73	163.98	26.08	3.25	6	19.37	83.80	115.71	26.87	2.72
2015	1	24	3	0.26	45.56	237.11	27.97	2.23	3	0.34	54.87	192.98	23.42	2.21	6	2.44	71.01	176.91	27.88	2.01
	2	66	3	0.44	54.75	176.82	26.57	3.98	8	1.33	56.87	200.20	30.27	3.42	1	4.88	75.60	163.62	28.95	2.52
	3	27	2	1.79	85.62	151.94	25.27	2.08	9	4.88	87.07	142.13	27.14	1.74	5	5.49	84.11	100.02	25.09	3.15
	4	6	2	2.13	78.12	179.29	26.77	4.22	1	5.84	82.57	108.83	26.40	5.88	9	6.14	80.27	165.54	28.24	3.21
	5	1	3	0.30	69.08	203.24	25.66	4.22	3	1.33	71.78	185.00	27.42	3.68	6	10.46	82.20	141.63	27.44	2.46
									La Niña											
1999	1	83	5	0.17	49.95	213.89	25.78	3.36	2	1.86	61.83	187.93	29.51	2.70	5	4.68	75.17	172.57	28.13	2.74
	2	6	5	0.11	59.90	219.48	22.68	1.87	2	1.16	65.64	196.45	27.46	1.62	5	6.73	89.27	145.45	24.64	1.87
	3	12	2	1.69	65.81	199.37	26.91	5.33	2	3.70	77.43	184.96	28.20	3.54	6	4.06	76.22	200.08	28.27	4.08
	4	22	1	4.61	89.13	103.48	24.90	2.97	3	4.77	87.38	140.23	25.08	2.39	8	5.96	88.42	148.28	26.67	1.71
	5	1	4	0.85	64.87	225.47	25.49	4.78	6	14.34	81.83	141.38	26.80	2.92	2	43.70	86.56	96.88	27.01	4.43
2007	1	61	6	2.22	64.76	181.64	27.62	3.10	2	2.33	63.08	181.16	27.88	3.04	4	2.38	64.51	180.27	27.76	2.98
	2	22	2	2.41	70.04	188.90	25.29	1.95	8	2.95	70.70	186.97	25.02	1.95	2	2.95	70.53	191.07	25.26	2.01
	3	12	6	2.12	70.71	197.18	28.12	4.51	1	2.83	69.34	211.15	26.99	4.00	5	7.74	78.87	163.87	27.76	3.76
	4	28	4	7.09	87.45	140.04	26.20	1.89	4	7.19	87.33	139.16	26.24	1.91	4	8.06	87.52	138.01	26.23	1.92
	5	1	1	0.06	64.49	215.43	24.41	5.29	4	2.51	72.80	196.31	25.93	4.26	7	15.47	82.66	136.28	26.99	2.52
2011	1	67	1	0.02	45.31	222.12	23.50	4.01	5	0.43	52.37	188.69	27.67	3.72	6	3.61	74.47	160.92	28.22	2.23
	2	15	2	0.03	50.35	210.01	21.59	2.51	4	0.27	54.41	210.47	25.88	2.16	6	4.47	84.04	153.64	25.37	1.63
	3	12	1	1.66	65.51	186.20	25.98	6.57	8	3.84	77.03	184.95	27.93	3.56	3	4.62	70.56	160.61	27.86	5.20
	4	29	1	3.77	84.22	117.80	24.59	3.30	8	5.00	87.95	142.08	26.45	1.69	3	7.42	86.86	110.50	26.10	2.50
	5	1	1	0.10	67.75	214.62	24.21	6.09	4	1.28	71.65	182.57	26.23	3.96	7	19.29	82.86	131.58	26.66	2.38

n—number of members, RF: rainfall, RH: relative humidity, SWD: solar downwelling, Temp: temperature, Wind: horizontal wind.

TC1 and TC2 in SC2, 1982 depicted a very dry season with average precipitation intensity of less than 2 mm/day. TC1 was composed of February–March. TC2 was December and January. TC3, on the other hand, was a wet season with an average precipitation of 4.51 mm/day for 8 months, April–November.

TC1 and TC2 in SC3, 1982 was a dry season for three months, December (TC1) and January–February (TC2), with an average precipitation of less than 2 mm/day, while TC3 was a wet season with an average precipitation of 4 mm/day or more for 9 months, March–November.

TCs in SC4 and SC5, 1982 were the same. TC1 was a dry season, January–February, with an average precipitation of less than 2 mm/day, while TC2 and TC3 were a wet season with an average precipitation of 2 mm/day or more for 10 months, December (TC2) and March–November (TC3).

TCs in each SC in La Niña were similar to those in El Niño. Nevertheless, there was higher average precipitation intensity in La Niña phenomenon, than those in El Niño phenomenon. Furthermore, the rainy season was a longer period in SC4 and SC5 for both events of ENSO.

4. Discussion

The highest average rainfall in 1982, 1997, and 2015 (5.64–11.06 mm/day) was less than that of in 1999, 2007, and 2011 (9.87–14.74 mm/day). This corresponds to the Oceanic Niño Index (ONI), showing that ONI in 1982, 1997, and 2015 was greater than 0.5 °C, meaning that El Niño occurred, and in 1999, 2007, and 2011 was below −0.5 °C, meaning that La Niña occurred [29].

Lower rainfall than usual was found, so there was a widespread drought in almost all regions of Thailand in 1982 and 1997, especially in Northeast [30]. There also was a severe El Niño effect in 2015, causing very low precipitation across the country ($\bar{x}$ = 2.28 mm/day).

Five spatial clusterings were formed. SC5 with the highest average precipitation was formed by only one station in Khlong Yai District, Trat Province, in every year whether there was an El Niño or La Niña phenomenon ($\bar{x}$ = 5.64 − 14.74 mm/day). The topography of Khlong Yai District is a coastline fully influenced by the southwest monsoon from the Gulf of Thailand; consequently, it has abundant rainfall for most of the year. This is consistent with the Trat Agricultural Meteorological Document that reports that Khlong Yai District, Trat Province, is the wettest area in Thailand [31].

There were approximately 80 stations in SC1 and SC2 with low average precipitation and especially low in 2015, mostly in the Central, North, and Northeast. It was consistent with a report that rainfall in these three regions when El Niño occurred was less than the average 30 years of rainfall of normal years.

There were three TCs in each SC. When the El Niño phenomenon occured, Thailand rainfall tended to be lower than normal, especially during the summer and early rainy season (mid-February–June). The dry season in El Niño was longer and less than average rainfall than TCs for the La Niña phenomena.

Most stations in the south were clustered into SC3 and SC4 with moderate and high rainfall, respectively, for both El Niño and La Niña phenomena. Usually, rainfall in Thailand, especially in the southeast coast, is high during October–December. In addition, some parts of Thailand were not affected by the ENSO phenomenon (El Niño and La Niña), such as Trat in SC5 with the highest rainfall, and Tak, Chiang Rai, Chiang Mai, Phayao, and Lampang in SC1 with the least rainfall. This may be due to their topography.

There are 35 provinces with more than one meteorological station of TMD. Of these, stations in 34 provinces were grouped into different SCs. This may be due to their topography affecting a different climate.

Spatial clusters were similar for both El Niño and La Niña except in 2015, when severe El Niño occurred. This might be the Euclidian distance matrix tending to cluster the samples with climate variables having similar mean. This suggests that other similarity matrices, such as correlation, may be possible to group samples based on trends and variation over time [11].

5. Conclusions

This paper employed multivariate cluster analysis with the average linkage to analyze the spatial and temporal grouping, using climate factors which are rainfall, relative humidity, average temperature, maximum temperature, lowest temperature, solar radiation, and wind speed at 124 locations over Thailand from CCAM (10 km), for the years 1982, 1997, and 2015 (El Niño) and 1999, 2007, and 2011 (La Niña).

Five SCs with a distance between a cluster of 12.5 were compared. It was observed that SCs were similar for both El Niño and La Niña except in 2015, when severe El Niño occurred. This indicated the more severe El Niño, the more spatial variation. The main difference between SC1–SC5 was the ascending amount of precipitation, where SC1 had the least amount of rainfall and SC5 had the heaviest rainfall.

In addition, three TC patterns in each SC were similar for both El Niño and La Niña. Nevertheless, the average precipitation intensity in La Niña was higher than that in El Niño.

This paper implements cluster analysis on atmospheric panel data. Even multivariable panel data is more complicated, but it is practical to cluster. Cluster results arealso more realistic than cross-sectional data and avoid information loss.

Future studies may focus on using future climate factors from the weather forecast models for clustering to study the spatial and temporal distributions. Other than the correlation distance suggested, the robust distance, for example the absolute distance or the Canberra distance to deal with outliers, should be further studied. Furthermore, as there might be extreme whether events in the ENSO phenomenon, for example less or abundant precipitation, which may affect the clustering, outliers should be detected and handled prior.

Author Contributions: Conceptualization, P.D.; methodology, P.D.; software, P.D., N.J., P.Y. and S.P.; validation, P.D.; formal analysis, P.D., N.J., P.Y. and S.P.; investigation, P.D.; resources, P.D., N.J., P.Y. and S.P.; data curation, P.D., N.J., P.Y. and S.P.; writing—original draft preparation, P.D., N.J., P.Y. and S.P.; writing—review and editing, P.D., U.H. and S.I.; visualization, P.D.; supervision, P.D.; project administration, P.D., U.H. and S.I. All authors have read and agreed to the published version of the manuscript.

Funding: This research project was supported by the Agricultural Research De-velopment Agency (ARDA) [PRP6405031190] and Thailand Sciece Research and Innovation (TSRI). Basic Research Fund: Fiscal year 2022 under project number FRB650048/0164.

Conflicts of Interest: The authors declare no conflict of interest.

References

1. TMD, Thai Meteorological Department, Enso Phenomenon. Available online: https://www.tmd.go.th/info/info.php?FileID=19 (accessed on 27 September 2020).
2. TMD, Thai Meteorological Department, El Niño. Available online: https://www.tmd.go.th/info/info.php?FileID=18 (accessed on 27 September 2020).
3. Office of Agricultural Economics, The Ministry of Agriculture and Cooperatives. *The Agriculture Development Plan under the Twelfth National Economic and Social Development Plan (2017–2021)*; The Ministry of Agriculture and Cooperatives: Bangkok, Thailand, 2016.
4. Tao, F.; Yokozawa, M.; Zhang, Z.; Hayashi, Y.; Grass, H.; Fu, C. Variability in climatology and agricultural production in China in association with the East Asian summer monsoon and El Niño Southern Oscillation. *Clim. Res.* **2004**, *28*, 23–30. [CrossRef]
5. Roberts, M.G.; Dawe, D.; Falcon, W.P.; Naylor, R.L. El Niño Southern Oscillation impacts on rice production in Luzon, the Philippines. *J. Appl. Meteorol. Climatol.* **2009**, *48*, 1718–1724. [CrossRef]
6. Xiangzheng, D.; Jikun, H.; Fangbin, Q.; Naylor, R.L.; Falcon, W.P.; Burke, M.; Rozelle, S.; Battisti, D. Impacts of El Niño-Southern Oscillation events on China's rice production. *J. Geogr. Sci.* **2010**, *20*, 3–16.
7. Bhuvaneswari, K.; Geethalakshmi, V.; Lakshmanan, A.; Srinivasan, R.; Sekhar, N.U. The impact of El Niño/Southern Oscillation on hydrology and rice productivity in the Cauvery Basin, India: Application of the soil and water assessment tool. *Weather Clim. Extrem.* **2013**, *2*, 39–47. [CrossRef]

8. Iizumi, T.; Luo, J.J.; Challinor, A.J.; Sakurai, G.; Yokozawa, M.; Sakuma, H.; Brown, M.E.; Yamagata, T. Impacts of El Niño Southern Oscillation on the global yields of major crops. *Nat. Commun.* **2014**, *5*, 3712. [CrossRef] [PubMed]
9. Limsakul, A. Impacts of El Niño-Southern Oscillation (ENSO) on Rice Production in Thailand during 1961–2016. *Environ. Nat. Resour. J.* **2019**, *17*, 30–42. [CrossRef]
10. Mayuening, E. Statistical Methods for Analyzing Rainfall in Thailand and the Southern Oscillation Index. Doctoral Dissertation, Prince of Songkla University, Pattani Campus, Pattani, Thailand, 2015.
11. Santos, C.A.G.; Brasil Neto, R.M.; da Silva, R.M.; Costa, S.G.F. Cluster Analysis Applied to Spatiotemporal Variability of Monthly Precipitation over Paraíba State Using Tropical Rainfall Measuring Mission (TRMM) Data. *Remote Sens.* **2019**, *11*, 637. [CrossRef]
12. Pinidluek, P.; Konyai, S.; Sriboonlue, V. Regionalization of rainfall in Northeastern Thailand. *GEOMATE J.* **2020**, *18*, 135–141. [CrossRef]
13. Moonsri, K.; Pochanart, P. The effect of climate change on sugarcane productivity in Northeastern Thailand. *J. Environ. Manag.* **2019**, *15*, 46–61.
14. Iyigun, C.; Türkeş, M.; Batmaz, İ.; Yozgatligil, C.; Purutçuoğlu, V.; Koç, E.K.; Öztürk, M.Z. Clustering current climate regions of Turkey by using a multivariate statistical method. *Theor. Appl. Climatol.* **2013**, *114*, 95–106. [CrossRef]
15. Zheng, B.; Li, S. Multivariable panel data cluster analysis and its application. *Comput. Model. New Technol.* **2014**, *18*, 553–557.
16. Akay, Ö.; Yüksel, G. Hierarchical clustering of mixed variable panel data based on new distance. *Commun. Stat. Simul. Comput.* **2021**, *50*, 1695–1710. [CrossRef]
17. Ramadhan, R.; Awalluddin, A.S.; Cahyandari, R. Multivariable Panel Data Cluster Analysis using Ward Method Gross Enrollment Ratio (GER) Data in West Java in the Year 2015–2018. In Proceedings of the International Conference on Science and Engineering, Yogyakarta, Indonesia, 28 November 2019. [CrossRef]
18. Wang, W.; Lu, Y. Application of Clustering Analysis of Panel Data in Economic and Social Research Based on R Software. *Acad. J. Bus. Manag.* **2021**, *3*, 98–104. [CrossRef]
19. Null, J. El Niño and La Niña Years and Intensities Based on Oceanic Niño Index (ONI). Available online: https://ggweather.com/enso/oni.htm (accessed on 11 April 2022).
20. Katzfey, J.; Nguyen, K.; McGregor, J.; Hoffmann, P.; Ramasamy, S.; Nguyen, H.V.; Khiem, M.V.; Nguyen, T.V.; Truong, K.B.; Vu, T.V.; et al. High-Resolution for Vietnam—Methodology and Evolution of Current Climate. *Asia Pac. J. Atmos. Sci.* **2016**, *52*, 91–106. [CrossRef]
21. Thevakaran, A.; McGregor, J.L.; Katzfey, J.; Hoffmann, P.; Suppiah, R.; Sonnadara, D.U.J. An Assessment of CSIRO Conformal Cubic Atmospheric model Simulation over Sri Lanka. *Clim. Dyn.* **2016**, *46*, 1861–1875. [CrossRef]
22. Pham, T.H.; Hoang, D.D.; Pham, Q.M.; Katzfey, J.; McGregor, J.L.; Nguyen, K.C.; Tran, Q.D.; Nguyen, M.L.; Phan, V.T. Implementation of Tropical Cyclone Detection Scheme to CCAM model for Seasonal Tropical Cyclone Prediction over the Vietnam East Sea. *VNU J. Sci. Earth Environ. Sci.* **2019**, *35*, 49–60.
23. Gareth, J.; Daniela, W.; Trevor, H.; Robert, T. *An Introduction to Statistical Learning: With Applications in R*; Springer: Berlin/Heidelberg, Germany, 2013.
24. Sokal, R.R.; Michener, C.D. A statistical method for evaluating systematic relationships. *Univ. Kans. Sci. Bull.* **1958**, *38*, 1409–1438.
25. Kaufman, L.; Rousseeuw, P.J. *Findings Groups in Data: An Introduction to Cluster Analysis*; John Wiley and Sons: Hoboken, NJ, USA, 2005.
26. RStudio Team. *RStudio: Integrated Development for R*; RStudio: Boston, MA, USA, 2020.
27. Rousseeuw, P.J. Silhouettes: A graphical aid to the interpretation and validation of cluster analysis. *J. Comput. Appl. Math.* **1987**, *20*, 53–65. [CrossRef]
28. Maechler, M.; Rousseeuw, P.J.; Struyf, A.; Hubert, M.; Hornik, K.; Studer, M.; Roudier, P.; Gonzalez, J.; Kozlowski, K.; Schubert, E.; et al. "Finding Groups in Data": Cluster Analysis Extended Rousseeuw et al. Available online: https://cran.r-project.org/web/packages/cluster/cluster.pdf (accessed on 20 February 2020).
29. National Oceanic and Atmospheric Administration, Cold & Warm Episodes by Season. Available online: https://origin.cpc.ncep.noaa.gov/products/analysis_monitoring/ensostuff/ONI_v5.php (accessed on 20 February 2020).
30. Amatayakul, P.; Chomtha, T. *Agricultural Meteorology to Know for Trat. Technical Document No. 551.6593-07-2016*; Agrometeorological Division, Meteorological Development Bureau, Thai Meteorological Department: Bangkok, Thailand, 2016.
31. Prapertchob, P.; Bhandari, H.; Pandy, S. Economic cost of drought and rice farmers' coping mechanisms in Northeast Thailand. In *Economic Costs of Drought and Rice Farmers' Coping Mechanisms: A Cross-Country Comparative Analysis*; Pandey, S., Bhandari, H., Hardy, B., Eds.; International Rice Research Institute: Los Baños, Philippines, 2007; pp. 113–148.

Article

Applications of the Sine Modified Lindley Distribution to Biomedical Data

Lishamol Tomy [1], Veena G [2] and Christophe Chesneau [3,*]

1. Department of Statistics, Deva Matha College, Kuravilangad 686633, Kerala, India; lishatomy@gmail.com
2. Department of Statistics, St.Thomas College, Palai 686574, Kerala, India; veenagpillai@hotmail.com
3. Laboratoire de Mathématiques Nicolas Oresme (LMNO), Université de Caen Normandie, Campus II, Science 3, 14032 Caen, France
* Correspondence: christophe.chesneau@gmail.com

Abstract: In this paper, the applicability of the sine modified Lindley distribution, recently introduced in the statistical literature, is highlighted via the goodness-of-fit approach on biological data. In particular, it is shown to be beneficial in estimating and modeling the life periods of growth hormone guinea pigs given tubercle bacilli, growth hormone treatment for children, and the size of tumors in cancer patients. We anticipate that our model will be effective in modeling the survival times of diseases related to cancer. The R codes for the figures, as well as information on how the data are processed, are provided.

Keywords: goodness-of-fit; biomedical data; Lindley distribution; trigonometric function; continuous distribution

1. Introduction

When people are diagnosed with cancer, COVID-19, or any other severe condition, or when clinical trials of a new treatment are conducted, survival is a major concern. Carcinoma is a general term that refers to a variety of diseases that can affect any part of the body. One of the risk factors is the abrupt development of aberrant cells that grow beyond their usual bounds, allowing them to infect nearby portions of the body and travel to other organs, which is one of the risk factors; this is known as metastasis. Widespread metastases are the leading cause of cancer death.

Doctors seek to regulate the growth and size of these tumors in the places where they arise in order to protect human lives. The tumor stage and survival times are two aspects, as they aid doctors in determining the best treatment for their patients. As a result, determining the probability distribution of tumor size and survival durations is crucial for selecting the best treatment option.

In order to analyze the numbers of people who are diagnosed with and die from severe diseases each year, the number of people who are currently living after the diagnosis of a disease, the mean age at which a disease was diagnosed, and the number of people who are still alive at a given time after diagnosis, statistics can be used. It also gives an idea of the differences among groups defined by age, sex, racial/ethnic group, geographic location, and other categories.

One such way of analyzing the properties of the survival data or the size of the tumor is by modeling the data. Data modelling related to biological science is of utmost importance to understanding the data statistically. Over the years, many researchers have developed discrete as well as continuous distributions that help in modelling biological data. Ref. [1] developed the Marshall–Olkin Inverse Lomax distribution (MO-ILD), which is used in modeling cancer stem cells. Ref. [2] studied the weighted generalized Quasi Lindley distribution, which was studied to model COVID-19 data from Algeria and Saudi Arabia, and Ref. [3] modeled the survival times of guinea pigs infected with virulent tubercle

bacilli using the Sine Half-Logistic Inverse Rayleigh distribution. With this motivation in mind, we use the existing sine-modified Lindley (S-ML) distribution developed by [4] in modelling data related to different types of cancer. We also provide optimized open source S-ML distribution codes for practitioners to use.

This paper is structured as follows. Section 2 covers a review of the existing S-ML distribution. Section 3 includes the application of the distribution to cancer data, as well as various visual presentations to back up the numbers, and Section 4 concludes the study.

2. The S-ML Distribution

In this section, a brief review of the definitions and properties of the sine generated (S-G or Sin-G in some references) family of distributions, the modified Lindley distribution, and the S-ML distribution is implemented. Due to their application and operating capability in a range of contexts, the families defined by "trigonometric transformations" have sparked a lot of interest in recent years. The sinusoidal transformation that contributes to the S-G family was initially studied by [5].

2.1. S-G Family of Distributions

The corresponding basic definitions of the associated distribution function (DF) and PDF given, respectively, by

$$F_{S-G}(y;\gamma) = \sin\left[\frac{\pi}{2}G(y;\gamma)\right], \quad y \in \mathbb{R}$$

and

$$f_{S-G}(y;\eta) = \frac{\pi}{2}g(y;\gamma)\cos\left[\frac{\pi}{2}G(y;\gamma)\right], \quad y \in \mathbb{R}$$

where $G(y;\gamma)$ and $g(y;\gamma)$ are the DF and PDF of a certain continuous distribution with parameter vector denoted by γ, respectively. These functions are linked to a reference or parent distribution that the practitioner determines ahead of time based on the study's goals. The S-G family is well-known as a potential parent family alternative. Without introducing extra parameters, the following stochastic ordering holds: $G(y;\gamma) \leq F_{S-G}(y;\gamma)$ for every $y \in \mathbb{R}$. The S-G family provides the capability to develop flexible statistical models that can handle a variety of data. The recent works on the S-G family include the sine Lindley and the sine exponential distribution introduced by [6], the transformed S-G family studied by [7], the sine Topp Leone-G family of distributions developed by [8], sine Kumaraswamy-G family introduced by [9], the sine extended odd Fréchet-G family of distributions studied by [10], and the sine power Lomax model by [11].

Ref. [4] improved the S-G family's performance by applying it to a specific one-parameter distribution established by [12]: the modified Lindley (ML) distribution. The S-ML distribution was developed as a result.

2.2. Modified Lindley Distribution

The ML distribution proposed by [12] is made possible by applying the tuning function $e^{-\beta y}$, $\beta > 0$ to the Lindley distribution with the goal of boosting its capabilities in a variety of domains. As a result, the ML distribution is defined by the DF expressed as follows:

$$G_{ML}(y;\beta) = 1 - \left[1 + \frac{\beta y}{1+\beta}e^{-\beta y}\right]e^{-\beta y}, \quad y > 0.$$

The PDF is given by

$$g_{ML}(y;\beta) = \frac{\beta}{1+\beta}e^{-2\beta y}\left[(1+\beta)e^{\beta y} + 2\beta y - 1\right], \quad y > 0,$$

respectively, with $\beta > 0$, and $G_{ML}(y;\beta) = g_{ML}(y;\beta) = 0$ for $y \leq 0$.

The ML distribution adapts to rising, reverse bathtub, and constant hazard rates and is a mixture of the exponential and gamma distributions with parameters β and $(2, 2\beta)$.

The practical benefit is very significant; for the three data sets shown in [12], the ML model outperforms the Lindley and exponential models. The wrapped modified Lindley distribution proposed by [13] and the inverted modified Lindley distribution proposed by [14] are two examples of improvements to the ML distribution.

2.3. S-ML Distribution

The corresponding DF and PDF of the S-ML distribution, respectively,

$$F_{S\text{-}ML}(y;\beta) = \cos\left[\frac{\pi}{2}\left(1 + e^{-\beta y}\frac{y\beta}{1+\beta}\right)e^{-\beta y}\right], \quad y > 0$$

and

$$f_{S\text{-}ML}(y;\beta) = \frac{\pi}{2}\frac{\beta}{1+\beta}e^{-2\beta y}\left[(1+\beta)e^{\beta y} + 2y\beta - 1\right]$$

$$\sin\left[\frac{\pi}{2}\left(1 + e^{-\beta y}\frac{y\beta}{1+\beta}\right)e^{-\beta x}\right], \quad y > 0,$$

with $\beta > 0$, and $F_{S\text{-}ML}(y;\beta) = f_{S\text{-}ML}(y;\beta) = 0$ for $y \leq 0$.

By varying the value of β, different variants of $f_{S\text{-}ML}(y;\beta)$ can be obtained. Figure 1 depicts the most representative of them.

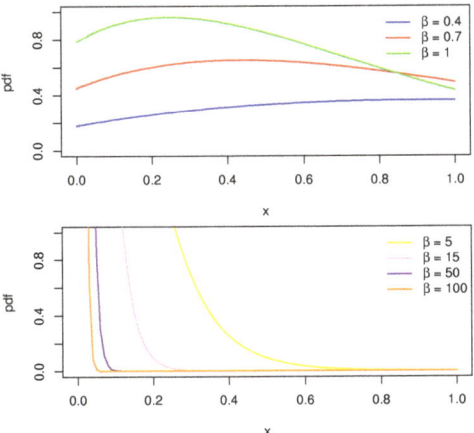

Figure 1. Illustration of $f_{S\text{-}ML}(y;\beta)$ with selected values of β.

We can imply from Figure 1, that for

smaller values of β, local increasing shape are seen; the distribution is unimodal, larger values of β, the plot of $f_{S\text{-}ML}(y;\beta)$ decreases and is leptokurtic in shape.

The shapes of the S-ML probability density function (PDF) are found to be adaptable to different shapes, being unimodal, decreasing, and right-skewed.

The S-ML distribution has also been shown to exhibit a non-monotonic hazard rate function (HRF), depicting an increasing-reverse bathtub-constant shape. The distribution's applicability and adaptability make it very appealing for modeling data from various fields and [4] has proved that the model stands strong against twelve other competent distributions, such as the generalized beta type 2 distribution introduced by [15], the Lomax distribution studied by [16], and the lognormal distribution developed by [17] in modelling data related to weather and engineering.

3. Applications

In the statistical literature on life-testing experiments, numerous distributions have been developed. Some of which can be used to model the increase or decrease in failure rates, while others can model bathtub and upside-down bathtub failure rates, and still others can do both. We have examined a few distributions in this case, which include the S-ML distribution against the sine-Lindley distribution (S-Lindley) defined by [18], the sine-exponential (S-Expo) distribution studied by [6], the inverse Lindley distribution (IL) introduced by [19], and the exponential (Expo) distribution as seen in [20].

The PDF and DF of the competing models used against the S-ML model are displayed in Table 1.

Table 1. DF and PDF of the competitive models used against the S-ML model.

Model	DF	PDF
S-Lindley	$\cos\left[\frac{\pi}{2}\left(1+\frac{y\beta}{1+\beta}\right)e^{-\beta y}\right]$	$\frac{\pi}{2}\frac{\beta^2}{1+\beta}(1+y)e^{-\beta y}\sin\left[\frac{\pi}{2}\left(1+\frac{\beta y}{1+\beta}\right)e^{-\beta y}\right]$
S-Expo	$\cos\left(\frac{\pi}{2}e^{-\beta y}\right)$	$\frac{\pi}{2}\beta\sin\left(\frac{\pi}{2}e^{-\beta y}\right)e^{-\beta y}$
IL	$\left(1+\frac{\beta}{y(1+\beta)}\right)e^{-\beta/y}$	$\frac{\beta^2}{1+\beta}\left(\frac{1+y}{y^3}\right)e^{-\beta/y}$
Expo	$1-e^{-\beta y}$	$\beta e^{-\beta y}$

3.1. Methodology

- We begin by investigating the descriptive measures of the modeled data-sets, which include the mean (μ), median (M), standard deviation (σ), skewness (γ_1) and the kurtosis (γ_2).
- A statistical analysis is conducted on the data-sets with the help of the statistical software [21]. The statistical analysis includes evaluating the estimate ($\hat{\beta}$) of the data by the method of maximum likelihood estimation, the related standard error (SE), and other statistical measures such as the goodness-of-fit (GOF) test statistics including Akaike Information criterion (AIC), Bayesian information criterion (BIC) along with Anderson Darling statistic (A^*), Cramér-von Mises statistic (w^*) and Kolmgrov–Smirnov statistic (D_n) with its correspondig p-value. The AIC is defined to be

$$AIC = 2k - 2ll,$$

the BIC is given by

$$BIC = k\log(n) - 2ll,$$

where ll denotes the log-likelihood function taken at the maximum likelihood estimate, n denotes the number of data and k represents the numver of model parameters. The model with the highest p-value and the lowest values for D_n, w^*, and A^*, as well as the AIC and BIC values, is the best fit for the data. It will be highlighted in the coming numerical tables with the blue color. The software R is used to conduct the estimation.
- Finally, for a visual representation, the empirical probability density function (EPDF) plots and the empirical cumulative density function (ECDF) plots, accompanied by the box plot and total time on test (TTT) plot, are displayed. The box plot gives a visual representation of the descriptive measures of the data and the TTT plot, proved useful for gaining information about the hazard form of the data. In many real-world situations, there is qualitative information about the shape of the failure rate function that might help in the selection of a particular distribution. The TTT plot has a convex shape for decreasing HRF and a concave shape for increasing HRF.

3.2. Survival Times of Growth Hormone Medication

The first data set consists of the estimated time from growth hormone medication until the children reached the target age in the Programa Hormonal de Secretaria de Saude de Minas Gerais in 2009, as reported in [22].

A summary of the measures of descriptive statistics is provided in Table 2 with the box and TTT plots plotted in Figure 2.

Table 2. Descriptive statistics of survival times of growth hormone medication.

μ	M	σ	γ_1	γ_2
5.979	5.260	2.810	0.851	3.119

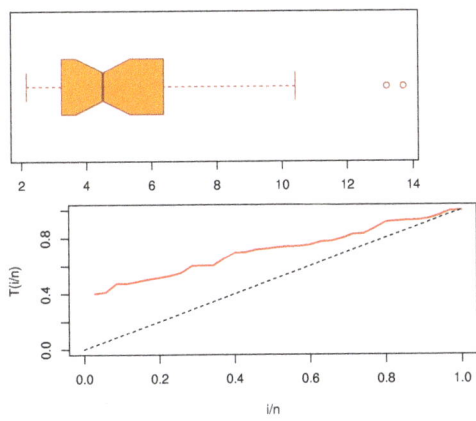

Figure 2. Box plot and TTT plot for the survival times of growth hormone medication.

Table 3 provides $\hat{\beta}$, the SE and the GOF metrics of the survival times of growth hormone medication.

Table 3. $\hat{\beta}$, SE and GOF metrics of the survival times of growth hormone medication.

Distribution	$\hat{\beta}$	SE	AIC	BIC	A^*	w^*	D_n	p-Value
S-ML	0.14870	0.01672	170.8874	172.4428	1.5424	0.23195	0.1978	0.1291
S-Lindley	0.22660	0.02426	173.38	174.9353	1.866	0.29474	0.22065	0.06620
S-Expo	0.10780	0.01691	185.8088	187.364	4.0824	0.77177	0.31926	0.00159
IL	4.9096	0.7251	186.918	188.4733	4.6262	0.8702	0.2905	0.00542
Exp	0.18848	0.03186	188.8149	190.3703	4.4891	0.85831	0.33317	0.00084

Statistical Analysis—Based on the information in Table 2, we can conclude that the data are positively skewed and mesokurtic, as evidenced by the box plot in Figure 2. The TTT plot of the survival times of the data set is displayed in Figure 2. It shows an increasing HRF plot. In addition, analysis of the data set shows that the evaluated model (S-ML) is the best model throughout all elements of the model selection criteria, such as the increasing hazard function. The S-ML model has a higher p-value and minimum values for the test statistics including the AIC, BIC, A^*, w^*, and D_n values, as shown in Table 3. The EPDF and ECDF plots are given in Figure 3.

The plots in Figure 3 display that the S-ML and S-Lindley models give a better fit to the data set than the S-Lindley, S-Expo, IL, and Expo models.

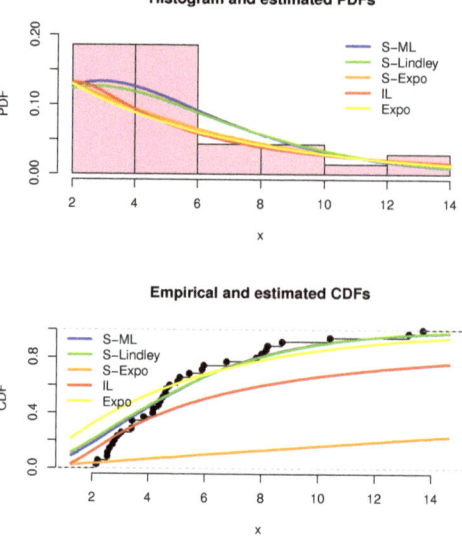

Figure 3. The EPDF and ECDF plot for the survival times of growth hormone medication.

3.3. Survival Times of Guinea Pigs Data

This data set was originally studied by [23], which has also been analyzed previously by [24]. The data set represents the survival times of $n = 72$ guinea pigs injected with different doses of tuberculosis bacilli. The main concern of this data set is to predict the survival times of the guinea pigs because they have a high susceptibility to human tuberculosis.

A summary of measures of descriptive statistics is provided in Table 4 with the box and TTT plots displayed in Figure 4.

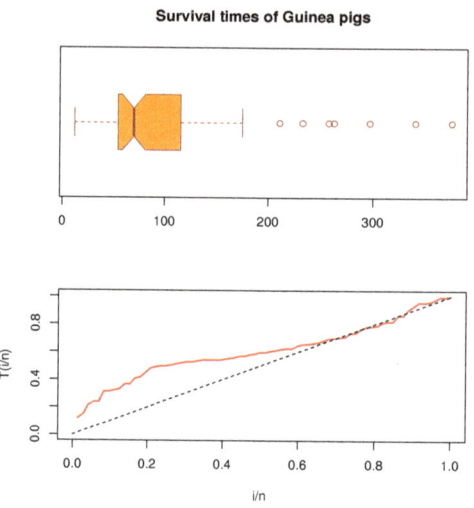

Figure 4. Box plot and TTT plot for the survival times of guinea pigs.

Table 4. Descriptive statistics of the survival times of guinea pigs.

μ	M	σ	γ_1	γ_2
99.82	70.00	81.11	1.796245	5.614

Table 5 displays $\hat{\beta}$, the SE and the GOF metrics for the survival times of guinea pigs.

Table 5. $\hat{\beta}$, SE and GOF metrics for the survival times of guinea pigs.

Distribution	$\hat{\beta}$	SE	AIC	BIC	A^*	w^*	D_n	p-Value
S-ML	0.0082780	0.0006544	791.9648	794.2415	2.022	0.3638	0.14389	0.1014
S-Lindley	0.013329	0.001003	795.0701	797.3468	2.6594	0.5020	0.16629	0.03728
S-Expo	0.00567	0.0006084	805.7007	807.9773	3.8741	0.68284	0.19629	0.0077
IL	61.066	7.084	807.3371	809.6137	4.590	0.8316	0.1845	0.01479
Expo	0.010018	0.001169	808.8843	811.1609	4.472	0.8059	0.2115	0.00317

Statistical Analysis—Table 4 informs us that the data are right-skewed and leptokurtic, as demonstrated by a graphical representation of the box plot in Figure 4. Figure 4 also illustrates the TTT plot of this data set. It displays an increasing HRF plot. Moreover, analysis of the data set implies that the S-ML distribution is the best model among the other competitive models, when statistical GOF criteria and the increasing HRF are considered. We can observe from Table 5, that the S-ML distribution has minimum values for the test statistics with a higher p-value and least values for GOF metrics. The EPDF and ECDF plots are displayed in Figure 5.

From Figure 5, we can also confirm this suitability behavior, as the plots of S-ML and S-Lindley distribution trace the shape of the data very well. We can conclude from Table 5 and Figure 5 that the S-ML model perfectly describes the survival times of guinea pigs.

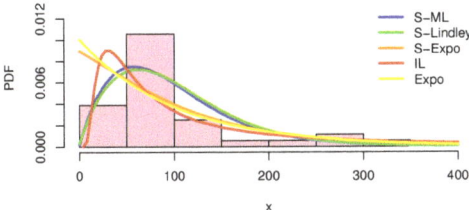

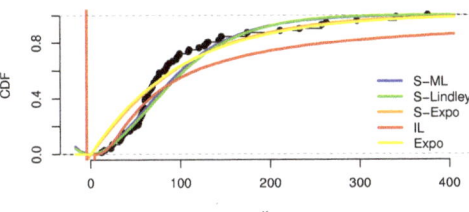

Figure 5. EPDF and ECDF plots for the survival times of Guinea pigs.

3.4. Size of Tumors in Lung Cancer Patients

A swelling or tumor arises when the cells in the lungs expand at an abnormally fast rate, which can lead to lung cancer. It is possible to identify that and see if its spread to other organs based on a variety of indicators. One of these characteristics is tumor stage, which aids doctors in determining the best treatment for their patients. The tumor size is used to determine the staging system. The data show the tumor size of 76 lung cancer patients at Tanta University's chest hospital, sixty of whom are in stage I, seven in stage II, and the rest in stage III.

A summary of measures of descriptive statistics is provided in Table 6 with the box and TTT plots plotted in Figure 6.

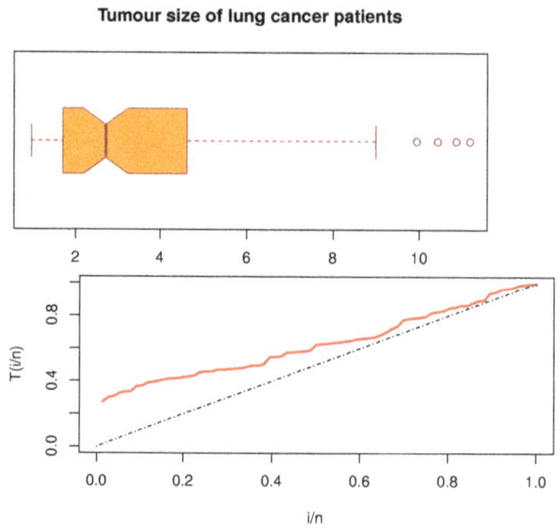

Figure 6. Box plot and TTT plot of the tumor size of the lung cancer patients.

Table 6. Descriptive statistics of the tumor size of the lung cancer patients.

μ	M	σ	γ_1	γ_2
3.531	2.700	2.570	1.442	4.269

Table 7 displays $\hat{\beta}$, the SE and the GOF metrics of the tumor size of the lung cancer patients.

Table 7. $\hat{\beta}$, SE and GOF metrics of the tumor size of the lung cancer patients.

Distribution	$\hat{\beta}$	SE	AIC	BIC	A^*	w^*	D_n	p-Value
S-ML	0.21916	0.02031	324.8085	327.1393	1.9164	0.2733	0.1239	0.1934
S-Lindley	0.3201	0.02373	329.8057	332.1365	2.3303	0.3321	0.15054	0.06381
S-Expo	0.16063	0.01724	341.6504	343.9812	4.5324	0.7384	0.2303	0.00063
IL	2.9640	0.2832	337.1246	339.4554	5.3844	0.9086	0.1823	0.01278
Expo	0.28313	0.03248	345.8022	348.133	5.2671	0.8884	0.2461	0.0002

Statistical Analysis—From Table 6, we see that the data are right-skewed and leptokurtic. This is proved in a graphical display of the box plot in Figure 6. Figure 6 shows the TTT plot of this data set. It illustrates an increasing HRF plot. From Table 7, the S-ML model has minimum values for D_n and higher p-value with least values for AIC and BIC. The EPDF and ECDF plots are illustrated in Figure 7.

The plots in Figure 7 show that S-ML distribution captures the shape of the histogram of the data set. We can conclude from Table 7 and Figure 7 that the S-ML distribution can be used to model this data set related to the size of tumors in lung cancer patients.

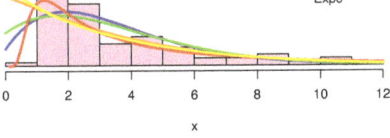
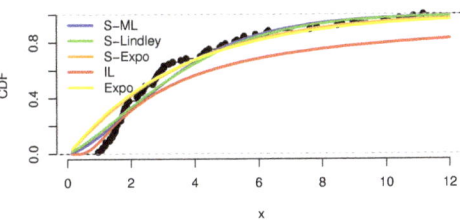

Figure 7. The EPDF and ECDF plot for the size of the tumor of the lung cancer patients.

4. Conclusions

In this paper, we have extended the applications of the sine-modified Lindley (S-ML) distribution developed by [4] to model biomedical data. The distribution yields the benefits of both the modified Lindley and S-G distributional functionalities. It was used to investigate the distribution of tumor size, patients diagnosed with cancer's survival durations, and medications provided. The AIC, BIC, and test statistics such as A^*, w^*, and D_n with their associated p-values are used to select the best-fitting model. These metrics are supported by a visual representation of how well the S-ML model fits the data, such as a box plot or a TTT plot. We believe the findings are superior to other competing distributions for modeling biomedical data and can be used to model a range of other biological data. We have also included the data sets and R codes for all of the figures in the paper, as well as all of the estimations, and the tests carried out. We refer readers to the Appendix A for these R codes.

Author Contributions: Conceptualization, L.T., V.G. and C.C.; methodology, L.T., V.G. and C.C.; software, L.T., V.G. and C.C.; validation, L.T., V.G. and C.C.; formal analysis, L.T., V.G. and C.C.; investigation, L.T., V.G. and C.C.; resources, L.T., V.G. and C.C.; data curation, L.T., V.G. and C.C.; writing—original draft preparation, L.T., V.G. and C.C.; writing—review and editing, L.T., V.G. and C.C.; visualization, L.T., V.G. and C.C. All authors have read and agreed to the published version of the manuscript.

Funding: This research received no external funding.

Acknowledgments: We would like to thank the two referees for the constructive comments on the paper.

Conflicts of Interest: The authors declare no conflict of interest.

Appendix A

In this section, we have included the code to analyze data set 1, using the software R. The codes for the graphs in the data analysis are also plotted.

Appendix A.1. Data Sets

Data set 1

(2.15, 2.20, 2.55, 2.56, 2.63, 2.74, 2.81, 2.90, 3.05, 3.41, 3.43, 3.43, 3.84, 4.16, 4.18, 4.36, 4.42, 4.51, 4.60, 4.61, 4.75, 5.03, 5.10, 5.44, 5.90, 5.96, 6.77, 7.82, 8.00, 8.16, 8.21, 8.72, 10.40, 13.20, 13.70)

Data set 2

(12, 15, 22, 24, 24, 32, 32, 33, 34, 38, 38, 43, 44, 48, 52, 53, 54, 54, 55, 56, 57, 58, 58,59, 60, 60, 60, 60, 61, 62, 63, 65, 65, 67, 68, 70, 70, 72, 73, 75, 76, 76, 81, 83, 84, 85, 87, 91, 95, 96,98, 99, 109, 110, 121, 127, 129, 131, 143, 146, 146, 175, 175, 211, 233, 258, 258, 263, 297, 341, 341, 376)

Data set 3

(0.96, 1.06, 1.09, 1.16, 1.19, 1.20, 1.32, 1.33, 1.40, 1.42, 1.46, 1.49, 1.51, 1.52, 1.54, 1.57, 1.59, 1.68, 1.70, 1.70, 1.76, 1.76, 1.77, 1.80, 1.81, 1.86, 1.89, 1.89, 1.94, 2.20, 2.20, 2.22, 2.36, 2.36, 2.39, 2.41, 2.45, 2.69, 2.71, 2.73, 2.77, 2.80, 2.83, 2.87, 2.94, 2.98, 3.03, 3.04, 3.19, 3.31, 3.57, 3.73, 4.17, 4.27, 4.30, 4.36, 4.45, 4.79, 4.85, 4.97, 5.26, 5.33, 5.53, 5.55, 5.91, 6.25, 6.31, 7.62, 7.84, 8.49, 8.63, 8.99, 9.94, 10.43, 10.86, 11.18)

Appendix A.2. Graphics for the PDF of S-ML Distribution

```
x= 0:10
f= function(x,p) #defining the pdf of S-ML model
{
(((pi/2)*(p/(1+p))*exp(-2*p*x))*(((1+p)*exp(p*x))+(2*p*x)-1)*sin((pi/2)*
(1+(exp(-p*x)*((p*x)/(1+p))))*exp(-p*x)))
}

curve(f(x,p= 5),col="yellow",xlab="x", ylim=c(0,1),ylab="pdf",lwd=2 )
curve(f(x,p= 15),col="pink",   lwd=2,  add= TRUE)
curve(f(x,p= 50),col="purple", lwd=2,  add= TRUE)
curve(f(x,p=100),col="orange", lwd=2,  add= TRUE)

legend("topright",legend=c(expression(paste(beta," = ",5)),
expression(paste(beta," = ",15)),
expression(paste(beta, " = ",50)),
expression(paste(beta, " = ",100))),
ncol=1, col=c("yellow","pink","purple", "orange"),
lwd=c(2,2,2,2), cex=c(1,1,1,1),text.width = 0.1, inset=0.011, bty ="n")
### In the same way, we can plot the pdf for other beta~values.
```

Appendix A.3. Parameter Estimate along with GOF Metrics

```
install.packages (c("EstimationTools", "MASS", "plyr" ))
library(EstimationTools)
library(MASS)
library(plyr)

# Data set 1
st = c(2.15, 2.20, 2.55, 2.56, 2.63, 2.74, 2.81, 2.90, 3.05, 3.41, 3.43,
   3.43, 3.84, 4.16, 4.18, 4.36, 4.42, 4.51, 4.60, 4.61, 4.75, 5.03, 5.10,
   5.44, 5.90, 5.96, 6.77, 7.82, 8.00, 8.16, 8.21, 8.72, 10.40, 13.20, 13.70)

# S-ML distribution
dSml = function(x, p, log = FALSE)   #log of pdf of S-ML model
{
n=count(x)
loglik <- (log(pi/2)+log(p)-log(1+p)-(2*p*x)+log((1+p)*exp(x*p)*(2*p*x)-1)+
log(sin((pi/2)*(1+exp(-p*x)*(p*x)/(1+p))*exp(-x*p))))
if ( log == FALSE)
```

```r
density <- exp(loglik)
else density <- loglik
return(density)
}

theta <- maxlogL(x =st, dist = "dSml",start = 0.59)

summary(theta)

# S-Lindley~distribution

dSL = function(x, p, log = FALSE) #log of pdf of S-Lindley model
{
n=count(x)
loglik <- (log(pi/2)+log(p^2)+log(1+x)-log(1+p)-(p*x)+
log(sin((pi/2)*(1+((x*p)/(1+p)))*exp(-x*p))))
if ( log == FALSE)
density <- exp(loglik)
else density <- loglik
return(density)
}

theta <- maxlogL(x =st, dist = "dSL",start = 0.59)
summary(theta)

# SE distribution
dSE = function(x, p, log = FALSE)   #log of pdf of SE model
{
n=count(x)
loglik <- (log(pi/2)+log(p)-(p*x)+log(sin((pi/2)*exp(-x*p))))
if ( log == FALSE)
density <- exp(loglik)
else density <- loglik
return(density)
}
theta <- maxlogL(x =st, dist = "dSE",start = 0.59)

summary(theta)

#IL distribution
dIL = function(x,p, log = FALSE)
{
n=count(x)
loglik <- (2*log(p))-log(1+p)-(p/x)+log(1+x)-(3*log(x))
if ( log == FALSE)
density <- exp(loglik)
else density <- loglik
return(density)
}

theta <- maxlogL(x =st, dist = "dIL",start = 0.65)

summary(theta)
```

```
#Exp distribution
dE = function(x,p, log = FALSE)  #log of pdf of Expo model
{
n=count(x)
loglik <- log(dexp(x, p, log = FALSE))
if ( log == FALSE)
density <- exp(loglik)
else density <- loglik
return(density)
}

st = c(2.15, 2.20, 2.55, 2.56, 2.63, 2.74, 2.81, 2.90, 3.05, 3.41, 3.43,
  3.43, 3.84, 4.16, 4.18, 4.36, 4.42, 4.51, 4.60, 4.61, 4.75, 5.03, 5.10,
  5.44, 5.90, 5.96, 6.77, 7.82, 8.00, 8.16, 8.21, 8.72, 10.40, 13.20, 13.70)

theta <- maxlogL(x =st, dist = "dE",start =  0.6)

summary(theta)
```

Appendix A.4. KS Test Statistic, p-Value and Other Test Statistics

```
install.packages ("goftest")
library(goftest)
y = c(2.15, 2.20, 2.55, 2.56, 2.63, 2.74, 2.81, 2.90, 3.05, 3.41, 3.43,
  3.43, 3.84, 4.16, 4.18, 4.36, 4.42, 4.51, 4.60, 4.61, 4.75, 5.03, 5.10,
  5.44, 5.90, 5.96, 6.77, 7.82, 8.00, 8.16, 8.21, 8.72, 10.40, 13.20, 13.70)

# S-ML~CDF

pSml = function(x,p)
{
p =   0.14870
cos((pi/2)*(1+(exp(-p*x)*(x*p)/(1+p)))*exp(-p*x))
}

ks1=ad1=cvm1=NULL
ks1=ks.test(y,pSml)
ad1=ad.test(y,pSml)
cvm1=cvm.test(y,pSml)
result1=c(ks1$statistic,ks1$p.value,ad1$statistic,ad1$p.value,
cvm1$statistic,cvm1$p.value)

# S-Lindley~CDF

pSL = function(x,p)
{
p =   0.22660
cos((pi/2)*(1+((x*p)/(1+p)))*exp(-p*x))
}

ks1=ad1=cvm1=NULL
ks1=ks.test(y,pSL)
ad1=ad.test(y,pSL)
cvm1=cvm.test(y,pSL)
result1=c(ks1$statistic,ks1$p.value,ad1$statistic,ad1$p.value,
```

```
cvm1$statistic,cvm1$p.value)

# S-Expo CDF
pSe = function(x,p)
{
p =    0.10780
cos((pi/2)*exp(-p*x))
}

ks1=ad1=cvm1=NULL
ks1=ks.test(y,pSe)
ad1=ad.test(y,pSe)
cvm1=cvm.test(y,pSe)
result1=c(ks1$statistic,ks1$p.value,ad1$statistic,ad1$p.value,
cvm1$statistic,cvm1$p.value)

#IL distribution
pIL = function(x,p)
{
p =    4.9096
(1 + (p/((1+p)*x)))*exp(-p/x)
}
ks1=ad1=cvm1=NULL
ks1=ks.test(y,pIL)
ad1=ad.test(y,pIL)
cvm1=cvm.test(y,pIL)
result1=c(ks1$statistic,ks1$p.value,ad1$statistic,ad1$p.value,
cvm1$statistic,cvm1$p.value)

# Expo~CDF

pEx = function(x,p)
{
p =    0.18848
pexp(x,p)
}

ks1=ad1=cvm1=NULL
ks1=ks.test(y,pEx)
ad1=ad.test(y,pEx)
cvm1=cvm.test(y,pEx)
result1=c(ks1$statistic,ks1$p.value,ad1$statistic,ad1$p.value,
cvm1$statistic,cvm1$p.value)
```

Appendix A.5. Graphics—To Plot the EPDF for the First Data Set

```
x = c(2.15, 2.20, 2.55, 2.56, 2.63, 2.74, 2.81, 2.90, 3.05, 3.41, 3.43,
3.43, 3.84, 4.16, 4.18, 4.36, 4.42, 4.51, 4.60, 4.61, 4.75, 5.03, 5.10,
5.44, 5.90, 5.96, 6.77, 7.82, 8.00, 8.16, 8.21, 8.72, 10.40, 13.20, 13.70)

hist(x,prob=T,main="Histogram and estimated PDFs",
col="pink", ylab = "PDF",ylim=c(0,0.05), bty ="n")

p = 0.14870 ## parameter estimate of S-ML model
```

```
curve(((pi/2)*(p/(1+p))*(exp(-2*p*x))*((1+p)*exp(p*x)+(2*x*p)-1)*
(sin((pi/2)*(1+exp(-p*x)*((x*p)/(1+p)))*exp(-x*p)))),col="blue",lwd=3,
add=T)

p = 0.22660 # parameter estimate of S-Lindley
curve(((pi/2)*((p^2)/(1+p))*(1+x)*exp(-p*x)*(sin((pi/2)*
(1+((x*p)/(1+p)))*exp(-x*p)))), col="green",lwd = 3, add=T)

p = 0.10780 # parameter estimate of S-Expo
curve(((pi/2)*(p*exp(-p*x))*(sin((pi/2)*exp(-x*p)))), col="orange",
lwd = 3, add=T)

p = 4.9096 # parameter estimate of IL
curve((((p^2)/(1+p))*((1+x)/x^3)*exp(-p/x)),col="red",lwd = 3, add=T)

p =  0.18848 # parameter estimate of Expo
curve(dexp(x,p), col="yellow", lwd = 3, add = T)

legend("topright",legend = c("S-ML","S-Lindley","S-Expo","IL","Expo"),
ncol = 1,
col= c("blue","green","orange","red","yellow"),lty =1,lwd=3,
text.width = 2.5 , inset= 0.00005, bty ="n")
```

Appendix A.6. Graphics—To Plot the ECDF for the First Data Set

```
y = c(2.15, 2.20, 2.55, 2.56, 2.63, 2.74, 2.81, 2.90, 3.05, 3.41, 3.43,
3.43, 3.84, 4.16, 4.18, 4.36, 4.42, 4.51, 4.60, 4.61, 4.75, 5.03, 5.10,
5.44, 5.90, 5.96, 6.77, 7.82, 8.00, 8.16, 8.21, 8.72, 10.40, 13.20, 13.70)

plot(ecdf(y) , verticals=TRUE, main="Empirical and estimated CDFs",
 ylab="CDF", xlab="x", bty ="n")

p = 0.14870 # parameter estimate of S-ML
curve((cos((pi/2)*(1+(exp(-p*x)*((x*p)/(1+p))))*exp(-p*x))),col="blue",
lwd=3, add=T)

p = 0.22660 # parameter estimate of S-Lindley
curve((cos((pi/2)*(1+((x*p)/(1+p)))*exp(-p*x))), col="green",lwd = 3,
add=T)

p = 0.010780 # parameter estimate of S-Expo
curve((cos((pi/2)*exp(-p*x))),col="orange",lwd = 3, add=T)

p = 4.9096 # parameter estimate of IL
curve(((1 + (p/((1+p)*x)))*exp(-p/x)),col="red",lwd = 3, add=T)

p = 0.18848 # parameter estimate of Expo
curve(pexp(x,p), col="yellow", lwd = 3, add = T)

legend("topleft",legend = c("S-ML","S-Lindley","S-Expo","IL","Expo"),
ncol = 1,
col= c("blue","green","orange","red","yellow"),lty =1,lwd=3,
text.width = 2.5 , inset= 0.00005, bty ="n")
#######
```

Appendix A.7. Graphics: Bar Plot and TTT Plot for First Data Set

```
###Bar plot
x = c(2.15, 2.20, 2.55, 2.56, 2.63, 2.74, 2.81, 2.90, 3.05, 3.41, 3.43,
3.43, 3.84, 4.16, 4.18, 4.36, 4.42, 4.51, 4.60, 4.61, 4.75, 5.03, 5.10,
5.44, 5.90, 5.96, 6.77, 7.82, 8.00, 8.16, 8.21, 8.72, 10.40, 13.20, 13.70)

boxplot(x,main = "Tumour size of lung cancer patients",
 col = "orange",  border="brown",horizontal = TRUE,notch = TRUE)

###TTT plot
install.packages ("AdequacyModel")
library(AdequacyModel)

TTT(x, lwd = 2, lty 2, col = "red", grid=FALSE)

#############################
```

References

1. Maxwell, O.; Chukwu, A.U.; Oyamakin, O.S.; Khaleel, M.A. The Marshall–Olkin inverse Lomax distribution (MO-ILD) with application on cancer stem cell. *J. Adv. Math. Comput. Sci.* **2019**, *33*, 1–12. [CrossRef]
2. Benchiha, S.; Al-Omari, A.I.; Alotaibi, N.; Shrahili, M. Weighted generalized quasi Lindley distribution: Different methods of estimation, applications for COVID-19 and engineering data. *AIMS Math.* **2021**, *6*, 11850–11878. [CrossRef]
3. Shrahili, M.; Elbatal, I.; Elgarhy, M. Sine Half-Logistic Inverse Rayleigh Distribution: Properties, Estimation, and Applications in Biomedical Data. *J. Math.* **2021**, *2021*, 4220479. [CrossRef]
4. Tomy, L.; G, V.; Chesneau, C. The sine modified Lindley distribution. *Math. Comput. Appl.* **2021**, *26*, 81. [CrossRef]
5. Souza, L.; Junior, W.R.O.; de Brito, C.C.R.; Chesneau, C.; Ferreira, T.A.E.; Soares, L. On the Sin-G class of distributions: Theory, model and application. *J. Math. Model.* **2019**, *7*, 357–379.
6. Kumar, D.; Singh, U.; Singh, S.K. A new distribution using sine function—Its application to bladder cancer patients data. *J. Stat. Appl. Probab.* **2015**, *4*, 417.
7. Jamal, F.; Chesneau, C.; Bouali, D.L.; Ul Hassan, M. Beyond the Sin-G family: The transformed Sin-G family. *PLoS ONE* **2021**, *16*, e0250790. [CrossRef] [PubMed]
8. Al-Babtain, A.A.; Elbatal, I.; Chesneau, C.; Elgarhy, M. Sine Topp-Leone-G family of distributions: Theory and applications. *Open Phys.* **2020**, *18*, 574–593. [CrossRef]
9. Chesneau, C.; Jamal, F. The sine Kumaraswamy-G family of distributions. *J. Math. Ext.* **2020**, *15*, 1–33.
10. Jamal, F.; Chesneau, C.; Aidi, K. The sine extended odd Fréchet-G family of distribution with applications to complete and censored data. *Math. Slovaca* **2021**, *71*, 961–982. [CrossRef]
11. Nagarjuna, V.B.; Vardhan, R.V.; Chesneau, C. On the accuracy of the sine power Lomax model for data fitting. *Modelling* **2021**, *2*, 78–104. [CrossRef]
12. Chesneau, C.; Tomy, L.; Gillariose, J. A new modified Lindley distribution with properties and applications. *J. Stat. Manag. Syst.* **2021**, *24*, 1383–1403. [CrossRef]
13. Chesneau, C.; Tomy, L.; Jose, M. Wrapped modified Lindley distribution. *J. Stat. Manag. Syst.* **2021**, *24*, 1025–1040. [CrossRef]
14. Chesneau, C.; Tomy, L.; Gillariose, J.; Jamal, F. The inverted modified Lindley distribution. *J. Stat. Theory Pract.* **2020**, *14*, 46. [CrossRef]
15. Kalbfleisch, J.D.; Prentice, R.L. *The Statistical Analysis of Failure Time Data*; Wiley: New York, NY, USA, 1980.
16. Lomax, K.S. Business Failures: Another Example of the Analysis of Failure Data. *J. Am. Stat. Assoc.* **1954**, *49*, 847–852. [CrossRef]
17. Aitchison, J.; Brown, J.A.C. *The Lognormal Distribution*; Cambridge University Press: Cambridge, UK, 1957.
18. Kumar, D.; Singh, U.; Singh, S.K.; Chaurasia, P.K. Statistical properties and application of a lifetime model using sine function. *Int. J. Creat. Res. Thoughts (IJCRT)* **2018**, *6*, 993–1002.
19. Sharma, V.K.; Singh, S.K.; Singh, U.; Agiwal, V. The inverse Lindley distribution: A stress-strength reliability model with application to head and neck cancer data. *J. Ind. Prod. Eng.* **2015**, *32*, 162–173. [CrossRef]
20. Johnson, N.L.; Kotz, S.; Balakrishnan, N. *Continuous Univariate Distributions*; John Wiley & Sons: Hoboken, NJ, USA, 1995; Volume 2.
21. R Development Core Team. *R: A Language and Environment for Statistical Computing*; R Foundation for Statistical Computing: Vienna, Austria, 2005; ISBN 3-900051-07-0.
22. Alizadeh, M.; Bagheri, S.; Bahrami, S.E.; Ghobadi, S.; Nadarajah, S. Exponentiated power Lindley power series class of distributions: Theory and applications. *Commun. Stat. Simul. Comput.* **2018**, *47*, 2499–2531. [CrossRef]

23. Bjerkedal, T. Acquisition of resistance in guinea pigs infected with different doses of virulent tubercle bacilli. *Am. J. Hyg.* **1960**, *72*, 130–148. [PubMed]
24. Gupta, R.C.; Kannan, N.; RayChoudhuri, A. Analysis of lognormal survival data. *Math. Biosci.* **1997**, *139*, 103–115. [CrossRef]

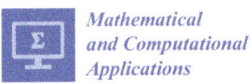

Article

Small Area Estimation of Zone-Level Malnutrition among Children under Five in Ethiopia

Kindie Fentahun Muchie [1,2,*], Anthony Kibira Wanjoya [3] and Samuel Musili Mwalili [3]

1. Institute for Basic Sciences, Technology and Innovation, Pan African University, Nairobi 62000-00200, Kenya
2. Department of Epidemiology and Biostatistics, Bahir Dar University, Bahir Dar 6000, Ethiopia
3. Department of Statistics and Actuarial Sciences, Jomo Kenyatta University of Agriculture and Technology, Nairobi 62000-00200, Kenya; awanjoya@gmail.com (A.K.W.); samuel.mwalili@gmail.com (S.M.M.)
* Correspondence: muchie.kindie@students.jkuat.ac.ke

Abstract: Child undernutrition is one of the 10 most significant public health problems worldwide. There is a rapidly growing demand to produce reliable estimates at the micro administrative level with small sample sizes. In this research, the authors employed small area estimation techniques to estimate the prevalence of malnutrition at the zonal level among children under five in Ethiopia. The small area estimation concept was sought for by linking the most recent possible survey data and census data in Ethiopia. The results show that there is spatial variation of stunting, wasting and being underweight across the zone level, showing different locations facing different challenges or different extents.

Keywords: undernutrition; prevalence; hierarchical Bayesian; spatial analysis; small area estimation; Markov chain Monte Carlo

1. Introduction

Malnutrition is defined as an imbalance in the quantity of protein, calories, and other nutrients consumed, usually including either undernutrition or overnutrition. Undernutrition is usually characterised by stunting, wasting, or being underweight. Child undernutrition can have an immediate impact on child mortality and morbidity, or it can have a long-term influence on the labour market and health consequences in adults.

Globally, in 2020, it was estimated that 149 million children under the age of 5 were stunted, 45 million were wasted, and 38.9 million were overweight or obese. Undernutrition among children is a significant public health issue in developing nations, as evidenced by the fact that undernutrition is ranked as the first priority among the world's 10 most important challenges. Ethiopia is one of the countries in the world with the highest rates of childhood undernutrition. Despite significant progress toward eliminating undernutrition in Ethiopia, between 2005 and 2019, the proportion of underweight children decreased from 33% to 21%, the proportion of stunted children decreased from 51% to 37%, and the proportion of wasted children decreased from 12% to 7% [1].

Both the World Health Assembly and the Sustainable Development Goals (SDGs) papers clearly emphasise the necessity of member countries implementing nutrition policies that prioritise maternal and child nutrition [2,3]. Implementation of these ambitious nutritional objectives outlined in the national and global agreements needs to be backed up with an unceasing stream of up-to-date evidence. Actions to eliminate malnutrition are also crucial for reaching the diet-related objectives of the global strategy for women's, children's, and adolescent's health for 2016–2030 [4] and the 2030 agenda for sustainable development [5]. The World Health Organization (WHO) likewise envisions a world free of all types of malnutrition in which all people attain good health and well-being. The WHO collaborates with partners and member countries to achieve universal access to healthy diets and effective nutrition interventions derived from resilient and sustainable food systems,

Citation: Muchie, K.F.; Wanjoya, A.K.; Mwalili, S.M. Small Area Estimation of Zone-Level Malnutrition among Children under Five in Ethiopia. *Math. Comput. Appl.* **2022**, *27*, 44. https://doi.org/10.3390/mca27030044

Academic Editor: Paweł Olejnik

Received: 7 April 2022
Accepted: 18 May 2022
Published: 22 May 2022

Publisher's Note: MDPI stays neutral with regard to jurisdictional claims in published maps and institutional affiliations.

Copyright: © 2022 by the authors. Licensee MDPI, Basel, Switzerland. This article is an open access article distributed under the terms and conditions of the Creative Commons Attribution (CC BY) license (https://creativecommons.org/licenses/by/4.0/).

according to the nutrition strategy of 2016–2025 [6]. The government of Ethiopia has taken several steps toward reducing undernutrition in the country. The recently endorsed 2019 Food and Nutrition Policy aims to achieve an optimal nutritional status throughout the life cycle via coordinated implementation of nutrition-specific and nutrition-sensitive interventions. In addition, through the Seqota Declaration, Ethiopia has committed to ending undernutrition in children under the age of 2 by 2030.

According to worldwide data, many diverse interventions enhance undernutrition outcomes, yet comparable interventions have varying effects in various situations and places [7]. In Ethiopia, several correlations and interventions have also been found to be important for undernutrition outcomes, including food aid and shocks [8,9], maternal nutritional and educational status [10–13], access to educated and trained health workers [14], access to feeding practices, and safe water [15,16]. The prevalence of various forms of undernutrition varies by geography in Ethiopia, showing that different geographic areas confront distinct problems with undernutrition [17].

The national-level Demographic and Health Survey (DHS) is the main source of official statistics in developing countries where there is no vital registration. The DHS data help generate a variety of relevant statistics at the macro level (administrative level and national levels). These days, there is a rapidly growing demand for micro-level statistics. However, the DHS data cannot be utilised directly to generate valid estimates at the micro level because of small sample sizes. Hence, employing small area estimation (SAE) is of paramount importance.

Even though there was a study conducted using 2014 Ethiopian Mini DHS in combination with census data to find small area estimates at the woreda level, it is not appropriate to estimate at the woreda level as the survey data are shifted spatially, where the shifting guarantees that the clusters will not be out of the zonal level [17]. It is also important to use the latest possible data: Ethiopian Mini DHS (EMDHS) 2019 data. In this work, SAE approaches were employed to obtain model-based estimates of the prevalence of malnutrition at the zone level in Ethiopia by linking data from the EMDHS 2019 and the Ethiopian Census 2007.

2. Methods and Materials

2.1. Study Setting and Design

Ethiopia is organised into four administrative levels: region, zone, woreda, and kebele. Ethiopia's first administrative division is the region, also known as a kilil or, alternatively, a regional state. The regions of Ethiopia are defined by ethno-linguistic areas. Currently, in 2022, there are 11 regions (Afar, Amhara, Oromia, Benishangul-Gumuz, Somali, Gambela, Harari, Sidama, Southern, South West, and Tigray) and 2 independently administrative cities (Addis Ababa and Dire Dawa). Zones are created by subdividing regions. Zones are administrative subdivisions in Ethiopia where DHS shifting guarantees that no survey clusters are outside of the zone. Zones are further subdivided into woredas, and woredas are also subdivided into kebeles. Going back over time within the country, woredas are generally stable administrative entities. Kebeles are the lowest administrative units or divisions of Ethiopia.

This study is a further analysis of secondary data: the Ethiopian Mini DHS (EMDHS) 2019 and the Ethiopian Census 2007. The Ethiopian Census 2007 is the country's third population and housing census and was conducted on 28 May 2007 for all regions except Afar and Somali, which were enumerated on 28 November 2007. The EMDHS 2019 was designed to represent national, urban-rural, and regional estimates of health and demographic outcomes. The samples for the EMDHS 2019 were chosen using stratified and two-stage cluster sampling procedures. Sketch maps were drawn for each of the clusters, and all conventional households were listed.

The EMDHS 2019 was a nationwide survey that included a nationally representative sample of 9150 randomly selected households. In the selected households, all children under the age of five were eligible for measurements of weight and height. The survey was designed to generate reliable estimates of important indicators for urban and rural areas, for each of the regions, and at the national level in Ethiopia. Figure 1 depicts the clusters included in the 2019 EMDHS.

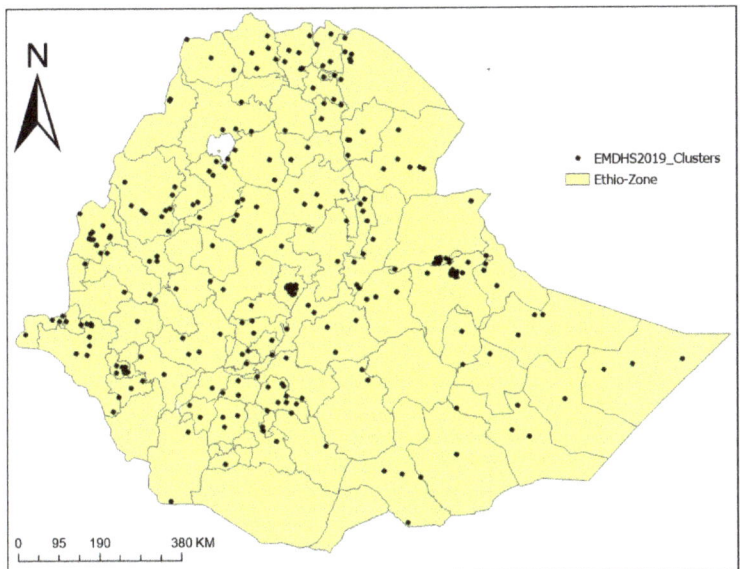

Figure 1. Study clusters of the EMDHS 2019.

2.2. Data Sources and Procedures

The analysis of the current study was based on the most recent available data: the EMDHS 2019 and the Ethiopian Census 2007. The EMDHS 2019 is a nationally representative cross-sectional household-based sample survey designed to provide information about several health and nutritional indicators in Ethiopia. The principal objective of the EMDHS 2019 was to provide up-to-date regional- and national-level estimates of the indicators. Specifically, the nutritional statuses of children under the age of five was assessed by measuring their weights and heights.

The EMDHS 2019 was carried out with the use of standardised data collection procedures and survey design. The EMDHS 2019 used a stratified cluster sampling technique to choose census enumeration areas (EAs) based on probability proportional to the enumeration area sizes. Following that, a random sample of households within the selected EAs was chosen. Data for the survey were gathered through face-to-face interviews with questionnaires administered to household heads and chosen household members who consented to being interviewed. Data collection for the EMDHS 2019 took place from 21 March to 28 June 2019. The EMDHS 2019 data for the current study were extracted from the Major DHS ((http://dhsprogram.com, accessed on 5 February 2021).

The 2007 Census of Ethiopia is one of the biggest and most recent data sources in Ethiopia, providing information on a wide variety of demographic, socioeconomic, and educational characteristics and the migration statuses of people at a disaggregated level. For the 2007 Census, short- and long-form questionnaires were created for use. The long form was administered to 20% of the randomly chosen households, which covered both housing and population topics. Specifically, the long form was composed of questions about internal migration and geographical characteristics, population characteristics, family and household characteristics, social and demographic characteristics, mortality and fertility, educational and literacy characteristics, disability characteristics, and economic characteristics. The long form was rich in terms of data, with demographic information such as assets, housing characteristics, education, fertility, and mortality.

The short format covered only basic demographics, and 80% of the households received the short format. All the questions in the short form were included in the long form. As a result, the short form was developed to gather information from the whole population. To collect information from individuals, households, and institutions, both the short and long forms were employed. Only the short-form questionnaire was used for the resident foreigners and the homeless. The 2007 population and housing census designated 86,805 enumeration areas (in all regions) with 17,363 urban areas and 69,462 rural areas. The current study's census data was obtained from the Integrated Public Use Microdata Series (IPUMS) (https://international.ipums.org, accessed on 12 May 2021).

All available shape files were collected. We used woreda-level geographic boundary mapping within regions to join data from different sources because woredas are relatively stable government structures, compared with structures below or above the woreda, during political or government changes.

2.3. Response Variable

This study applied the recent possible available survey—the EMDHS 2019—unlike the EMDHS 2014, which was used by a previous study [18].

The outcome variables considered were childhood stunting, wasting, and being underweight, which are binary at the individual level. According to WHO criteria, children with height-for-age, weight-for-height, and weight-for-age z-scores of less than -2 standard deviations (SDs) were leveled as stunted, wasting, and underweight, respectively. The parameter of interest was estimating the zone level prevalence of stunting, wasting, and being underweight among children under five. Notably, the EMDHS 2019 was not designed to provide zone-level estimates for key maternal or child health indicators, including childhood undernutrition, and therefore this study did not use the SAE technique to estimate the prevalence of stunting, wasting, and being underweight. The study restricted our analysis to children under five, as the EMDHS 2019 collected information on key indicators of child health and development only for those who were born in the five years prior to the survey.

2.4. Auxiliary Covariates

In this study, the auxiliary variables were taken from the 2007 Population and Housing Census of Ethiopia. The chosen covariates were at the district level and zonal level. Literacy is widely acknowledged to benefit the individual and society and is associated with a number of positive outcomes for health and nutrition. We considered household characteristics, including education, as an auxiliary variable, as they were used from a study conducted in Ethiopia using the EMDHS 2014 data [18]. As the number of auxiliary variables we can consider should be small in number, we selected a few of the available variables in the census data, namely those which explained the outcome variable more. Hence, we summarised both the woreda- and zone-level summary values (percentages) of the auxiliary variables, including the percent of literacy and percent of access to an improved water supply.

2.5. Data Processing and Analysis

The analysis was estimated first by mapping the census data at the woreda-level and zone-level areas. Then, the data sources were mapped by subnational states before processing the data in the analysis. We overlaid the population grid on a current shapefile and then aggregated the population within each area to generate woreda-level and zone-level population estimates for the country. Zone-level estimates of prevalence of stunting, wasting, and being underweight were compiled in the country.

Let $Y_i(A_r)$ be a binary response variable for the ith individual in the rth area A_r, where $i = 1, \ldots, N_r$, $r = 1, \ldots, n$ and $\sum_{r=1}^{n} N_r = N$, with n referring the number of small areas and N_r referring the number of individuals in the rth area. We considered $y_i(A_r)$ as taking values one (with probabilities p_i) and zero (with probabilities $1 - p_i$), being a realisation of a random variable $Y_i(A_r)$ following a Bernoulli distribution (i.e., $y_i(A_r) \sim ber[p_i(A_r)]$).

For comparison purposes, we also computed the zone-level estimated direct prevalence of malnutrition and its corresponding variance. We used sampling weights to compute the estimated weighted direct prevalence. Similarly, variances were computed using Taylor series linearisation [19] to estimate the variance, considering the sampling weight.

Spatial analysis (both global and local spatial autocorrelation) were used so as to check the importance of including the spatial effect in the modeling. The survey GPS coordinates were combined with the weighted prevalence of stunting, wasting, and being underweight in each of the EMDHS 2019 clusters. As a result, the cluster level weighted prevalence was used to depict the hot and cold spots of clusters. Geographic variation in stunting, wasting, and underweight prevalence among the EMDHS 2019 clusters was identified using spatial analysis [20]. Geographic variation of a significant high prevalence or low prevalence of stunting, wasting, and being underweight was computed for each cluster using Moran's I statistic [20]. Maps depicting the distribution and variations of stunting, wasting, and being underweight throughout the country were constructed.

Regarding the small area estimation, we adapted the Bayesian approach of modeling for its ability to combine information from several sources [21]. The approach also simplifies computation of the measures of accuracy in the SAE, which produces realisations of the posterior distribution of the target quantities [22]. Empirical Bayesian (EB) and hierarchical Bayesian (HB) methods are Bayesian approaches which are more generally applicable in the sense of handling models for binary and count data [23]. The ELL method [24], which is an EB estimation method used by some authors [18,25], assumes a nested error model on the transformed variables [23]. Though the ELL method can handle data from survey and census sources, its nested error modeling nature requires individual-level auxiliary variables, which we could not get from the census data for the individuals in the survey. Under the HB framework, there are a number of developed models for discrete outcome variables [26–28].

Accordingly, we used a spatial hierarchical Bayesian small area model for the binary response variable [28], which enabled us to use area-level auxiliary data from the census and individual-level data from the survey. We considered zone-level classification as small area two, which we wanted to estimate, whereas woreda-level classification was classified as small area one. Furthermore, 72 knot points within two resolutions were considered in the spatial dimension reduction. Further details of the modeling can be found in a published study elsewhere [28].

Weakly informative priors were considered for the model parameters. All the priors used being proper guaranteed that the appropriateness of the posterior distribution. Markov chain Monte Carlo (MCMC) simulation was used to generate posterior samples from the conditional distributions of the parameters of the model. Inspection of the plots (trace plots, density plots, and autocorrelation plots) and formal tests (Geweke's test) were considered in order to check the convergence of the simulated sequences in the models.

Measures of precision play a crucial role in small area estimation. Consider R as the number of MCMC samples after removing the burnin period followed by thinning. By the ergodic theorem for the Markov chains [29], $\hat{p}_i$ converges to $E(p_i|y)$ and $\hat{V}(p_i|y)$ to $V(p_i|y)$ as $R \longrightarrow \infty$. Checks of the convergence of the MCMC were used to guarantee the ergodic theorem. Hence, the estimate of p_i and its corresponding posterior variance for the ith area are obtained directly from the predictive distribution of $y(A_k^*)$ accordingly:

$$\hat{p}_i \approx \frac{1}{R} \sum_{k=1}^{R} \hat{p}_i^{(k)} = \hat{p}_i^{(\cdot)}$$

and

$$V(\hat{p}_i) \approx \frac{1}{R-1} \sum_{k=1}^{R} \left(\hat{p}_i^{(k)} - \hat{p}_i^{(\cdot)}\right)^2$$

Benchmarking is important in that the model-based estimators do not benchmark against the direct survey estimate for large areas [30]. To avoid possible overshrinkage and model misspecification, the model-based HB estimates $\hat{p}_i$ were benchmarked so that the benchmarked HB (BHB) estimates added up to the direct large area (regional level in this study case) estimate. The posterior mean squared error (PMSE) was used to measure the variability of the BHB estimators. The PMSE is the sum of a bias correction term and the posterior variance.

We take $\hat{p}_i^{BHB}$ as the benchmarked HB (BHB) estimator of p_i such that $\hat{p}_i^{BHB}$ is a function of the HB estimators $\hat{p}_i^{HB}$ (i.e., $\hat{p}_i^{BHB} = f(\hat{p}_1^{HB}, \ldots, \hat{p}_n^{HB})$) for some function $f(\cdot)$, satisfying the benchmark property [30] $\sum_{i=1}^{n} \hat{p}_i^{BHB} = \sum_{i=1}^{n} \hat{p}_i^{Direct}$, where $i = 1, \ldots, n$ and $\hat{p}_i^{Direct}$ is the direct survey estimator. Hence, the BHB estimator can be obtained as follows:

$$\hat{p}_i^{BHB} = \hat{p}_i^{HB} \frac{\sum_{k=1}^{n} \hat{p}_k^{Direct}}{\sum_{k=1}^{n} \hat{p}_k^{HB}}$$

To obtain a measure of variability associated with the BHB estimator $\hat{p}_i^{BHB}$, we use the following posterior mean squared error (PMSE):

$$\text{PMSE}\left(\hat{p}_i^{BHB}\right) = \left(\hat{p}_i^{BHB} - \hat{p}_i^{HB}\right)^2 + V\left(\hat{p}_i^{HB}\right).$$

Thus, the PMSE of $\hat{p}_i^{BHB}$ is simply the sum of the posterior variance $V(p_i \mid y)$ and a bias correction term $\left(\hat{p}_i^{BHB} - \hat{p}_i^{HB}\right)^2$.

3. Results

3.1. Data Description

Both census and survey data were considered in this study. We obtained 10% of the census 2007 data, with 7,434,086 individuals covering all 720 woredas at the time of the census. Out of these, 6,132,270 had short-form data. Accordingly, the remaining 1,301,816 participants with the long-form data type were considered to compute the woreda- and zone-level auxiliary variables for our analysis. We summarised the woreda- and zone-level weighted averages and percentages of the auxiliary variables.

From the EMDHS 2019, 4552 children under 5 with complete information of their anthropometric measurements (height, weight, and age) were considered in the analysis. Furthermore, the global positioning system (GPS) data (position of enumeration areas) of the EMDHS 2019 and shapefiles were overlaid to demonstrate the estimates visually.

3.2. Direct Estimates of Malnutrition

The weighted prevalence of malnutrition was computed as a direct estimate at the woreda level and the estimates are given in Table 1. The corresponding 95% confidence intervals were computed using variance estimates from Taylor series linearisation.

Table 1. Weighted direct prevalence of indicators of malnutrition among children under 5 in Ethiopia in 2019 (n = 4552).

Indicator	Prevalence	SE	95% CI
Stunted	38.9	1.29	[36.38, 41.45]
Wasting	22.5	1.36	[19.81, 25.14]
Underweight	6.9	0.53	[5.82, 7.89]

From Figure 2, we can see the number of clusters included and the direct estimate for the weighted prevalence of malnutrition. Specifically, there were a few zones with zero clusters included in the survey (Figure 2a), and hence direct estimates for these zones could not be found (Figure 2c–e). Hence, small area estimation is of paramount importance to obtaining estimates for zonal level administrative classification. That aside, it also important to check whether taking the spatial effect into account in the small area estimation process improves the estimates or not. Spatial autocorrelation analysis and spatial pattern analysis helped us to check this.

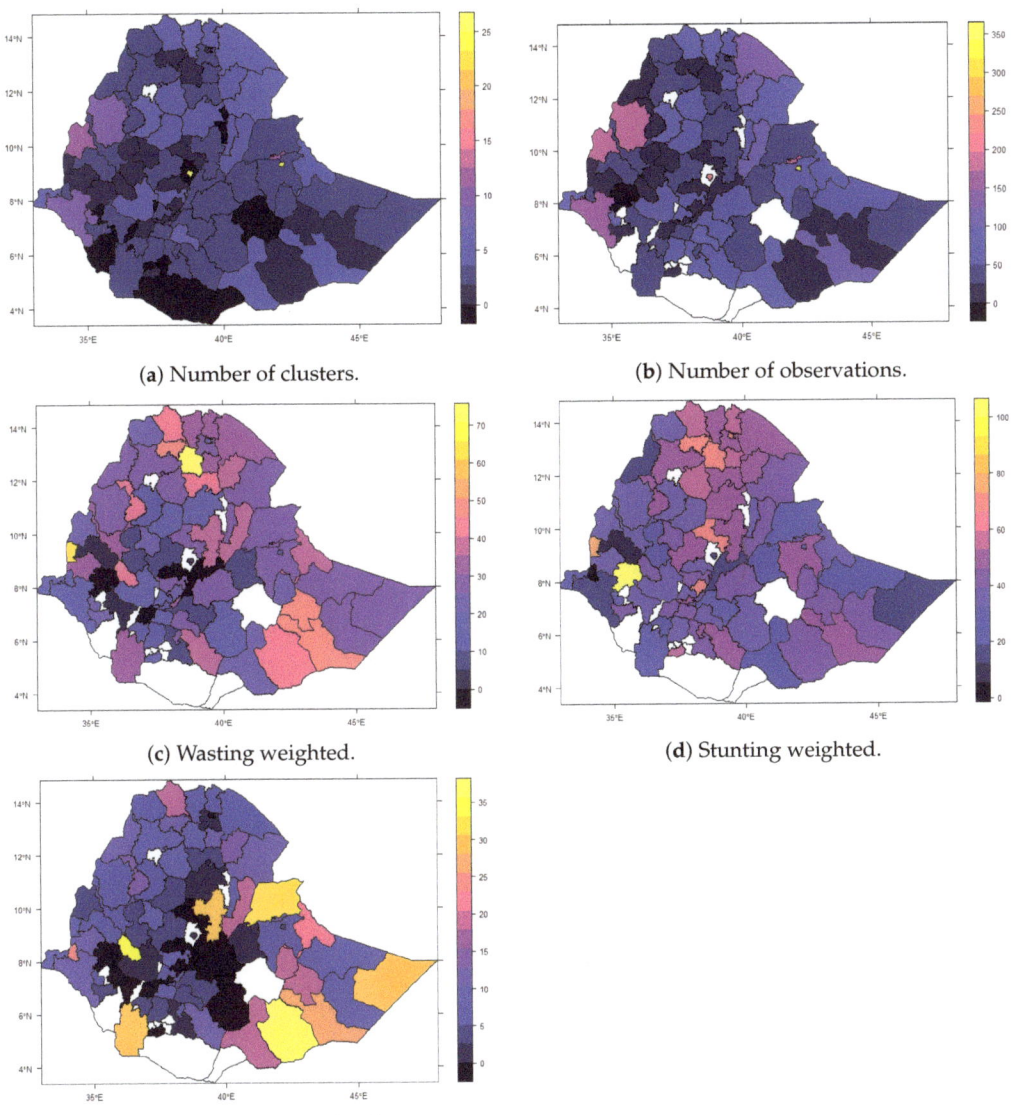

Figure 2. Number of clusters and weighted prevalence of malnutrition.

In general, we can see the spatial structure in the undernutrition estimates, namely in that two areas that are neighbors have more similar risks than two areas that are far apart. Specifically, spatial pattern analysis showed that there is spatial effect for being underweight, wasting, and stunting among children. The Global Moran's I test results are given in Table 2, showing the existence of significant spatial autocorrelation. Similarly, Anselin local Moran's I analysis and pattern (cluster and hotspot) analysis (Figure 3) showed the existence of significant clusters of a high as well as low prevalence of malnutrition. Hence, taking the spatial random effect in modeling evidently will have a great role. Therefore, the spatial random effect was taken into account in the small area estimation.

Table 2. Global Moran's I summary statistics for the malnutrition indicators.

Measure	Underweight	Stunting	Wasting
Moran's Index	0.218227	0.395968	0.372278
Expected Index	−0.003289	−0.003289	−0.003289
Variance	0.002098	0.002122	0.002118
z-score	4.836414	8.667563	8.160307
p-value	0.000001	0.000000	0.000000

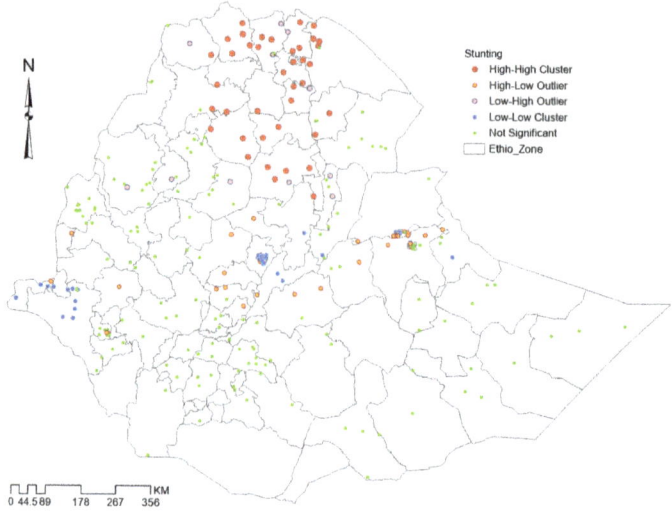

(**a**) Stunting.

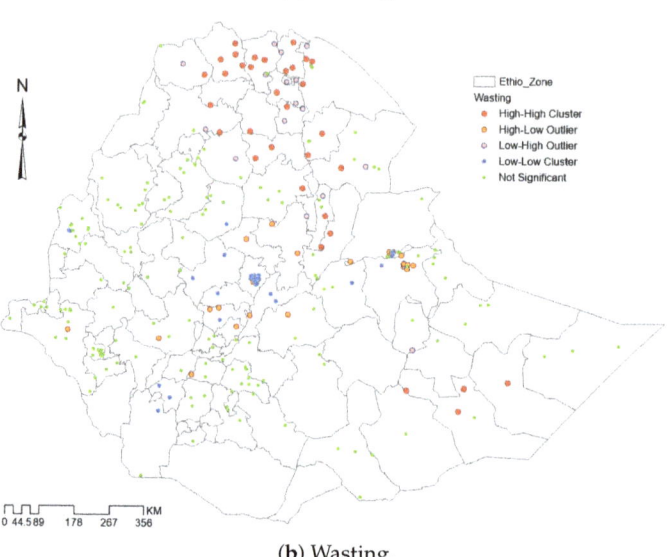

(**b**) Wasting.

Figure 3. *Cont.*

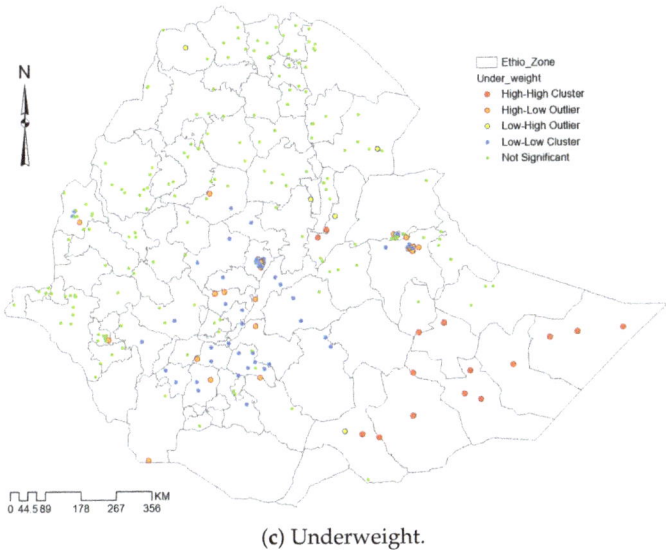

(c) Underweight.

Figure 3. Cluster and hotspot analysis of prevalence of malnutrition.

3.3. Small Area Estimates of Malnutrition

Some of the second administrative level in the country (zones) were not represented at all in the EMDHS 2019. These include the Oromiya Zone in Amhara, Yem Special, Sheka, Konta Special, Mirab Omo, Basketo, Alle, Derashe, Amaro and Burji in SNNP, Borena, East Bale in Oromia, and Daawa in Somali. Meanwhile, no woredas, the third adminstrative level in the country, in the unrepresented zones nor some more woredas from represented zones were represented at all, and a few of them had small sample sizes. Accordingly, the application of small area estimation is of paramount importance at the zonal level and woreda level. Aside from that, to comply with the assumption of the model, we considered the zone as secondary small area (SA2) and woreda as primary small area (SA1). The 2 resolutions with 72 knot points were considered for dimension reduction.

A total of 20,000 MCMC samples were generated from the posterior distribution. Considering a burnin period of 6000 and thinning for every third, 4667 samples were retained for the final process. The convergence and independence of the samples were confirmed from the trace plots (Figure 4), density plots (Figure 5), autocorrelation plots (Figure 6), and Geweke's test of convergence (Table 3). All these show that there is no evidence of assumption violations; that is, the samples were a realisation of stationary distribution.

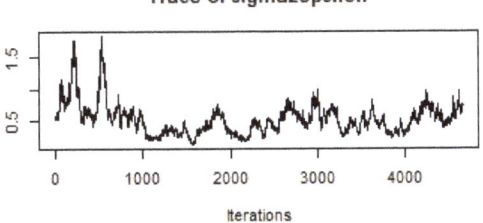
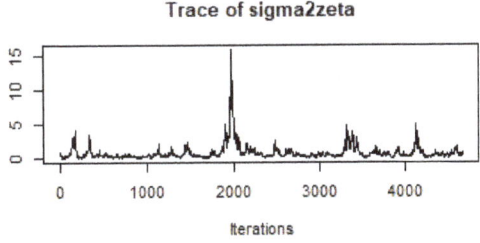

Figure 4. Cont.

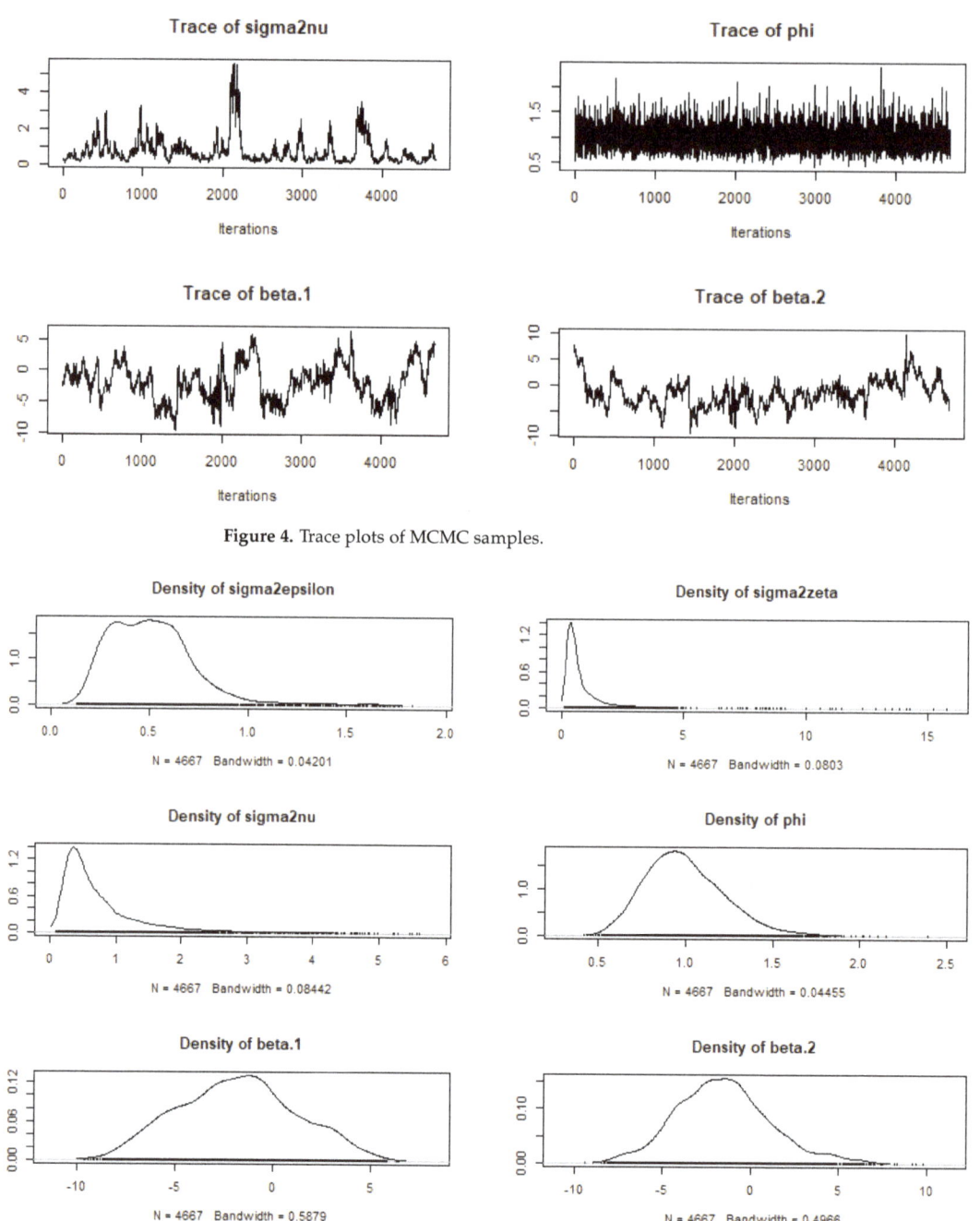

Figure 4. Trace plots of MCMC samples.

Figure 5. Kernel density plots of MCMC samples.

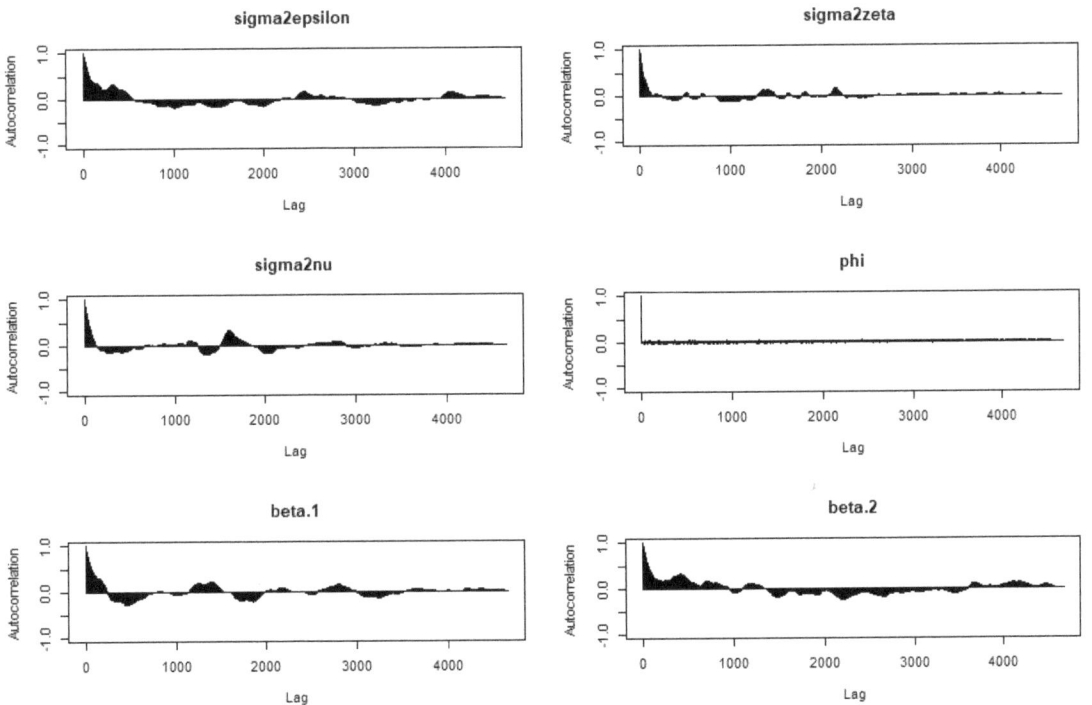

Figure 6. Autocorrelation function plots of MCMC samples.

Table 3. Geweke's test of convergence for application.

Parameters	σ_ϵ^2	σ_ζ^2	σ_ν^2	ϕ	β_1	β_2
Z-value	1.71403	0.32273	0.03037	-0.05075	0.46410	0.19037

The PMSE plots were generated, showing the variability of the benchmarked HB estimates at the zonal level in Ethiopia (Figure 7).

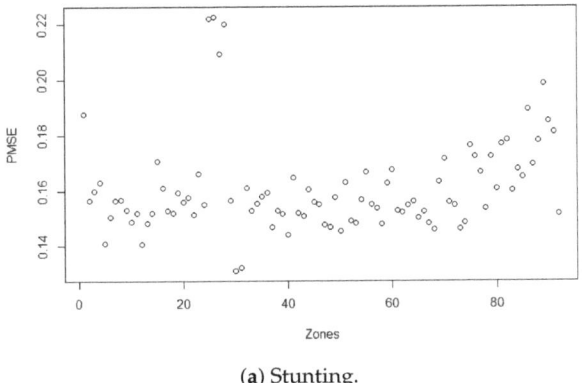

(**a**) Stunting.

Figure 7. *Cont.*

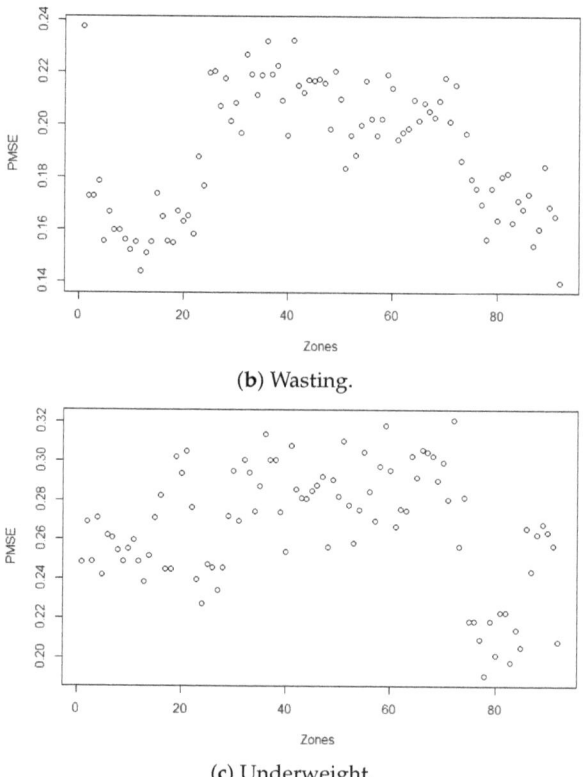

(b) Wasting.

(c) Underweight.

Figure 7. PMSE for benchmarked HB small area estimates.

The predictive posterior variance of prevalence of undernutrition and correction bias due to benchmarking are given in Figure 8.

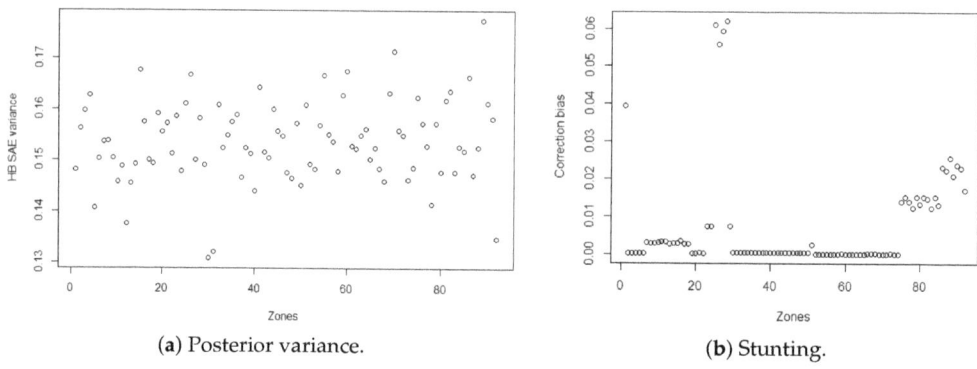

(a) Posterior variance.

(b) Stunting.

Figure 8. *Cont.*

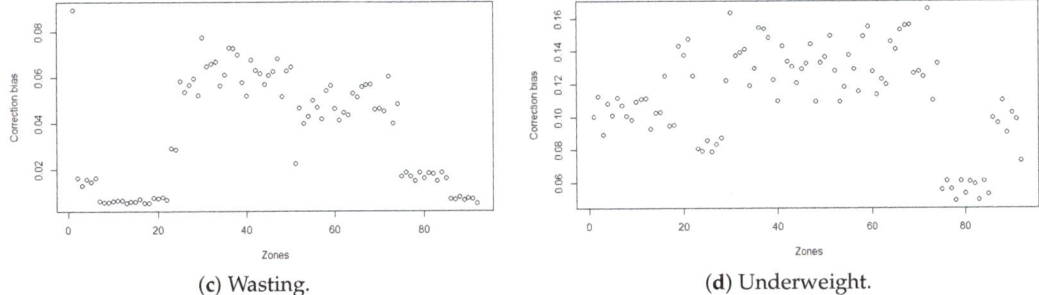

(c) Wasting. (d) Underweight.

Figure 8. Posterior variance of HB SAE and benchmarking correction bias for undernutrition.

The estimates of malnutrition and its corresponding 95% credible interval using the proportion of literate persons and proportion of individuals from improved sources of drinking water as auxiliary variables are given. Accordingly, the plot for stunting generated from small area estimation is given in Figure 9. From the figure, we can understand the distribution of the burden of stunting among children under five at the zonal level in the country. The highest small area estimated prevalence of stunting was observed in the southwestern part of the country of Ethiopia. The corresponding measures of precision using the 95% credible interval is given in Figure 9, which shows the SAE-provided precise estimate.

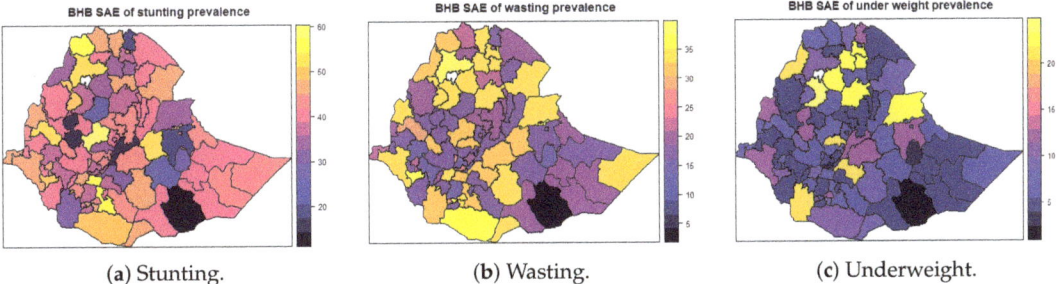

(a) Stunting. (b) Wasting. (c) Underweight.

Figure 9. Prevalence of undernutrition from benchmarked HB small area estimation at zonal level in Ethiopia.

4. Discussion

The national level estimates of stunting (38.9%), wasting (22.5%), and being underweight (6.9%) indicate that stunting is still a severe public problem in the country, followed by wasting. This is in line with the related studies.

In addition to the national estimates, the local-level prevalence of malnutrition in Ethiopia was estimated. The design-based estimates were not adequate for estimation in the lower-level administrative areas, as the survey was not representative for lower-level administrative areas. There was a significant overall spatial autocorrelation, as well as hotspot areas with a high prevalence of undernutrition among children under five in Ethiopia. It was determined that including a spatial random effect in the estimation process was crucial in Ethiopia.

Accordingly, the hierarchical Bayesian spatial small area model was applied to estimate the prevalence of stunting, wasting, and being underweight. The estimate showed the existence of spatial variation of undernutrition at the zonal level.

In the midst of significant advancements in global economic growth, issues connected to child malnutrition have consistently posed a major challenge in low- and middle-income nations [2,31,32]. Globally, hunger and malnutrition diminish a country's gross domestic product (GDP) by USD 1.4–2.1 trillion every year. Malnutrition costs the 54 African countries between 3 and 16 percent of their annual GDPs, with Ethiopia accounting for 16.5 percent, Malawi accounting for 10.3 percent, Rwanda accounting for 11.5 percent, and Burkina Faso accounting for 7.7 percent [3,31,33].

Increased agricultural productivity, girls' education promotion, immunisation, integrated management of neonatal and childhood illnesses, improved access to water and sanitation, and skilled birth delivery could all help to reduce the burden of undernutrition among Ethiopian children under the age of five [34].

The spatial variation of undernutrition at the zonal level could be related with economical variation, drought, food insecurity, and variation in cultivation [35]. This might suggest the need for the design and implementation of effective public health interventions at the zonal level to reduce undernutrition among children under five in Ethiopia.

This study is not without limitations. It is probable that there are other auxiliary variables that impact children's nutritional conditions, but the current study did not evaluate those variables due to a lack of information in the census data. Furthermore, the census data used were 15 years old, which may have influenced the results. Hence, readers are advised to take these limitations into account.

5. Conclusions

The prevalence of undernutrition among children under five in Ethiopia was estimated at the zone level using small area estimation techniques. The small area estimation concept was sought out by linking the most recent possible survey data and census data in Ethiopia. In Ethiopia, undernutrition had significant spatial variations across the country. The results specifically show that there is spatial variation in stunting, wasting, and being underweight across the zone level, showing that different locations faced different challenges and to different extents of undernutrition. Therefore, public health interventions that reduce undernutrition among children and enhance women's awareness toward undernutrition in zones with a high prevalence of undernutrition are crucial, and the Ethiopian Federal Ministry of Health (FMOH) should design tailored nutritional intervention for children under five who are living in zones with a high prevalence of undernutrition.

Author Contributions: Conceptualisation, K.F.M., A.K.W. and S.M.M.; methodology, K.F.M., A.K.W. and S.M.M.; software, K.F.M.; validation, K.F.M., A.K.W. and S.M.M.; formal analysis, K.F.M.; investigation, K.F.M.; resources, K.F.M.; data curation, K.F.M.; writing—original draft preparation, K.F.M.; writing—review and editing, K.F.M., A.K.W. and S.M.M.; visualisation, K.F.M.; supervision, A.K.W. and S.M.M.; project administration, K.F.M. All authors have read and agreed to the published version of the manuscript.

Funding: This research received no external funding.

Institutional Review Board Statement: The analysis presented in this study was based on the EMDHS 2019 and Census 2007, which are publicly available data sets with no identifiable information on the participants in the survey. Standard survey and census procedures for all ethical issues, like informed consent, were followed strictly in the EMDHS 2019 and Ethiopian Census 2007, respectively. Accordingly, for this study, no separate informed consent or ethical approval was required. However, we have received a grant of permission from Major DHS (http://dhsprogram.com, accessed on 5 February 2021) and IPUMS (https://international.ipums.org, accessed on 12 May 2021) to use survey data and census data, respectively.

Data Availability Statement: Minimal data that support the findings of this study can be accessed from the correspondence upon reasonable request. The data are not publicly available due to privacy or ethical restrictions.

Acknowledgments: We would like to express our appreciation to the Pan African University Institute of Basic Sciences, Technology and Innovation for the support. We also extend our appreciation to Major DHS and IPUMS for granting us permission to use the data. The authors wish to acknowledge the statistical office that provided the underlying data making this research possible: Central Statistical

Agency, Ethiopia. We also would like to acknowledge four anonymous reviewers for their insightful comments which helped us to improve the paper.

Conflicts of Interest: The authors declare no conflict of interest.

Abbreviations

ACF: autocorrelation Function; AIDS: acquired immunodeficiency syndrome; BHB: benchmarked hierarchical Bayesian; CSA: Central Statistics Agency; DHS: Demographic and Health Survey; EDHS: Ethiopia DHS; EMDHS: Ethiopian Mini DHS; EPHI: Ethiopian Public Health Institute; FMoH: Federal Ministry of Health; GDP: gross domestic product; HB: hierarchical Bayesian; HH: household; IPUMS: Integrated Public Use Microdata Series; MCMC: Markov chain Monte Carlo; PMSE: posterior mean squared error; RBHB: ratio-benchmarked hierarchical Bayesian; SA1: small area one; SA2: small area two; SAE: small area estimation; SD: standard deviation; SNNP: Southern Nations, Nationalities and People; SNNPR: Southern Nations Nationalities and People's Region; U5: under five; WHO: World Health Organization.

References

1. Ethiopia Mini Demographic and Health Survey 2019. Available online: https://www.dhsprogram.com/pubs/pdf/FR363/FR363.pdf (accessed on 6 April 2022).
2. Branca, F.; Grummer-Strawn, L.; Borghi, E.; Blössner, M.; Onis, M. Extension of the WHO maternal, infant and young child nutrition targets to 2030. *SCN News* **2015**, *41*, 55–58.
3. Federal Democratic Republic of Ethiopia. *National Nutrition Program 2016–2020*; Federal Democratic Republic of Ethiopia: Addis Ababa, Ethiopia, 2016.
4. WHO. *Global Strategy for Women's, Children's and Adolescents' Health (2016–2030)*; WHO: Geneva, Switzerland, 2016; Volume 20, pp. 4–103.
5. Transforming Our World: The 2030 Agenda for Sustainable Development. Available online: https://sdgs.un.org/2030agenda (accessed on 6 April 2022).
6. WHO. *Ambition and Action in Nutrition: 2016–2025*; World Health Organization: Geneva, Switzerland, 2017.
7. Ainsworth, M.; Ambel, X.; Martin, G.; Sinha, S.; Huppi, M. *What Can We Learn from Nutrition Impact Evaluations*; Independent Evaluation Group, the World Bank: Washington, DC, USA, 2010.
8. Quisumbing, A.R. Food aid and child nutrition in rural Ethiopia. *World Dev.* **2003**, *31*, 1309–1324. [CrossRef]
9. Yamano, T.; Alderman, H.; Christiaensen, L. Child growth, shocks, and food aid in rural Ethiopia. *Am. J. Agric. Econ.* **2005**, *87*, 273–288. [CrossRef]
10. AlemayehuAzeze, A.; Huang, W. Maternal education, linkages and child nutrition in the long and short-run: Evidence from the Ethiopia Demographic and Health Surveys. *Int. J. Afr. Dev.* **2014**, *1*, 3.
11. Negash, C.; Whiting, S.J.; Henry, C.J.; Belachew, T.; Hailemariam, T.G. Association between maternal and child nutritional status in Hula, rural Southern Ethiopia: A cross sectional study. *PLoS ONE* **2015**, *10*, e0142301. [CrossRef] [PubMed]
12. Dereje, N. Determinants of severe acute malnutrition among under five children in Shashogo Woreda, southern Ethiopia: A community based matched case control study. *J. Nutr. Food Sci.* **2014**, *4*, 300. [CrossRef]
13. Ayele, D.G.; Zewotir, T.T.; Mwambi, H.G. Structured additive regression models with spatial correlation to estimate under-five mortality risk factors in Ethiopia. *BMC Public Health* **2015**, *15*, 268. [CrossRef] [PubMed]
14. Mekonnen, A.; Jones, N.; Tefera, B. *Tackling Child Malnutrition in Ethiopia: Do the Sustainable Development Poverty Reduction Programme's Underlying Policy Assumptions Reflect Local Realities?* Young Lives: London, UK, 2005.
15. Bantamen, G.; Belaynew, W.; Dube, J. Assessment of factors associated with malnutrition among under five years age children at Machakel Woreda, Northwest Ethiopia: A case control study. *J. Nutr. Food Sci.* **2014**, *4*, 1.
16. Asfaw, M.; Wondaferash, M.; Taha, M.; Dube, L. Prevalence of undernutrition and associated factors among children aged between six to fifty nine months in Bule Hora district, South Ethiopia. *BMC Public Health* **2015**, *15*, 41. [CrossRef] [PubMed]
17. Rajkumar, A.S.; Gaukler, C.; Tilahun, J. *Combating Malnutrition in Ethiopia: An Evidence-Based Approach for Sustained Results*; World Bank Publications: Washington, DC, USA, 2011.
18. Sohnesen, T.P.; Ambel, A.A.; Fisker, P.; Andrews, C.; Khan, Q. Small area estimation of child undernutrition in Ethiopian woredas. *PLoS ONE* **2017**, *12*, e0175445. [CrossRef] [PubMed]
19. Wolter, K.M.; Wolter, K.M. Taylor Series Methods. In *Introduction to Variance Estimation*; Springer: New York, NY, USA, 2007; pp. 226–271. Available online: https://link.springer.com/chapter/10.1007/978-0-387-35099-8_6 (accessed on 6 April 2022).
20. Anselin, L. Local indicators of spatial association—LISA. *Geogr. Anal.* **1995**, *27*, 93–115. [CrossRef]
21. Sahu, S.K.; Gelf, A.E.; Holl, D.M. Fusing point and areal level space–time data with application to wet deposition. *J. R. Stat. Soc. Ser. Appl. Stat.* **2010**, *59*, 77–103. [CrossRef]
22. Pfeffermann, D. New Important Developments in Small Area Estimation. *Stat. Sci.* **2013**, *28*, 40–68. [CrossRef]
23. Rao, J.N.; Molina, I. *Small Area Estimation*; John Wiley & Sons, Inc.: Hoboken, NJ, USA, 2015.

24. Elbers, C.; Lanjouw, J.O.; Lanjouw, P. Micro-level estimation of poverty and inequality. *Econometrica* **2003**, *71*, 355–364. [CrossRef]
25. Betti, G.; Dabalen, A.; Ferré, C.; Neri, L. Updating poverty maps between Censuses: A case study of Albania. In *Poverty and Exclusion in the Western Balkans*; Springer: Berlin/Heidelberg, Germany, 2013; pp. 55–70.
26. Ghosh, M.; Natarajan, K.; Stroud, T.W.F.; Carlin, B.P. Generalized Linear Models for Small-Area Estimation. *J. Am. Stat. Assoc.* **1998**, *93*, 273–282. [CrossRef]
27. Bakar, K.S.; Biddle, N.; Kokic, P.; Jin, H. A Bayesian spatial categorical model for prediction to overlapping geographical areas in sample surveys. *J. R. Stat. Soc. Ser. Stat. Soc.* **2020**, *183*, 535–563. [CrossRef]
28. Muchie, K.F.; Wanjoya, A.K.; Mwalili, S.M. A Simulation Study of Hierarchical Bayesian Fusion Spatial Small Area Model for Binary Outcome under Spatial Misalignment. *Open J. Stat.* **2021**, *11*, 993–1009. [CrossRef]
29. Orey, S. An ergodic theorem for Markov chains. *Z. Wahrscheinlichkeitstheorie Verwandte Geb.* **1962**, *1*, 174–176. [CrossRef]
30. You, Y.; Rao, J.; Dick, J. Benchmarking hierarchical Bayes small area estimators with application in census undercoverage estimation. In Proceedings of the Survey Methods Section, 2002 Annual Meeting in Hamilton, Hamilton, ON, Canada, 26–29 May 2002; pp. 86–90.
31. WHO. *The Double Burden of Malnutrition-Policy Brief*; WHO: Geneva, Switzerland, 2016.
32. WHO. *The State of Food Security and Nutrition in the World 2018: Building Climate Resilience for Food Security and Nutrition*; Food and Agriculture Organization: Rome, Italy, 2018.
33. African Union Commission; NEPAD Planning and Coordinating Agency; UN Economic Commission for Africa; UN World Food Programme. *The Cost of Hunger in Africa: Social and Economic Impact of Child Undernutrition in Egypt, Ethiopia, Swaziland and Uganda*; United States Agency for International Development: Washington, DC, USA, 2013.
34. *National Nutrition Programme June 2013–June 2015*; Government of Federal Democratic Republic of Ethiopia: Addis Ababa, Ethiopia, 2013. Available online: https://www.medbox.org/pdf/5e148832db60a2044c2d2ccb (accessed on 6 April 2022).
35. Swinburn, B.A.; Kraak, V.I.; Allender, S.; Atkins, V.J.; Baker, P.I.; Bogard, J.R.; Brinsden, H.; Calvillo, A.; De Schutter, O.; Devarajan, R.; et al. The global syndemic of obesity, undernutrition, and climate change: The Lancet Commission report. *Lancet* **2019**, *393*, 791–846. [CrossRef]

Article

Morlet Cross-Wavelet Analysis of Climatic State Variables Expressed as a Function of Latitude, Longitude, and Time: New Light on Extreme Events

Jean-Louis Pinault

Independent Researcher, 96, Rue du Port David, 45370 Dry, France; jeanlouis_pinault@hotmail.fr

Abstract: This study aims to advance our knowledge in the genesis of extreme climatic events with the dual aim of improving forecasting methods while clarifying the role played by anthropogenic warming. Wavelet analysis is used to highlight the role of coherent Sea Surface Temperature (SST) anomalies produced from short-period oceanic Rossby waves resonantly forced, with two case studies: a Marine Heatwave (MHW) that occurred in the northwestern Pacific with a strong climatic impact in Japan, and an extreme flood event that occurred in Germany. Ocean–atmosphere interactions are evidenced by decomposing state variables into period bands within the cross-wavelet power spectra, namely SST, Sea Surface Height (SSH), and the zonal and meridional modulated geostrophic currents as well as precipitation height, i.e., the thickness of the layer of water produced during a day, with regard to subtropical cyclones. The bands are chosen according to the different harmonic modes of the oceanic Rossby waves. In each period band, the joint analysis of the amplitude and the phase of the state variables allow the estimation of the regionalized intensity of anomalies versus their time lag in relation to the date of occurrence of the extreme event. Regarding MHWs in the northwestern Pacific, it is shown how a warm SST anomaly associated with the northward component of the wind resulting from the low-pressure system induces an SST response to latent and sensible heat transfer where the latitudinal SST gradient is steep. The SST anomaly is then shifted to the north as the phase becomes homogenized. As for subtropical cyclones, extreme events are the culmination of exceptional circumstances, some of which are foreseeable due to their relatively long maturation time. This is particularly the case of ocean–atmosphere interactions leading to the homogenization of the phase of SST anomalies that can potentially contribute to the supply of low-pressure systems. The same goes for the coalescence of distinct low-pressure systems during cyclogenesis. Some avenues are developed with the aim of better understanding how anthropogenic warming can modify certain key mechanisms in the evolution of those dynamic systems leading to extreme events.

Keywords: wavelet analysis; extreme subtropical cyclones; climate change; sea surface temperature anomalies; oceanic Rossby waves; Marine Heatwaves

Citation: Pinault, J.-L. Morlet Cross-Wavelet Analysis of Climatic State Variables Expressed as a Function of Latitude, Longitude, and Time: New Light on Extreme Events. *Math. Comput. Appl.* **2022**, *27*, 50. https://doi.org/10.3390/mca27030050

Academic Editor: Leonardo Trujillo

Received: 28 April 2022
Accepted: 2 June 2022
Published: 4 June 2022

Publisher's Note: MDPI stays neutral with regard to jurisdictional claims in published maps and institutional affiliations.

Copyright: © 2022 by the author. Licensee MDPI, Basel, Switzerland. This article is an open access article distributed under the terms and conditions of the Creative Commons Attribution (CC BY) license (https://creativecommons.org/licenses/by/4.0/).

1. Introduction

Although they seem distant, Marine Heatwaves (MHWs) and extreme subtropical cyclones have a common origin, the resonant forcing of oceanic Rossby waves at midlatitudes. The present research is focused on those Rossby waves whose period varies from a few days to a few months. At mid-latitudes, they form preferentially where the western boundary currents move away from the continents to re-enter the subtropical gyres [1]. These Rossby waves induce very active convergent or divergent geostrophic currents in the formation of positive or negative Sea Surface Temperature (SST) anomalies. They appear as harmonics of an annual fundamental Rossby wave resonantly forced by the declination of the sun.

While the climatic impact of Rossby waves is well known, their interaction with the atmosphere still presents some mysteries. However, behind this natural cause, there is a reality: these extreme events are becoming more and more frequent, as numerous studies show. There is therefore a compelling need to elucidate how anthropogenic warming intervenes in the genesis of these extreme events in order to better understand the ocean–atmosphere interactions involved as well as to better anticipate them.

The proposed method consists in representing, in contiguous bands of periods, the amplitude and the time lag, with respect to the date of occurrence of the extreme event, of each of the climatic state variables. For this, both the amplitude and the phase of the climatic state variables are mapped. The amplitude and the phase are deduced from the cross-wavelet power spectra of these state variables expressed as a function of longitude and latitude, and from a reference time series representative of the evolution of the extreme event. In order to optimize the temporal resolution of the dynamics of the observed phenomena, the cross-wavelet power spectra are both scale-averaged over the bandwidths and time-averaged over a time interval bracketing the date of occurrence of the extreme event, the length of which is equal to the bandwidth.

1.1. Marine Heatwaves

MHWs are observed in all oceans. They have impacted fishery resources and the occurrence of harmful algal blooms where rich marine ecosystems are at risk [2]. For example, recent MHW events in the Tasman Sea have had dramatic impacts on the ecosystems, fisheries, and aquaculture off Tasmania's east coast [3]. Similar damages have been investigated in the South China Sea where MHWs were strongly regulated by El Niño–Southern Oscillation (ENSO) [4]. The high latitudes are not spared: Alaska was impacted in 2016 [5]. The economic impact of these events, little known until the recent past, has given rise to much research in recent years. However, our understanding of the large-scale drivers and potential predictability of MHW events is still in its infancy.

The dynamic processes related to the initiation of an advective MHW were investigated in continental shelves, namely the Middle Atlantic Bight of the Northwest Atlantic [6], the North West Australia [7], the Indonesian-Australian Basin and areas including the Timor Sea and Kimberley shelf [8], and in the Pacific shelf waters off southeast Hokkaido, Japan [9]. Favorable climatic conditions are mentioned for driving cross-isobath intrusions of warm, salty offshore water onto the continental shelf.

Long-term temperature changes under the influence of human-induced greenhouse gas-forcing drive coastal MHW trends globally. Cross-shore gradients of MHW and SST changes are reported in the Chilean coast region [10], in mid-latitude coasts like the Mediterranean Sea, Japan Sea, and Tasman Sea, as well as in the northeastern coast of the United States [11], along the Australian coastlines [12], in the Tasman Sea [13], in Canada's British Columbia coastal waters, from Queen Charlotte Strait to the Strait of Georgia [14], in the Coastal Zone of Northern Baja California [15], in the Southern California Bight [16], and in the Oyashio region [17–22].

Studies focused on MHWs have reported conditions favoring the warming of surface waters caused by increased solar radiation because of reduced cloud cover, namely in summer MHWs in the South China Sea [18], in the East China Sea, and the South Yellow Sea [19]. The genesis and trend of MHWs in the Indian Ocean and their role in modulating the Indian summer monsoon have been investigated [20], as well as the role of oceanic Rossby waves forced in the interior South Pacific on observed MHW occurrences off southeast Australia [3].

Finally, intense MHWs occurred at the sea surface over extensive areas of the northwestern Pacific Ocean, including the entire Sea of Japan and part of the Sea of Okhotsk [21,22]. An extreme event due to its extension and intensity, occurred in July–August 2021 [21]. In this article, we will attempt to highlight the role played by oceanic Rossby waves in the genesis of this event, the conditions of formation of which have not yet been fully elucidated.

1.2. Extreme Subtropical Cyclones

The intensity of the heaviest extreme precipitation events is known to increase with global warming [23–27] almost everywhere in the world [28,29]. Particularly impacted are regions subject to subtropical cyclones [30]. At mid-latitudes, these regions are easily identifiable by their precipitation pattern in the 5–10 year band, while they only show a weak seasonality [31,32]. The main areas subject to rainfall oscillation in the 5–10 year band are: (a) Southwest North America, (b) Texas, (c) Southeastern North America, (d) Northeastern North America, (e) Southern Greenland, (f) Europe and Central and Western Asia, (g) the region of the Río de la Plata, (h) Southwestern and Southeastern Australia, and (i) Southeast Asia.

Global warming is projected to lead to a higher intensity of precipitation and longer dry periods in North America [33–35] and Europe [36–41]. Extreme floods during the recent decades in Europe are more frequent compared to the last 500 years [42]. For Germany, the number of people exposed to flood risks could more than triple and damages more than quadruple by the end of the century [43,44]. In summer, an increase is also projected in most parts of Europe, although decreases are projected for some regions in southern and southwestern Europe, partly due to a projected decrease in cyclone frequency in the Mediterranean [45].

In spite of potentially large societal impacts, mechanisms involved in changes in frequency and intensity of heavy precipitation are much less explored. The purpose of this article is to improve techniques for predicting these extreme precipitation events and to advance our knowledge of the possible mechanisms whose incidence and intensity are linked to global warming. For this, we will analyze in detail the different phases of hydroclimatic mechanisms that led to an extreme precipitation event in Germany in July 2021, that is, a region reputed not to be floodable, causing many casualties.

1.3. Oceanic Rossby Waves at Mid-Latitudes

Oceanic Rossby waves have a well-known effect on the climate. The role of Rossby waves in local air-sea interactions over the tropical Indian Ocean and in remote forcing from the tropical Pacific Ocean has been investigated during El Niño and positive Indian Ocean Dipole years [46,47]. High-resolution subsurface observations have provided insight into equatorial oceanic Rossby wave activity forced by Madden-Julian Oscillation events [48].

However, the role played by the oceanic Rossby waves on the climate is not limited to the tropical oceans. The Rossby waves that develop where the western boundary currents leave the continents to re-enter the subtropical gyres have a strong impact on climate [31]. Located at the same latitudes as the subtropical jet streams, they thus participate in the cyclogenesis of mid-latitude eddy systems (anticyclones and depressions) then moving under these powerful air currents.

Oceanic Rossby waves propagate westward. Being approximately non-dispersive, their phase velocity given by the dispersion relation only depends on the latitude [49]. The phase velocity decreases when the latitude increases. At mid-latitudes, it is lower than the velocity of the eastward propagating wind-driven current of the gyre resulting from Ekman pumping associated with the wind curl. Rossby waves are driven by the circulation of the gyre.

Based on the momentum equations of Rossby waves, these baroclinic waves are forced by changes in solar irradiance induced by solar and orbital cycles [50,51]. This property is specific to Rossby waves at mid-latitudes because, in tropical oceans, they are mainly driven by the wind, in this case the trade winds. Under the effect of radiative forcing, in addition to a Sea Surface Height (SSH) anomaly, the propagation of Rossby waves along the subtropical gyre induces a zonal and a meridional modulated current. The meridional current is in phase with the forcing while the zonal current and the SSH perturbation, i.e., the ridge of the Rossby wave, are in quadrature. During the ascending phase of the zonal ridge, the meridional modulated currents converge toward the ridge.

The convergence causes the thermocline to deepen due to the inflow of warm water from the surface of the ocean. The affected ocean surface extends well beyond the gyre due to the meridional currents. A quarter of a period later, the zonal modulated current reaches its maximum at the same time as the ridge. The zonal currents are in opposite phase on either side of the ridge, causing the zonal propagation of the thermocline wave.

Both the meridional and the zonal modulated current change direction every half-period. Note that the speed of the zonal current is expressed in a relative way because the westward propagating Rossby wave is carried by the eastward propagating wind-driven current of the gyre. Its absolute speed is obtained by adding it to that of the steady wind-driven current. Thus, the zonal current of the gyre periodically accelerates, slows down, and sometimes even reverses direction.

During its ascending phase, the Rossby wave behaves like a heat sink while, during its recession phase, the upwelling which occurs along the ridge causes the Rossby wave to release heat that has been stored when the thermocline was lowering. This explains the climatic impact of Rossby waves at mid-latitudes; sometimes they favor high-pressure systems, sometimes low-pressure.

2. Materials and Methods

2.1. The Caldirola–Kanai Oscillator

Several Rossby waves of different periods overlap along the gyre. Sharing the same zonal and meridional modulated currents, these Rossby waves behave like coupled oscillators with inertia. The equation of the Caldirola–Kanai (CK) oscillator, which is a fundamental model of dissipative systems that is usually used to develop a phenomenological single-particle approach for the damped harmonic oscillator [52], can be expressed by considering the conditions of durability of the dynamic system. For that, the equation of the CK oscillator is formulated to express the mode of coupling between N Rossby waves that share the same modulated geostrophic currents [50]:

$$\mathcal{M}_i \ddot{u}_i + \gamma \mathcal{M}_i \dot{u}_i + \sum_{j=1}^{N} J_{ij}(u_i - u_j) = I_i \cos(\Omega t) \qquad (1)$$

where u_i is the zonal geostrophic current velocity of the ith oscillator along the gyre, $\mathcal{M}_i$ the mass of water displaced during a cycle resulting from the quasi-geostrophic motion of the ith oscillator, γ the Rayleigh friction, and J_{ij} the measure of the coupling strength between the oscillators i and j. The right-hand side represents the periodic driving on the ith oscillator with frequency Ω and amplitude I_i, that is, Coriolis and pressure gradient forces. The restoring force simply depends on the difference in velocity of zonal geostrophic currents between the oscillators. So, it vanishes when the velocities are equal $u_i = u_j$ which, in the absence of friction, removes any interaction between the oscillators i and j. On the other hand, the strength of the interaction increases as the difference in velocities increases. The coupling of Rossby waves is exercised by the fact that the velocities u_i are common at the convergence of the modulated geostrophic currents of the resonant oscillatory system.

In order to ensure the durability of that dynamic system, the coupled oscillators have to form oscillatory subsystems so that the resonance conditions are defined recursively:

$$\tau_i = \frac{1}{n_i} \tau_{i-1} \text{ with } \tau_0 = T \qquad (2)$$

where $n_i = 2$ or 3. T is the period of the fundamental wave, that is, one year.

The CK oscillators resonate in subharmonic and harmonic modes of the annual fundamental wave. The apparent eastward propagation velocity of this fundamental wave depends on the latitude of the gyre where the western boundary current leaves the continent, and to the velocity of the steady wind-driven current. In the case of a pseudo-periodic forcing, its apparent wavelength is adjusted to the forcing period when the average forcing frequency is present in the frequency spectrum of the dynamic system. Natural frequencies close to the forcing frequency are favored, while those far from it are dampened because of friction so that the fundamental wave is resonantly forced by the variations in solar irradiance resulting from the declination of the sun. This is by far the primary source of temperature variability in surface and subsurface waters of the oceans at mid-latitudes.

2.2. Data

Daily gridded data (1/4 degree × 1/4 degree) of SSH, geostrophic currents [53], and SST [54–57] are used. SSH and geostrophic current data begin 15 March 2019. Although starting earlier, SST data is used over the same time interval as SSH. The last update was on 17 October 2021.

Data of precipitation is produced as part of the Global Precipitation Climatology Project (GPCP) Climate Data Record (CDR) Daily analysis, which spans the time period October 1996 to the near present [58,59]. The algorithm to produce the daily 1° GPCP product takes inputs from several different sources and merges them to create the most consistent and accurate daily precipitation estimates [60].

2.3. Wavelet Analysis

2.3.1. Marine Heatwaves

The problem that we are going to tackle, which relates to the genesis of MHWs at mid-latitudes, consists in highlighting the evolution of brief SST anomalies at different time scales, reflecting the driving role of oceanic Rossby waves. A Morlet cross-wavelet analysis is performed to estimate the amplitude of variations in characteristic period bands of four state variables, that is, SSH, modulated geostrophic currents, and SST, as well as their phase compared to a reference time series [61]. Presently, SST averaged along the parallel 34.125° N, between 145.625° E and 148.125° E, is used as the time reference. The average of the SST data over a short segment of the parallel makes it possible to specify the evolution of the heat wave over time by reducing the noise without significantly harming the spatial resolution from which the location of the reference is defined.

As we will see later, SST anomalies observed on 5 January 2020 and on 23 July 2021 are representative of a phenomenon that led to a "marine cold wave" in the first case and a "marine heatwave" in the second. This time reference is chosen so that it unambiguously reflects those two extreme events, both being defined as a sharp surface temperature anomaly (the extremum does not last more than a day), positive or negative as the case may be.

Under these conditions, the square root of the wavelet power applied to the state variable time series, scale-averaged over the period bands, is the regionalized amplitude of anomalies, whatever their date of occurrence. The cross-wavelet power applied to both the state variable time series and the time reference, scale-averaged over the period bands, is the regionalized phase of anomalies. It is the time-lag between the extrema of the anomalies and the date of occurrence of the extreme event, namely the marine cold wave or the MHW [62]. Consequently, for each state variable and for each period band a paired map of the amplitude and the phase of anomalies is obtained.

The wavelet analysis of the state variables is carried out over short periods of time framing the date of occurrence of the extreme events. In this way, for each band, both the amplitude and the phase of the anomalies are time-averaged over a time interval coinciding with the width of the band, centered on the date of occurrence of the extreme events.

The choice of each period band is guided by the properties of the CK oscillator considered as a prototype of coupled Rossby waves. Harmonics of the CK oscillator are identifiable in Figure 1b that represents the Wavelet Fourier spectrum of SSH at 34.125° N, 140.125° E located on the north Pacific gyre, 0.75° south of the Pacific shelf off the southeastern region of Japan. The richness of the Fourier spectrum is probably attributable to the proximity of the coasts of Japan facing the Pacific Ocean. This suggests a local resonance of Rossby waves, which strengthens harmonics. The Fourier spectrum distinctly shows the annual fundamental wave, the amplitude of which is predominant, which gives rise to harmonics whose main periods are 1/3 yr, 1/6 yr, 1/12 yr, and 1/24 yr.

Rossby waves are subject to very large fluctuations as attested by the width of the peaks in the Fourier spectrum. Only the amplitudes of the harmonics whose periods are 1/12 yr and 1/24 yr are known with a level of confidence greater than 95% (the lack of precision of the amplitude of the annual Rossby wave results from the short duration of the observation period, which was barely 3 years).

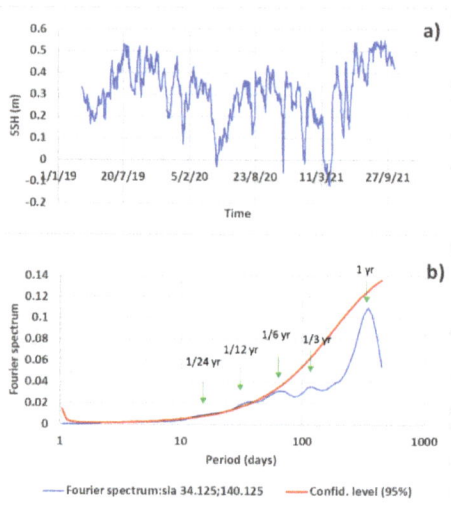

Figure 1. The SSH anomaly at 34.125° N, 140.125° E–(**a**) the raw signal–(**b**) the Wavelet Fourier spectrum (adimentional) and the main harmonics. SSH data is provided by the National Oceanic and Atmospheric Administration (NOAA) https://coastwatch.noaa.gov/pub/socd/lsa/rads/sla/daily/nrt/ (accessed on 27 April 2022).

In Table 1 the period bands are chosen in accordance with (2). Bandwidths are deduced from the mean period τ of harmonics. Lower and upper limits are $0.75 \times \tau$ and $1.5 \times \tau$, respectively, so that the bands are contiguous, since the periods are halved from one harmonic to another. The progression of the bands is continued beyond the periods analyzed (mean periods 1/48 and 1/96 years).

Table 1. Properties of observed or presumed harmonics and bandwidths.

Harmonic	n_i from (1)	Mean Period (Days)	Lower Limit (Days)	Upper Limit (Days)
1	–	365.2	–	–
1/3	3	121.7	91.3	182.6
1/6	2	60.9	45.7	91.3
1/12	2	30.4	22.8	45.7
1/24	2	15.2	11.4	22.8
1/48	2	7.6	5.7	11.4
1/96	2	3.8	2.9	5.7

2.3.2. Subtropical Cyclones

The phenomenological study of climatic phenomena leading to extreme precipitation at mid-latitudes is performed in the same way. Three state variables are jointly analyzed, the precipitation height, i.e., the thickness of the layer of water produced during a day, SST, and SSH. When a positive SST anomaly is locally in phase with the extreme precipitation event while being within the perimeter of the cyclonic low-pressure system, this means that the water vapor evaporated from the ocean is involved in the cycle of cyclogenesis by providing latent heat during the condensation process. This concomitance results from the fact that the atmospheric phenomena leading to the transport of water vapor within the low-pressure system are very rapid compared to the oceanic processes at the origin of the SST anomalies. As will be justified later, the supply of the cyclonic system from the free surface of the ocean occurs in less than a day, whereas the maturation of a large surface and uniform phase SST anomaly generally takes at least ten days.

3. Results

3.1. Marine Heatwaves

As shown in Figure 2c,d sudden positive or negative SST anomalies may occur where Rossby waves are resonantly forced [1]. Two major positive anomalies occurred during the time of observation, namely from 1 January 2019 to 27 September 2021. The first positive anomaly occurred on 30 May 2019, the second on 23 July 2021. The first is one month ahead of the corresponding negative SSH anomaly, while the second is 2 weeks in advance (Figure 2a,c). Anticipation of SST anomalies means that they occurred while the thermocline was lifting. One major negative SST anomaly occurred on 5 January 2020, one week behind the corresponding positive SSH anomaly, while the thermocline was deepening (Figure 2b,d).

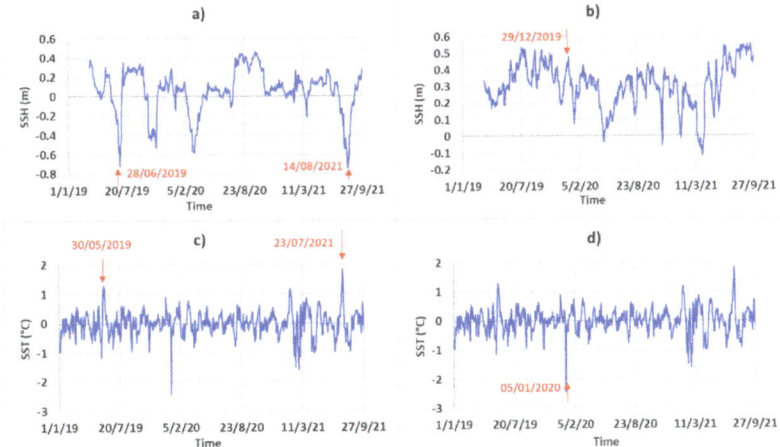

Figure 2. Abrupt events highlighted by SSH at 34.125° N, 148.125° E in (**a**) and at 34.125° N, 140.125° E in (**b**), and by SST averaged along the parallel 34.125° N between 145.625° E and 148.125° E, and filtered in the band of 1–68 days to emphasize the rapid variations while attenuating the annual variations. This series is used as the time reference in the wavelet analysis of data. (**a,c**) are referring to warm events, (**b,d**) to cold events. SST data is provided by NOAA https://www.ncei.noaa.gov/data/sea-surface-temperature-optimum-interpolation/v2.1/access/avhrr/ (accessed on 27 April 2022).

Each SST anomaly corresponds to an opposite SSH anomaly. The reverse is not true; some SSH anomalies do not produce significant SST anomalies. This suggests strong ocean–atmosphere interactions are required for the Rossby waves to produce coherent SST anomalies, with a threshold effect.

3.1.1. The Marine Heatwave That Occurred on 21 July 2021

The climatic impact of this heatwave was significant. One of the most notable records in July 2021 was registered in Asahikawa 43°46′ N, 142°22′ E [63]. The city registered 36.2 °C on 27 July, breaking the previous record of 36 °C set on 7 August 1989.

3.1.2. Wavelet Analysis of Climatic State Variables

As shown in Figure 3a,c,e,g,i, the harmonics of SSH are visible along the North-Pacific gyre, mainly between latitudes 25° N and 35° N, and between longitudes 130° E and 180°. The longitudinal and meridional extensions of Rossby waves increases with period. With regarding to the harmonic 1/6, the SSH anomaly in Figure 3i,j extends over areas of the northwestern Pacific Ocean, including the Yellow Sea, and the entire Sea of Japan.

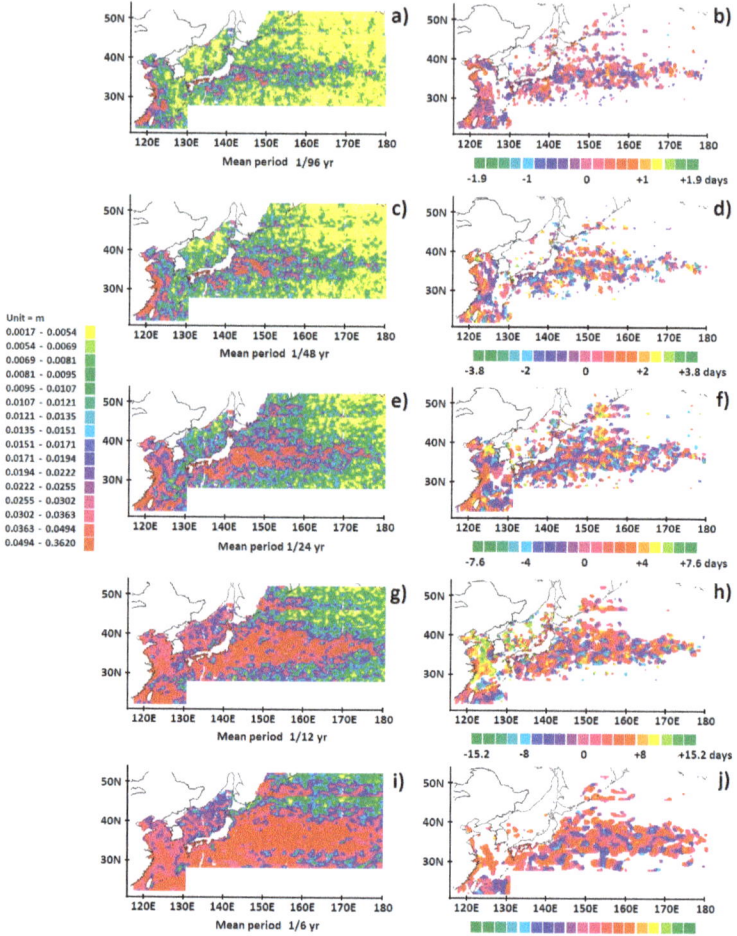

Figure 3. The amplitude (**a,c,e,g,i**) and the phase (**b,d,f,h,j**) of the harmonics of SSH. The periods are 1/6 yr in (**a**), (**b**) 1/12 yr in (**c**), (**d**) 1/24 yr in (**e**), (**f**) 1/48 yr in (**g**), and (**h**) 1/96 yr in (**i,j**). The amplitudes are expressed in 16 classes, each containing the same number of individuals (quantiles). The color of the bar associated with the phase represents an angle varying from −180° to + 180° [61] (each class corresponds to 20°). This angle is reflected by a segment of time of one period, hence the coincidence of the colors at the ends. Time lags in (**b,d,f,h,j**) are relative to 23 July 2021. The time reference is the SST anomaly averaged along the parallel 34.125° N between 145.625° E and 148.125° E. The SSH anomaly is negative when the time lag is zero (the SSH anomaly is negatively correlated with the SST anomaly and late compared to SST). Only the phase corresponding to the 37.5% quantile of the highest values of the amplitude is displayed. Same data sources as in previous figures.

In Figure 3b,d,f,h,j, the phase of the SSH anomaly clearly shows a succession of ridges and troughs in phase opposition. The momentum equations applied to a quasi-geostrophic motion of oceanic Rossby waves show that the meridional geostrophic current V is in phase with the forcing while both the zonal current U and SSH anomalies are in quadrature. However, the phase is more precise for the zonal current (Figure 4) and the meridional current (Figure 5) than for SSH in Figure 3. Indeed, the estimation of geostrophic current velocities from SSH anomalies has a filtering effect because it involves surrounding measurements of SSH. This has the effect of reducing noise and making the interpolated values more representative than the raw measurements of SSH.

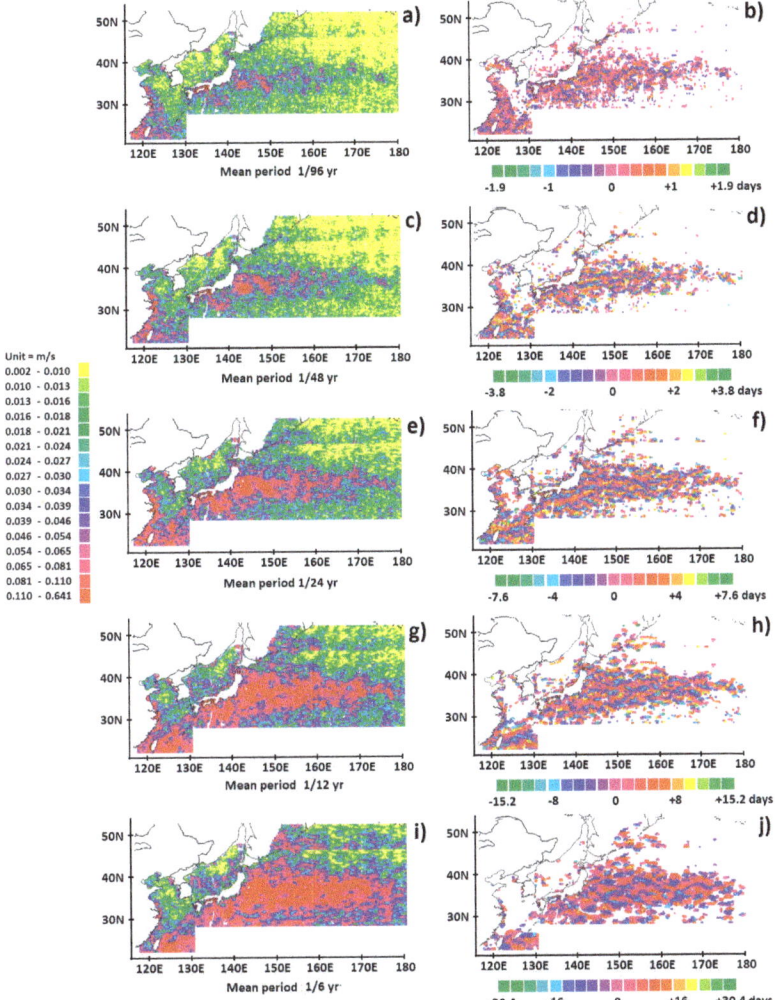

Figure 4. Same as Figure 3 for the amplitude (**a,c,e,g,i**) and the phase (**b,d,f,h,j**) of the zonal geostrophic current U oriented to the east when the time lag is zero. Same data sources as in previous figures.

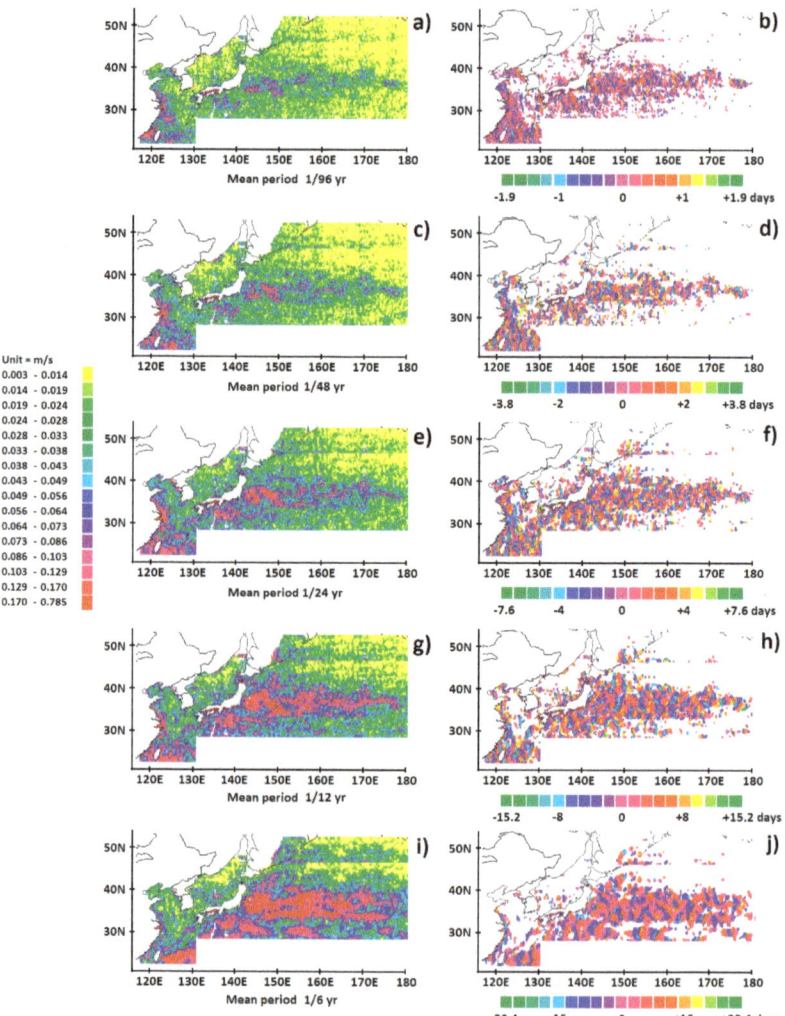

Figure 5. Same as Figure 3 for the amplitude (**a,c,e,g,i**) and the phase (**b,d,f,h,j**) of the meridional geostrophic current V oriented to the south when the time lag is zero. Same data sources as in previous figures.

The modulated geostrophic currents change direction every apparent half-wavelength of Rossby waves. Thus, in Figure 4, the zonal current U shows a succession of regions in phase opposition whose size corresponds to an apparent half wavelength of Rossby waves. These regions form a mosaic of cells in which the zonal geostrophic currents converge or diverge when the cell is translated longitudinally by half of a wavelength. This alternation is still observable for meridional current V (Figure 5), but this time convergence or divergence occur in the North-South direction.

Figures 4 and 5 confirm the previous observations regarding the longitudinal and meridional extensions of Rossby waves along the gyre from SSH as the period increases. This also applies to the speed of modulated geostrophic currents. However, the anomalies of modulated geostrophic currents remain localized along the gyre as the period increases, without stretching to the Yellow Sea, and the Sea of Japan, as does SSH. This difference in the behavior of Rossby waves in the semi-closed seas suggests that these seas are not large enough to allow the formation of perceptible geostrophic currents.

Anomalies in opposite phase also widen with the period, consistent with the increase in apparent Rossby wavelength. The anomalies of the zonal component of the geostrophic current U extend longitudinally with the period while the anomalies of the meridional component V extend latitudinally as shown in Figures 4j and 5j.

Downwelling that occurs in convergent cells means that the thermocline lowers, the intake of warm water resulting from geostrophic currents. On the contrary, upwelling that occurs in divergent cells makes the thermocline rise, restoring warm water under the effect of geostrophic currents. The alternation of convergent or divergent cells throughout the gyre at mid-latitudes highlights the determining role of these cells regarding their climatic impacts. These privileged ocean–atmosphere interactions along the gyre occur at all time scales extending from the annual, seasonal cycles to time intervals not exceeding a few days.

These ocean–atmosphere interactions induce atmospheric baroclinic instabilities as suggested by the variations in SST at the rate of the different periods of the Rossby waves, as shown in Figure 6. The transient SST anomalies occur along the gyre from which the Kuroshio leaves the Asian continent to a longitude close to 180°. Regarding the harmonic 1/6, the SST anomaly in Figure 6i,j is translated over extensive areas of the northwestern Pacific Ocean, including the Yellow Sea, the entire Sea of Japan, and part of the Sea of Okhotsk, as does SSH.

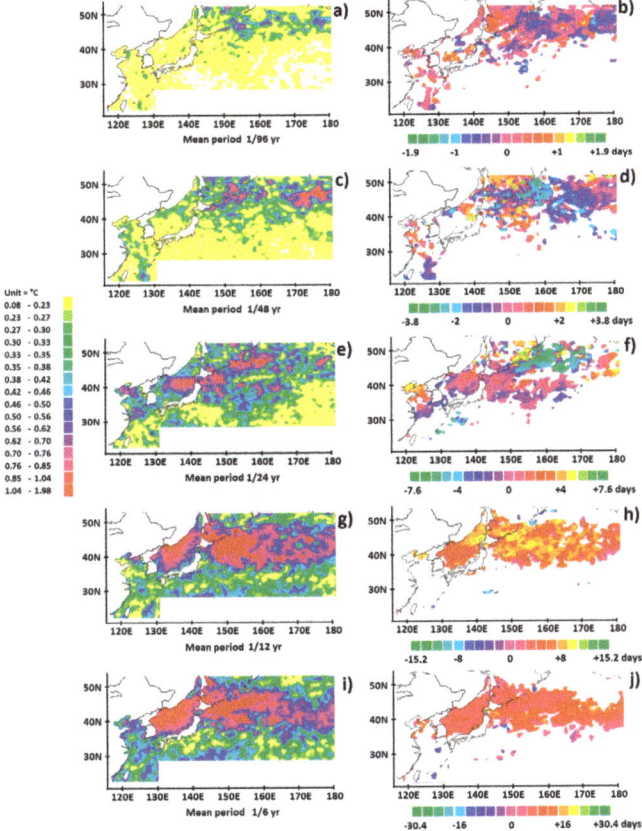

Figure 6. Same as Figure 3 for the amplitude (a,c,e,g,i) and the phase (b,d,f,h,j) of SST whose anomalies are positive when the time lag is zero. Same data sources as in previous figures.

Compared to SSH, SST anomalies are translated to the north while widening (Figure 6). This translation that appears especially during the first 3 periods is of short duration, which suggests the role of the atmosphere. Highly contrasted during the first 3 periods, the phases of SST anomalies become uniform as the period increases. As shown in the Figure 6j, uniformity of the phase is achieved for the harmonic 1/6, which confirms that the lifetime of the SST anomaly is very short compared to the period close to 2 months.

3.2. The Marine Cold Wave That Accurred on 5 January 2020

Figure 7 shows the amplitude and the phase of SST anomalies during the cold event. Anomalies are little translated toward the north, which suggests the weakness of the SST response to the meridional component of the wind resulting from a high-pressure system initiated by the negative SST anomaly of the gyre. Here again, the phase shows a mosaic of convergent and divergent cells characterized by the inversion of geostrophic currents (Figure 7d,f). But contrary to what happens for MHWs, the phase does not homogenize when the period increases, reflecting the SSH anomaly. This suggests the weakness of the ocean–atmosphere interactions, hence the weak climatic impact of marine cold waves.

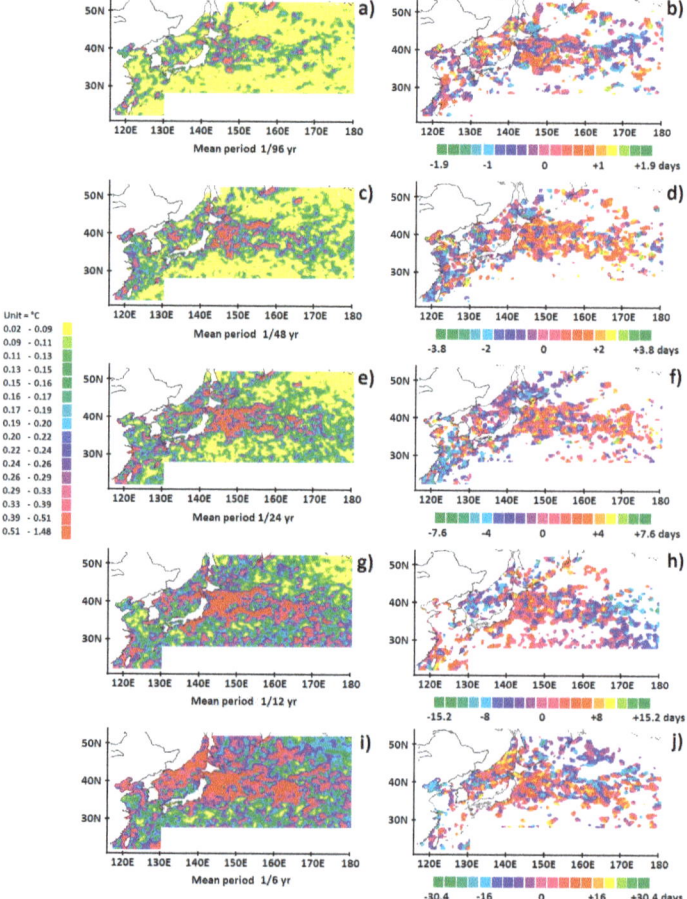

Figure 7. Same as Figure 3 for the amplitude (a,c,e,g,i) and the phase (b,d,f,h,j) of SST whose anomalies are negative when the time lag is zero. Time lags in (b,d,f,h,j) are expressed in relation to 5 January 2020. Same data sources as in previous figures.

3.3. Subtropical Cyclones

Subtropical cyclones develop at mid-latitudes around a stationary front due to an upper-level disturbance, generally an upper-level trough downstream of a strong westerly jet [63,64]. Cyclogenesis results from the combination of vorticity advection and thermal advection created by the latitudinal temperature gradient, a low-pressure center causing upward motion around the low [65]. This rotational flow will push polar air equatorward west of the low via its cold front, and warmer air will push poleward low via the warm front.

3.3.1. An Extreme Precipitation Event, Germany, July 2021

During one week in July 2021, severe flooding occurred across Europe due to dangerous thunderstorms and rain, hitting Germany the hardest. This country experienced up to 182 mm of rain within 72 h. More than 170 people have lost their lives and entire communities have been destroyed. The number of victims of this flood disaster exceeds that of all previous inland floods in Germany since 1900 combined [66].

In mid-July 2021, a pronounced high altitude low shifted from France to the Alps and southern Germany. On its front, very warm and humid air masses were directed to the north and east of Germany, concomitantly with fresher Atlantic air to the south and south-west of Germany, causing record rainfall in parts of North Rhine-Westphalia and Rhineland-Palatinate.

3.3.2. Wavelet Analysis of State Variables

Precipitation height is represented in Figure 8. Here again, the low-pressure system is decomposed into the 5 period bands, the time shift of precipitation areas being relative to the date of occurrence of the extreme rainfall event on 14 July 2021.

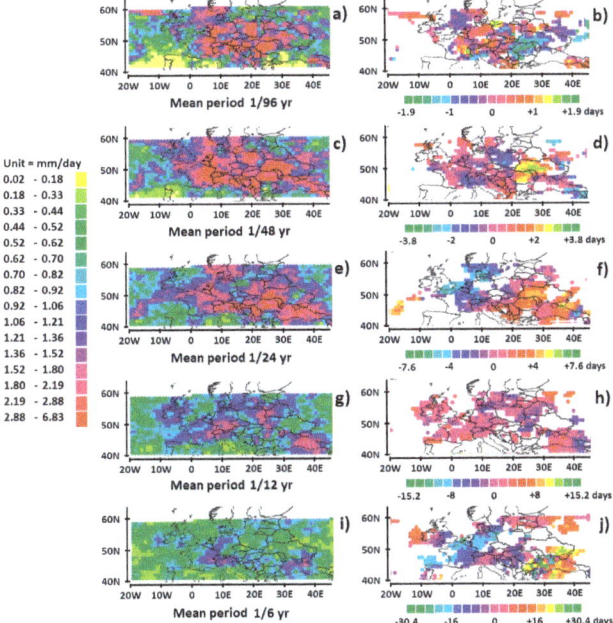

Figure 8. Same as Figure 3 for the amplitude (**a,c,e,g,i**) and the phase (**b,d,f,h,j**) of the harmonics of the precipitation height in western Europe. Time lags in (**b,d,f,h,j**) are expressed in relation to 14 July 2021. The time reference is the rainfall height in Germany at 47° N, 18° E. Daily precipitation data is provided by NOAA https://www.ncei.noaa.gov/data/global-precipitation-climatology-project-gpcp-daily/access/ (accessed 27 April 2022).

The two main precipitation areas represented in Figure 8i,j, i.e., within the band centered on the period 1/6 yr, are independent since they are strongly out of phase with each other. In contrast, Figure 8g,h highlights a coherent low-pressure system at the synoptic scale within the band centered on the period 1/12 yr. The phase of the three main rainfall areas over central and western Europe are indeed only slightly shifted.

According to Figure 8e,f, the rotation of the low-pressure system occurs within the period band centered on 1/24 yr. This deduction is based on the presence of two rainfall areas in phase opposition on both sides of the disaster area, which confirms the hypothesis that the different rainfall areas belong to the same dynamic system at a synoptic scale. The cyclonic flow is fed mainly by the Atlantic west of the coasts of western Europe (Figure 9g,h) and the Baltic Sea (Figure 9e,f).

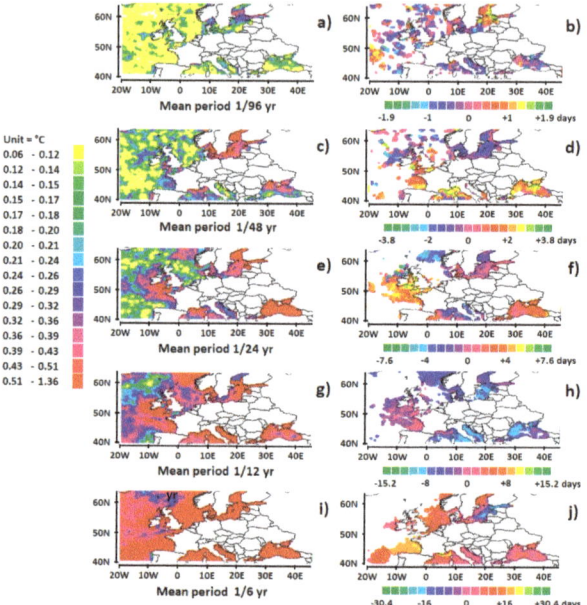

Figure 9. Same as Figure 8 for the amplitude (a,c,e,g,i) and the phase (b,d,f,h,j) of SST. SST anomalies are positive when the time lag is zero. Same data sources as in previous figures.

An SST anomaly over the Atlantic west of the coasts of southern England, Ireland, and France does indeed occur within the period band centered on 1/12 yr (Figure 9g,h), reaching more than 1 °C. The phase of this anomaly is close to zero, showing that ocean–atmosphere interactions are occurring while the SST anomaly is peaking. With regard to the Baltic Sea, the SST anomalies peak within the period bands centered on 1/48 and 1/24 yr, as shown in Figure 9c–f. The phase of the anomaly is close to zero within the period band centered on 1/24 yr, while it is slightly shifted negatively within the period band centered on 1/48 yr, but it nevertheless contributes significantly to the feeding of the low-pressure system by peaking the day before the extreme event occurred.

In the Baltic Sea, SST increases when SSH decreases, i.e., when the thermocline rises. This phenomenon is mainly observable within the period band centered on 1/24 yr (Figures 9e,f and 10e,f) where the phase of the SST anomaly is close to zero. The phases of both SSH and SST anomalies are uniform in seas bordered by coasts, which modifies the apparent wavelength of Rossby waves. It is elongated, in this case, in the absence of a strong current flowing east in which Rossby waves would be embedded. The latter result from the declination of the sun and the variation in solar irradiance during the year, which induces the motion of the thermocline. The westward propagating Rossby waves and their harmonics remain confined in these seas. Convection processes occur in subsurface water, favoring the warming of surface water. These conditions are conducive to the formation of baroclinic instabilities in the atmosphere as a result of increased evaporation.

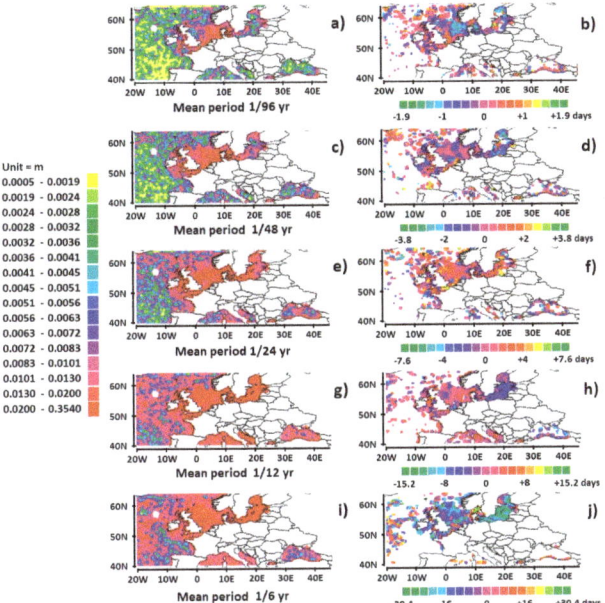

Figure 10. Same as Figure 8 for the amplitude (**a,c,e,g,i**) and the phase (**b,d,f,h,j**) of SSH. SSH anomalies are negative when the time lag is zero. Same data sources as in previous figures.

With regard to the Atlantic Ocean, SSH anomalies are weak off the coasts of western Europe, whereas the amplitude of SST anomalies is high. This suggests that this temperature anomaly results from atmospheric phenomena that translate the SST anomalies developing along the North Atlantic gyre toward the east, a process that leads to baroclinic instabilities in the atmosphere.

As shown in Figure 8a–f, the size of the cyclonic flow reduces as the mean period decreases from 1/24 to 1/96 yr, remaining centered in Germany while the rotation accelerates. Within the period band centered on 1/96 yr, the precipitation area is concentrated between latitudes 45° N and 55° N, and longitudes 3° E and 20° E. East of 20° E the precipitation does not contribute to the genesis of the extreme event since it is strongly out of phase (Figure 8b). In addition, the phase is uniform, close to zero, within sampling errors (the step is daily).

Figure 8b,d,f show that several atmospheric layers are rotating simultaneously. They concentrate around the disaster zone as the rotation accelerates. The half-period of rotation passes from the order of 4 days (Figure 8f) to a few hours (Figure 8b). Since the precipitation areas concentrate around the axis of rotation of the low-pressure system while the rotation period decreases, this suggests that the rotation is accelerating as the layer rises, driven by the upward flow of the cyclonic system. In this way, the uppermost layer is fed by the lower layers. Its phase is uniform so that the rotation period is less than the duration of the extreme precipitation event. The water vapor contained in the different atmospheric layers condenses when they rise due to the lowering of the temperature, which leads to heavy precipitation.

However, the concentration of precipitation, which occurs during cycles of shorter periods, cannot be approached using the same data, which is beyond the scope of this article. Here, the spatial and temporal resolution of the rainfall data [60] are suited to highlighting the various stages leading to the deepening of the low-pressure system, namely the merging of the various low-pressure systems at the synoptic scale, and the feeding of the cyclonic flow from the Atlantic Ocean off the coasts of western Europe and the Baltic Sea.

4. Discussion

4.1. Marine Heatwaves

Regarding MHWs, uniformization of the phase as the SST anomaly migrates north only becomes mature in the 1/6 harmonic mode. In the 1/12 mode, the maturation of the SST anomaly is not complete, a time shift of the order of one week remaining within the anomaly (Figure 6h).

With regard to short cycles corresponding to harmonic modes 1/24, 1/48, and 1/96, the northernmost fringe of the SST anomaly whose phase is heterogeneous is transient. Indeed, it disappears completely during long cycles, the SST anomaly concentrating around a zonal midline at approximately 42° N (Figure 6j). This suggests that the warm, humid air from the low-pressure system warms the sea surface as it migrates north, inducing convective processes in subsurface water while the SST becomes increasingly cold. This promotes the creation of a vertical profile of convection/evaporation tending toward an equilibrium between the thermocline and the surface of the ocean. But stratification of the subsurface water leading to this vertical profile seems unstable and does not occur systematically, as shown in Figure 2.

This strong ocean–atmosphere interaction which causes the thermocline to rise, could explain the uniformization of the phases of the SST anomalies at latitudes where the northward thermal gradient of surface water is steep. Uniformization of the phases then amounts to assuming an overall movement of the thermocline during the longest cycle, hence the brief but intense heatwave which appeared around 27 July 2021.

The northward translation of the SST anomaly is only significant in the case of heatwaves due to the low-pressure system that forms above the gyre before developing into a synoptic cyclonic system. This enhances the SST response to latent and sensible heat fluxes directed to the north. This sudden SST response to atmospheric transfers has already been observed, which sparked interest in this research [21]. According to the authors, MHW observed at the sea surface in the summer of 2021 was the largest in extent and intensity since the beginning of satellite measurements of global SST in 1982, with a strong societal impact.

Other works reported such MHWs in the northwestern Pacific [9,22]. Ref. [22] reported the positive SST anomaly that occurred in August 2020 in subtropical waters in the surroundings of the gyre 120° E–180° E, 20° N–35° N, which was attributed to anthropogenic forcing. Further investigations seem necessary to validate such a hypothesis. Indeed, this positive SST anomaly does not seem distinguishable from internal variability in the context of the present study (Figure 2c).

In [17] the SST of the Oyashio region abruptly increased in the summer of 2010, and a high summertime SST repeated every year until 2016. This was attributed to the strengthening of the Kuroshio water influence. In [9], extreme weather and MHWs are reported; these occurred simultaneously around the Pacific shelf off southeastern Hokkaido, Japan. In these two cases, the influence of the western boundary current was presumably involved, in conjunction with extreme weather. Based on recent works relying on the properties of Rossby waves at mid-latitudes, the present paper proposes a common cause for these intriguing phenomena.

4.2. Subtropical Cyclones

A low-pressure system is forming at the synoptic scale, the result of the merger of several low-pressure subsystems. To achieve this merger, dew-point fronts have to be formed, separating moist air masses found ahead of the dry line from drier air masses found behind it. The drier air behind dew-point fronts lifts up the moist air ahead, triggering strong moist convection. A barometric trough gradually forms, which creates a convergence zone in the lower layers of the atmosphere and upper-level divergence.

The increasingly rapid rotation of cyclonic flows in the various atmospheric layers as they rise produces an extreme rainfall event. The rapid cycles of cyclogenesis contrast with the slowly maturing phenomena without which the cyclonic system could not have developed with such magnitude. They may lead to SST anomalies concomitant with the extreme rainfall event, which occur within the period bands centered on 1/12 and 1/24 yr. Monitoring these maturation processes could help predict the occurrence of devastating climatic phenomena.

The analysis of the different stages leading to subtropical low-pressure systems makes it possible to address an essential problem that relates to the presumed impact of anthropogenic forcing. One mechanism for the increase in such transient events discussed in the literature is related to the slowing of the predominant westerly wind circulation evident in observational data [66,67], due to a strong warming of the Arctic as a result of global warming [68]. Such a slowdown has been linked to observed increases in the persistence of weather systems [69,70].

By influencing the rapid cycles of cyclogenesis, such a mechanism could contribute to explaining the increase in the frequency of extreme rainfall events observed during the last decades in the northern hemisphere, in particular in the North America. But the ubiquity of the increase in the frequency as well as the intensity of extreme rainfall events also suggest an evolution in the mechanisms favoring the development of cyclonic flows at the synoptic scale. This hypothesis is corroborated by the fact that extreme rainfall events occur in places deemed not to be flood-prone, causing numerous victims, as happened in Germany in July 2021, thus deceiving the vigilance of weather-watch systems.

The development of coherent SST anomalies, the main driver of synoptic-scale subtropical cyclones, is unambiguously linked to the propagation of oceanic Rossby waves. These result from solar forcing, independent of anthropogenic forcing. In contrast, other mechanisms related to global warming appear to be decisive in the context of slow cycles during which the coalescence of low-pressure systems occurs. Such mechanisms are strengthened by a temperature increase of ocean surface water associated with an overall increase in atmospheric humidity, which lowers the dew point and favors the formation of fronts. In return, the extension of the low-pressure system at the synoptic scale centered on a continental low favors the feeding of the cyclonic flow by overlapping over surrounding SST anomalies. Owing to the accumulated latent heat, with regard to their internal energy these low-pressure systems promote upper-level lows, favoring blocks. This may explain the record precipitations observed during the last decades when pouring over regions deemed not to be flood prone, as has occurred in many places in Western and Central Europe.

5. Conclusions

The wavelet analysis of high temporal and spatial resolution data, namely SSH, geostrophic currents, and SST in the northwestern Pacific, allowed the highlighting of the formation of a mosaic of convergent and divergent cells along the north Pacific gyre from where the Kuroshio leaves the Asian continent to nearly 180°. Upwelling and downwelling are associated with Rossby waves of short apparent wavelengths embedded in the wind-driven current of the gyre. The driver of the fundamental Rossby wave and the harmonics is the declination of the sun. Sudden SSH anomalies may occur, some of them producing abrupt extensive positive or negative SST anomalies, opposite in sign to SSH anomalies from which they originated. This phenomenon is general and is observable along the subtropical gyres where the western boundary currents move away from the continents.

Regarding MHWs in the northwestern Pacific, a warm SST anomaly associated with the northward component of the wind resulting from the low-pressure system induces an SST response to latent and sensible heat transfer where the latitudinal SST gradient is steep. The SST anomaly is then shifted north while the phases become uniform.

The wavelet analysis of high temporal and spatial resolution of SSH, SST, and rainfall height in the North Atlantic, the Baltic Sea, and northwest Europe has made it possible to highlight the evolution of an extratropical cyclone in northwestern Europe, of exceptional intensity, at different time scales. Intensification of subtropical cyclones as well as the increase in their frequency appear to be mainly related to the evolution of conditions favoring the formation of low-pressure systems at the synoptic scale. These conditions are probably exacerbated by anthropogenic warming which promotes the maturation of the mechanisms leading to the coalescence of lows. In these conditions, the interactions between the atmosphere and the coherent positive SST anomalies on the surrounding ocean play a major role in feeding the cyclonic flow centered on a continental low. Owing to the accumulated latent heat, extreme subtropical cyclones induce upper-level lows that favor the persistence of the cyclonic flow.

The innovative nature of this study is based on the dynamics of the various systems implicated in the formation of extreme climatic events. These events are the culmination of exceptional circumstances, some of which are foreseeable due to their relatively long maturation time. Some avenues are developed with the aim of better understanding how anthropogenic warming can modify certain key mechanisms in the evolution of the dynamic system at the interface between the oceans and the atmosphere.

Future work will focus on the role played by the anthropogenic forcing in the formation of extensive MHWs. On the other hand, by taking advantage of high-resolution data on geostrophic currents, a systematic study of short-period Rossby waves developing where the western boundary currents leave the continents to re-enter the subtropical gyres would be rich in teaching how to specify their climatic impacts, including the conditions of formation of MHWs and extreme rainfall events. Using the same method of investigation, other case studies focusing in particular on the southern hemisphere are required with the aim of generalizing these investigations.

Funding: This research received no external funding.

Data Availability Statement: Only public data duly referenced are used.

Conflicts of Interest: The author declares no conflict of interest.

References

1. Pinault, J.-L. A Review of the Role of the Oceanic Rossby Waves in Climate Variability. *J. Mar. Sci. Eng.* **2022**, *10*, 493. [CrossRef]
2. Yao, Y.; Wang, J.; Yin, J.; Zou, X. Marine heatwaves in China's marginal seas and adjacent offshore waters: Past, Present, and Future. *J. Geophys. Res. Oceans* **2020**, *125*, e2019JC015801. [CrossRef]
3. Li, Z.; Holbrook, N.J.; Zhang, X.; Oliver, E.C.J.; Cougnon, E.A. Remote Forcing of Tasman Sea Marine Heatwaves. *J. Clim.* **2020**, *33*, 5337–5354. [CrossRef]
4. Liu, K.; Xu, K.; Zhu, C.; Liu, B. Diversity of Marine Heatwaves in the South China Sea Regulated by ENSO Phase. *J. Clim.* **2022**, *35*, 877–893. [CrossRef]
5. Walsh, J.E.; Thoman, R.L.; Bhatt, U.S.; Bieniek, P.A.; Brettschneider, B.; Brubaker, M.; Danielson, S.; Lader, R.; Fetterer, F.; Holderied, K.; et al. The High Latitude Marine Heat Wave of 2016 and Its Impacts on Alaska. *Bull. Am. Meteor. Soc.* **2018**, *99*, S39–S43. [CrossRef]
6. Chen, K.; Gawarkiewicz, G.; Yang, J. Mesoscale and Submesoscale Shelf-Ocean Exchanges Initialize an Advective Marine Heatwave. *J. Geophys. Res. Oceans* **2021**, *127*, e2021JC017597. [CrossRef]
7. Maggiorano, A.; Feng, M.; Wang, X.H.; Ritchie, L.; Stark, C.; Colberg, F.; Greenwood, J. Hydrodynamic drivers of the 2013 marine heatwave on the North West Shelf of Australia. *J. Geophys. Res. Oceans* **2021**, *126*, e2020JC016495. [CrossRef]
8. Benthuysen, J.A.; Oliver, E.C.J.; Feng, M.; Marshall, A.G. Extreme marine warming across tropical Australia during austral summer 2015–2016. *J. Geophys. Res. Oceans* **2018**, *123*, 1301–1326. [CrossRef]
9. Kuroda, H.; Taniuchi, Y.; Kasai, H.; Nakanowatari, T.; Setou, T. Co-occurrence of marine extremes induced by tropical storms and an ocean eddy in summer 2016: Anomalous hydrographic conditions in the Pacific shelf waters off southeast Hokkaido, Japan. *Atmosphere* **2021**, *12*, 888. [CrossRef]
10. Marin, M.; Bindoff, N.L.; Feng, M.; Phillips, H.E. Slower long-term coastal warming drives dampened trends in coastal marine heatwave exposure. *J. Geophys. Res. Oceans* **2021**, *126*, e2021JC017930. [CrossRef]
11. Marin, M.; Feng, M.; Phillips, H.E.; Bindoff, N.L. A global, multiproduct analysis of coastal marine heatwaves: Distribution, characteristics and long-term trends. *J. Geophys. Res. Oceans* **2021**, *126*, e2020JC016708. [CrossRef]

12. Feng, M.; Zhang, X.; Oke, P.; Monselesan, D.; Chamberlain, M.; Matear, R.; Schiller, A. Invigorating Ocean boundary current systems around Australia during 1979–2014: As simulated in a near-global eddy-resolving ocean model. *J. Geophys. Res. Oceans* **2016**, *121*, 3395–3408. [CrossRef]
13. Perkins-Kirkpatrick, S.E.; King, A.D.; Cougnon, E.A.; Holbrook, N.J.; Grose, M.R.; Oliver EC, J.; Lewis, S.C.; Pourasghar, F. The Role of Natural Variability and Anthropogenic Climate Change in the 2017/2018 Tasman Sea Marine Heatwave. *Bull. Am. Meteorol. Soc.* **2019**, *100*, S105–S110. [CrossRef]
14. Dosser, H.V.; Waterman, S.; Jackson, J.M.; Hannah, C.G.; Evans, W.; Hunt, B.P.V. Stark physical and biogeochemical differences and implications for ecosystem stressors in the Northeast Pacific coastal ocean. *J. Geophys. Res. Oceans* **2021**, *126*, e2020JC017033. [CrossRef]
15. Delgadillo-Hinojosa, F.; Félix-Bermúdez, A.; Torres-Delgado, E.V.; Durazo, R.; Camacho-Ibar, V.; Mejía, A.; Ruiz, M.C.; Linacre, L. Impacts of the 2014–2015 warm-water anomalies on nutrients, chlorophyll-a and hydrographic conditions in the coastal zone of northern Baja California. *J. Geophys. Res. Oceans* **2020**, *125*, e2020JC016473. [CrossRef]
16. Fumo, J.T.; Carter, M.L.; Flick, R.E.; Rasmussen, L.L.; Rudnick, D.L.; Iacobellis, S.F. Contextualizing marine heatwaves in the Southern California bight under anthropogenic climate change. *J. Geophys. Res. Oceans* **2020**, *125*, e2019JC015674. [CrossRef]
17. Miyama, T.; Minobe, S.; Goto, H. Marine heatwave of sea surface temperature of the Oyashio region in summer in 2010–2016. *Front. Mar. Sci.* **2021**, *7*, 576240. [CrossRef]
18. Yao, Y.; Wang, C. Variations in summer marine heatwaves in the South China Sea. *J. Geophys. Res. Oceans* **2021**, *126*, e2021JC017792. [CrossRef]
19. Gao, G.; Marin, M.; Feng, M.; Yin, B.; Yang, D.; Feng, X.; Ding, Y.; Song, D. Drivers of marine heatwaves in the East China Sea and the South Yellow Sea in three consecutive summers during 2016–2018. *J. Geophys. Res. Oceans* **2020**, *125*, e2020JC016518. [CrossRef]
20. Saranya, J.S.; Roxy, M.K.; Dasgupta, P.; Anand, A. Genesis and trends in marine heatwaves over the tropical Indian Ocean and their interaction with the Indian summer monsoon. *J. Geophys. Res. Oceans* **2020**, *127*, e2021JC017427. [CrossRef]
21. Kuroda, H.; Setou, T. Extensive Marine Heatwaves at the Sea Surface in the Northwestern Pacific Ocean in Summer 2021. *Remote Sens.* **2021**, *13*, 3989. [CrossRef]
22. Hayashi, M.; Shiogama, H.; Emori, S.; Ogura, T.; Hirota, N. The northwestern Pacific warming record in August 2020 occurred under anthropogenic forcing. *Geophys. Res. Lett.* **2021**, *48*, e2020GL090956. [CrossRef]
23. Myhre, G.; Alterskjær, K.; Stjern, C.W.; Hodnebrog, Ø.; Marelle, L.; Samset, B.H.; Sillmann, J.; Schaller, N.; Fischer, E.; Schulz, M.; et al. Frequency of extreme precipitation increases extensively with event rareness under global warming. *Sci. Rep.* **2019**, *9*, 16063. [CrossRef] [PubMed]
24. Sillmann, J.; Kharin, V.V.; Zwiers, F.W.; Zhang, X.; Bronaugh, D. Climate Extremes Indices in the CMIP5 Multimodel Ensemble. Part 2: Future Climate Projections. *J. Geophys. Res. Atmos.* **2013**, *118*, 2473–2493. [CrossRef]
25. Giorgi, F.; Coppola, E.; Raffaele, F. A Consistent Picture of the Hydroclimatic Response to Global Warming from Multiple Indices: Models and Observations. *J. Geophys. Res. Atmos.* **2014**, *119*, 11695–11708. [CrossRef]
26. Lehmann, J.; Coumou, D.; Frieler, K. Increased record-breaking precipitation events under global warming. *Clim. Chang.* **2015**, *132*, 501–515. [CrossRef]
27. Fischer, E.; Knutti, R. Detection of spatially aggregated changes in temperature and precipitation extremes. *Geophys. Res. Lett.* **2014**, *41*, 547–554. [CrossRef]
28. Dong, S.; Sun, Y.; Li, C.; Zhang, X.; Min, S.K.; Kim, Y.H. Attribution of extreme precipitation with updated observations and CMIP6 simulations. *J. Clim.* **2021**, *34*, 871–881. [CrossRef]
29. Sun, Q.; Zhang, X.; Zwiers, F.; Westra, S.; Alexander, L.V. A global, continental, and regional analysis of changes in extreme precipitation. *J. Clim.* **2021**, *34*, 243–258. [CrossRef]
30. Hawcroft, M.; Walsh, E.; Hodges, K.; Zappa, G. Significantly increased extreme precipitation expected in Europe and North America from subtropical cyclones. *Environ. Res. Lett.* **2018**, *13*, 124006. [CrossRef]
31. Pinault, J.-L. Regions Subject to Rainfall Oscillation in the 5–10 Year Band. *Climate* **2018**, *6*, 2. [CrossRef]
32. Pinault, J.-L. Global warming and rainfall oscillation in the 5–10 year band in Western Europe and Eastern North America. *Clim. Chang.* **2012**, *114*, 621–650. [CrossRef]
33. Easterling, D.R.; Arnold, J.R.; Knutson, T.; Kunkel, K.E.; LeGrande, A.N.; Leung, L.R.; Vose, R.S.; Waliser, D.E.; Wehner, M.F. Precipitation Change in the United States. In *Climate Science Special Report: Fourth National Climate Assessment, Volume I*; Wuebbles, D., Fahey, J.D.W., Hibbard, K.A., Eds.; U.S. Global Change Research Program: Washington, DC, USA, 2017.
34. Feng, Z.; Leung, L.R.; Hagos, S.; Houze, R.A.; Burleyson, C.D.; Balaguru, K. More frequent intense and long-lived storms dominate the springtime trend in central US rainfall. *Nat. Commun.* **2016**, *7*, 13429. [CrossRef] [PubMed]
35. Changnon, S.A.; Westcott, N.E. Heavy rainstorms in Chicago: Increasing frequency, altered impacts and future implications. *J. Am. Water Resour. Assoc.* **2002**, *38*, 1467–1475. [CrossRef]
36. *Managing the Risks of Extreme Events and Disasters to Advance Climate Change Adaptation. Special Report of the Intergovernmental Panel on Climate Change*; Cambridge University Press: Cambridge, UK, 2012. Available online: https://www.ipcc.ch/report/managing-the-risks-of-extreme-events-and-disasters-to-advance-climate-change-adaptation/ (accessed on 27 April 2022).
37. Rajczak, J.; Pall, P.; Schär, C. Projections of Extreme Precipitation Events in Regional Climate Simulations for Europe and the Alpine Region. *J. Geophys. Res. Atmos.* **2013**, *118*, 3610–3626. [CrossRef]

38. Rajczak, J.; Schär, C. Projects of Future Precipitation Extremes Over Europe: A Multimodel Assessment of Climate Simulations. *J. Geophys. Res. Atmos.* **2017**, *122*, 10773–10800. [CrossRef]
39. Zeder, J.; Fischer, E.M. Observed extreme precipitation trends and scaling in Central Europe. *Weather Clim. Extrem.* **2020**, *29*, 100266. [CrossRef]
40. Ali, H.; Fowler, H.J.; Lenderink, G.; Lewis, E.; Pritchard, D. Consistent Large-Scale Response of Hourly Extreme Precipitation to Temperature Variation Over Land. *Geophys. Res. Lett.* **2021**, *48*, e2020GL090317. [CrossRef]
41. Blöschl, G.; Hall, J.; Viglione, A.; Perdigão, R.A.; Parajka, J.; Merz, B.; Lun, D.; Arheimer, B.; Aronica, G.T.; Bilibashi, A. Changing climate both increases and decreases European river floods. *Nature* **2019**, *573*, 108–111. [CrossRef]
42. Blöschl, G.; Kiss, A.; Viglione, A.; Barriendos, M.; Böhm, O.; Brázdil, R.; Coeur, D.; Demarée, G.; Llasat, M.C.; Macdonald, N. Current European flood-rich period exceptional compared with past 500 years. *Nature* **2020**, *583*, 560–566. [CrossRef]
43. Willner, S.N.; Levermann, A.; Zhao, F.; Frieler, K. Adaptation required to preserve future high-end river flood risk at present levels. *Sci. Adv.* **2018**, *4*, eaao1914. [CrossRef] [PubMed]
44. Dottori, F.; Szewczyk, W.; Ciscar, J.C.; Zhao, F.; Alfieri, L.; Hirabayashi, Y.; Bianchi, A.; Mongelli, I.; Frieler, K.; Betts, R.A.; et al. Increased human and economic losses from river flooding with anthropogenic warming. *Nat. Clim. Chang.* **2018**, *8*, 781–786. [CrossRef]
45. Jacob, D.; Petersen, J.; Eggert, B.; Alias, A.; Christensen, O.B.; Bouwer, L.M.; Braun, A.; Colette, A.; Déqué, M.; Georgievski, G.; et al. EURO-CORDEX: New High-Resolution Climate Change Projections for European Impact Research. *Reg. Environ. Chang.* **2014**, *14*, 563–578. [CrossRef]
46. Chakravorty, S.; Gnanaseelan, C.; Chowdary, J.S.; Luo, J.-J. Relative role of El Niño and IOD forcing on the southern tropical Indian Oceanic Rossby waves. *J. Geophys. Res. Oceans* **2014**, *119*, 8. [CrossRef]
47. Sprintall, J.; Cravatte, S.; Dewitte, B.; Du, Y.; Gupta, A.S. ENSO Oceanic Teleconnections. In *El Niño Southern Oscillation in a Changing Climate*; McPhaden, M.J., Santoso, A., Cai, W., Eds.; American Geophysical Union: Washington, DC, USA, 2020. [CrossRef]
48. Webber, B.G.; Matthews, A.J.; Heywood, K.J.; Kaiser, J.; Schmidtko, S. Seaglider observations of equatorial Indian Oceanic Rossby waves associated with the Madden-Julian Oscillation. *J. Geophys. Res. Oceans* **2014**, *119*, 6. [CrossRef]
49. Gill, A.E. *Atmosphere-Ocean Dynamics, International Geophysics Series 30*; Academic Press: Cambridge, MA, USA, 1982; 662p.
50. Pinault, J.-L. Resonantly Forced Baroclinic Waves in the Oceans: Subharmonic Modes. *J. Mar. Sci. Eng.* **2018**, *6*, 78. [CrossRef]
51. Pinault, J.-L. Resonantly Forced Baroclinic Waves in the Oceans: A New Approach to Climate Variability. *J. Mar. Sci. Eng.* **2021**, *9*, 13. [CrossRef]
52. Choi, M.Y.; Thouless, D.J. Topological interpretation of subharmonic mode locking in coupled oscillators with inertia. *Phys. Rev. B* **2001**, *64*, 014305. [CrossRef]
53. Sea Level Anomaly and Geostrophic Currents, Multi-Mission, Global, Optimal Interpolation, Gridded, Are Provided by the National Oceanic and Atmospheric Administration (NOAA). Available online: https://coastwatch.noaa.gov/pub/socd/lsa/rads/sla/daily/nrt/ (accessed on 19 November 2021).
54. Daily Sea Surface Temperature Is Provided by NOAA. Available online: https://www.ncei.noaa.gov/data/sea-surface-temperature-interpolation/v2.1/access/avhrr/ (accessed on 8 January 2022).
55. Reynolds, R.W.; Smith, T.M.; Liu, C.; Chelton, D.B.; Casey, K.S.; Schlax, M.G. Daily High-Resolution-Blended Analyses for Sea Surface Temperature. *J. Clim.* **2007**, *20*, 5473–5496. [CrossRef]
56. Banzon, V.; Smith, T.M.; Chin, T.M.; Liu, C.; Hankins, W. A long-term record of blended satellite and in situ sea-surface temperature for climate monitoring, modeling and environmental studies. *Earth Syst. Sci. Data* **2016**, *8*, 165–176. [CrossRef]
57. Huang, B.; Liu, C.; Banzon, V.; Freeman, E.; Graham, G.; Hankins, B.; Smith, T.; Zhang, H.M. Improvements of the Daily Optimum Interpolation Sea Surface Temperature (DOISST) Version v2.1. *J. Clim.* **2021**, *34*, 2923–2939. [CrossRef]
58. Global Daily Precipitation Is Provided by NOAA. Available online: https://www.ncei.noaa.gov/data/global-precipitation-climatology-project-gpcp-daily/access/ (accessed on 25 March 2022).
59. Report of Global Precipitation Climatology Project (GPCP). Available online: https://www.ncei.noaa.gov/data/global-precipitation-climatology-project-gpcp-daily/doc/CDRP-ATBD-0913%20Rev%200%20Precipitation%20-%20GPCP%20Daily%20C-ATBD%20(01B-35)%20(DSR-1159).pdf (accessed on 25 March 2022).
60. Huffman, G.J.; Adler, R.F.; Morrissey, M.M.; Bolvin, D.T.; Curtis, S.; Joyce, R.; McGavock, B.; Susskind, J. Global Precipitation at One-Degree Daily Resolution from Multisatellite Observations. *J. Hydrometeorol.* **2001**, *2*, 36–50. [CrossRef]
61. Torrence, C.; Compo, G.P. A Practical Guide to Wavelet Analysis. *Bull. Am. Meteorol. Soc.* **1998**, *79*, 61–78. [CrossRef]
62. Pinault, J.-L.; Pereira, L. What Speleothems Tell Us about Long-Term Rainfall Oscillation throughout the Holocene on a Planetary Scale. *J. Mar. Sci. Eng.* **2021**, *9*, 853. [CrossRef]
63. The Watchers. Available online: https://watchers.news/2021/07/28/record-breaking-heatwave-hokkaido-japan-july-2021/ (accessed on 7 January 2021).
64. Wash, C.H.; Heikkinen, S.H.; Liou, C.S.; Nuss, W.A. A Rapid Cyclogenesis Event during GALE IOP 9. *Mon. Weather Rev.* **1990**, *118*, 234–257. [CrossRef]
65. Wallace, J.M.; Hobbs, P.V. *Atmospheric Science: An Introductory Survey*; University of Washington: Seattle, WA, USA, 2006.
66. The July 2021 Floods in Germany and the Climate Crisis—A Statement by Members of Scientists for Future. Available online: https://info-de.scientists4future.org/the-july-2021-floods-in-germany-and-the-climate-crisis/ (accessed on 22 January 2022).

67. Coumou, D.; Lehmann, J.; Beckmann, J. The weakening summer circulation in the Northern Hemisphere mid-latitudes. *Science* **2015**, *348*, 324–327. [CrossRef] [PubMed]
68. Kornhuber, K.; Tamarin-Brodsky, T. Future Changes in Northern Hemisphere Summer Weather Persistence Linked to Projected Arctic Warming. *Geophys. Res. Lett.* **2021**, *48*, e2020GL091603. [CrossRef]
69. Pfleiderer, P.; Coumou, D. Quantification of temperature persistence over the Northern Hemisphere land-area. *Clim. Dyn.* **2018**, *51*, 627–637. [CrossRef]
70. Harvey, B.J.; Cook, P.; Shaffrey, L.C.; Schiemann, R. The Response of the Northern Hemisphere Storm Tracks and Jet Streams to Climate Change in the CMIP3, CMIP5, and CMIP6 Climate Models. *J. Geophys. Res. Atmos.* **2020**, *125*, e2020JD032701. [CrossRef]

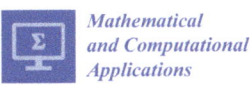

Article

A Note on Gerber–Shiu Function with Delayed Claim Reporting under Constant Force of Interest

Kokou Essiomle and Franck Adekambi *

School of Economics, University of Johannesburg, Johannesburg 2006, South Africa; 201900490@student.uj.ac.za
* Correspondence: fadekambi@uj.ac.za

Abstract: In this paper, we analyze the Gerber–Shiu discounted penalty function for a constant interest rate in delayed claim reporting times. Using the Poisson claim arrival scenario, we derive the differential equation of the Laplace transform of the generalized Gerber–Shiu function and show that the differential equation can be transformed to a Volterra equation of the second kind with a degenerated kernel. In the case of an exponential claim distribution, a closed-expression for the Gerber–Shiu function is obtained via sequence expansion. This result allows us to calculate the absolute (relative) ruin probability. Additionally, we discuss a method of solving the Volterra equation numerically and provide an illustration of the ruin's probability to support the finding.

Keywords: Gerber–Shiu function; constant force of interest; Volterra equation; absolute ruin; delayed reporting times

1. Introduction

The classical ruin problem formulated by Lundberg [1] and Cramér [2] was reconsidered by Taylor [3], who incorporated inflationary conditions into it. He determined that the inflation rate should be taken into consideration and increased. This result is perhaps not surprising since the presence of inflation means that if the effects of inflation are not offset by the premium adjustment, for example, it is the insurer who will suffer prejudice when claims are ultimately initiated. To generalize the problem of ruin, Gerber and Shiu [4] introduced the expected discounted penalty function, which allowed them to analyze the joint distribution of the time of ruin, the surplus immediately before ruin, and the deficit at ruin through the Laplace transform of the defective renewal equation. Cai [5] built upon Gerber and Shiu [4]'s work by including a constant rate of interest and allowing the initial surplus to be negative. Based on these assumptions, he studied absolute ruin using the defective renewal equation of the expected discounted penalty function. Originally studied in the insurance industry, the ruin problem and discounted penalty functions have been found to be applicable to the financial sector as well. We refer readers to Adekambi and Essiomle [6] and Gerber et al. [7] and the references therein for further information on the use of ruin theory in the banking sector.

Sundt and Teugels [8] extended the classical compound Poisson model by assuming that both the premium and the initial reserve are invested in a risk-free asset with a constant interest rate. They derived a differential equation of the ruin probability using the Poisson risk process. In addition, they discussed two special cases where there is no initial reserve and assumed that the claim amounts are exponentially distributed. In the same vein, Kalashnikov and Konstantinides [9] extended Taylor [3]'s work, under a subexponential claims distribution assumption. Similarly, Kalashnikov and Konstantinides [9] and Yuen [10] studied the Gerber–Shiu discounted penalty function when they extended Taylor [3]'s work by incorporating a force of interest and a constant barrier into the classic compound risk model.

Li and Lu [11], in turn, extended Yuen [10]'s work, when they considered a risk process, which incorporates credit and debt interest, and studied the generalized Gerber–Shiu

Citation: Essiomle, K.; Adekambi, F. A Note on Gerber–Shiu Function with Delayed Claim Reporting under Constant Force of Interest. *Math. Comput. Appl.* **2022**, *27*, 51. https://doi.org/10.3390/mca27030051

Received: 7 May 2022
Accepted: 13 June 2022
Published: 20 June 2022

Publisher's Note: MDPI stays neutral with regard to jurisdictional claims in published maps and institutional affiliations.

Copyright: © 2022 by the authors. Licensee MDPI, Basel, Switzerland. This article is an open access article distributed under the terms and conditions of the Creative Commons Attribution (CC BY) license (https://creativecommons.org/licenses/by/4.0/).

penalty function. They derived the exact expressions for the penalty functions when credit and debt interest rates are equal and when claim sizes are exponentially distributed.

Several studies have shown that the assumption of independence between claim sizes and inter-claim times is unrealistic and is created purely for mathematical tractability. Scholars assumed that the risk would have a certain dependence structure to make the model more realistic. Furthermore, Adekambi and Essiomle [12] and Essiomle and Adekambi [13] examined the asymptotic tail probability and the ruin probability of the discounted claim. In their study, they developed a closed analytical form of the ruin probability that assumes a specific distribution of claims.

Cheung et al. [14] examined the structure of various Gerber–Shiu functions using Sparre Andersen's model and allowed a dependence structure between claim sizes and inter-claim times. Cheung et al. [14] analyzed Lundberg's fundamental equation and the generalized adjustment coefficient using the defective renewal equation. Based on Cheung et al. [14]'s research, Willmot and Woo [15] assumed a more general type of dependency; they analyzed the expected discounted penalty function, the surplus prior to ruin, the deficit at ruin, and the surplus after the second claim prior to the moment of ruin.

Furthermore, Gerber argues that when interest yield is present, the surplus can fall below zero and then rise above zero. As a result, the ruin may not occur. As a result of this scenario, Gerber defined what he termed the absolute ruin time, which states that a ruin occurs if the surplus falls below a certain threshold, say $-\frac{c}{\delta}$, where c represents the constant premium rate and $\delta > 0$ represents the force of interest. In alignment with this idea, Konstantinides et al. [16] assumed a constant force of interest and studied the probabilities of finite and infinite absolute ruin time. Their study derived explicit asymptotic expressions for the finite and infinite absolute ruin probabilities for the compound Poisson model.

In most of the actuarial literature, it is assumed that claims are not delayed, which means they are reported or settled as soon as they occur. As a result, the model is unrealistic since the insured reported the claim after some days of the event. Furthermore, due to some legislative constraints (for example, claim investigation by experts), the insurers may take some months to settle claims. To address these observations, Cai [17] examined the effects of of the timing of claim payments and interest rate on ruin probability. In his study, the rate of interest was assumed to be dependent on the auto-regressive structure, as well as on the Lundberg inequality. The same perspective was adopted by Zou and Jie [18], who assumed that the claim occurrence may be delayed, and so, the total loss within the time horizon can be divided into two sub-claims: the main claim and the by-claim. In their analyze, they used the mean of Lagrange interpolation and the defective renewal equation of the discounted penalty function. Through a compound geometric distribution, they were able to establish an exact representation of the solution.

This paper incorporates the reporting delay time (when the claim occurs, it takes a while for the insured to declare that claim or for the insurance company to report that claim). Put differently, we assume that the exact amount of claim can only be determined after a reporting delay. We extend Li and Lu [11]'s work by assuming that the loss sizes are only known after a random delay time, during which the insurer continues to receive the premium and investment return. With these assumptions, we establish an integro-differential equation of the discounted penalty function and find that the solution to this equation can be represented by an infinite series. We derive an explicit solution for the exponential claim distribution.

To our best knowledge, there is no such study in the literature. Thus, our result contributes to the body of knowledge on the effects of delayed reporting times. The rest of the paper is structured as follows: Section 2 provides an overview of the model and a brief explanation of its assumptions. In Section 3, we present the main results of our work and discuss some special cases. The following section provides some numerical illustrations. We conclude the paper in Section 5 with suggested ways in which the work can be extended.

2. Model Setting and Rationality

Suppose that the premiums are continuously received and invested in the market with a constant force of interest ($\delta > 0$) and that $\mathbf{U}_\delta(t)$ is the surplus at time t. The initial reserve is also subject to investment and generates a constant force of interest. Let us further assume that the n-th claim occurs at $T_n = \sum_{k=0}^{n} \tau_k$ and it takes V_n time for the exact claim size or for the claim to be reported. Then, the surplus process $(\mathbf{U}_\delta(t), t \geq 0)$ is as follows:

$$\mathbf{U}_\delta(t) = u e^{\delta t} + c \int_0^t e^{\delta(t-y)} dy - \int_0^t e^{\delta(t-y)} dS_y, \quad t \geq 0, \qquad (1)$$

where $S_t = \sum_{k=1}^{N(t)} X_k \mathbb{1}_{\{T_k + V_k \leq t\}}$ and:

- $\{\tau_k, k \in \mathbb{N}\}$ is a sequence of continuous positive independent and identically distributed (i.i.d.) random variables, such that $\tau_k, k \geq 0$ represent the inter-occurrence time between the $(k-1)$-th and k-th claims;
- $\{T_k, k \in \mathbb{N}\}$ represent the occurrence time of the claims received by the insurer;
- $\{V_k, k \in \mathbb{N}\}$ is a sequence of continuous positive independent and identically distributed (i.i.d.) random variables, such that V_k is the time from T_k taken by the insurer to report the k-th claim;
- $\{X_k, k \in \mathbb{N}\}$ is a sequence of positive i.i.d. random variables, dependent on T_k and V_k such that X_k represents the claim amount paid by the insurer;
- $\{N(t), t \geq 0\}$ is the renewal process generated by the inter-occurrence times $\{\tau_k, k \in \mathbb{N}\}$;
- The random variables X_k, T_k, and V_k may exhibit some dependence structures.

Making things simple, we assume that T_k and V_k are independent, an assumption that can be justified by the fact that the delay time may depend mostly on the claim size. Thus, the joint distribution of the claim size, the claim arrival time, and the delay time are given by

$$f_{X_k, T_k, V_k}(x, t, z) = f_{T_k}(t) f_{V_k}(z) f_{X_k | T_k, V_k}(x).$$

Under some circumstances, the claim reporting and settlement process may take more time due to bad economics in the company (insufficient liquidity during wartime or natural disasters). In addition, the delay time is more relevant in a health insurance contract because one has to determine if the illness began within the period that the insurance covers before payments can be made. Additionally, during that holding period, the company will continue to receive premiums from policyholders, along with a compounded yield on the initial reserve. By delaying the claim payment time, companies can partially resolve the liquidity problem. Thus, they will minimize the absolute ruin probability, and therefore reduce the bankruptcy rate.

Gerber and Shiu [4] and Cai [5] argue that when the surplus drops below $-\frac{c}{\delta}$, the surplus cannot be positive due to the debts of the insurer, because the present value at that time is less than or equal to c, which is the present value of all premium income available after that time, and the delay in reporting and payment. A good investment strategy may be required in some types of contracts; for example, when the insurer covers only low risks and invests heavily in high asset returns (we refer to this scenario as a $-\frac{c}{\delta}$ small enough risk pool), the surplus may be positive as the company will continue to receive premiums and returns at the time of the ruin. In light of this, the time to absolute ruin defined by Gerber and Shiu [4] and Cai [5] is considered to be an absolute (relative) time to ruin. Throughout this paper, we assume that $\frac{c}{\delta}$ is small enough. That is, the insurer covers only low risk and invests in high return assets ($\delta \gg c$).

3. Main Results

We present our main results in this section. To start with, we use the Gerber–Shiu function and the formula for the absolute (relative) time to ruin. The absolute ruin probability T_δ is calculated as follows:

$$T_\delta = \begin{cases} \inf\left\{ t \mid \mathbf{U}_\delta(t) < -\dfrac{c}{\delta} \right\} \\ \infty \quad \text{if } \mathbf{U}_\delta(t) \geq -\dfrac{c}{\delta} \end{cases} \quad \text{for all } t \geq 0; \tag{2}$$

Here, $c_0 = -\dfrac{c}{\delta}$ and is defined through the Gerber–Shiu function by

$$\Phi_{\alpha,\delta}(u, c_0) = \mathbb{E}\left[e^{-\alpha T_\delta} w(\mathbf{U}_\delta(T_\delta^-), c_0 - \mathbf{U}_\delta(T_\delta)) \mathbf{1}_{\{T_\delta < \infty\}} \right], \tag{3}$$

α is interpreted as the discounted rate, and the function $w()$ represents the penalty function and is assumed to be bounded.

3.1. Integro-Differential Equation of the Gerber–Shiu Function

We obtain the following proposition.

Proposition 1. *Assume from the financial surplus given by Equation (1) that the delay time is exponentially distributed with parameter β. Then, the Gerber–Shiu function satisfies the integro-differential equation below:*

$$(\delta u + c)^2 \Phi''_{\alpha,\delta}(u, c_0) = \eta_0 (\delta u + c) \Phi'_{\alpha,\delta}(u, c_0) - \eta_1 \Phi_{\alpha,\delta}(u, c_0) \\ + \lambda \beta \int_0^{u-c_0} \Phi_{\alpha,\delta}(u - x, c_0) dF_{X|\tau,V}(x) + \lambda \beta \omega(u), \tag{4}$$

where $\eta_0 = 2\alpha + \beta + \lambda - \delta$, $\eta_1 = (\lambda + \alpha)(\alpha + \beta)$, $c_0 = -\dfrac{c}{\delta}$, λ is the claim number process rate in a Poisson process, and

$$\omega(u) = \int_{u-c_0}^{\infty} w(u, c_0 + x - u) dF_{X|\tau,V}(x).$$

Proof. Let $s(t, v) = u e^{\delta(t+v)} + c \dfrac{e^{\delta(t+v)} - 1}{\delta}$ and $c_0 = -\dfrac{c}{\delta}$.

By conditioning on the first claim occurrence time, the corresponding claim size, and the delay time, we have the equation:

$$\begin{aligned}
\Phi_{\delta,\alpha}(u, c_0) &= \int_0^\infty \int_0^\infty \int_0^{s(t,v)-c_0} e^{-\alpha(t+v)} \Phi_{\delta,\alpha}(s(t,v) - x, c_0) dF_{X|\tau,V}(x) dF_\tau(t) dF_V(v) \\
&+ \int_0^\infty \int_0^\infty \int_{s(t,v)-c_0}^\infty e^{-\alpha(t+v)} w(s(t,v), c_0 + x - s(t,v)) dF_{X|\tau,V}(x) dF_\tau(t) dF_V(v) \\
&= \int_0^\infty \int_0^\infty \int_0^{s(t,v)-c_0} e^{-\alpha(t+v)} \beta e^{-\beta t} \lambda e^{-\lambda t} \Phi_{\delta,\alpha}(s(t,v) - x, c_0) dF_{X|\tau,V}(x) dt dv \\
&+ \int_0^\infty \int_0^\infty \int_{s(t,v)-c_0}^\infty e^{-\alpha(t+v)} \beta e^{-\beta t} \lambda e^{-\lambda t} w(s(t,v), c_0 + x - s(t,v)) dF_{X|\tau,V}(x) dt dv.
\end{aligned}$$

Let $z = s(t, v)$. Then,

$$e^t = \left(\dfrac{\delta z + c}{\delta u + c} \right)^{1/\delta} e^{-v}; \text{ and}$$

$$dt = \dfrac{1}{\delta z + c} dz.$$

Plugging these relations in the above equation yields

$$
\begin{aligned}
\Phi_{\delta,\alpha}(u,c_0) &= \int_0^\infty \int_{s(0,v)}^\infty \int_0^{z-c_0} \lambda\beta \left(\frac{\delta z + c}{\delta u + c}\right)^{-(\alpha+\lambda)/\delta} e^{-(\beta-\lambda)v} \frac{1}{\delta z + c} \Phi_{\delta,\alpha}(z-x,c_0) \\
&\quad \times dF_{X|\tau,V}(x) dz dv \\
&\quad + \int_0^\infty \int_{s(0,v)}^\infty \int_{z-c_0}^\infty \lambda\beta \left(\frac{\delta z + c}{\delta u + c}\right)^{-(\alpha+\lambda)/\delta} e^{-(\beta-\lambda)} \frac{1}{\delta z + c} W(z, c_0 + x - z) \\
&\quad \times dF_{X|\tau,V}(x) dz dv, \\
&= \lambda\beta(\delta u + c)^{(\alpha+\lambda)/\delta} \int_0^\infty \int_{s(0,v)}^\infty (\delta z + c)^{-\frac{\alpha+\lambda}{\delta}-1} e^{-(\beta-\lambda)v} \\
&\quad \times \left[\int_0^{z-c_0} \Phi_{\delta,\alpha}(z-x,c_0) dF_{X|\tau,V}(x) + \int_{z-c_0}^\infty W(z, c_0 + x - z) dF_{X|\tau,V}(x) \right] dz dv.
\end{aligned} \quad (5)
$$

Now, we define $\omega(z) := \int_{z-c_0}^\infty W(z, c_0 + x - z) dF_{X|\tau,V}(x)$ and $y = s(0,v)$ so that

$$
e^v = \left(\frac{\delta y + c}{\delta u + c}\right)^{1/\delta} \implies dv = \frac{1}{\delta y + c} dy.
$$

Then, inverting the above relations into Equation (5) gives

$$
\begin{aligned}
\Phi_{\delta,\alpha}(u,c_0) &= \lambda\beta(\delta u + c)^{(\alpha+\lambda)/\delta} \int_u^\infty \int_y^\infty (\delta z + c)^{-\frac{\alpha+\lambda}{\delta}-1} \frac{1}{\delta y + c} \left(\frac{\delta y + c}{\delta u + c}\right)^{-(\beta-\lambda)/\delta} \\
&\quad \times \left[\int_0^{z-c_0} \Phi_{\delta,\alpha}(z-x,c_0) dF_{X|\tau,V}(x) + \omega(z) \right] dz dy, \\
&= \lambda\beta(\delta u + c)^{(\alpha+\beta)/\delta} \int_u^\infty \int_y^\infty (\delta z + c)^{-\frac{\alpha+\lambda}{\delta}-1} (\delta y + c)^{-\frac{\beta-\lambda}{\delta}-1} \\
&\quad \times \left[\int_0^{z-c_0} \Phi_{\delta,\alpha}(z-x,c_0) dF_{X|\tau,V}(x) + \omega(z) \right] dz dy, \\
&= \lambda\beta(\delta u + c)^{(\alpha+\beta)/\delta} \int_u^\infty (\delta y + c)^{-\frac{\beta-\lambda}{\delta}-1} \int_y^\infty (\delta z + c)^{-\frac{\alpha+\lambda}{\delta}-1} \\
&\quad \times \left[\int_0^{z-c_0} \Phi_{\delta,\alpha}(z-x,c_0) dF_{X|\tau,V}(x) + \omega(z) \right] dz dy.
\end{aligned} \quad (6)
$$

Taking the first derivative of Equation (6):

$$
\begin{aligned}
\Phi'_{\delta,\alpha}(u,c_0) &= \frac{\alpha+\beta}{\delta u + c} \Phi_{\delta,\alpha}(u,c_0) - \lambda\beta \frac{(\delta u + c)^{(\alpha+\lambda)/\delta}}{\delta u + c} \int_u^\infty (\delta z + c)^{-\frac{\alpha+\lambda}{\delta}-1} \\
&\quad \times \left[\int_0^{z-c_0} \Phi_{\delta,\alpha}(z-x,c_0) dF_{X|\tau,V}(x) + \omega(z) \right] dz, \\
(\delta u + c)\Phi'_{\delta,\alpha}(u,c_0) &= (\alpha+\beta)\Phi_{\delta,\alpha}(u,c_0) - \lambda\beta(\delta u + c)^{(\alpha+\lambda)/\delta} \int_u^\infty (\delta z + c)^{-\frac{\alpha+\lambda}{\delta}-1} \\
&\quad \times \left[\int_0^{z-c_0} \Phi_{\delta,\alpha}(z-x,c_0) dF_{X|\tau,V}(x) + \omega(z) \right] dz.
\end{aligned} \quad (7)
$$

Upon rearranging Equation (7), we obtain:

$$
\begin{aligned}
&- \lambda\beta(\delta u + c)^{(\alpha+\lambda)/\delta} \int_u^\infty (\delta z + c)^{-\frac{\alpha+\lambda}{\delta}-1} \left[\int_0^{z-c_0} \Phi_{\delta,\alpha}(z-x,c_0) dF_{X|\tau,V}(x) + \omega(z) \right] dz \\
&= (\delta u + c)\Phi'_{\delta,\alpha}(u,c_0) - (\alpha+\beta)\Phi_{\delta,\alpha}(u,c_0).
\end{aligned} \quad (8)
$$

By taking the derivative of (7), we have

$$
\begin{aligned}
(\delta u+c)\Phi_{\delta,\alpha}''(u,c_0)+\delta\Phi_{\delta,\alpha}'(u,c_0) &= (\alpha+\beta)\Phi_{\delta,\alpha}'(u,c_0)-(\alpha+\lambda)\frac{\lambda\beta}{\delta u+c}(\delta u+c)^{(\alpha+\lambda)/\delta}\\
&\quad \times \int_u^\infty (\delta z+c)^{-\frac{\alpha+\lambda}{\delta}-1}\left[\int_0^{z-c_0}\Phi_{\delta,\alpha}(z-x,c_0)\mathrm{d}F_{X|T,V}(x)+\omega(z)\right]\mathrm{d}z\\
&\quad + \frac{\lambda\beta}{\delta u+c}\left[\int_0^{u-c_0}\Phi_{\delta,\alpha}(u-x,c_0)\mathrm{d}F_{X|T,V}(x)+\omega(u)\right],\\
(\delta u+c)^2\Phi_{\delta,\alpha}''(u,c_0) &= (\alpha+\beta-\delta)(\delta u+c)\Phi_{\delta,\alpha}'(u,c_0)-(\alpha+\lambda)\lambda\beta(\delta u+c)^{(\alpha+\lambda)/\delta}\\
&\quad \times \int_u^\infty (\delta z+c)^{-\frac{\alpha+\lambda}{\delta}-1}\left[\int_0^{z-c_0}\Phi_{\delta,\alpha}(z-x,c_0)\mathrm{d}F_{X|T,V}(x)+\omega(z)\right]\mathrm{d}z\\
&\quad + \lambda\beta\left[\int_0^{u-c_0}\Phi_{\delta,\alpha}(u-x,c_0)\mathrm{d}F_{X|T,V}(x)+\omega(u)\right]. \quad (9)
\end{aligned}
$$

Finally, inserting Equation (8) into (9), the equation:

$$
\begin{aligned}
(\delta u+c)^2\Phi_{\delta,\alpha}''(u,c_0) &= (2\alpha+\lambda+\beta-\delta)(\delta u+c)\Phi_{\delta,\alpha}'(u,c_0)-(\lambda+\alpha)(\alpha+\beta)\Phi_{\delta,\alpha}(u,c_0)\\
&\quad + \lambda\beta\left[\int_0^{u-c_0}\Phi_{\delta,\alpha}(u-x,c_0)\mathrm{d}F_{X|T,V}(x)+\omega(u)\right],
\end{aligned}
$$

which proves the statement by setting $\eta_0=(2\alpha+\lambda+\beta-\delta)$ and $\eta_1=(\lambda+\alpha)(\alpha+\beta)$. □

Remark 1. *In deriving their results, Cai [5] and Li and Lu [11] distinguished between $\Phi_{\alpha,\delta}^+(u)$ for the return interest rate when the initial surplus exceeds zero and $\Phi_{\alpha,\delta}^-(u)$ for debt interest when the initial surplus drops below zero. However, as pointed out by Cai [5]'s work, when the initial surplus approaches zero, then both quantities are equalized. In this paper, we assume that $\frac{c}{\delta}$ approaches 0, and thanks to the delayed reporting time, it is reasonable to assume that in the interval $[-\frac{c}{\delta},0]$, $\Phi_{\alpha,\delta}^-(u)\approx\Phi_{\alpha,\delta}^+(u)$.*

Equation (4) is difficult to solve. In the following, we use the Laplace transform to derive a more practical equation for the Gerber–Shiu function.

Let us define as in Li and Lu [11] the following transformation:

$$
Y_{\delta,\alpha}(u,c_0)=\begin{cases}\dfrac{\Phi_{\delta,\alpha}(c_0,c_0)-\Phi_{\delta,\alpha}(u,c_0)}{\Phi_{\delta,\alpha}(c_0,c_0)}, & \text{if } u\geq c_0\\ 0 & \text{if } u<c_0.\end{cases} \quad (10)
$$

Clearly, $Y_{\delta,\alpha}(c_0,c_0)=0$. If in addition, we assume that the penalty function $w(x,y)$ is bounded, then $\lim_{u\to\infty}\Phi_{\delta,\alpha}(u,c_0)=0$ (see Li and Lu [11]), which implies that $\lim_{u\to\infty}Y_{\delta,\alpha}(u,c_0)=1$.

Theorem 1. *In the model given by Equation (1), the Laplace transform of the discounted penalty function satisfies the following non-homogeneous differential equation.*

$$
\tilde{y}_{\delta,\alpha}''(s)+f_1(s)\tilde{y}_{\delta,\alpha}'(s)+f_0(s)\tilde{y}_{\delta,\alpha}(s)=g(s), \quad (11)
$$

where $Y_{\delta,\alpha}(u,c_0)$ is given by Equation (10),

$$
\begin{aligned}
f_1(s) &= \frac{\eta_2+2}{\delta}\times\frac{1}{s}+2c_0;\quad \eta_2=\eta_0+2\delta;\quad \eta_3=\eta_1+\delta\eta_2,\\
f_0(s) &= \frac{c^2}{\delta^2}+\frac{1}{s^2}\left(\frac{\eta_3}{\delta^2}-\frac{\lambda\beta}{\delta^2}\tilde{f}(s)\right)-\frac{1}{s}\left(\frac{2c}{\delta}+\frac{\eta_2 c}{\delta^2}\right),\\
g(s) &= -\frac{1}{(\delta s)^2}\left[\frac{e^{-sc_0}}{s}(\lambda\beta\tilde{f}(s)-\eta_1)+\frac{\lambda\beta}{\Phi_{\delta,\alpha}(c_0)}\tilde{w}(s)\right].
\end{aligned}
$$

$\tilde{y}_{\delta,\alpha}(s)$, $\tilde{f}(s)$ and $\tilde{w}(s)$ are respectively the Laplace transforms of $Y_{\delta,\alpha}(u)$, $dF_{X|\tau,V}(x)$, and $\omega(u)$.

Proof. Hereafter, let $dF(x) := dF_{\tau,V}(x)$.

Integrating Equation (5) on both sides from c_0 to u yields

$$\begin{aligned}
\int_{c_0}^{u}(\delta t+c)^2 \Phi''_{\delta,\alpha}(t,c_0)dt &= \eta_0 \int_{c_0}^{u}(\delta t+c)\Phi'_{\delta,\alpha}(t,c_0)dt - \eta_1 \int_{c_0}^{u}\Phi_{\delta,\alpha}(t,c_0)dt \\
&+ \lambda\beta \int_{c_0}^{u}\int_{0}^{t-c_0}\Phi_{\delta,\alpha}(t-x,c_0)dF(x)dt + \lambda\beta \int_{c_0}^{u}\omega(t)dt.
\end{aligned}$$

That is,

$$\begin{aligned}
(\delta u+c)^2 \Phi'_{\delta,\alpha}(u,c_0) &= \eta_2 \int_{c_0}^{u}(\delta t+c)\Phi'_{\delta,\alpha}(t,c_0)dt - \eta_1 \int_{c_0}^{u}\Phi_{\delta,\alpha}(t,c_0)dt \\
&+ \lambda\beta \int_{0}^{u-c_0}dF(x)\left(\int_{c_0}^{u-x}\Phi_{\delta,\alpha}(y)dy\right) + \lambda\beta \int_{c_0}^{u}\omega(t)dt.
\end{aligned}$$

so that,

$$\begin{aligned}
(\delta u+c)^2 \Phi'_{\delta,\alpha}(u,c_0) &= \eta_2(\delta u+c)\Phi_{\delta,\alpha}(u,c_0) - \eta_3 \int_{c_0}^{u}\Phi_{\delta,\alpha}(t,c_0)dt \quad (12)\\
&+ \lambda\beta \int_{0}^{u-c_0}dF(x)\left(\int_{c_0}^{u-x}\Phi_{\delta,\alpha}(y,c_0)dy\right) + \lambda\beta \int_{c_0}^{u}\omega(t)dt.
\end{aligned}$$

Plugging Equation (10) into (12) gives

$$\begin{aligned}
-(\delta u+c)^2 \Phi_{\delta,\alpha}(c_0,c_0) Y'_{\delta,\alpha}(u,c_0) &= \eta_2(\delta u+c)[\Phi_{\delta,\alpha}(c_0,c_0) - \Phi_{\delta,\alpha}(c_0,c_0)Y_{\delta,\alpha}(u,c_0)] \\
&- \eta_3 \int_{c_0}^{u}[\Phi_{\delta,\alpha}(c_0,c_0) - \Phi_{\delta,\alpha}(c_0,c_0)Y_{\delta,\alpha}(t,c_0)]dt \\
&+ \lambda\beta \int_{c_0}^{u}\omega(t)dt + \lambda\beta \int_{0}^{u-c_0}dF(x) \\
&\times \left(\int_{c_0}^{u-x}[\Phi_{\delta,\alpha}(c_0,c_0) - \Phi_{\delta,\alpha}(c_0,c_0)Y_{\delta,\alpha}(y,c_0)]dy\right).
\end{aligned}$$

Dividing both sides by $\Phi_{\delta,\alpha}(c_0,c_0)$ yields

$$\begin{aligned}
-(\delta u+c)^2 Y'_{\delta,\alpha}(u,c_0) &= \eta_2(\delta u+c) - \eta_2(\delta u+c)Y_{\delta,\alpha}(u,c_0) - \eta_3(u-c_0) + \eta_3 \int_{c_0}^{u}Y_{\delta,\alpha}(t,c_0)dt \\
&+ \frac{\lambda\beta}{\Phi_{\delta,\alpha}(c_0,c_0)} \int_{c_0}^{u}\omega(t)dt + \lambda\beta \int_{0}^{u-c_0}(u-c_0-x)dF(x) \\
&- \lambda\beta \int_{0}^{u-c_0}dF(x)\left(\int_{c_0}^{u-x}Y_{\delta,\alpha}(y,c_0)dy\right), \\
-(\delta u+c)^2 Y'_{\delta,\alpha}(u,c_0) &= -\eta_3 \int_{c_0}^{u}dt - \eta_2(\delta u+c)Y_{\delta,\alpha}(u,c_0) + \eta_3 \int_{c_0}^{u}Y_{\delta,\alpha}(t,c_0)dt \\
&+ \frac{\lambda\beta}{\Phi_{\delta,\alpha}(c_0,c_0)} \int_{c_0}^{u}\omega(t)dt + \lambda\beta \int_{0}^{u-c_0}(u-c_0-x)dF(x) \\
&- \lambda\beta \int_{0}^{u-c_0}dF(x)\left(\int_{c_0}^{u-x}Y_{\delta,\alpha}(y,c_0)dy\right).
\end{aligned}$$

Finally, we obtain

$$\begin{aligned}-\delta^2 Y'_{\delta,\alpha}(u,c_0) &= 2\delta c Y'_{\delta,\alpha}(u,c_0) + c^2 Y'_{\delta,\alpha}(u,c_0) - \eta_1 \int_{c_0}^u dt - \eta_2 \delta u Y_{\delta,\alpha}(u,c_0) \\ &\quad - \eta_2 c Y_{\delta,\alpha}(u,c_0) + \eta_3 \int_{c_0}^u Y_{\delta,\alpha}(t,c_0)dt + \frac{\lambda\beta}{\Phi_{\delta,\alpha}(c_0,c_0)} \int_{c_0}^u w(t)dt \\ &\quad + \lambda\beta \int_0^{u-c_0}(u-c_0-x)dF(x) - \lambda\beta \int_0^{u-c_0} dF(x)\left(\int_{c_0}^{u-x} Y_{\delta,\alpha}(y,c_0)dy\right).\end{aligned} \tag{13}$$

Taking the Laplace transform of Equation (13) gives

$$\begin{aligned}-\delta^2\left(s\tilde{y}''_{\delta,\alpha}(s) + 2\tilde{y}'_{\delta,\alpha}(s)\right) &= -2\delta c\left(s\tilde{y}'_{\delta,\alpha}(s) + \tilde{y}_{\delta,\alpha}(s)\right) + c^2 s\tilde{y}_{\delta,\alpha}(s) \\ &\quad + \eta_2\delta \tilde{y}'_{\delta,\alpha}(s) - \eta_2 c \tilde{y}_{\delta,\alpha}(s) + \frac{\eta_3}{s}\tilde{y}_{\delta,\alpha}(s) + \frac{\lambda\beta}{\Phi_{\delta,\alpha}(c_0)}\frac{\tilde{w}(s)}{s} \\ &\quad - \eta_1\frac{e^{-sc_0}}{s^2} + \frac{\lambda\beta}{s^2}e^{-sc_0}\tilde{f}(s) - \frac{\lambda\beta}{s}\tilde{f}(s)\tilde{y}_{\delta,\alpha}(s),\end{aligned}$$

which completes the proof. □

The differential equation given in (11) is not straightforward to solve. Thus, one may think of transforming this equation to a well-known equation, which can be easily solved analytically or numerically.

For simplicity, let

$$\tilde{y}_{\delta,\alpha}(c_0,c_0) = b_0 = \frac{e^{-c_0^2}}{c_0} - \frac{\Phi_{\delta,\alpha}(c_0,c_0)}{\Phi_{\delta,\alpha}(c_0,c_0)}, \text{ and } \tilde{y}'_{\delta,\alpha}(c_0,c_0) = b_1 = -\left[\frac{\Phi'_{\delta,\alpha}(c_0,c_0)}{\Phi_{\delta,\alpha}(c_0,c_0)} + e^{-c_0^2}\left(1 + \frac{1}{c_0^2}\right)\right].$$

In the following corollary, we transform Equation (11) into the Volterra equation of the second kind with a degenerate kernel.

Corollary 1. *Under the assumption of Theorem 1, Equation (11) becomes*

$$\mathcal{U}(s) - \int_{c_0}^s K(s,t)\mathcal{U}(t)dt = h(s), \tag{14}$$

where $\mathcal{U}(s)$, $K(s,t)$, and $h(s)$ are given by

$$\begin{aligned}q(s) &= \int_{c_0}^s (s-t)\mathcal{U}(t)dt, \\ \tilde{y}_{\delta,\alpha}(s) &= q(s) + b_1(s-c_0) + b_0, \\ K(s,t) &= -[f_1(s) + f_0(s)(s-t)], \\ h(s) &= g(s) - [b_0 + b_1(s-c_0)f_0(s) + b_1 f_1(s)].\end{aligned}$$

Proof. Consider Equation (11) and introduce the function $q(s)$ such that $\tilde{y}_{\delta,\alpha}(s) = q(s) + b_1(s-c_0) + b_0$. Clearly, $q(c_0) = q'(c_0) = 0$ and

$$\tilde{y}''_{\delta,\alpha}(s) = q''(s), \quad \tilde{y}'_{\delta,\alpha}(s) = q'(s) + b_1.$$

Plugging these expressions into Equation (11) yields

$$q''(s) + f_1(s)q'(s) + f_0(s)q(s) = h(s), \tag{15}$$

where $h(s) = g(s) - [b_0 + b_1(s-c_0)f_0(s) + b_1 f_1(s)]$. Let us further introduce another auxiliary function $\mathcal{U}(t)$ such that $q(s) = \int_{c_0}^s (s-t)\mathcal{U}(t)dt$.

$\mathcal{U}(t)$ satisfies the following Volterra equation:

$$\mathcal{U}(t) + \int_{c_0}^{s} f_1(s)\mathcal{U}(t)\mathrm{d}t + \int_{c_0}^{s} f_0(s)(s-t)\mathcal{U}(t)\mathrm{d}t = h(s),$$

and this completes the proof. □

The mathematical literature is replete with several papers that try to solve Equation (14). The book of Polyanin and Manzhirov [19] presents an overview of the analytical solution of (14), which gives a particular non-trivial solution of Equation (15) without the second member. The following remark presents the analytical solution in a general form given that we know a non-trivial solution $(q_1(s))$.

Remark 2. *Let $q_1(s)$ be a non-trivial particular solution of Equation (15) without the second member. Then, the other solution of Equation (15) without the second member is in the form*

$$q_2(s) = q_1(s) \int_{c_0}^{s} \frac{\left|\frac{t}{c_0}\right|^{\frac{\eta_2+2}{\delta}} \exp\left[\frac{2c}{\delta}(c_0-t)\right]}{q_1(t)^2} \mathrm{d}t,$$

from which the final solution for Equation (15) is given by

$$q(s) = q_2(s) \int_{c_0}^{s} \frac{q_1(t)}{\left|\frac{t}{c_0}\right|^{\frac{\eta_2+2}{\delta}} \exp\left[\frac{2c}{\delta}(c_0-t)\right]} h(t)\mathrm{d}t - q_1(s) \int_{c_0}^{s} \frac{q_2(t)}{\left|\frac{t}{c_0}\right|^{\frac{\eta_2+2}{\delta}} \exp\left[\frac{2c}{\delta}(c_0-t)\right]} h(t)\mathrm{d}t.$$

Although this solution is interesting, the main problem is the determination of $q_1(s)$, which is not straightforward. An alternative way to solve this problem is to find the solution numerically. In recent years, scholars in applied mathematics have investigated the numerical solution of Equation (14). For an overview of these method, the readers are referred to Maleknejad et al. [20] and Nadir and Rahmoune [21] and the references therein. The following remark presents the numerical way to solve (14) by Simpson's quadrature rule as given in Nadir and Rahmoune [21].

Remark 3. *Let us consider the transformed Equation (14) given below:*

$$\mathcal{U}(s) + \int_{c_0}^{s} K_0(s,t)\mathcal{U}(t)\mathrm{d}t = h(s), \qquad (16)$$

where $K_0(s,t) = f_1(s) + f_0(s)(s-t)$.
For $s_0 = c_0 < s_1 < \cdots < s_{2n}$ and, moreover, for notation simplicity, we set $\mathcal{U}(2j) = \mathcal{U}(s_{2j})$, $K_0(2j,2j) = K_0(s_{2j},s_{2j})$, $h(2j) = h(s_{2j})$. Hence, for $j = 1, \cdots, n$, it was proven in Nadir and Rahmoune [21] that

$$\mathcal{U}(2j)\left(1 - \frac{h}{3}\left[2K_0(2j,2j-1) + K_0(2j,2j)\right]\right) = h(2j) + \frac{h}{3}\left(K_0(2j,0) + 2K_0(2j,1)\right)\mathcal{U}(c_0) + \frac{2}{3}h$$
$$\times \sum_{i=1}^{j-1}\left(K_0(2j,2i-1) + K_0(2j,2i) + K_0(2j,2i+1)\right)$$
$$\times \mathcal{U}(2i),$$

where h is the step of subdivision, $\mathcal{U}(2j+1) = \dfrac{\mathcal{U}(2j) + \mathcal{U}(2j+2)}{2}$, $\mathcal{U}(c_0) = h(c_0)$.

This formula provides us with a powerful tool to approximate the Gerber–Shiu function. Section 4 aims to investigate the numerical solution of $\tilde{y}_{\delta,\alpha}(s)$ and, thus, its inverse Laplace transform in order to find the Gerber–Shiu function in particular cases where a given distribution of conditional claim sizes is assumed with a particular penalty function.

3.2. Exponentially Distributed Conditional Claim

Equation (4) can be transformed to a non-homogeneous differential equation of order three under an exponentially distributed conditional claim. The following proposition states the form of the differential equation satisfied by the discounted penalty function under this assumption.

Proposition 2. *In the model given in Equation* (1), *under exponentially distributed conditional claims with parameter* ρ, *Equation* (4) *becomes*

$$(u-c_0)^2 \Phi'''_{\delta,\alpha}(u,c_0) = \left[\mu_0(u-c_0) - \rho(u-c_0)^2\right]\Phi''_{\delta,\alpha}(u,c_0) + \frac{\lambda\beta}{\delta^2}\left[\rho\omega(u) - \omega'(u)\right] \\ + \mu_3 \Phi_{\delta,\alpha}(u,c_0) + [\mu_1 + \mu_2(u-c_0)]\Phi'_{\delta,\alpha}(u,c_0), \quad (17)$$

where

$$\mu_0 = \frac{\eta_0}{\delta} - 2; \quad \mu_1 = \frac{\eta_0\delta - \eta_1}{\delta^2}, \quad \mu_2 = \frac{\rho\eta_0}{\delta} \text{ and } \mu_3 = \frac{\rho}{\delta^2}(\lambda\beta - \eta_1).$$

Proof. Differentiating Equation (4) gives:

$$(\delta u + c)^2 \Phi'''_{\delta,\alpha}(u,c_0) = (\eta_0 - 2\delta)(\delta u + c)\Phi''_{\delta,\alpha}(u,c_0) + (\eta_0\delta - \eta_1)\Phi'_{\delta,\alpha}(u,c_0) \\ + \lambda\beta \frac{d}{du}\left[e^{-\rho(u)}\int_{c_0}^{u}\Phi_{\delta,\alpha}(y)\rho e^{\rho y}dy\right] + \lambda\beta\omega'(u). \quad (18)$$

That is,

$$(\delta u + c)^2 \Phi'''_{\delta,\alpha}(u,c_0) = (\eta_0 - 2\delta)(\delta u + c)\Phi''_{\delta,\alpha}(u,c_0) + (\eta_0\delta - \eta_1)\Phi'_{\delta,\alpha}(u,c_0) + \lambda\beta\omega'(u) \\ + \lambda\beta\rho\left(\Phi_{\delta,\alpha}(u,c_0) - \int_0^{u-c_0}\Phi_{\delta,\alpha}(u-x,c_0)\rho e^{-\rho x}dx\right). \quad (19)$$

From Equation (4), we have

$$-\lambda\beta \int_0^{u-c_0} \Phi_{\delta,\alpha}(u-x,c_0)\rho e^{-\rho x}dx = -(\delta u + c)^2\Phi''_{\delta,\alpha}(u,c_0) + \eta_0(\delta u + c)\Phi'_{\delta,\alpha}(u,c_0) \\ -\eta_1\Phi_{\delta,\alpha}(u,c_0) + \lambda\beta\omega(u). \quad (20)$$

Plugging Equation (20) into (19) yields

$$(\delta u + c)^2 \Phi'''_{\delta,\alpha}(u,c_0) = \left[(\eta_0 - 2\delta)(\delta u + c) - \rho(\delta u + c)^2\right]\Phi''_{\delta,\alpha}(u,c_0) \\ + \lambda\beta[\rho\omega(u) + \omega'(u)] + [(\eta_0\delta - \eta_1) + \rho\eta_0(\delta u + c)]\Phi'_{\delta,\alpha}(u,c_0) \\ - \rho\eta_1\Phi_{\delta,\alpha}(u,c_0) + \lambda\beta\rho\Phi_{\delta,\alpha}(u,c_0).$$

Dividing both sides of the above formula by δ^2 completes the proof. □

Under some assumptions on the penalty function (a penalty function depends only on the deficit at ruin), Equation (17) can be solved via the sequence expansion technique.

Theorem 2. *Suppose that the penalty function depends only on the deficit at ruin. Then, Equation* (17), *with initial condition* $\Phi_{\delta,\alpha}(c_0,c_0)$, *has the following solution:*

$$\Phi_{\delta,\alpha}(u,c_0) = \Phi_{\delta,\alpha}(c_0,c_0) \sum_{n=0}^{\infty}\left[\prod_{i=1}^{n}(v)_i\right](u-c_0)^n, \quad (21)$$

where $\prod_{i=1}^{0} = 1$ *and*

$$(v)_n = \frac{(n-1)[\mu_2 - \rho(n-2)] + \mu_3}{n[(n-1)(n-2-\mu_0) - \mu_1]}.$$

Proof. Under the assumption of Theorem 2, Equation (17) becomes

$$(u-c_0)^2 \Phi'''_{\delta,\alpha}(u,c_0) = \left[\mu_0(u-c_0) - \rho(u-c_0)^2\right]\Phi''_{\delta,\alpha}(u,c_0) \qquad (22)$$
$$+ [\mu_1 + \mu_2(u-c_0)]\Phi'_{\delta,\alpha}(u,c_0) + \mu_3 \Phi_{\delta,\alpha}(u,c_0).$$

Let $\Phi_{\delta,\alpha}(u,c_0)$ be defined by

$$\Phi_{\delta,\alpha}(u,c_0) := \sum_{n=0}^{\infty} a_n(u-c_0)^n, \text{ hence}$$

$$\Phi^{(i)}_{\delta,\alpha}(u,c_0) = \sum_{n=i}^{\infty} n(n-1)\cdots(n-i+1)a_n(u-c_0)^{n-i}, \qquad (23)$$

where $\Phi^{(i)}_{\delta,\alpha}(u,c_0)$ is the i-th derivative of $\Phi_{\delta,\alpha}(u,c_0)$, for $i = 1,2,3$.

Plugging Equation (23) into (22) yields

$$\sum_{n=1}^{\infty} n(n-1)(n-2)a_n(u-c_0)^{n-1} = \mu_0 \sum_{n=1}^{\infty} n(n-1)a_n(u-c_0)^{n-1}$$
$$- \rho \sum_{n=0}^{\infty} n(n-1)a_n(u-c_0)^n + \mu_3 \sum_{n=0}^{n} a_n(u-c_0)^n$$
$$+ \mu_1 \sum_{n=1}^{\infty} na_n(u-c_0)^{n-1} + \mu_2 \sum_{n=0}^{\infty} na_n(u-c_0)^n.$$

Letting n be $n+1$, then

$$\sum_{n=0}^{\infty} (n+1)(n)(n-1)a_{n+1}(u-c_0)^n = \mu_0 \sum_{n=0}^{\infty} (n+1)(n)a_{n+1}(u-c_0)^n$$
$$- \rho \sum_{n=0}^{\infty} n(n-1)a_n(u-c_0)^n + \mu_3 \sum_{n=0}^{n} a_n(u-c_0)^n$$
$$+ \mu_1 \sum_{n=0}^{\infty} (n+1)a_{n+1}(u-c_0)^n + \mu_2 \sum_{n=0}^{\infty} na_n(u-c_0)^n.$$

Thus,

$$\sum_{n=0}^{\infty} \left[(n+1)(n)(n-1) - \mu_0(n+1)(n) - \mu_1(n+1)\right]a_{n+1}(u-c_0)^n$$
$$= \sum_{n=0}^{\infty} \left[\mu_5 + \mu_2(n) - \rho(n)(n-1)\right]a_n(u-c_0)^n,$$

from which it follows that

$$a_{n+1} = \frac{n(\mu_2 - \rho(n-1)) + \mu_3}{(n+1)[n((n-1) - \mu_0) - \mu_1]} a_n.$$

By expanding the above relation recursively, we have

$$a_n = a_0 \prod_{i=1}^{n} \left[\frac{(i-1)(\mu_2 - \rho(i-2)) + \mu_3}{i[(i-1)((i-2) - \mu_0) - \mu_1]}\right]$$
$$= a_0 \prod_{i=1}^{n} (v)_i,$$

which proves the statement by inserting this relation back into Equation (23) and by the fact that $\Phi_{\delta,\alpha}(c_0, c_0) = a_0$. □

Remark 4. In Proposition 1, if we let $\alpha = 0$, we obtain the integro-differentiating of the (relative) absolute ruin. Under the exponential claim distribution, one obtains the explicit expression of the (relative) absolute ruin probability by letting $\alpha = 0$ in Theorem 2.

4. Numerical Illustration

This section describes how to solve the Volterra equation given in Equation (16) and from there calculate the ruin probability using the inverse Laplace transform of Equation (10).

As outlined by Nadir and Rahmoune [21], we first find the solution to Equation (16) by numerical approximation. Moreover, since we must integrate the latter function to obtain the function $q(s)$ given by Equation (14) and, hence, the Laplace transform function $\tilde{y}_{\delta,\alpha}(s)$, we approximate the solution to Equation (16) by a function using spline approximation (the spline approximation was chosen over polynomial, cubic spline, or Lagrange polynomial, linear approximations because it yields the best accuracy). Although the rational approximation is also a good approximation, we prefer the late one since the rational approximation produces a singular function that does not allow the integral to converge. $Y_{\delta,\alpha}(u)$ is then obtained by applying the inverse Laplace to the function $\tilde{y}_{\delta,\alpha}(s)$ derived above.

In the absence of real data, we chose our parameters arbitrarily. For comparison purposes, we varied the distribution parameter of the reporting delay time to examine its impact on the (relative) absolute ruin. Numerical calculations show that the ruin probability decreases as the reporting delay increases on average, regardless of the initial capital, which confirms that the delay in claim reporting can positively affect the ruin probability.

The numerical setting assumes a Gamma distribution with parameter $a = 10.2$, $b = 0.008$, which means that each claim average is 1275.

Table 1, as well as Table 2 show that the ruin probability is impacted by the initial surplus; however, the ruin probability is normally not influenced by only the initial surplus (increasing the initial reserve will not always decrease the ruin probability), as this may depend on the delay in reporting or settlement times as well. In comparing Tables 1 and 2, it turns out that an increase in the average reporting delay decreases the ruin probability. This finding potentially provides risk managers and actuaries with a powerful instrument to manage risks, in particular the risk of insolvency.

Table 1. Ruin probability for $\delta = 0.02$, $\lambda = 0.01$, $c = 0.1\%$, $\beta = 0.05$.

u	10.20	24.50	40.80	51.02	65.30	81.63	100
$\psi_\delta(u)$	0.5729536	0.5060346	0.5046236	0.4992245	0.5000019	0.5001424	0.5008247

Table 2. Ruin probability for $\delta = 0.02$, $\lambda = 0.01$, $c = 0.1\%$, $\beta = 0.02$.

u	10.20	24.50	40.80	51.02	65.30	81.63	100
$\psi_\delta(u)$	0.5751914	0.5058820	0.5051878	0.5007443	0.4998882	0.5001839	0.5007640

5. Discussion and Conclusions

This study analyzed the impact of the reporting delay time on the Gerber–Shiu discounted penalty function under an interest rate environment, where the claim effective's cost distribution is determined by the occurrence of the claim and the reporting delays (this somehow incorporated indirectly the inflation on the claim cost). We showed that the Gerber–Shiu function follows a degenerate Volterra equation with an arbitrary distribution of conditional claim sizes. A numerical illustration of our result using an approximation method (prescribed by Nadir and Rahmoune [21]) confirmed the effect of the delay in reporting time on the ruin probability. In addition, we derived a closed analytical expression of the Gerber–Shiu function under the exponential conditional claim distribution assumption. The numerical results indicated that the ruin probability as a function of the initial surplus is not a monotonic function as one usually expects (increases of the initial surplus decrease the ruin probability).

Although this study contributes to the existing literature, it has some drawbacks in that we assumed that when the surplus belongs to $\left[-\frac{c}{\delta},\ 0\right]$, the insurer could continue to operate without borrowing (in contrast with Gerber and Shiu [4] and Cai [5]), which may be the case when $\frac{c}{\delta}$ is small enough and there is flexibility in the delayed reporting times. As a result, insurers must seek assets with high risk-free rates and protect only low-risk contracts ($\delta >> c$). A good distribution of the reporting delayed time must be bounded from above, due to regulation constraints (anti-selection ban and maximum accepting reporting time). In future studies, we will extend our results by analyzing the discounted penalty function with stochastic interest rates, since evidence in finance studies has demonstrated that there is no such rate as a risk-free rate for reporting.

Author Contributions: The authors contributed equally to this work from the Introduction to Conclusions. All authors have read and agreed to the published version of the manuscript.

Funding: This research was funded by the Global Excellence Stature Fellowship 4.0 and NRF INCENTIVE GRANT from the University of Johannesburg.

Conflicts of Interest: The authors declare no conflict of interest.

References

1. Lundberg, F. Framställning av Sannolikehetsfunktionen. In *Återförsäkering av Kollektivrisker*; Almqvist & Wiksell: Uppsala, Sweden, 1903.
2. Cramér, C.H. On the mathematical theory of risk. In *Forsakringsaktiebolaget Skandias Festskrift*; Centraltryckeriet: Stockholm, Sweden, 1930.
3. Taylor, G.C. Probability of ruin under inflationary conditions or under experience rating. *ASTIN Bull. J. IAA* **1979**, *10*, 149–162. [CrossRef]
4. Gerber, H.U.; Shiu, E.S.W. On the Time Value of Ruin. *N. Am. Actuar. J.* **1998**, *2*, 48–72. [CrossRef]
5. Cai, J. On the Time Value of Absolute Ruin with Interest. *Adv. Appl. Probab.* **2007**, *39*, 343–357. [CrossRef]
6. Adékambi, F.; Essiomle, K. Ruin Probability for Stochastic Flows of Financial Contract under Phase-Type Distribution. *Risks* **2020**, *8*, 53. [CrossRef]
7. Gerber, H.U.; Shiu, E.S.W.; Yang, H. The Omega model: From bankruptcy to occupation times in the red. *Eur. Actuar. J.* **2012**, *2*, 259–272. [CrossRef]
8. Sundt, B.; Teugels, J.L. Ruin estimates under interest force. *Insur. Math. Econ.* **1995**, *16*, 7–22. [CrossRef]
9. Kalashnikov, V.; Konstantinides, D. Ruin under interest force and subexponential claims: A simple treatment. *Insur. Math. Econ.* **2000**, *17*, 145–149. [CrossRef]
10. Yuen, K.C.; Wang, G.; Li, W.K. The Gerber–Shiu expected discounted penalty function for risk processes with interest and a constant dividend barrier. *Insur. Math. Econ.* **2007**, *40*, 104–112. [CrossRef]
11. Li, S.; Lu, Y. On the generalized Gerber–Shiu discounted penalty function for surplus processes with interest. *Insur. Math. Econ.* **2013**, *52*, 127–134. [CrossRef]
12. Adékambi, F.; Essiomle, K. Asymptotic Tail Probability of the Discounted Aggregate Claims under Homogeneous, Non-Homogeneous and Mixed Poisson Risk Model. *Risks* **2021**, *9*, 122. [CrossRef]
13. Essiomle, K.; Adékambi, F. Ruin formulas for a delay renewal risk model with general dependence. *J. Appl. Probab. Stat.* **2020**, *15*, 147–165.
14. Cheung, E.C.; Landriault, D.; Willmot, G.E.; Woo, J.K. Structural properties of Gerber–Shiu functions in dependent Sparre Andersen models. *Insur. Math. Econ.* **2010**, *46*, 117–126. [CrossRef]
15. Willmot, G.E.; Woo, J.-K. On the analysis of a general class of dependent risk processes. *Insur. Math. Econ.* **2012**, *51*, 134–141. [CrossRef]
16. Konstantinides, D.G.; Ng, K.W.; Tang, Q. The probabilities of absolute ruin in the renewal risk model with constant force of interest. *J. Appl. Probab.* **2010**, *47*, 323–334. [CrossRef]
17. Cai, J. Ruin probabilities with dependent rates of interest. *J. Appl. Probab.* **2002**, *39*, 312–323. [CrossRef]
18. Zou, W.; Xie, J. On the Gerber–Shiu discounted penalty function in a risk model with delayed claim. *Insur. Math. Econ.* **2012**, *46*, 117–126. [CrossRef]
19. Polyanin, A.D.; Manzhirov, A. *Handbook of Integral Equations*; Chapman & Hall: London, UK, 2008.

20. Maleknejad, K.; Sohrabi, S.; Rostami, Y. Numerical solution of nonlinear Volterra equations of the second kind by using Chebyshev polynomials. *Appl. Math. Comput.* **2007**, *188*, 123–128. [CrossRef]
21. Nadir, M.; Rahmoune, A. Modified method for solving linear Volterra integral equations of the second kind using Simpson's rule. *Int. J. Math. Manuscr.* **2007**, *1*, 141–146.

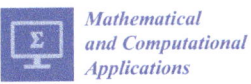

Article

The Generalized Odd Linear Exponential Family of Distributions with Applications to Reliability Theory

Farrukh Jamal [1,*], Laba Handique [2], Abdul Hadi N. Ahmed [3], Sadaf Khan [1], Shakaiba Shafiq [1] and Waleed Marzouk [3]

1. Department of Statistics, The Islamia University of Bahawalpur, Bahawalpur 63100, Pakistan; smkhan6022@gmail.com (S.K.); shakaiba.hashmi@gmail.com (S.S.)
2. Department of Statistics, Darrang College, Gauhati University, Tezpur 784001, India; handiquelaba@gmail.com
3. Department of Mathematical Statistics, Institute of Statistical Studies and Research, Cairo University, Giza 12613, Egypt; dr.hadi@cu.edu.eg (A.H.N.A.); wgm-74@hotmail.com (W.M.)
* Correspondence: farrukh.jamal@iub.edu.pk

Abstract: A new family of continuous distributions called the generalized odd linear exponential family is proposed. The probability density and cumulative distribution function are expressed as infinite linear mixtures of exponentiated-F distribution. Important statistical properties such as quantile function, moment generating function, distribution of order statistics, moments, mean deviations, asymptotes and the stress–strength model of the proposed family are investigated. The maximum likelihood estimation of the parameters is presented. Simulation is carried out for two of the mentioned sub-models to check the asymptotic behavior of the maximum likelihood estimates. Two real-life data sets are used to establish the credibility of the proposed model. This is achieved by conducting data fitting of two of its sub-models and then comparing the results with suitable competitive lifetime models to generate conclusive evidence.

Keywords: generalized odd linear distribution; hazard rate function; moments; residual analysis; maximum likelihood estimation; Monte Carlo simulation

1. Introduction

Analysis of lifetime data is an important subject in many fields, including reliability, social sciences, biomedical, engineering and other fields. In practice, it has been observed that many phenomena do not follow any of the classical distributions; for this reason, many efforts have been made in the last few decades to introduce new generators or families of distributions that extend these classical distributions to provide considerable flexibility in modeling data in diverse spectrums. Many authors have suggested new generators or families in the literature, for example, and not exclusively: Marshall and Olkin (1997) [1] introduced the Marshall–Olkin family, Gupta et al. (1998) [2] introduced the exponentiated-G family, Eugene et al. (2002) [3] proposed the beta-G family, Cordeiro and Castro (2011) [4] suggested the Kumaraswamy-G family, Alexander et al. (2012) [5] presented the McDonald-G family, Alzaatreh et al. (2013) [6] proposed the transformed-transformer (T-X) family, Bourguignon et al. (2014) [7] presented the Weibull-G family, Tahir et al. (2015) [8] studied the odd generalized exponential-G family, Cordeiro et al. (2016) [9] discussed the Zografos Balakrishnan odd log-logistic family, Gomes-Silva et al. (2017) [10] presented the odd Lindley-G family, Alizadeh et al. (2017) [11] provided the Gompertz-G family and Jamal et al. (2017) [12] defined the odd Burr-III family, among others. For a clearer understanding of the odds ratio to define new G-classes, we motivate the readers to Khan et al. (2021) [13], in which the authors adopted a unique odd function to propose an alternate generalized odd generalized exponential-G family.

The linear exponential or (linear failure rate) distribution is the distribution of the minimum of two independent random variables Z_1 and Z_2 having exponential (a) and Rayleigh (b) (Sen and Bhattacharyya, 1995 [14]). Therefore, the variables have exponential and Rayleigh distributions as special cases, which are well-known distributions for modeling lifetime data in reliability and medical studies. The linear exponential distribution is used to model phenomena with linearly increasing failure rates, but it does not provide a reasonable fit for modeling phenomena with decreasing, non-linear increasing, or non-monotonic failure rates, which include the bathtub and upside-down bathtub, among others. These phenomena are common in reliability and biological studies. This motivated us to introduce generalizations of linear, exponential distribution so that their goodness of fit measures may improve the tail properties. Our motivations and the main goals of this paper are to propose a random variable that follows the linear exponential distribution as a new generator to introduce new models which can yield all types of the hazard rate functions with improved goodness of fit properties for real-life data.

2. The Generalized Odd Linear Exponential (GOLE-F) Family

Suppose the random variable Z has a linear exponential distribution with parameters $a, b \geq 0$ where $a + b > 0$, then its cumulative distribution function (CDF) and probability density function (PDF) are, respectively,

$$R(z) = 1 - e^{-(az + \frac{b}{2}z^2)}, \; z \geq 0 \tag{1}$$

$$r(z) = (a + bz)e^{-(az + \frac{b}{2}z^2)}, \; z > 0. \tag{2}$$

Adopting the T-X framework defined by the authors in [6], for any power parameter $c > 0$, we define the CDF of a new wider family called the generalized odd linear exponential ("GOLE-F" for short) family by

$$G(x; a, b, c, \boldsymbol{\phi}) = \int_0^{\frac{F(x;\boldsymbol{\phi})^c}{1 - F(x;\boldsymbol{\phi})^c}} (a + bz)e^{-(az + \frac{b}{2}z^2)}dz = 1 - \exp\left[-\left(\frac{aF(x;\boldsymbol{\phi})^c}{1 - F(x;\boldsymbol{\phi})^c} + \frac{b}{2}\left(\frac{F(x;\boldsymbol{\phi})^c}{1 - F(x;\boldsymbol{\phi})^c}\right)^2\right)\right], \tag{3}$$

where $W[F(x)] = \frac{F(x;\boldsymbol{\phi})^c}{1 - F(x;\boldsymbol{\phi})^c}$ is the link function with $F(x; \boldsymbol{\phi})$ as the baseline CDF of an absolutely continuous distribution with parameter vector $\boldsymbol{\phi}$ and pdf $f(x; \boldsymbol{\phi})$.

The PDF of GOLE-F corresponding to the CDF in Equation (3) is provided by

$$g(x; a, b, c, \boldsymbol{\phi}) = \left[\frac{cf(x;\boldsymbol{\phi})F(x;\boldsymbol{\phi})^{c-1}(a + (b-a)F(x;\boldsymbol{\phi})^c)}{(1 - F(x;\boldsymbol{\phi})^c)^3}\right] \times \exp\left[-\left(\frac{aF(x;\boldsymbol{\phi})^c}{1 - F(x;\boldsymbol{\phi})^c} + \frac{b}{2}\left(\frac{F(x;\boldsymbol{\phi})^c}{1 - F(x;\boldsymbol{\phi})^c}\right)^2\right)\right]. \tag{4}$$

Henceforth, for any parent model, we will simply write $F(x) = F(x; \boldsymbol{\phi})$ as the distribution function and $f(x) = f(x; \boldsymbol{\phi})$ as the density function. Further, any random variable X with density function (4) is denoted by $X \sim GOLE - F(a, b, c, \boldsymbol{\phi})$.

The hazard rate function (HRF) and reversed hazard rate function (RHRF) of the random variable X are, respectively,

$$h(x; a, b, c, \boldsymbol{\phi}) = \frac{cf(x)F(x)^{c-1}(a + (b-a)F(x)^c)}{(1 - F(x)^c)^3}, \tag{5}$$

and

$$\tau(x; a, b, c, \boldsymbol{\phi}) = \frac{\{cf(x)F(x)^{c-1}(a + (b-a)F(x)^c)\}e^{-\left(\frac{aF(x)^c}{1-F(x)^c} + \frac{b}{2}\left(\frac{F(x)^c}{1-F(x)^c}\right)^2\right)}}{(1 - F(x)^c)^3\left\{1 - e^{-\left(\frac{aF(x)^c}{1-F(x)^c} + \frac{b}{2}\left(\frac{F(x)^c}{1-F(x)^c}\right)^2\right)}\right\}}. \tag{6}$$

The quantile function of the random variable X can be obtained by inverting Equation (3), and hence the GOLE-F distribution can be simulated easily from the following Equation.

$$X = Q(U) = F^{-1}\left(\left[\frac{-a+\sqrt{a^2-2b\log(1-U)}}{b-a+\sqrt{a^2-2b\log(1-U)}}\right]^{1/c}\right), \quad (7)$$

where U has a uniform distribution over the interval (0,1), in particular, if $u = 1/2$ we obtain the median of the random variable X as follows:

$$M = Q\left(\frac{1}{2}\right) = F^{-1}\left(\left[\frac{-a+\sqrt{a^2-2b\log(1-1/2)}}{b-a+\sqrt{a^2-2b\log(1-1/2)}}\right]^{1/c}\right). \quad (8)$$

3. Special Model of the GOLE-F Family

In this section, we provide two extended distributions as special models of the GOLE-F family and display their plots of density and hazard rate functions.

3.1. The Generalized Odd Linear Exponential-Weibull (GOLE-W) Distribution

Consider the Weibull distribution with density and distribution functions $f(x;\lambda,\beta) = \lambda\beta x^{\beta-1}e^{-\lambda x^\beta}$ and $F(x;\lambda,\beta) = 1 - e^{-\lambda x^\beta}$, respectively, where $\lambda, \beta > 0$ and $x \geq 0$. Then, the GOLE-W distribution has (PDF) provided by

$$g(x;a,b,c,\lambda,\beta) = \left[\frac{c\lambda\beta x^{\beta-1}e^{-\lambda x^\beta}\left(1-e^{-\lambda x^\beta}\right)^{c-1}\left(a+(b-a)\left(1-e^{-\lambda x^\beta}\right)^c\right)}{\left(1-\left(1-e^{-\lambda x^\beta}\right)^c\right)^3}\right]$$
$$\times \exp\left[-\left(\frac{a\left(1-e^{-\lambda x^\beta}\right)^c}{1-\left(1-e^{-\lambda x^\beta}\right)^c} + \frac{b}{2}\left(\frac{\left(1-e^{-\lambda x^\beta}\right)^c}{1-\left(1-e^{-\lambda x^\beta}\right)^c}\right)^2\right)\right].$$

Figure 1a show a wealth of possible shapes of the distribution once different choices of the parameters are made. For example, the shape can be U and inverted-U, right-skewed, reversed-J shape or symmetrical. Additionally, Figure 1b reveal that the HRF of the GOLE-W distribution can be increasing–constant, constant–monotone–increasing or monotone–increasing shapes.

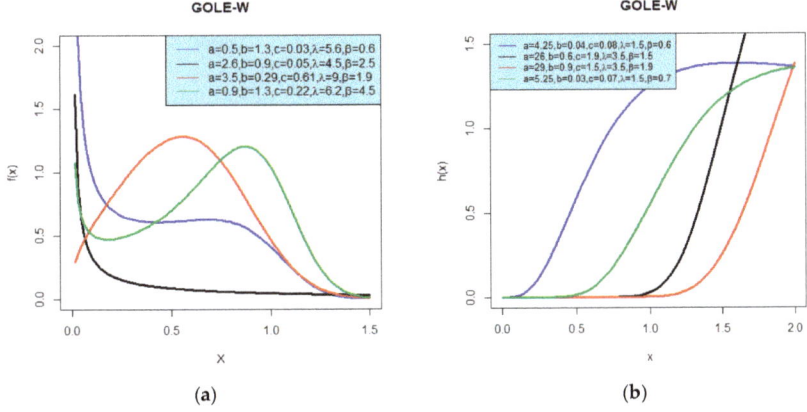

Figure 1. (a) Density function and (b) hazard rate plots of the GOLE-W distribution for different parameter values.

3.2. The Generalized Odd Linear Exponential-Exponential (GOLE-E) Distribution

Consider the Exponential distribution with density and distribution functions $f(x;\lambda) = \lambda e^{-\lambda x}$ and $F(x;\lambda) = 1 - e^{-\lambda x}$, respectively, where $\lambda > 0$ and $x \geq 0$. Then, the GOLE-E distribution has (PDF) provided by

$$g(x;a,b,c,\lambda) = \left[\frac{c\lambda e^{-\lambda x}(1-e^{-\lambda x})^{c-1}\left(a+(b-a)(1-e^{-\lambda x})^c\right)}{(1-(1-e^{-\lambda x})^c)^3}\right] \times \exp\left[-\left(\frac{a(1-e^{-\lambda x})^c}{1-(1-e^{-\lambda x})^c}+\frac{b}{2}\left(\frac{(1-e^{-\lambda x})^c}{1-(1-e^{-\lambda x})^c}\right)^2\right)\right].$$

Figure 2a show possible shapes of the GOLE-E distribution for different choices of the parameters. The shapes of pdf can be right-skewed, or symmetrical. Further, Figure 2b reveal that the HRF of the GOLE-E distribution can be decreasing–constant, monotone–increasing or bathtub shape. The PDF and HRF of the GOLE-W and GOLE-E distributions for some selected values of the parameters indicate the flexibility of the new family.

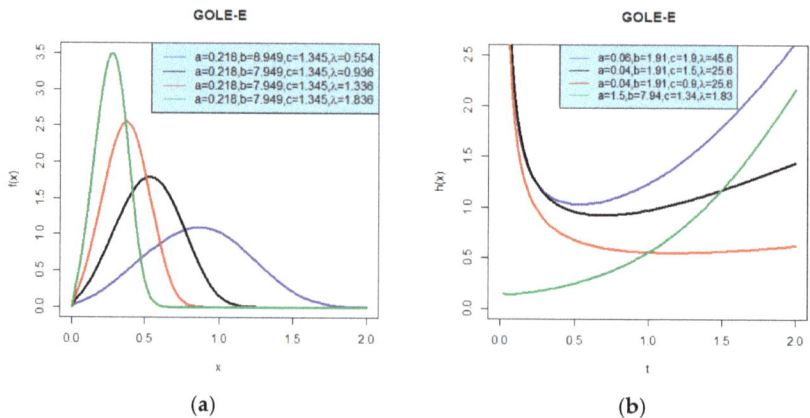

Figure 2. Plots of (**a**) density function and (**b**) hazard rate of the GOLE-E distribution for different parameter values.

4. Mathematical Properties of the GOLE-F Family

In this section, some mathematical properties of the GOLE-F family are obtained.

4.1. Asymptotic Behavior of GOLE-F Family

First of all, for the statements of the following results, we recall that $F(x)$ is the CDF of an absolutely continuous distribution with pdf $f(x)$.

Proposition 1. *The asymptotes corresponding to Equations (3)–(5) when $x \to -\infty$ are provided by*

$$G(x) \sim a\, F(x)^c, \tag{9}$$

$$g(x) \sim c\, a\, f(x) F(x)^{c-1}, \tag{10}$$

$$h(x) \sim c\, a\, f(x) F(x)^{c-1}. \tag{11}$$

Proposition 2. *The asymptotes corresponding to Equations (3)–(5) when $x \to \infty$ are provided by*

$$1 - G(x) \sim 1 - e^{-\left(\frac{a}{1-F(x)^c}+\frac{b}{2}\left\{\frac{1}{1-F(x)^c}\right\}^2\right)}, \tag{12}$$

$$g(x) \sim \frac{bcf(x)}{(1-F(x)^c)^3} e^{-\left(\frac{a}{1-F(x)^c}+\frac{b}{2}\left\{\frac{1}{1-F(x)^c}\right\}^2\right)} \tag{13}$$

$$h(x) \sim \frac{bcf(x)}{(1-F(x)^c)^3}. \tag{14}$$

For detail see Appendix A.

4.2. Useful Expansions for CDF and PDF of the New Family

Using the power series for the exponential function and the generalized binomial expansion

$$e^{-z} = \sum_{i=0}^{\infty} \frac{(-1)^i z^i}{i!},$$

and

$$(1-v)^n = \sum_{i=0}^{\infty} (-1)^i \binom{n}{i} v^i,$$

respectively, where $|v| < 1$ and n is any real number, we can rewrite the CDF of the GOLE-F family as follows:

$$G(x; a, b, c, \phi) = 1 - \sum_{i=0}^{\infty} \sum_{j=0}^{i} \sum_{k=0}^{\infty} \frac{(-1)^i b^j a^i \binom{i}{j}\binom{i+j+k-1}{k}}{i! 2^j} F(x)^{c(i+j+k)}. \tag{15}$$

Again, based on the binomial expansion, we find

$$F(x)^{c(i+j+k)} = (1 - (1 - F(x))^{c(i+j+k)} = \sum_{m=0}^{\infty} \sum_{l=m}^{\infty} (-1)^{l+m} \binom{l}{m}\binom{c(i+j+k)}{l} F(x)^m. \tag{16}$$

From (15) and (16), we obtain

$$G(x; a, b, c, \phi) = 1 - \sum_{m=0}^{\infty} \omega_m F(x)^m, \tag{17}$$

where

$$\omega_m = \sum_{l=m}^{\infty} \sum_{i=0}^{\infty} \sum_{j=0}^{i} \sum_{k=0}^{\infty} \rho_{i,j,k,l}(a,b,c),$$

and

$$\rho_{i,j,k,l}(a,b,c) = \frac{(-1)^{i+l+m} b^j a^i \binom{i}{j}\binom{l}{m}\binom{i+j+k-1}{k}\binom{c(i+j+k)}{l}}{i! 2^j}.$$

Now, we can write the CDF of the GOLE-F family in Equation (17), as

$$G(x; a, b, c, \phi) = \sum_{m=0}^{\infty} \delta_m F(x)^m, \tag{18}$$

where $\delta_0 = 1 - \omega_0$, and $\delta_m = -\omega_m$ for $m = 1, 2, \ldots$ By differentiating Equation (18), we obtain the expansion of the density function of the GOLE-F family as an infinite linear mixture of exp-F densities in the following form

$$g(x; a, b, c, \phi) = \sum_{m=0}^{\infty} \delta_{m+1} \pi_{m+1}(x), \tag{19}$$

where $\pi_{m+1}(x) = (m+1)f(x)F(x)^m$ is the exp-F density function with power parameter $(m+1)$. Now, if the random variable Y_{m+1} has the density function $\pi_{m+1}(x)$, then many mathematical properties of the random variable X, including the ordinary and incomplete moments and moment generating function can easily be obtained based on the exp-F distribution.

4.3. Moments

Suppose that the random variable Y_{m+1} has the density function $\pi_{m+1}(x)$ in (19), then the nth moment of the random variable X can be obtained from

$$\mu'_n = E(X^n) = \sum_{m=0}^{\infty} \delta_{m+1} E(Y^n_{m+1}). \tag{20}$$

A second alternative formula for μ'_n in terms of the baseline qf. $Q_F(u)$ can be obtained as

$$\mu'_n = \sum_{m=0}^{\infty} \delta_{m+1}(m+1) \int_0^1 Q_F(u)^n u^m du, \tag{21}$$

where $Q_F(u) = F^{-1}(u)$ is the qf of the parent distribution and $u \in (0,1)$.

The incomplete moments have an important role in measuring inequality, for example, income quantiles, the mean deviations and Lorenz and Bonferroni curves. The nth incomplete moment of X is provided by

$$\eta_n(z) = \sum_{m=0}^{\infty} \delta_{m+1}(m+1) \int_0^{F(z)} Q_F(u)^n u^m du. \tag{22}$$

The last integral can be computed analytically or numerically for most baseline distributions. Bonferroni and Lorenz curves have applications in many different areas such as economics to study income and poverty, reliability, demography, insurance and medicine. For a random variable X, the Bonferroni and Lorenz curves are defined by $B(p) = \eta_1(q)/pE(X)$ and $L(p) = \eta_1(q)/E(X)$, respectively, where p is a given probability, $q = Q(p)$ and $\eta_1(q)$ is the first incomplete moment that can be calculated from the above Equation with $r = 1$ at q. Table 1 display the mean, variance, skewness and kurtosis of the GOLE-E distribution for some choices values of the parameters. We note from Table 1 that the skewness of the GOLE-E distribution is always positive, whereas the kurtosis of the GOLE-E distribution varies only in the interval (1.0571, 2.6112).

Table 1. Mean, variance, skewness and kurtosis of the GOLE-E distribution with different values of a, b, c and $\lambda = 1$.

a	b	c	Mean	Variance	Skewness	Kurtosis
0.5	0.5	1	0.6704	0.1330	1.2979	1.8368
		2	1.1667	0.2186	1.1776	1.4755
		5	1.9517	0.3004	1.0960	1.2512
		10	2.5987	0.3351	1.0635	1.1656
		20	3.2683	0.3541	1.0440	1.1146
		50	4.1704	0.3660	1.0288	1.0753
		100	4.8587	0.3701	1.0218	1.0571
1	1	1	0.4614	0.0824	1.3869	2.1392
		2	0.8995	0.1587	1.2180	1.5987
		5	1.6411	0.2408	1.1103	1.2923
		10	2.2721	0.2776	1.0699	1.1839
		20	2.9334	0.2982	1.0467	1.1226
		50	3.8303	0.3114	1.0294	1.0774
		100	4.5169	0.3159	1.0218	1.0574
2	1.5	1	0.3091	0.0488	1.5119	2.6112
		2	0.6860	0.1130	1.2703	1.7688
		5	1.3790	0.1919	1.1269	1.3424
		10	1.9914	0.2297	1.0768	1.2044
		20	2.6430	0.2514	1.0494	1.1308
		50	3.5339	0.2654	1.0299	1.0793
		100	4.2185	0.2702	1.0217	1.0574

4.4. Generating Function

Here, we provide three formulae for the mgf $M(t) = E\left(e^{tX}\right)$ of the random variable X. The first one is provided by

$$M(t) = \sum_{n=0}^{\infty} \frac{t^n}{n!} \mu'_n, \qquad (23)$$

where μ'_n is the nth moment of the random variable X. A second formula for $M(t)$ comes from (19) as

$$M(t) = \sum_{m=0}^{\infty} \delta_{m+1} M_{m+1}(t), \qquad (24)$$

where $M_{m+1}(t)$ is the mgf of the random variable $Y_{m+1} \sim$ exp-F $(m+1)$. A third formula for $M(t)$ can also be derived based on (19) in terms of the baseline qf. $Q_F(u)$ as

$$M(t) = \sum_{m=0}^{\infty} \delta_{m+1}(m+1) \int_0^1 \exp(tQ_F(u)) u^m du, \qquad (25)$$

where $Q_F(u) = F^{-1}(u)$ is the qf of the baseline distribution and $u \in (0,1)$.

4.5. Mean Deviations

The amount of scattering in a population is evidently measured to some extent by the totality of deviations from the mean and median. These are known as the mean deviation about the mean and the mean deviation about the median. These measures can be calculated using the following relationships:

$\delta_1(X) = 2\mu G(\mu) - 2\int_{-\infty}^{\mu} xg(x)dx$ and $\delta_2(X) = \mu - 2\int_{-\infty}^{M} xg(x)dx$, respectively, where $\mu = E(X)$ and $M = Q\left(\frac{1}{2}\right)$.

4.6. Order Statistics

Let $X_1, X_2, \ldots, X_n$ be a random sample from the GOLE-F family with CDF and PDF defined in Equations (3) and (4), respectively. Suppose $X_{1:n}, X_{2:n}, \ldots, X_{n:n}$ denote the order statistics obtained from this sample and $X_{r:n}$ is the ith order statistic, then the density function of the rth order statistic is provided by

$$g_{r:n}(x) = \frac{n!}{(r-1)!(n-r)!} \sum_{s=0}^{n-r} (-1)^s \binom{n-r}{s} g(x) G(x)^{r+s-1}. \qquad (26)$$

From (17), we determine

$$G(x)^{r+s-1} = \left[\sum_{m=0}^{\infty} \delta_m F(x)^m\right]^{r+s-1} = \sum_{m=0}^{\infty} d_{r+s-1,m} F(x)^m, \qquad (27)$$

where

$d_{r+s-1,0} = \delta_0^{r+s-1}$ and $d_{r+s-1,m} = (m\delta_0)^{-1} \sum_{q=1}^{m} [q(r+s) - m]\delta_q d_{r+s-1,m-q}$.

By replacing t instead of m in Equation (19), we obtain

$$g(x) = \sum_{t=0}^{\infty} \delta_{t+1}(t+1) f(x) F(x)^t. \qquad (28)$$

By substituting (26) in (27) and (28), we determine the PDF of the rth order statistic $X_{r:n}$ as

$$g_{r:n}(x) = \sum_{t,m=0}^{\infty} \pi_{t,m} h_{t+m+1}(x), \qquad (29)$$

where $h_{t+m+1}(x)$ denotes the PDF of exp-F distribution with power parameter $(t+m+1)$, and

$$\pi_{t,m} = \sum_{s=0}^{n-r} \frac{(-1)^s \binom{n-r}{s} n! \delta_{t+1}(t+1) d_{r+s-1,m}}{(r-1)!(n-r)!(t+m+1)}.$$

Based on Equation (29), several mathematical properties of these order statistics such as ordinary and incomplete moments, factorial moments, moment generating function, mean deviations and several others, can be obtained.

4.7. Stochastic Orderings

Stochastic orders and inequalities are used in many different areas of probability and statistics. Such areas include reliability theory, survival analysis, economics, insurance, actuarial science, queuing theory, biology, operations research, management science, etc. For more detail regarding stochastic ordering, see (Shaked et al., 1994 [15]). Given two random variables X and Y, we say that X is smaller than Y in the:

1. usual stochastic order, denoted by $X \leq_{st} Y$, if $G_X(x) \geq G_Y(x)$, for all x;
2. hazard rate order, denoted by $X \leq_{hr} Y$, if $h_X(x) \geq h_Y(x)$, for all x;
3. reversed hazard rate order, denoted by $X \leq_{rh} Y$, if $G_X(x)/G_Y(x)$, is decreases in x;
4. mean residual life order, denoted by $X \leq_{mrl} Y$, if $m_X(x) \leq m_Y(x)$, for all x;
5. likelihood ratio order, denoted by $X \leq_{lr} Y$, if $g_X(x)/g_Y(x)$, is decreases in x.

For all the previous orders, we determine the following chains of implications:

$$X \leq_{lr} Y \Rightarrow X \leq_{hr} Y \Rightarrow X \leq_{st} Y$$

and

$$X \leq_{lr} Y \Rightarrow X \leq_{rh} Y \Rightarrow X \leq_{st} Y,$$

also

$$X \leq_{hr} Y \Rightarrow X \leq_{mrl} Y.$$

For the proposed GOLE-F family, the following theorem provides the stochastic comparison results with respect to the above orderings.

Theorem 1. Let $X \sim GOLE(a_1, b_1, c_1, \phi)$ and $Y \sim GOLE(a_2, b_2, c_2, \phi)$. If $a_1 \geq a_2$. and $b_1 \geq b_2$ and $c_1 \leq c_2$, then $X \leq_{st} Y$.

Proof. If $c_1 \leq c_2$, then

$$\frac{F(x)^{c_1}}{1-F(x)^{c_1}} \geq \frac{F(x)^{c_2}}{1-F(x)^{c_2}}. \tag{30}$$

Hence, if $a_1 \geq a_2$ and $b_1 \geq b_2$, then

$$\frac{a_1 F(x)^{c_1}}{1-F(x)^{c_1}} + \frac{b_1}{2}\left(\frac{F(x)^{c_1}}{1-F(x)^{c_1}}\right)^2 \geq \frac{a_2 F(x)^{c_2}}{1-F(x)^{c_2}} + \frac{b_2}{2}\left(\frac{F(x)^{c_2}}{1-F(x)^{c_2}}\right)^2. \tag{31}$$

Therefore,

$$\left[-\left(\frac{a_1 F(x)^{c_1}}{1-F(x)^{c_1}} + \frac{b_1}{2}\left(\frac{F(x)^{c_1}}{1-F(x)^{c_1}}\right)^2\right)\right] \leq \left[-\left(\frac{a_2 F(x)^{c_2}}{1-F(x)^{c_2}} + \frac{b_2}{2}\left(\frac{F(x)^{c_2}}{1-F(x)^{c_2}}\right)^2\right)\right]. \tag{32}$$

Thus,

$$1 - \exp\left[-\left(\frac{a_1 F(x)^{c_1}}{1-F(x)^{c_1}} + \frac{b_1}{2}\left(\frac{F(x)^{c_1}}{1-F(x)^{c_1}}\right)^2\right)\right]$$
$$\geq 1 - \exp\left[-\left(\frac{a_2 F(x)^{c_2}}{1-F(x)^{c_2}} + \frac{b_2}{2}\left(\frac{F(x)^{c_2}}{1-F(x)^{c_2}}\right)^2\right)\right]. \tag{33}$$

That means $G_X(x) \geq G_Y(x)$ and $X \leq_{st} Y$. □

Theorem 2. Let $X \sim GOLE(a_1, b_1, c, \phi)$ and $Y \sim GOLE(a_2, b_2, c, \phi)$. If $a_1 > a_2$ and $b_1 = b_2$, then $X \leq_{lr} Y$.

Proof. We determine

$$\frac{g_X(x)}{g_Y(x)} = \frac{[(a_1 + (b_1 - a_1)F(x)^c)] \times \exp\left[-\left(\frac{a_1 F(x)^c}{1-F(x)^c} + \frac{b_1}{2}\left(\frac{F(x)^c}{1-F(x)^c}\right)^2\right)\right]}{[(a_2 + (b_2 - a_2)F(x)^c)] \times \exp\left[-\left(\frac{a_2 F(x)^c}{1-F(x)^c} + \frac{b_2}{2}\left(\frac{F(x)^c}{1-F(x)^c}\right)^2\right)\right]}. \quad (34)$$

Thus,

$$\log\left(\frac{g_X(x)}{g_Y(x)}\right) = \log\left[(a_1 + (b_1 - a_1)F(x)^c)\right] - \left(\frac{a_1 F(x)^c}{1-F(x)^c} + \frac{b_1}{2}\left(\frac{F(x)^c}{1-F(x)^c}\right)^2\right) \\ - \log[(a_2 + (b_2 - a_2)F(x)^c)] + \left(\frac{a_2 F(x)^c}{1-F(x)^c} + \frac{b_2}{2}\left(\frac{F(x)^c}{1-F(x)^c}\right)^2\right). \quad (35)$$

By differentiating the last Equation and after some simplifications, we obtain

$$\frac{d}{dx}\left(\log\left(\frac{g_X(x)}{g_Y(x)}\right)\right) = \frac{(a_2 b_1 - a_1 b_2) c f(x) F(x)^{c-1}}{(a_1 + (b_1 - a_1)F(x)^c)(a_2 + (b_2 - a_2)F(x)^c)} + \frac{(a_2 - a_1) c f(x) F(x)^{c-1}}{(1-F(x)^c)^2} + \frac{(b_2 - b_1) c f(x) F(x)^{2c-1}}{(1-F(x)^c)^3}. \quad (36)$$

Now, if $a_1 > a_2$ and $b_1 = b_2$, then $\frac{d}{dx}\left[\log\left(\frac{g_X(x)}{g_Y(x)}\right)\right] < 0$, and hence $g_X(x)/g_Y(x)$ is decreases in x. This implies that $X \leq_{lr} Y$. □

4.8. Stress-Strength Model

The stress–strength model defines the life of an element which has a random strength Y that is subjected to an accidental stress X. The component fails at the instant that the stress applied to it exceeds the strength, and the component will function suitably whenever $X < Y$. Hence, $R = P(X < Y)$ is a measure of component reliability (Kotz et al., 2003 [16]). It has many applications, especially in reliability engineering. We derive the reliability R when Y and X are two independent continuous random variables from the GOLE-F (a_1, b_1, c_1, ϕ_1) and GOLE-F (a_2, b_2, c_2, ϕ_2) distributions, respectively. The reliability is defined by

$$R = \int_0^\infty g_Y(x) G_X(x) dx. \quad (37)$$

Using the PDF in (19) and the CDF in (18), we obtain

$$R = \sum_{m,t=0}^\infty \delta_{m+1} \delta_t R_{m+1,t}, \quad (38)$$

where
$R_{m+1,t} = \int_0^\infty \pi_{m+1}(x, \phi_1) \Pi_t(x, \phi_2) dx$, and
$\pi_{m+1}(x, \phi_1) = (m+1) f(x, \phi_1) F(x, \phi_1)^m$, $\Pi_t(x, \phi_2) = F(x, \phi_2)^t$.
The constants δ_t, δ_{m+1} are defined as:

$$\delta_t = \sum_{l=t}^\infty \sum_{i=0}^\infty \sum_{j=0}^i \sum_{k=0}^\infty \frac{(-1)^{i+l+t+1} b_2^j a_2^i \binom{i}{j}\binom{l}{t}\binom{i+j+k-1}{k}\binom{c_2(i+j+k)}{l}}{i! 2^j},$$

for $t \geq 1$. For $t = 0$, then

$$\delta_0 = 1 - \sum_{l=0}^{\infty}\sum_{i=0}^{\infty}\sum_{j=0}^{i}\sum_{k=0}^{\infty} \frac{(-1)^{i+l+m}b_2^j a_2^i \binom{i}{j}\binom{i+j+k-1}{k}\binom{c_2(i+j+k)}{l}}{i!2^j},$$

and

$$\delta_{m+1} = \sum_{l=m+1}^{\infty}\sum_{i=0}^{\infty}\sum_{j=0}^{i}\sum_{k=0}^{\infty} \frac{(-1)^{i+l+m+2}b_1^j a_1^i \binom{i}{j}\binom{l}{m+1}\binom{i+j+k-1}{k}\binom{c_1(i+j+k)}{l}}{i!2^j},$$

for $m \geq 0$.

If $\phi_1 = \phi_2$, then the model reduces to

$$R = \sum_{m,t=0}^{\infty} \frac{\delta_{m+1}\delta_t(m+1)}{m+t+1}. \tag{39}$$

5. Estimation and Simulation

5.1. Estimation of the Parameters

Here, we find the maximum likelihood estimates (MLEs) of the parameters of the new family of distributions from complete samples only. Let $x_1, x_2, \ldots, x_n$ be observed values from the GOLE − F family with parameters a, b, c and ϕ. Let $\xi = (a, b, c, \phi)^T$ be the parameters vector. The total log-likelihood function for ξ is obtained by

$$l(\xi) = n\log c + \sum_{i=1}^{n}\log f(x_i;\phi) + (c-1)\sum_{i=1}^{n}\log F(x_i;\phi) + \sum_{i=1}^{n}\log\left(a + (b-a)F(x_i;\phi)^c\right) \\ -3\sum_{i=1}^{n}\log(1 - F(x_i;\phi)^c) - a\sum_{i=1}^{n}H(x_i,\phi) - \frac{b}{2}\sum_{i=1}^{n}H(x_i,\phi)^2, \tag{40}$$

where $H(x_i,\phi) = \frac{F(x_i;\phi)^c}{1-F(x_i;\phi)^c}$ and $H(x_i,\phi)^2 = \left(\frac{F(x_i;\phi)^c}{1-F(x_i;\phi)^c}\right)^2$.

The components of the score vector $U(\xi)$ are obtained by

$$U_a = \sum_{i=1}^{n} \frac{1 - F(x_i;\phi)^c}{a + (b-a)F(x_i;\phi)^c} - \sum_{i=1}^{n} H(x_i,\phi), \tag{41}$$

$$U_b = \sum_{i=1}^{n} \frac{F(x_i;\phi)^c}{a + (b-a)F(x_i;\phi)^c} - \frac{1}{2}\sum_{i=1}^{n} H(x_i,\phi)^2, \tag{42}$$

$$U_c = \frac{n}{c} + \sum_{i=1}^{n}\log F(x_i;\phi) + \sum_{i=1}^{n}\frac{(b-a)F(x_i;\phi)^c \log F(x_i;\phi)}{a + (b-a)F(x_i;\phi)^c} + 3\sum_{i=1}^{n}\frac{F(x_i;\phi)^c \log F(x_i;\phi)}{1 - F(x_i;\phi)^c} \\ -a\sum_{i=1}^{n}\frac{F(x_i;\phi)^c \log F(x_i;\phi)}{(1-F(x_i;\phi)^c)^2} - b\sum_{i=1}^{n}\frac{F(x_i;\phi)^{2c}\log F(x_i;\phi)}{(1-F(x_i;\phi)^c)^3}, \tag{43}$$

and

$$U_{\phi_k} = \sum_{i=1}^{n}\frac{\frac{\partial f(x_i;\phi)}{\partial \phi_k}}{f(x_i;\phi)} + (c-1)\sum_{i=1}^{n}\frac{\frac{\partial F(x_i;\phi)}{\partial \phi_k}}{F(x_i;\phi)} + \sum_{i=1}^{n}\frac{c(b-a)F(x_i;\phi)^{c-1}\frac{\partial F(x_i;\phi)}{\partial \phi_k}}{a + (b-a)F(x_i;\phi)^c} \\ + 3\sum_{i=1}^{n}\frac{cF(x_i;\phi)^{c-1}\frac{\partial F(x_i;\phi)}{\partial \phi_k}}{1 - F(x_i;\phi)^c} - a\sum_{i=1}^{n}\frac{\partial H(x_i,\phi)}{\partial \phi_k} - b\sum_{i=1}^{n} H(x_i,\phi)\frac{\partial H(x_i,\phi)}{\partial \phi_k}. \tag{44}$$

Setting U_a, U_b, U_c and U_ϕ equal to zero, and solving the equations simultaneously, yields the MLE $\hat{\xi} = \left(\hat{a}, \hat{b}, \hat{c}, \hat{\phi}\right)^T$ of $\xi = (a, b, c, \phi)^T$. These equations cannot be solved analytically, and statistical software can be used to solve them numerically using iterative methods such as the Newton–Raphson type algorithms.

5.2. Simulation Study

In this section, a graphical Monte Carlo simulation study is conducted to compare the performance of the different estimators of the unknown parameters for the GOLE-E (a, b, c, λ) distribution. All the computations in this section are conducted using the R program. We generate $N = 1000$ samples of size $n = 20, 25, \ldots, 500$ from the GOLE-W and GOLE-E distributions. The true parameter values for GOLE-W ($\lambda = 1$) are $a = 1.8, b = 0.5, c = 1.7$ and $\beta = 2.8$, and those for GOLE-E are $a = 2, b = 1.5, c = 2$ and $\lambda = 2.5$, respectively. We also calculate the bias and mean square error (MSE) of the MLEs empirically. The bias and MSE are computed by

$$\hat{Bias}_h = \frac{1}{N} \sum_{i=1}^{N} \left(\hat{h}_i - h \right), \quad \hat{MSE}_h = \frac{1}{N} \sum_{i=1}^{N} \left(\hat{h}_i - h \right)^2.$$

For $h = a, b, c, \lambda$, respectively.

We provide the results of this simulation study in Figures 3–6. From these figures, we can perceive that when the sample size increases, the empirical biases and MSEs approach zero in all cases for the two models.

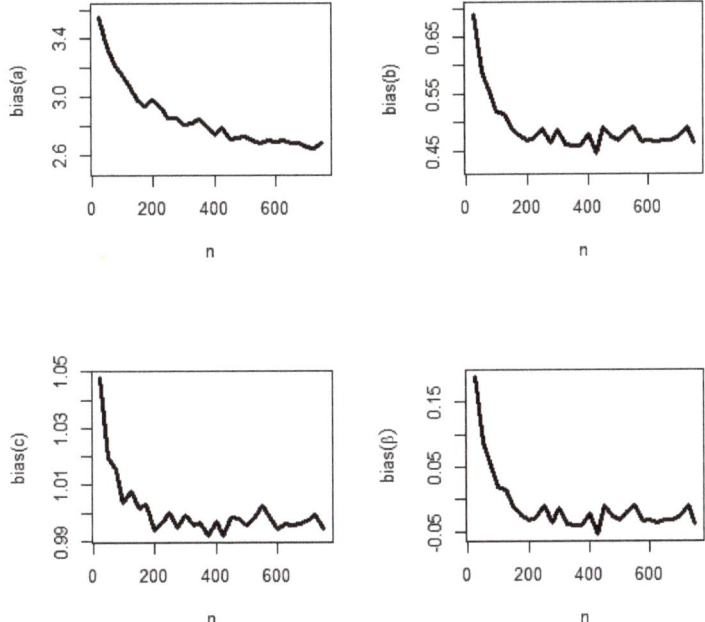

Figure 3. The biases of the estimates of parameters of the GOLE-W distribution.

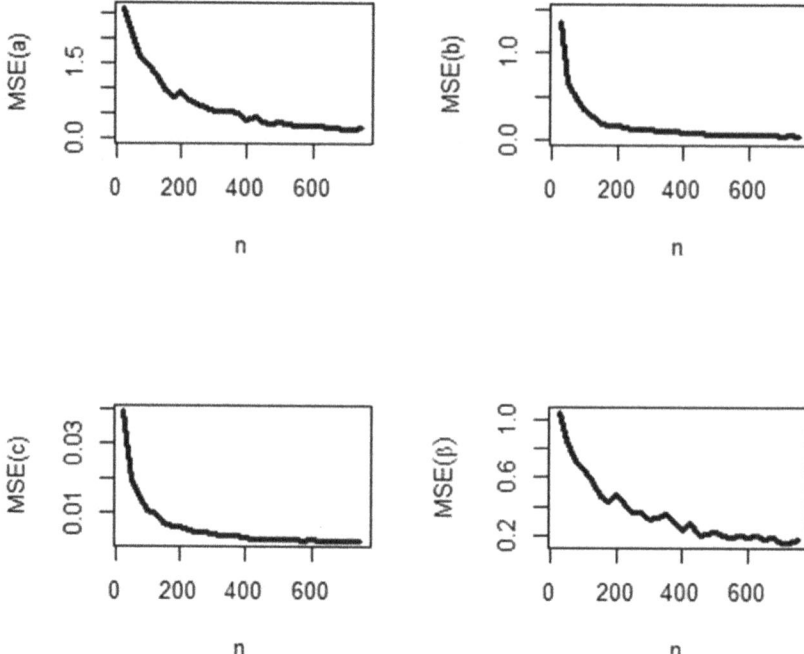

Figure 4. The *MSEs* of the estimates of parameters of the GOLE-W distribution.

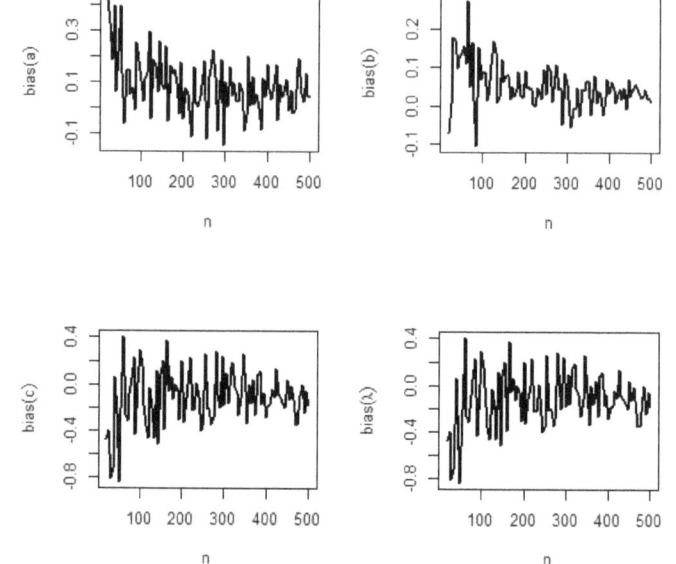

Figure 5. The biases of the estimates of parameters of the GOLE-E distribution.

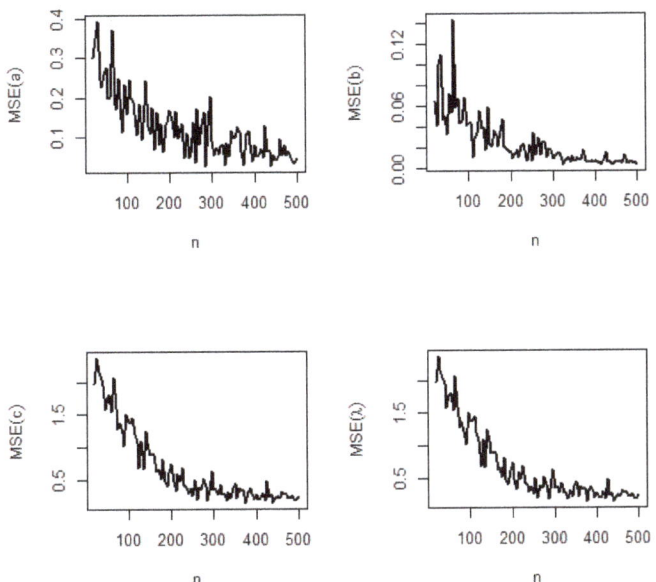

Figure 6. The *MSE* of the estimates of parameters of the GOLE-E distribution.

6. Applications on Real-Life Data Sets

In this section, we illustrate the suitability of the proposed family by fitting two real data sets on the special models viz-a-viz GOLE – W(a, b, c, λ, β) and GOLE – E (a, b, c, λ), arising due to this family with PDF mentioned in Sections 3.1 and 3.2, respectively. The comparison is conducted with some of the existing models via numerical maximizations of log-likelihood functions using the method of a limited memory quasi-Newton code for bound–constrained maximization (L-BFGS-B). We determine the log-likelihood function adjudicated at the MLEs by estimating the parameters.

Data I: The first data set is related to the measurements of nicotine levels in 346 cigarettes. [https://arxiv.org/ftp/arxiv/papers/1509/1509.08108.pdf, accessed on 19 May 2022]. Data II: The second data set consists of 74 observations of gauge lengths of 20 mm of single carbon fibers pertaining to failure stresses. (Kundu and Raqab, 2009 [17]). The descriptive statistics related to this data sets are given in Table 2.

Table 2. Descriptive Statistics for the data set I and data set II.

Data Sets	Min.	Mean	Median	S.D.	Skewness	Kurtosis	1st Q.	3rd Q.	Max.
I	0.10	0.85	0.90	0.33	0.17	0.29	0.60	1.10	2.00
II	1.312	2.477	2.513	0.487	−0.151	−0.127	2.150	2.816	3.5

The total time on test (TTT) plot proposed by Aarset (1987) [18] is a technique to extract information about the shape of the hazard function. This is drawn by plotting $T(i/n) = \{(\sum_{r=1}^{i} y_{(r)}) + (n-i)y_{(i)}\}/\sum_{r=1}^{n} y_{(r)}$, where $i = 1, 2, \ldots, n$ and $y_{(r)}$ where $r = 1, 2, \ldots, n$ is the order statistics of the sample against (i/n). The constant hazard plot is a straight diagonal, while for decreasing (increasing) hazards, it is convex (concave), respectively. The TTT plots for the data sets in Figure 7 indicate that the data sets have an increasing hazard rate.

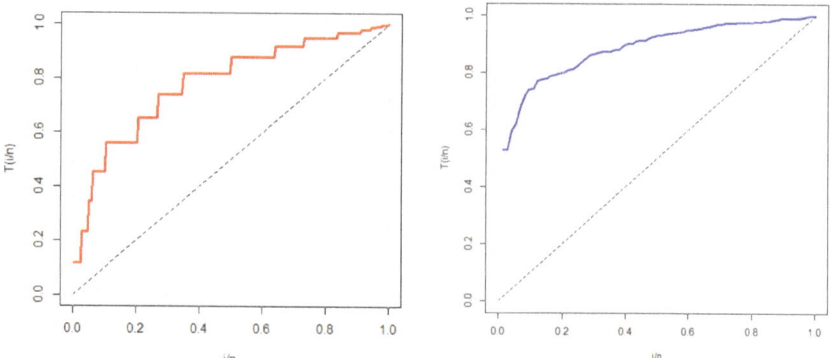

Figure 7. TTT plots of the data set I and II.

The best model is chosen on the basis of information criteria such as AIC (Akaike Information Criterion), CAIC (Consistent Akaike Information Criterion), BIC (Bayesian Information Criterion) and HQIC (Hannan–Quinn Information Criterion) with the goodness of fit measures as A* (Anderson–Darling criterion), W* (Cramér–von Mises criterion) and Kolmogorov–Smirnov (K-S) tests with p-values. The model with minimum values for these statistics could be chosen as the best model to fit the data except for the KS *p*-value, whose maximum value is the desired outcome. Asymptotic standard errors and 95% confidence intervals of the MLEs of the parameters for each competing model are also computed. For visual comparison, the fitted PDFs and the fitted CDFs are plotted with the corresponding observed histograms and ogives.

6.1. Application of GOLE-E

The GOLE-E (a, b, c, λ) distribution is compared with some models, namely exponential (E), moment exponential (ME) (Dara and Ahmad, 2012 [19]), exponentiated moment exponential (EM-E) (Hasnain et al., 2015 [20]), exponentiated exponential (E-E) (Gupta and Kundu, 2001 [21]), beta exponential (B-E) (Nadarajah and Kotz, 2006 [22]) and Kumaraswamy exponential (Kw-E) (Cordeiro and de Castro, 2011 [4]) distributions for all data sets.

In Tables 3–6, the MLEs, standard errors (SEs) and confidence interval (in parentheses) of the parameters from all the fitted distributions along with the AIC, BIC, CAIC and HQIC for the two data sets are presented. From Tables 3–6, it is evident that for the data sets, the GOLE-E distribution is the best model with the lowest values of the AIC, BIC, CAIC, HQIC, A*, W* and highest *p*-value of the K-S statistics. Hence, it is a better model than some recently introduced models, namely exponential (E), moment exponential (ME), exponentiated moment exponential (EM-E), exponentiated exponential (E-E), beta exponential (B-E) and Kumaraswamy exponential (Kw-E) distribution, for the two data sets. More information is provided for a visual comparison in the form of histograms, ogives or cumulative frequency curves of the observed data with the fitted densities and fitted cdfs displayed in Figures 8 and 9. These plots show that the proposed distributions provide the closest fit to all the observed data sets.

Table 3. MLEs, standard error (in parentheses), confidence interval values [in brackets] for the data set I.

Models	$\hat{a}$	$\hat{b}$	$\hat{c}$	$\hat{\lambda}$
GOLE-E (a,b,c,λ)	0.218 (0.315) [0, 0.84]	8.949 (3.246) [2.58, 15.31]	1.345 (0.237) [0.88, 1.81]	0.554 (0.083) [0.39, 0.72]
Kw-E (a,b,λ)	3.020 (0.163) [2.70, 3.34]	105.575 (38.348) [30.41, 180.73]	-	0.252 (0.045) [0.160.34]
B-E (a,b,λ)	4.922 (0.364) [4.21, 5.64]	17.433 (8.216) [1.32, 33.54]	-	0.298 (0.128) [0.05, 0.55]
E-E (b,λ)	-	5.526 (0.514) [4.52, 6.53]	-	2.726 (0.128) [2.475, 2.98]
EM-E (a,b)	2.574 (0.229) [2.13, 3.02]	0.284 (0.012) [0.26, 0.31]	-	-
M-E (b)	-	0.406 (0.016) [0.37, 0.44]	-	-
E (λ)	-	-	-	1.173 (0.063) [1.04, 1.29]

Table 4. The AIC, BIC, CAIC, HQIC, A*, W* and KS (p-value) values for data set I.

Models	AIC	BIC	CAIC	HQIC	A*	W*	KS (p-Value)
GOLE-E (a,b,c,λ)	232.14	247.54	232.28	238.30	2.67	0.47	0.25 (0.29)
Kw-E (a,b,λ)	236.92	248.46	236.99	241.51	3.37	0.58	0.12 (0.03)
B-E (a,b,λ)	276.04	287.59	276.11	280.66	6.48	1.09	0.24 (0.16)
E-E (b,λ)	302.44	310.14	302.47	305.52	9.42	1.59	0.22 (0.23)
EM-E (a,b)	290.62	298.32	290.65	293.70	8.48	1.46	0.24 (0.20)
M-E (b)	388.70	392.55	388.71	390.24	6.49	1.09	0.23 (0.008)
E (λ)	583.66	587.51	583.67	585.20	6.54	1.11	0.34 (0.002)

Table 5. MLEs, standard error (in parentheses) and confidence interval values [in brackets] for data set II.

Models	$\hat{a}$	$\hat{b}$	$\hat{c}$	$\hat{\lambda}$
GOLE-E (a,b,c,λ)	0.365 (0.160) [0.05, 0.68]	1.299 (0.657) [0.01, 2.59]	4.091 (1.248) [1.64, 6.54]	2.748 (0.531) [1.71, 3.78]
Kw-E (a,b,λ)	12.473 (3.939) [4.75, 20.19]	24.773 (23.936) [0, 71.68]	-	0.559 (0.194) [0.17, 0.93]
B-E (a,b,λ)	26.259 (5.838) [14.81, 37.70]	14.354 (17.832) [0, 49.30]	-	0.421 (0.376) [0, 1.16]
E-E (b,λ)	-	89.394 (32.458) [25.77, 153.01]	-	2.018 (0.171) [1.68, 2.35]
EM-E (a,b)	32.319 (10.705) [11.33, 53.30]	0.418 (0.032) [0.35, 0.48]	-	-
M-E (b)	-	1.238 (0.101) [1.04, 1.44]	-	-
E (λ)	-	-	-	0.403 (0.046) [0.31, 0.49]

Table 6. The AIC, BIC, CAIC, HQIC, A*, W* and KS (p-value) values for data set II.

Models	AIC	BIC	CAIC	HQIC	A*	W*	KS (p-Value)
GOLE-E (a,b,c,λ)	107.90	116.20	108.48	111.58	0.43	0.04	0.06 (0.83)
Kw-E (a,b,λ)	112.66	119.56	113.00	115.36	0.52	0.06	0.07 (0.79)
B-E (a,b,λ)	116.82	123.72	117.16	119.52	0.62	0.09	0.08 (0.71)
E-E (b,λ)	121.60	126.20	121.76	123.40	1.04	0.16	0.01 (0.44)
EM-E (a,b)	119.90	124.50	120.07	121.70	0.63	0.10	0.09 (0.52)
M-E (b)	230.16	232.46	230.22	231.06	0.58	0.08	0.35 (0.002)
E (λ)	284.24	286.54	284.29	285.14	0.57	0.09	0.44 (0.01)

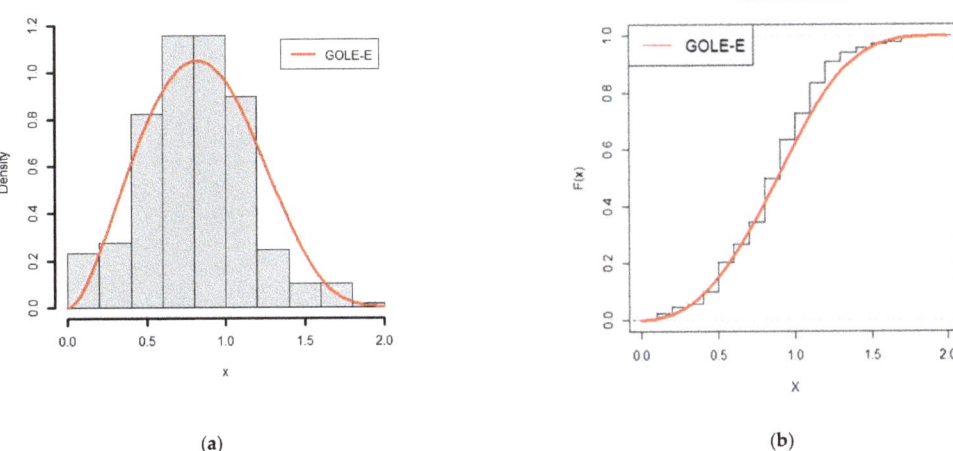

Figure 8. Plots of (**a**) the fitted PDF and (**b**) estimated CDF for the GOLE-E distribution for data set I.

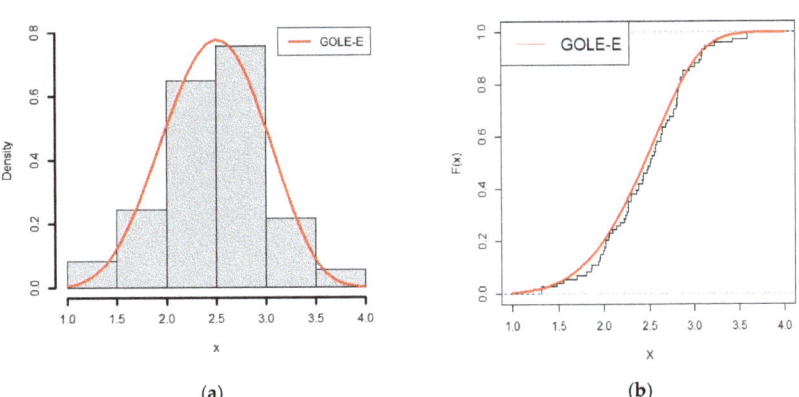

Figure 9. Plots of (**a**) the fitted PDF and (**b**) estimated CDF for the GOLE-E distribution for data set II.

6.2. Application of GOLE-W

The GOLE-W $(a, b, c, \lambda, \beta)$ distribution with $(\lambda = 1)$ is compared with some models, namely Weibull (W), moment exponential (ME), exponentiated Weibull (EW) (Mudholker and Srivastava, 1993 [23]), generalized Weibull (GW) (Lai 2014 [24]), beta Weibull (B-W) (Lee et al., 2007 [25]) and Kumaraswamy Weibull (Kw-W) (Cordeiro et al. 2010 [26]) distributions for all data sets.

Likewise, in Tables 7–10, the MLEs, standard errors (in parentheses) and confidence interval [in brackets] of the parameters from all the competitive models along with AIC, CAIC, BIC and HQIC for the two data sets are presented. From these tables, it is quite obvious that for the two data sets, GOLE-W distribution is the best model with the lowest values of AIC, BIC, CAIC, HQIC, A* , W* and highest p-value of the K-S statistics. Hence, it is worth emphasizing that the proposed GOLE-F provides a more useful generalization (with exponential and Weibull as special models) than the competitive models for both of the datasets. A much more useful depiction is presented in the form of a visual comparison in Figures 10 and 11, where the densities and distribution function of observed data are compared against the fitted models, respectively. These plots reveal that the proposed distributions provide the closest fit to all the observed data sets.

Table 7. MLEs, standard errors (in parentheses) and confidence interval [in brackets] values for data set I.

Models	$\hat{a}$	$\hat{b}$	$\hat{c}$	$\hat{\lambda}$	$\hat{\beta}$
GOLE-W (a,b,c,β)	2.3893 (2.1340) [0, 6.57]	114.6653 (50.1098) [16.45, 212.88]	4.8673 (2.1515) [0.65, 9.08]	-	0.506777 (0.1582) [0.20, 0.82]
Kw-W (a,b,λ,β)	0.7103 (0.0233) [0.66, 0.76]	0.2623 [0.23, 0.29]	-	3.0464 (0.0263) [2.99, 3.10]	3.8368 (0.0174) [3.80, 3.87]
B-W (a,b,λ,β)	0.7730 (0.0673) [0.64, 0.90]	0.2276 (0.0137) [0.20, 0.25]	-	3.0201 (0.0042) [3.01, 3.02]	4.3742 (0.0042) [4.37, 4.38]
E-W (b,λ,β)	-	0.8090 (0.1515) (4.52, 6.53)	-	3.068922 (0.3541) (2.475, 2.98)	0.9440 (0.1732) [0.60, 1.28]
G-W (a,λ,β)	0.5597 (11.2701) [0, 22.65]	-	-	2.7190 (0.1140) [2.50, 2.94]	2.0240 (0.7523) [0.55, 3.50]
M-E (b)	-	0.406 (0.016) [0.37, 0.44]	-	-	-
W (λ,β)	-	-	-	1.132926 (0.0623) [1.01, 1.26]	2.71898 (0.1140) [2.50, 2.94]

Table 8. The AIC, CAIC, BIC, HQIC, A*, W* and KS (p-value) values for data set I.

Models	AIC	CAIC	BIC	HQIC	A*	W*	KS (p-Value)
GOLE-W (a,b,c,β)	**230.11**	**230.23**	**245.49**	**236.23**	**2.51**	**0.44**	**0.10 (0.17)**
Kw-W (a,b,λ,β)	231.86	231.96	247.23	237.97	2.57	0.45	0.11 (0.03)
B-W (a,b,λ,β)	232.84	232.95	248.22	238.96	2.67	0.46	0.12 (0.01)
E-W (b,λ,β)	232.42	232.39	243.86	236.91	2.81	0.48	0.12 (0.000)
G-W (a,λ,β)	233.56	233.63	245.09	238.15	2.97	0.51	0.24 (0.000)
M-E (b)	388.70	392.55	388.71	390.24	6.49	1.09	0.23 (0.008)
W (λ,β)	231.56	231.88	239.25	234.62	2.97	0.51	0.14 (0.000)

Table 9. MLEs, standard errors (in parentheses), confidence interval values [in brackets] for data set II.

Models	$\hat{a}$	$\hat{b}$	$\hat{c}$	$\hat{\lambda}$	$\hat{\beta}$
GOLE-W (a,b,c,β)	6.9553 (6.6794) [0, 20.04]	114.6653 (50.1098) [0, 27.65]	20.1203 (6.6638) [7.06, 33.18]	-	0.8448 (0.31433) [0.23, 1.46]
Kw-W (a,b,λ,β)	1.6646 (0.8438) [0.01, 3.32]	1.0950 (0.5829) [0.23, 0.29]	-	4.3675 (2.8871) [0, 10.0262]	0.0187 (0.0192) [0, 0.0564]
B-W (a,b,λ,β)	1.7401 (1.3064) [0, 4.30]	0.9961 (0.9249) [0, 2.81]	-	4.2987 (2.0778) [0.23, 8.37]	0.0222 (0.0253) [0, 0.07]
E-W (b,λ,β)	-	1.7298 (0.7208) [0.32, 3.14]	-	4.3083 (0.9066) [2.53, 6.09]	0.9440 (0.1732) [0, 0.07]
G-W (a,λ,β)	1.5460 (0.9021) [0, 3.3142]	-	-	5.3816 (0.4906) [4.42, 6.34]	0.0033 (0.0008) [0.002, 0.005]
M-E (b)	-	0.406 (0.016) [0.37, 0.44]	-	-	-
W (λ,β)	-	-	-	0.0036 (0.0009) [0.0002, 0.0053]	5.7342 (0.2428) [5.26, 6.21]

Table 10. The AIC, CAIC, BIC, HQIC, A*, W* and KS (p-value) values for data set II.

Models	AIC	CAIC	BIC	HQIC	A*	W*	KS (p-Value)
GOLE-W (a,b,c,β)	**110.57**	**111.15**	**119.79**	**114.25**	0.25	0.031	**0.06 (0.93)**
Kw-W (a,b,λ,β)	111.06	112.83	120.08	114.98	2.27	0.037	0.08 (0.91)
B-W (a,b,λ,β)	111.32	112.90	120.13	115.99	0.26	0.038	0.07 (0.92)
E-W (b,λ,β)	118.33	118.72	122.89	120.68	0.31	0.075	0.098 (0.89)
G-W (a,λ,β)	113.84	113.13	123.02	119.15	0.30	0.052	0.08 (0.88)
M-E (b)	388.70	392.55	388.71	390.24	6.49	1.09	0.23 (0.008)
W (λ,β)	117.45	116.61	121.77	117.91	0.29	0.037	0.09 (0.87)

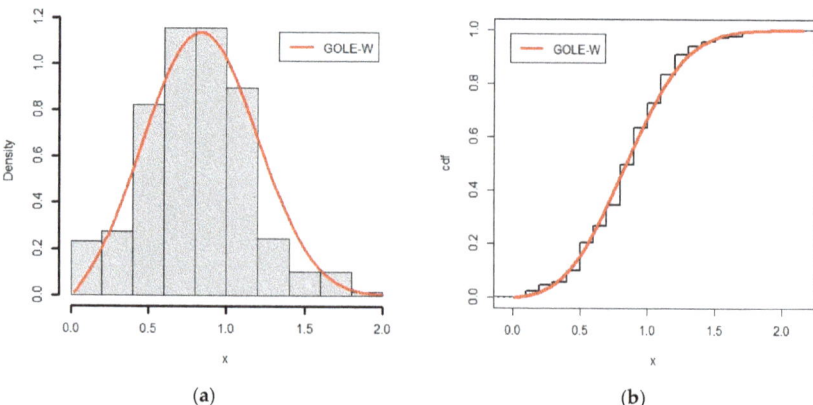

Figure 10. Plots of (**a**) the fitted PDF for the GOLE-W distribution and (**b**) estimated CDF for the GOLE-W distribution for data set I.

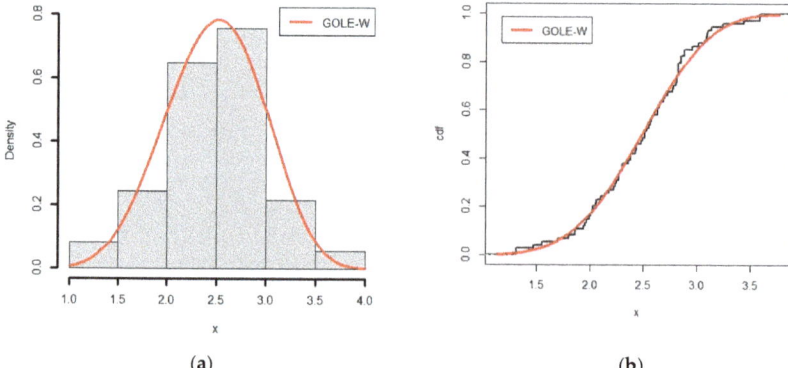

Figure 11. Plots of (**a**) the fitted PDF for the GOLE-W distribution and (**b**) estimated CDF for the GOLE-W distribution for data set II.

7. Conclusions

Through this paper, we provide a new general family of distributions to generalize any continuous baseline distribution. The main properties of the new family and other properties associated with the area of reliability are discussed. It was noted that the distributions generated by the new family are highly flexible in data modeling where we used one member to fit two real data to illustrate the importance of this family. This member provided consistently better fits than the other comparative distributions.

Author Contributions: Data curation, L.H.; Formal analysis, L.H.; Investigation, S.S.; Project administration, A.H.N.A.; Resources, S.K.; Supervision, W.M.; Writing—original draft, F.J. All authors have read and agreed to the published version of the manuscript.

Funding: This research received no external funding.

Acknowledgments: The authors are thankful to the Editor-in-Chief and the anonymous referees for their meticulous and thorough reading, which significantly enhanced the readability of this paper, and special thanks to Christophe Chesneau, Department of Mathematics, University of Caen-Normandie, LMNO, France, for their appreciable directions regarding Propositions 1 and 2.

Conflicts of Interest: The authors declare no conflict of interest.

Appendix A

Recalling Equations (3) and (4) and assigned new numbers as (A1) and (A2), respectively, as follows

$$G(x;a,b,c,\boldsymbol{\phi}) = 1 - \exp\left[-\left(\frac{aF(x;\boldsymbol{\phi})^c}{1-F(x;\boldsymbol{\phi})^c} + \frac{b}{2}\left(\frac{F(x;\boldsymbol{\phi})^c}{1-F(x;\boldsymbol{\phi})^c}\right)^2\right)\right]. \quad (A1)$$

$$g(x;a,b,c,\boldsymbol{\phi}) = \left[\frac{cf(x;\boldsymbol{\phi})F(x;\boldsymbol{\phi})^{c-1}(a+(b-a)F(x;\boldsymbol{\phi})^c)}{(1-F(x;\boldsymbol{\phi})^c)^3}\right] \times \exp\left[-\left(\frac{aF(x;\boldsymbol{\phi})^c}{1-F(x;\boldsymbol{\phi})^c} + \frac{b}{2}\left(\frac{F(x;\boldsymbol{\phi})^c}{1-F(x;\boldsymbol{\phi})^c}\right)^2\right)\right]. \quad (A2)$$

Proposition A1. *Given $x = F(x)$, by using the equivalence: $e^y \sim 1 + y$ when $y \to 0$ since $\lim_{x \to -\infty} F(x)^c \to 0$. Then, by the properties of the CDF in Equation (A1), we arrive at*

$$G(x) \sim \frac{aF(x)^c}{1-F(x)^c} + \frac{b}{2}\left(\frac{F(x)^c}{1-F(x)^c}\right)^2,$$

and, by asymptotic dominance, we obtain

$$G(x) \sim a\,F(x)^c. \quad (A3)$$

Using the same arguments, we obtain

$$g(x) \sim c\,a\,f(x)F(x)^{c-1}. \quad (A4)$$

In addition, the survival function is close to one; thus, the denominator in the hazard function is close to one. Then, using Equations (A3) and (A4), we obtain

$$h(x) \sim c\,a\,f(x)F(x)^{c-1}. \quad (A5)$$

Proposition A2. *Similarly, using the same arguments when $\lim_{x \to +\infty} F(x)^c \to 1$, we can prove that the survival function can be approximately reduced as follows*

$$1 - G(x) \sim 1 - e^{-\left(\frac{a}{1-F(x)^c} + \frac{b}{2}\left\{\frac{1}{1-F(x)^c}\right\}^2\right)}. \quad (A6)$$

Using the same arguments, we obtain

$$g(x) \sim \frac{bcf(x)}{(1-F(x)^c)^3}\, e^{-\left(\frac{a}{1-F(x)^c} + \frac{b}{2}\left\{\frac{1}{1-F(x)^c}\right\}^2\right)}. \quad (A7)$$

Using Equations (A6) and (A7), we obtain

$$h(x) \sim \frac{bcf(x)}{(1-F(x)^c)^3}.$$

This completes the proof.

References

1. Marshall, A.W. A new method for adding a parameter to a family of distributions with application to the exponential and Weibull families. *Biometrika* **1997**, *84*, 641–652. [CrossRef]
2. Gupta, R.C.; Gupta, P.L.; Gupta, R.D. Modeling failure time data by Lehman alternatives. *Commun. Stat.-Theory Methods* **1998**, *27*, 887–904. [CrossRef]

3. Eugene, N.; Lee, C.; Famoye, F. Beta-normal distribution and its applications. *Commun. Stat. Theory Methods* **2002**, *31*, 497–512. [CrossRef]
4. Cordeiro, G.M.; De Castro, M. A new family of generalized distributions. *J. Stat. Comput. Simul.* **2011**, *81*, 883–898. [CrossRef]
5. Alexander, C.; Cordeiro, G.M.; Ortega, E.M.; Sarabia, J.M. Generalized beta-generated distributions. *Comput. Stat. Data Anal.* **2012**, *56*, 1880–1897. Available online: https://EconPapers.repec.org/RePEc:eee:csdana:v:56:y:2012:i:6 (accessed on 19 May 2022). [CrossRef]
6. Alzaatreh, A.; Lee, C.; Famoye, F. A new method for generating families of continuous distributions. *Metron* **2013**, *71*, 63–79. [CrossRef]
7. Bourguignon, M.; Silva, R.B.; Cordeiro, G.M. The Weibull-G Family of Probability Distributions. *J. Data Sci.* **2014**, *12*, 53–68. Available online: http://www.jds-online.com/volume-12-number-1-january-2014 (accessed on 19 May 2022). [CrossRef]
8. Tahir, M.H.; Cordeiro, G.M.; Alizadeh, M.; Mansoor, M.; Zubair, M.; Hamedani, G.G. The odd generalized exponential family of distributions with applications. *J. Stat. Distrib. Appl.* **2015**, *2*, 1. [CrossRef]
9. Cordeiro, G.M.; Alizadeh, M.; Ortega, E.M.; Serrano, L.H.V. The Zografos-Balakrishnan odd log-logistic family of distributions: Properties and Applications. *Hacet. J. Math. Stat.* **2015**, *46*, 11781–11803. Available online: https://dergipark.org.tr/hujms/issue/43489/524407 (accessed on 19 May 2022). [CrossRef]
10. Gomes-Silva, F.; Percontini, A.; De Brito, E.; Ramos, M.W.; Venâncio, R.; Cordeiro, G.M. The Odd Lindley-G Family of Distributions. *Austrian J. Stat.* **2017**, *46*, 65–87. [CrossRef]
11. Alizadeh, M.; Cordeiro, G.M.; Pinho, L.G.B.; Ghosh, I. The Gompertz-G family of distributions. *J. Stat. Theory Pract.* **2016**, *11*, 179–207. [CrossRef]
12. Jamal, F.; Nasir, M.A.; Tahir, M.H.; Montazeri, N.H. The odd Burr-III family of distributions. *J. Stat. Appl. Probab.* **2017**, *6*, 105–122. Available online: http://www.naturalspublishing.com/files/published/4nk6g57u5e512l.pdf (accessed on 19 May 2022). [CrossRef]
13. Khan, S.; Balogun, O.S.; Tahir, M.H.; Almutiry, W.; Alahmadi, A.A. An Alternate Generalized Odd Generalized Exponential Family with Applications to Premium Data. *Symmetry* **2021**, *13*, 2064. [CrossRef]
14. Sen, A.; Bhattacharyya, G.K. Inference procedures for the linear failure rate model. *J. Stat. Plan. Inference* **1995**, *46*, 59–76. [CrossRef]
15. Shaked, M.; Shanthikumar, J.G. *Stochastic Orders and Their Applications*; Academic Press: San Diego, CA, USA, 2014. [CrossRef]
16. Kotz, S.; Pensky, M. *The Stress-Strength Model and Its Generalizations: Theory and Applications*; World Scientific: Singapore, 2003. [CrossRef]
17. Kundu, D.; Raqab, M.Z. Estimation of R = P (Y < X) for three-parameter Weibull distribution. *Stat. Probab. Lett.* **2009**, *79*, 1839–1846. [CrossRef]
18. Aarset, M.V. How to Identify a Bathtub Hazard Rate. *IEEE Trans. Reliab.* **1987**, *36*, 106–108. [CrossRef]
19. Dara, S.T.; Ahmad, M. *Recent Advances in Moment Distribution and Their Hazard Rates*; Lap Lambert Academic Publishing: Chisinau, Republic of Moldova, 2012.
20. Hasnain, S.A.; Iqbal, Z.; Ahmad, M. On exponentiated moment exponential distribution. *Pak. J. Stat.* **2015**, *31*, 267–280. Available online: https://www.statindex.org/journals/1313/31/2 (accessed on 19 May 2022).
21. Gupta, R.D.; Kundu, D. Exponentiated Exponential Family: An Alternative to Gamma and Weibull Distributions. *Biom. J.* **2001**, *43*, 117–130. [CrossRef]
22. Nadarajah, S.; Kotz, S. The beta exponential distribution. *Reliab. Eng. Syst. Saf.* **2006**, *91*, 689–697. [CrossRef]
23. Mudholkar, G.; Srivastava, D. Exponentiated Weibull family for analyzing bathtub failure-rate data. *IEEE Trans. Reliab.* **1993**, *42*, 299–302. [CrossRef]
24. Lai, C.D. Generalized Weibull Distributions. In *Generalized Weibull Distributions*; Springer Briefs in Statistics; Springer: Berlin/Heidelber, Germany, 2014. [CrossRef]
25. Lee, C.; Famoye, F.; Olumolade, O. Beta-Weibull Distribution: Some Properties and Applications to Censored Data. *J. Mod. Appl. Stat. Methods* **2007**, *6*, 173–186. [CrossRef]
26. Cordeiro, G.M.; Ortega, E.M.; Nadarajah, S. The Kumaraswamy Weibull distribution with application to failure data. *J. Frankl. Inst.* **2010**, *347*, 1399–1429. [CrossRef]

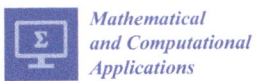

Article

Prony Method for Two-Generator Sparse Expansion Problem

Abdulmtalb Hussen [1,*] and Wenjie He [2]

[1] School of Engineering, Math and Technology Navajo Technical University, Lowerpoint Rd State Hwy 371, Crownpoint, NM 87313, USA
[2] Department of Computer Science, University of Missouri, St. Louis, MO 63121, USA; hew@umsl.edu
* Correspondence: ahussen@navajotech.edu

Abstract: In data analysis and signal processing, the recovery of structured functions from the given sampling values is a fundamental problem. Many methods generalized from the Prony method have been developed to solve this problem; however, the current research mainly deals with the functions represented in sparse expansions using a *single generating function*. In this paper, we generalize the Prony method to solve the sparse expansion problem for *two generating functions*, so that more types of functions can be recovered by Prony-type methods. The two-generator sparse expansion problem has some special properties. For example, the two sets of frequencies need to be separated from the zeros of the Prony polynomial. We propose a *two-stage least-square detection method* to solve this problem effectively.

Keywords: Prony method; exponential sums; eigenfunctions; eigenvalues; sparse expansion; generating function; Hankel matrix; short time Fourier transform; least-square method

1. Introduction

The Prony method is a popular tool used to recover the functions represented in sparse expansions using one generating function. For example, the function with the following form

$$f(x) = \sum_{j=1}^{M} c_j e^{ix\phi_j} \quad (1)$$

can be recovered from $2M$ equispaced sampling values $f(lh), l = 0, \ldots, 2M - 1$ for an appropriate positive constant h; however, in many real-world applications, we need to deal with the functions represented by more than one generating functions. For example, the *harmonic signals* with the form

$$f(x) = \sum_{j=1}^{M} \left(c_j \cos(\phi_j x) + d_j \sin(\beta_j x) \right), \quad (2)$$

are generated by two generating functions (or simply generators): $\cos(\phi x)$ and $\sin(\beta x)$, where ϕ and β are generic parameters used as the placeholders for the real parameters $\{\phi_j\}_{j=1}^{M}$ and $\{\beta_j\}_{j=1}^{M}$ to generate the specific terms in the expansion. In this system, we have two sets of coefficients $\{c_j\}_{j=1}^{M}$ and $\{d_j\}_{j=1}^{M}$ and two sets of frequencies $\{\phi_j\}_{j=1}^{M}$ and $\{\beta_j\}_{j=1}^{M}$. Analogous to the original Prony method, we expect to use $4M$ equispaced sampling values $f(lh), l = 0, \ldots, 4M - 1$ to recover those four sets of parameters.

There are some existing methods to solve this problem. The first one is to convert it to a single-generator problem by the following formulas

$$\cos x = \frac{1}{2}(e^{ix} + e^{-ix}) \quad \text{and} \quad \sin x = \frac{1}{2i}(e^{ix} - e^{-ix}),$$

which results in problem (1) (see [1]). Another way using the same idea is based on the *even/odd* properties for $\cos x$ and $\sin x$ (see [2]) as follows

$$f(x) + f(-x) = 2 \sum_{j=1}^{M} c_j \cos(\phi_j x). \tag{3}$$

However, this approach is very restrictive, because the chance that one can make this kind of conversion is very small. In this paper, we are interested in solving the *general* two-generator sparse expansion problem by a new way of generalized Prony method. More specifically, we study the functions with the following two-generator sparse expansion

$$f(x) = \sum_{j=1}^{M_1} c_j u(\phi_j x) + \sum_{l=1}^{M_2} d_l v(\beta_l x), \tag{4}$$

where $u(\phi x)$ and $v(\beta x)$ are two different functions used as the generators. In order to make the Prony method work, we need a critical condition for our special technique: *There exists a linear operator, such that $u(\phi x)$ and $v(\beta x)$ are both eigenfunctions of this operator.*

Another situation that could lead to the two-generator expansion problem is when we apply some special transforms on a sparse expansion. For example, when we apply the *short time Fourier transform* (STFT), i.e.,

$$\text{STFT}\{f(x)\}(\omega, \tau) = \int_{-\infty}^{\infty} f(x) w(x - \tau) e^{-i\omega x} dx \tag{5}$$

using the Gaussian window function $w(x) = \frac{1}{\sqrt{2\pi}} e^{-\frac{x^2}{2\sigma^2}}$ on the sparse cosine expansion

$$f(x) = \sum_{j=1}^{M} c_j \cos(\phi_j x), \tag{6}$$

we would obtain a two-generator sparse expansion as follows,

$$f(x) = \sum_{j=1}^{M} c_j e^{-\beta(\phi_j - x)^2} + \sum_{j=1}^{M} c_j e^{-\beta(\phi_j + x)^2}. \tag{7}$$

In this example, the two generators are $e^{-\beta(\phi - x)^2}$ and $e^{-\beta(\phi + x)^2}$ with $\beta \neq 0$. Actually, the original single-generator problem (6) can be solved directly. For example, one can convert $\cos(\phi x)$ to $\frac{1}{2}(e^{i\phi x} + e^{-i\phi x})$ (see [1]), or use a method based on the Chebyshev polynomials (see [3]). When we solve problem (6) directly, we use the sampling values in the time domain; when we solve the problem in the form of (7), we use the sampling values in the frequency domain. (See [4] for a discussion on sampling values in the frequency domain.) In this paper, we use this example to study the special properties of the two-generator sparse expansion problem.

Since the signals could take various forms, not necessarily in the exponential form studied in the classical Prony method, many researchers generalized the Prony method to handle different types of signals. For example, many results in [1,3,5–12] have been developed over the last few years. In particular, Peter and Plonka in [1,8] generalized the Prony method to reconstruct M-sparse expansions in terms of eigenfunctions of some special linear operators. In [3], Plonka and others reconstructed different signals by exploiting the generalized shift operator. These results provide us the building blocks for our method in this paper.

We organize our presentation in the remaining sections as follows. In Section 2, we quickly review the classical Prony method and one of its generalizations for the Gaussian generating function to establish the foundation of our method. In Section 3, we describe the details of our method using the example with two generators: cosine and sine functions. In Section 4, we apply our method on two different types of Gaussian generating functions, so that we can study an interesting property: *When the Hankel matrix for finding the coefficients of the Prony polynomial is singular, what does it really mean?* In Section 5, we show two examples that correspond to the two problems solved in Sections 3 and 4, respectively. Finally, we make conclusions in Section 6 and describe two related research problems to be solved in the future.

2. Review of the Prony Method and One of Its Generalizations

Our method is built on top of the Prony method and one of its generalizations. Before we present our technique, we review these basic methods.

2.1. Classical Prony Method

Let $f(x)$ be a function in the form of

$$f(x) = \sum_{j=1}^{M} c_j e^{-ix\phi_j} \tag{8}$$

with $M \geq 1$. Then the coefficients $\{c_j\}_1^M$ and the frequencies $\{\phi_j\}_1^M$ can be recovered from the sampling values $f(lh), l = 0, ..., 2M - 1$, where h is some positive constant. To solve this problem, a special polynomial called the Prony polynomial can help us convert the relatively hard *non-linear* problem (8) to two *linear* problems and a *simple non-linear* problem (finding zeros of a polynomial). The Prony polynomial for (8) is defined as

$$\Lambda(z) = \prod_{j=1}^{M}(z - e^{-ih\phi_j}) = \sum_{l=0}^{M} \lambda_l z^l, \tag{9}$$

where $\lambda_l, l = 0, ..., M$ are the coefficients of the monomial terms in (9) with the leading coefficient $\lambda_M = 1$. The technique is based on the following critical property:

$$\sum_{l=0}^{M} \lambda_l f(h(l+m)) = \sum_{l=0}^{M} \lambda_l \sum_{j=1}^{M} c_j e^{-ih(l+m)\phi_j} = \sum_{j=1}^{M} c_j e^{-ihm\phi_j} \underbrace{\sum_{l=0}^{M} \lambda_l e^{-ihl\phi_j}}_{=0} = 0 \tag{10}$$

for any $m = 0, 1, \ldots, M - 1$, which can be written as the following linear system

$$\left[f(h(l+m))\right]_{l,m=0}^{M-1} \begin{bmatrix} \lambda_0 \\ \vdots \\ \lambda_{M-1} \end{bmatrix} = -\begin{bmatrix} f(hM) \\ \vdots \\ f(h(2M-1)) \end{bmatrix}. \tag{11}$$

The coefficient vector $\boldsymbol{\lambda} = [\lambda_0, \lambda_1, \ldots, \lambda_{M-1}]^T$ can be calculated from the $2M$ sampling values $f(lh), l = 0, ..., 2M - 1$. The linear system (11) is guaranteed to have a unique solution under the condition that all ϕ_j's are distinct in $(-K, K) \subset \mathbb{R}$ for some $K > 0$ (with h in the range $0 < h < \frac{\pi}{K}$), and $c_1, \ldots, c_M$ are nonzero in $\mathbb{C}$, which is a natural requirement for problem (8). This property is a direct result of the following matrix factorization

$$\left[f(h(l+m))\right]_{l,m=0}^{M-1} = \boldsymbol{V}^T \text{diag}(c_1, ..., c_M) \boldsymbol{V}, \tag{12}$$

where $V := [e^{-ilh\phi_j}]_{l=0,j=1}^{l=M-1,j=M}$ is a Vandermonde matrix, which is non-singular for distinct ϕ_j's and $h\phi_j \in (-\pi, \pi]$ for $j = 1, ..., M$. The frequencies can be extracted from the zeros of $\Lambda(z)$ (in the form of $z_j = e^{-ih\phi_j}$) using the formula

$$\phi_j = \frac{-\text{Im}(\ln(z_j))}{h}, \quad j = 1, ..., M. \tag{13}$$

Finally, the coefficients $c_j, j = 1, ..., M$ can be determined by solving the following *overdetermined* linear system (with M unknowns and $2M$ equations)

$$f(lh) = \sum_{j=1}^{M} c_j e^{-ilh\phi_j}, \quad l = 0, ..., 2M-1. \tag{14}$$

The redundant equations in this overdetermined linear system will play a critical role in our two-generator method to help us separate the frequencies associated with the two generators (see Section 3).

2.2. Sparse Expansions on Shifted Gaussian

In order to solve the two-generator sparse expansion problem (7), we need to apply the technique presented in [3], which solves a single-generator sparse expansion problem with the following form

$$f(x) = \sum_{j=1}^{M} c_j e^{-\beta(x-\phi_j)^2}, \tag{15}$$

where $\beta \in \mathbb{C} \setminus \{0\}$. The technique relies on the following generalized shift operator

$$S_{K,h} f(x) = K(x,h) f(x+h), \tag{16}$$

where $h \neq 0$, and $K(\cdot, \cdot)$ has the property

$$K(x, h_1 + h_2) = K(x, h_1) K(x+h_1, h_2) = K(x, h_2) K(x+h_2, h_1).$$

The $K(x,h)$ function in (16) is chosen to be $e^{\beta h(2x+h)}$, so that we have the following critical property

$$(S_{K,h} e^{-\beta(\phi-\cdot)^2})(x) = e^{2\beta\phi h} e^{-\beta(\phi-x)^2}, \tag{17}$$

which means that $e^{-\beta(\phi_j-x)^2}$'s are eigenfunctions of $S_{K,h}$ for all $\phi_j \in \mathbb{R}$.

The sparse expansion $f(x)$ in (15) can be reconstructed using $2M$ sampling values $f(x_0 + hk), k = 0, ..., 2M-1$, and x_0 is an arbitrary real number. If $\text{Re}\,\beta \neq 0$, then $h \in \mathbb{R} \setminus \{0\}$; while if $\text{Re}\,\beta = 0$, then $0 < h \leq \frac{\pi}{2|\text{Im}\,\beta|L}$ with $\phi_j \in (-L, L)$ for $j = 1, ..., M$ for some given L. (See [3].) The Prony polynomial for the problem in (15) can be defined as:

$$\Lambda(z) := \prod_{j=1}^{M} (z - e^{2h\beta\phi_j}) = \sum_{l=0}^{M} \lambda_l z^l \tag{18}$$

with $\lambda_M = 1$. Then, we have the following linear system

$$\sum_{l=0}^{M-1} \lambda_l e^{\beta h(l+m)(2x_0 + h(l+m))} f(x_0 + h(l+m)) = -e^{\beta h(m+M)(2x_0 + h(m+M))} f(x_0 + h(m+M)) \tag{19}$$

for $m = 0, 1, ..., M-1$, which can be represented as an inhomogeneous system

$$H\lambda = -G, \tag{20}$$

where $G := [(\mathcal{S}_{K,(M+m)h} f)(x_0)]_{m=0}^{M-1}$, and $H := [(\mathcal{S}_{K,(l+m)h} f)(x_0 + (l+m)h)]_{l,m=0}^{M-1}$. This H matrix is a Hankel matrix, and it has the following structure

$$H := [(\mathcal{S}_{K,(l+m)h} f)(x_0 + (l+m)h)]_{l,m=0}^{M-1} = \left[K(x_0, (l+m)h) f(x_0 + (l+m)h)\right]_{l,m=0}^{M-1} \quad (21)$$

$$= V \mathrm{diag}(c_j e^{-\beta(\phi_j - x_0)^2}) V^T,$$

with the Vandermonde matrix

$$V := \begin{bmatrix} 1 & 1 & \cdots & 1 \\ e^{2\beta h \phi_1} & e^{2\beta h \phi_2} & \cdots & e^{2\beta h \phi_M} \\ \vdots & \vdots & \cdots & \vdots \\ e^{2(M-1)\beta h \phi_1} & e^{2(M-1)\beta h \phi_2} & \cdots & e^{2(M-1)\beta h \phi_M} \end{bmatrix}.$$

Thus, H is invertible for distinct ϕ_j's in $(-L, L) \subset \mathbb{R}$ for $L > 0$, and the vector of the coefficients $\lambda := [\lambda_0, ..., \lambda_{M-1}]^T$ are obtained by solving the system (20), which can be used to calculate the parameters $\{\phi_j\}$'s.

Finally, the coefficients c_j's in the expansion (15) can be computed by solving the following overdetermined linear system:

$$f(x_0 + lh) = \sum_{j=1}^{M} c_j e^{-\beta(x_0 - \phi_j + lh)^2}, \quad l = 0, ..., 2M - 1. \quad (22)$$

3. The Sparse Expansion Problem with Two Generators: Cosine and Sine

In this section, we present our method for solving the two-generator sparse expansion problem in the following form

$$f(x) = \sum_{j=1}^{M_1} c_j \cos(\phi_j x) + \sum_{l=1}^{M_2} d_l \sin(\beta_l x) \quad (23)$$

through a modified Prony method. We present our method in the following theorem.

Theorem 1. *Assume that a function $f(x)$ has the two-generator sparse expansion form of (23), where the number of terms for two generators M_1 and M_2 are known, but the two sets of coefficients in $\{c_1, ..., c_{M_1}\}$ and $\{d_1, ..., d_{M_2}\}$ and the two sets of frequencies in $\{\phi_1, ..., \phi_{M_1}\}$ and $\{\beta_1, ..., \beta_{M_2}\}$ are unknown. If $4(M_1 + M_2) - 1$ equispaced sampling values of the form $f(x_0 + kh)$ for $k = -2(M_1 + M_2) + 1, ..., -1, 0, 1, ..., 2(M_1 + M_2) - 1$ are provided, then the original function $f(x)$ can be uniquely reconstructed under the following conditions:*
1° *All the coefficients $\{c_1, ..., c_{M_1}, d_1, ..., d_{M_2}\}$ are nonzero in $\mathbb{C}$.*
2° *All the frequencies $\{\phi_1, ..., \phi_{M_1}, \beta_1, ..., \beta_{M_2}\}$ are distinct in a range $[0, K) \subset \mathbb{R}$ for some $K > 0$. Furthermore, h is selected from the range $0 < h < \frac{\pi}{K}$.*
3° *The value of $x_0 \in \mathbb{R}$ is selected to make the $(M_1 + M_2)$ numbers $\cos(\phi_1 x_0), ..., \cos(\phi_{M_1} x_0)$, $\sin(\beta_1 x_0), ..., \sin(\beta_{M_2} x_0)$ nonzero.*

Proof. First, we choose an appropriate linear operator, such that our two generating functions $\cos(\phi x)$ and $\sin(\beta x)$ in (23) are both the eigenfunctions of this operator. We consider the *symmetric shift operator* (see [3])

$$\mathcal{S}_{h,-h} f(x) := \left(\frac{\mathcal{S}_{-h} + \mathcal{S}_h}{2}\right) f(x) = \frac{f(x-h) + f(x+h)}{2}. \quad (24)$$

When we apply this operator on $\cos(\phi x)$ and $\sin(\beta x)$, we obtain

$$(\mathcal{S}_{h,-h}) \cos(\phi x) = \cos(\phi h) \cos(\phi x),$$
$$(\mathcal{S}_{h,-h}) \sin(\beta x) = \cos(\beta h) \sin(\beta x), \tag{25}$$

where $\cos(\phi h)$ and $\cos(\beta h)$ are the eigenvalues. Now we define the Prony polynomial for problem (23) using all the eigenvalues $\{\cos(\phi_j h)\}_{j=1}^{M_1}$ and $\{\cos(\beta_l h)\}_{l=1}^{M_2}$ as follows:

$$\Lambda(z) = \prod_{j=1}^{M_1}(z - \cos(h\phi_j)) \prod_{l=1}^{M_2}(z - \cos(h\beta_l)), \tag{26}$$

which can be written in terms of the Chebyshev polynomials as

$$\Lambda(z) = \sum_{k=0}^{M_1+M_2} \lambda_k T_k(z), \tag{27}$$

where $T_k(z) := \cos(k \cos^{-1}(z))$. (See [3] for more information on this technique.) Since the leading coefficient of the Chebyshev polynomial $T_k(z)$ is 2^{k-1}, we choose $\lambda_{M_1+M_2} = 2^{1-(M_1+M_2)}$, so that $\Lambda(z)$ in (27) has the leading coefficient 1. This Prony polynomial has the following critical property:

$$\sum_{k=0}^{M_1+M_2} \lambda_k T_k(\cos(\phi_j h)) = 0 \quad \text{and} \quad \sum_{k=0}^{M_1+M_2} \lambda_k T_k(\cos(\beta_l h)) = 0$$

for $j = 1, 2, \ldots, M_1$ and $l = 1, 2, \ldots, M_2$, respectively, which is essential to help us derive the following linear system.

To derive a linear system for $\{\lambda_k\}_{k=0}^{M_1+M_2-1}$, we need to calculate the following expression

$$\sum_{k=0}^{M_1+M_2} \lambda_k \left(\mathcal{S}_{kh,-kh} \mathcal{S}_{mh,-mh} f(x_0) \right),$$

which can be shown to be zero. That is,

$$\frac{1}{4} \sum_{k=0}^{M_1+M_2} \lambda_k \left(f(x_0 + (m+k)h) + f(x_0 - (m+k)h) + f(x_0 + (m-k)h) + f(x_0 - (m-k)h) \right) = 0 \tag{28}$$

for $m = 0, 1, \ldots, M_1 + M_2 - 1$. Indeed, using the right-hand-side expression in (23) for $f(x)$ in (28) and for a fixed $m \in \{0, 1, \ldots, M_1 + M_2 - 1\}$, we obtaining

$$\frac{1}{4} \sum_{k=0}^{M_1+M_2} \lambda_k \left[\sum_{j=1}^{M_1} 2c_j \left(\cos(\phi_j(x_0 + mh)) + \cos(\phi_j(x_0 - mh)) \right) \cos(\phi_j kh) \right]$$
$$+ \frac{1}{4} \sum_{k=0}^{M_1+M_2} \lambda_k \left[\sum_{l=1}^{M_2} 2d_l \left(\sin(\beta_l(x_0 + mh)) + \sin(\beta_l(x_0 - mh)) \right) \cos(\beta_l kh) \right]$$
$$= \sum_{j=1}^{M_1} c_j \cos(\phi_j x_0) \cos(\phi_j mh) \left(\sum_{k=0}^{M_1+M_2} \lambda_k \cos(\phi_j kh) \right) + \sum_{l=1}^{M_2} d_l \sin(\beta_l x_0) \cos(\beta_l mh) \left(\sum_{k=0}^{M_1+M_2} \lambda_k \cos \beta_l(kh) \right)$$
$$= \sum_{j=1}^{M_1} c_j \cos(\phi_j x_0) \cos(\phi_j mh) \underbrace{\left(\sum_{k=0}^{M_1+M_2} \lambda_k T_k(\cos(\phi_j h)) \right)}_{=0} + \sum_{l=1}^{M_2} d_l \sin(\beta_l x_0) \cos(\beta_l mh) \underbrace{\left(\sum_{k=0}^{M_1+M_2} \lambda_k T_k(\cos(\beta_l h)) \right)}_{=0}$$
$$= 0.$$

We can reformulate the system (28) as

$$\sum_{k=0}^{(M_1+M_2)-1} \lambda_k \Big(f(x_0 + (m+k)h) + f(x_0 - (m+k)h) + f(x_0 + (m-k)h) + f(x_0 - (m-k)h) \Big)$$
$$= -2^{1-(M_1+M_2)} \Big(f(x_0 + ((M_1+M_2)+m)h) + f(x_0 - ((M_1+M_2)+m)h) \qquad (29)$$
$$+ f(x_0 + ((M_1+M_2)-m)h) + f(x_0 - ((M_1+M_2)-m)h) \Big)$$

for $m = 0, 1, \ldots, M_1 + M_2 - 1$. To solve this system, we need $4(M_1 + M_2) - 1$ sampling values in the form of $f(x_0 + kh)$ for $k = -2(M_1 + M_2) + 1, \ldots, -1, 0, 1, \ldots, 2(M_1 + M_2) - 1$.

In order to see that the linear system in (29) has a unique solution, we study the $(M_1 + M_2) \times (M_1 + M_2)$ coefficient matrix in (29), which we denote as H. As in the classical Prony method, we can factorize H in the following structure

$$H := \Big[f(x_0 + (m+k)h) + f(x_0 - (m+k)h) + f(x_0 + (m-k)h) + f(x_0 - (m-k)h) \Big]_{m,k=0}^{(M_1+M_2)-1}$$
$$= 4 \Bigg[\sum_{j=1}^{M_1} c_j \cos(\phi_j x_0) \cos(\phi_j m h) \cos(\phi_j k h) + \sum_{l=1}^{M_2} d_l \sin(\beta_l x_0) \cos(\beta_l m h) \cos(\beta_l k h) \Bigg]_{m,k=0}^{(M_1+M_2)-1}$$
$$= 4 V_h D V_h^T,$$

where the Vandermonde Block matrix V_h can be written as

$$V_h := \begin{bmatrix} A & | & B \end{bmatrix}, \qquad (30)$$

with

$$A := \begin{bmatrix} 1 & \cdots & 1 \\ T_1(\cos \phi_1 h) & \cdots & T_1(\cos \phi_{M_1} h) \\ \vdots & \cdots & \vdots \\ T_{(M_1+M_2)-1}(\cos \phi_1 h) & \cdots & T_{(M_1+M_2)-1}(\cos \phi_{M_1} h) \end{bmatrix}_{(M_1+M_2) \times M_1} \qquad (31)$$

and

$$B := \begin{bmatrix} 1 & \cdots & 1 \\ T_1(\cos \beta_1 h) & \cdots & T_1(\cos \beta_{M_2} h) \\ \vdots & \cdots & \vdots \\ T_{(M_1+M_2)-1}(\cos \beta_1 h) & \cdots & T_{(M_1+M_2)-1}(\cos \beta_{M_2} h) \end{bmatrix}_{(M_1+M_2) \times M_2}, \qquad (32)$$

and the diagonal block matrix D can be written as

$$D := \begin{bmatrix} D1 & 0 \\ 0 & D2 \end{bmatrix} \qquad (33)$$

where

$$D1 := \begin{bmatrix} c_1 \cos(\phi_1 x_0) & & \\ & \ddots & \\ & & c_{M_1} \cos(\phi_{M_1} x_0) \end{bmatrix} \qquad (34)$$

and
$$D2 := \begin{bmatrix} d_1 \sin(\beta_1 x_0) & & \\ & \ddots & \\ & & d_{M_2} \sin(\beta_{M_2} x_0) \end{bmatrix}. \tag{35}$$

Thus, H is guaranteed to be invertible by the conditions 2° and 3° of the theorem. Then, we can find the unique solution for $\{\lambda_k\}_{k=0}^{M_1+M_2-1}$ from the linear system (29).

With these λ_k values for $\Lambda(z)$ as in (26), we can determine ϕ_j's and β_l's from the zeros of $\Lambda(z)$; however, this step is non-trivial, because we do not know what zeros correspond to ϕ_j's and what zeros correspond to β_l's. In order to resolve this ambiguity, we consider all the possible cases: Among $M_1 + M_2$ zeros of $\Lambda(z)$, M_1 of them correspond to ϕ_j's. Thus, there are a total $\binom{M_1+M_2}{M_1}$ possible choices for ϕ_j's, among which there is exactly one choice for the solution; however, how do we select the right one? We need to go to the next *overdetermined* linear system for the answer.

When we determine the coefficients c_j's and d_l's in (23), we have the following linear system

$$f(x_0 + hn) = \sum_{j=1}^{M_1} c_j \cos(\phi_j(x_0 + hn)) + \sum_{l=1}^{M_2} d_l \sin(\beta_l(x_0 + hn)) \tag{36}$$

for $n = -2(M_1 + M_2) + 1, \ldots, -1, 0, \ldots, 2(M_1 + M_2) - 1$ corresponding to all the sampling values, which has $4(M_1 + M_2) - 1$ equations and $M_1 + M_2$ unknowns. This *overdetermined* linear system gives us the extra information we need to select the true-solution case from the remaining non-solution cases.

Our method is based on an observation: The sampling values $\{f(x_0 + nh)\}_{n=-2(M_1+M_2)+1}^{2(M_1+M_2)-1}$ are calculated using the original ϕ_j's and β_l's (corresponding to the true-solution case), which means that all the $4(M_1 + M_2) - 1$ equations in (36) are completely satisfied for the true-solution case. In other words, the least-square solution of (36) for the true-solution case should have this property: Its error term is zero *theoretically* (or very close to zero due to rounding errors in computation). While the least-square solution for any non-solution case would have a *significant* (with respect to the rounding errors) nonzero error term, which makes the true solution stand out clearly. □

Our experiments have verified this phenomenon. Based on this observation, we develop a *two-stage least-square detection* method to minimize the computing cost, and in Section 5, we demonstrate the effectiveness of this method using a simple example.

Remark 1. *The overdetermined linear system (36) plays an important role in determining the true solution from certain number of possible cases. Typically, this situation happens in the multi-generator sparse expansion problem. For the single-generator case, we can select same number of linearly independent equations from the overdetermined system as the number of unknowns to find the solution; however, for the multi-generator case, the redundant equations are very useful in the least-square method.*

4. The Sparse Expansion Problem with Two Gaussian Generators

In this section, we solve another two-generator sparse expansion problem as in (7) that uses the two Gaussian generating functions, $e^{-\beta(\phi-x)^2}$ and $e^{-\beta(\phi+x)^2}$, in the form of

$$f(x) = \sum_{j=1}^{M} c_j e^{-\beta(\phi_j-x)^2} + \sum_{j=1}^{M} c_j e^{-\beta(\phi_j+x)^2} \tag{37}$$

for some constant $\beta \in \mathbb{C}\setminus\{0\}$. In order to recover the coefficients $c_j \in \mathbb{C}\setminus\{0\}$ and the parameters ϕ_j's, we need $4M$ sampling values $f(x_0 + kh), k = 0, \ldots, 4M - 1$, where $x_0 \in \mathbb{R}$, and h satisfies the same condition as in Section 2.2.

This two-generator sparse expansion problem has a special property: When $\phi_{j_0} = 0$ for some $j_0 \in \{1, ..., M\}$, the two functions $e^{-\beta(\phi_{j_0} - x)^2}$ and $e^{-\beta(\phi_{j_0}+x)^2}$ are the same. This property would cause some problem for our method presented in the previous section. In order to make the discussion easier, we separate these two cases, and consider the case that $\phi_j \in \mathbb{R} \setminus \{0\}$ for all $j = 1, ..., M$ first.

Theorem 2. *Assume that a function $f(x)$ has the two-generator sparse expansion form of* (37), *where the number of terms M and the constant $\beta \in \mathbb{C} \setminus \{0\}$ are known, but the coefficients in $\{c_1, ..., c_M\}$ and the parameters in $\{\phi_1, ..., \phi_M\}$ are unknown. If $4M$ equispaced sampling values of the form $f(x_0 + kh)$ for $k = 0, 1, ..., 4M - 1$ are provided, then the original function $f(x)$ can be uniquely reconstructed under the following conditions:*

1° *The coefficients $\{c_1, ..., c_M\}$ are nonzero in $\mathbb{C}$.*
2° *The parameters $\{\phi_1, ..., \phi_M\}$ are nonzero in $(-L, L) \subset \mathbb{R}$ for some $L > 0$, and they are distinct.*
3° *If $\operatorname{Re} \beta \neq 0$, then $h \in \mathbb{R} \setminus \{0\}$; while if $\operatorname{Re} \beta = 0$, then $0 < h \leq \frac{\pi}{2|\operatorname{Im}\beta|L}$.*

Proof. Our method relies on existence of some *critical linear operator*, such that both generating functions are its eigenfunctions. Here we use the operator $\mathcal{S}_{K,h}$ as defined in (16) with $K(x, h) := e^{\beta h(2x+h)}$, which has the following properties:

$$(\mathcal{S}_{K,h} e^{-\beta(\phi - \cdot)^2})(x) = e^{2\beta\phi h} e^{-\beta(\phi - x)^2}, \tag{38}$$
$$(\mathcal{S}_{K,h} e^{-\beta(\phi + \cdot)^2})(x) = e^{-2\beta\phi h} e^{-\beta(\phi + x)^2}.$$

Clearly $e^{-\beta(\phi_j - \cdot)^2}$ and $e^{-\beta(\phi_j + \cdot)^2}$ are eigenfunctions of $\mathcal{S}_{K,h}$ for all $\phi_j \in \mathbb{R} \setminus \{0\}$ with corresponding eigenvalues $e^{2\beta\phi_j h}$ and $e^{-2\beta\phi_j h}$, respectively, for $j = 1, ..., M$. Hence we can define the Prony polynomial using all these eigenvalues:

$$\Lambda(z) = \prod_{j=1}^{M}(z - e^{2h\beta\phi_j}) \prod_{j=1}^{M}(z - e^{-2h\beta\phi_j}) = \sum_{l=0}^{2M} \lambda_l z^l \tag{39}$$

with $\lambda_{2M} = 1$. Since the real number $\phi_j \neq 0$, we can assume that $\phi_j > 0$ for all $j = 1, ..., M$ based on the structure in (37) to improve the certainty without loss of generality.

Then for $m = 0, 1, ..., 2M - 1$, we calculate

$$\sum_{l=0}^{2M} \lambda_l (\mathcal{S}_{K,(l+m)h} f)(x_0) = \sum_{l=0}^{2M} \lambda_l e^{\beta h(l+m)(2x_0 + h(l+m))} f(x_0 + h(l+m))$$
$$= \sum_{l=0}^{2M} \lambda_l e^{\beta h(l+m)(2x_0+h(l+m))} \sum_{j=1}^{M} c_j e^{-\beta(\phi_j - (x_0+h(l+m)))^2} + \sum_{l=0}^{2M} \lambda_l e^{\beta h(l+m)(2x_0+h(l+m))} \sum_{j=1}^{M} c_j e^{-\beta(\phi_j + (x_0+h(l+m)))^2}$$
$$= \left(\sum_{j=1}^{M} c_j e^{-\beta(x_0+hm-\phi_j)^2} e^{\beta hm(2x_0+hm)} \right) \underbrace{\left(\sum_{l=0}^{2M} \lambda_l e^{2\beta h l \phi_j} \right)}_{=0} + \left(\sum_{j=1}^{M} c_j e^{-\beta(x_0+hm+\phi_j)^2} e^{\beta hm(2x_0+hm)} \right) \underbrace{\left(\sum_{l=0}^{2M} \lambda_l e^{-2\beta h l \phi_j} \right)}_{=0} = 0,$$

which can be written as the following linear system

$$\sum_{l=0}^{2M-1} \lambda_l e^{\beta h(l+m)(2x_0+h(l+m))} f(x_0 + h(l+m)) = -e^{\beta h(m+2M)(2x_0+h(m+2M))} f(x_0 + h(m+2M)) \tag{40}$$

for $m = 0, 1, ..., 2M − 1$. To solve this system, we need $4M$ sampling values: $f(x_0 + kh)$ for $= 0, 1, ..., 4M − 1$. To study existence of the solution for this linear system, we would like to simplify it with respect to the unknown vector $\lambda := [\lambda_0, ..., \lambda_{2M-1}]^T$ as follows,

$$H\lambda = -G, \tag{41}$$

with $G := [(\mathcal{S}_{K,(M+m)h} f)(x_0)]_{m=0}^{2M-1}$ and

$$H := [(\mathcal{S}_{K,(l+m)h} f)(x_0)]_{l,m=0}^{2M-1}. \tag{42}$$

The invertibility of H can be seen from the following matrix factorization:

$$\begin{aligned}
H &= \left[K(x_0, h(l+m)) f(x_0 + h(l+m)) \right]_{l,m=0}^{2M-1} \\
&= \left[\sum_{j=1}^{M} c_j e^{\beta h(l+m)(2x_0 + h(l+m))} e^{-\beta(\phi_j - (x_0 + h(l+m)))^2} + \sum_{j=1}^{M} c_j e^{\beta h(l+m)(2x_0 + h(l+m))} e^{-\beta(\phi_j + (x_0 + h(l+m)))^2} \right]_{l,m=0}^{2M-1} \\
&= \left[\sum_{j=1}^{M} c_j e^{-\beta(\phi_j - x_0)^2} e^{2\beta h(l+m)\phi_j} + \sum_{j=1}^{M} c_j e^{-\beta(\phi_j + x_0)^2} e^{-2\beta h(l+m)\phi_j} \right]_{l,m=0}^{2M-1} \\
&= V_h \operatorname{diag}\left(c_j e^{-\beta(\phi_j - x_0)^2} + c_j e^{-\beta(\phi_j + x_0)^2} \right) V_h^T \\
&= V_h D V_h^T
\end{aligned} \tag{43}$$

where the Vandermonde block matrix V_h has the following form

$$V_h := \begin{bmatrix} A & | & B \end{bmatrix} \tag{44}$$

with

$$A := \begin{bmatrix} 1 & \cdots & 1 \\ e^{2\beta h \phi_1} & \cdots & e^{2\beta h \phi_M} \\ \vdots & \cdots & \vdots \\ e^{2(2M-1)\beta h \phi_1} & \cdots & e^{2(2M-1)\beta h \phi_M} \end{bmatrix}_{(2M) \times M} \tag{45}$$

and

$$B := \begin{bmatrix} 1 & \cdots & 1 \\ e^{-2\beta h \phi_1} & \cdots & e^{-2\beta h \phi_M} \\ \vdots & \cdots & \vdots \\ e^{-2(2M-1)\beta h \phi_1} & \cdots & e^{-2(2M-1)\beta h \phi_M} \end{bmatrix}_{(2M) \times M}, \tag{46}$$

and the diagonal block matrix D is given by

$$D := \begin{bmatrix} D1 & 0 \\ 0 & D2 \end{bmatrix} \tag{47}$$

with

$$D1 := \begin{bmatrix} c_1 e^{\beta(\phi_1 - x_0)^2} & & \\ & \ddots & \\ & & c_M e^{\beta(\phi_M - x_0)^2} \end{bmatrix} \tag{48}$$

and
$$D2 := \begin{bmatrix} c_1 e^{\beta(\phi_1+x_0)^2} & & \\ & \ddots & \\ & & c_M e^{\beta(\phi_M+x_0)^2} \end{bmatrix}. \tag{49}$$

From this structure, we can see that the Vandermonde matrix V_h in (44) is invertible by conditions 2° and 3° of the theorem, and hence H in (42) is also invertible by condition 1°, which results in the unique solution for λ.

With all the λ_l values found from the above linear system, we can find all the ϕ_j values by calculating the zeros of the Prony polynomial of (39). In this case, we do not need to deal with the ambiguity that we encountered in the previous section due to the special structure of the pairs $(\phi_j, -\phi_j)$'s. Finally, the coefficients c_j's of the sparse expansion (37) can be computed by solving the following *overdetermined* linear system:

$$f(x_0 + lh) = \sum_{j=1}^{M} c_j \left(e^{-\beta(\phi_j + x_0 - lh)^2} + e^{-\beta(\phi_j + x_0 + lh)^2} \right) \tag{50}$$

for $l = 0, \ldots, 4M - 1$. □

Remark 2. *Our method above only works for the case when $\phi_j \neq 0$ for all j in $\{1, 2, \ldots, M\}$; however, in the real-world situation, when we solve a problem of (37) using $4M$ sampling values, how do we know if there exists any $\phi_j = 0$ in it or not? We need a detection method to tell us if all the ϕ_j's are nonzero before we apply the above method.*

Let us investigate the existence of a solution for the linear system (41), which is determined by the invertibility of H in (42). We notice that when $\phi_1 = 0$, the first column of (45) and the first column of (46) are the same, which causes the matrix V_h in (44) to be singular. Then, we conclude that H in (43) is singular if any $\phi_j = 0$. In other words, by checking the invertibility of H, we can tell if there is any $\phi_j = 0$ for problem (37). If H in (42) is singular, our current method does not work. Fortunately we can modify our method to solve the problem for this special situation.

Let us assume that $\phi_0 = 0$, and the remaining ϕ_j's are positive numbers. In this case, we modify (37) to

$$f(x) = c_0 e^{-\beta x^2} + \sum_{j=1}^{M} c_j e^{-\beta(\phi_j - x)^2} + \sum_{j=1}^{M} c_j e^{-\beta(\phi_j + x)^2}, \tag{51}$$

and its corresponding Prony polynomial is defined as

$$\Lambda(z) = (z-1) \prod_{j=1}^{M}(z - e^{2h\beta\phi_j}) \prod_{j=1}^{M}(z - e^{-2h\beta\phi_j}) = \sum_{l=0}^{2M+1} \lambda_l z^l \tag{52}$$

with $\lambda_{2M+1} = 1$. Since $\Lambda(1) = 0$, it leads to

$$\sum_{l=0}^{2M+1} \lambda_l = 0. \tag{53}$$

Then we can show that

$$\sum_{l=0}^{2M+1} \lambda_l (\mathcal{S}_{K,(l+m)h} f)(x_0) = 0, \quad \text{for } m = 0, 1, \ldots, 2M, \tag{54}$$

because we can split the above left-hand-side summation into the following three summations with zero value each:

$$\sum_{l=0}^{2M+1} \lambda_l e^{\beta h(l+m)(2x_0+h(l+m))} c_0 e^{-\beta(x_0+h(l+m))^2} = c_0 e^{-\beta x_0^2} \underbrace{\sum_{l=0}^{2M+1} \lambda_l}_{=0} = 0,$$

$$\sum_{l=0}^{2M+1} \lambda_l e^{\beta h(l+m)(2x_0+h(l+m))} \sum_{j=1}^{M} c_j e^{-\beta(\phi_j-(x_0+h(l+m)))^2} = \left(\sum_{j=1}^{M} c_j e^{-\beta(x_0+hm-\phi_j)^2} e^{\beta hm(2x_0+hm)}\right) \underbrace{\left(\sum_{l=0}^{2M+1} \lambda_l e^{2\beta hl\phi_j}\right)}_{=0} = 0,$$

and

$$\sum_{l=0}^{2M+1} \lambda_l e^{\beta h(l+m)(2x_0+h(l+m))} \sum_{j=1}^{M} c_j e^{-\beta(\phi_j+(x_0+h(l+m)))^2} = \left(\sum_{j=1}^{M} c_j e^{-\beta(x_0+hm+\phi_j)^2} e^{\beta hm(2x_0+hm)}\right) \underbrace{\left(\sum_{l=0}^{2M+1} \lambda_l e^{-2\beta hl\phi_j}\right)}_{=0} = 0.$$

The linear system (54) for $\lambda := [\lambda_0, ..., \lambda_{2M}]^T$ can be written as

$$H\lambda = -G, \tag{55}$$

with $G := [(\mathcal{S}_{K,(2M+m+1)h} f)(x_0)]_{m=0}^{2M}$ and

$$H := [(\mathcal{S}_{K,(l+m)h} f)(x_0)]_{l,m=0}^{2M}. \tag{56}$$

We use $(4M+2)$ sampling values: $f(x_0 + kh)$ for $k = 0, 1, \ldots, 4M+1$ to solve the system. Similar to (43), we still have

$$H = V_h D V_h^T,$$

but we need to modify V_h to

$$\begin{bmatrix} 1 & 1 & \cdots & 1 & 1 & \cdots & 1 \\ 1 & e^{2\beta h\phi_1} & \cdots & e^{2\beta h\phi_M} & e^{-2\beta h\phi_1} & \cdots & e^{-2\beta h\phi_M} \\ \vdots & \vdots & \cdots & \vdots & \vdots & \cdots & \vdots \\ 1 & e^{4M\beta h\phi_1} & \cdots & e^{4M\beta h\phi_M} & e^{-4M\beta h\phi_1} & \cdots & e^{-4M\beta h\phi_M} \end{bmatrix}_{(2M+1)\times(2M+1)},$$

which is invertible for positive distinct $\{\phi_1, \ldots, \phi_M\} \subset (0, L)$, and the diagonal block matrix D becomes

$$D = \begin{bmatrix} c_0 e^{-\beta x_0^2} & 0 & 0 \\ 0 & D1 & 0 \\ 0 & 0 & D2 \end{bmatrix}$$

with $D1$ and $D2$ maintaining the same forms of (48) and (49), respectively.

After we solve the linear system of (55), we obtain the Prony polynomial that contains one zero at $z = 1$ and the remaining zeros appear in pairs of (z_j, z_j^{-1})'s, which correspond to the parameter values 0 and $(\phi_j, -\phi_j)$ pairs. Finally, we will solve the following overdetermined linear system for $c_0, c_1, \ldots, c_M$ values

$$f(x_0 + lh) = c_0 e^{-\beta(x_0-lh)^2} + \sum_{j=1}^{M} c_j \left(e^{-\beta(\phi_j+x_0-lh)^2} + e^{-\beta(\phi_j+x_0+lh)^2}\right) \tag{57}$$

for $l = 0, \ldots, 4M+1$. From this example, we can see that the value of $\det(H)$ can give us some important information, that is, which of the two systems in (37) and (51) we should work on. This property could be useful when we consider a problem in which the M value in (51) is unknown, but restricted in certain range. (See discussion in Section 6).

5. Numerical Experiments

In this section, we use two simple examples to illustrate the implementation details of our method for the two-generator sparse expansion problem described in the previous sections. The first example is for version (23) in Section 3. The second example is for version (37) in Section 4.

Example 1. *We consider a function $f(x)$ (see Figure 1) that is a two-generator expansion with each generator producing 5 terms in the following form*

$$f(x) = \sum_{j=1}^{5} c_j \cos(\phi_j x) + \sum_{j=1}^{5} d_j \sin(\beta_j x), \quad (58)$$

and the 20 parameters we used are listed in the table below to generate the sampling values.

How to use the 39 equispaced sampling values (where 39 comes from $4(5+5) - 1$) in the form of $f(x_0 + kh), k = -19, \ldots, 0, \ldots, 19$ to recover the original parameters in Table 1?

Table 1. Original parameters of the function $f(x)$ in (58).

j	c_j	d_j	ϕ_j	β_j
1	-2	5	2	3
2	3	-6	4	5
3	-4	4	7	6
4	8	-3	8	9
5	7	2	10	11

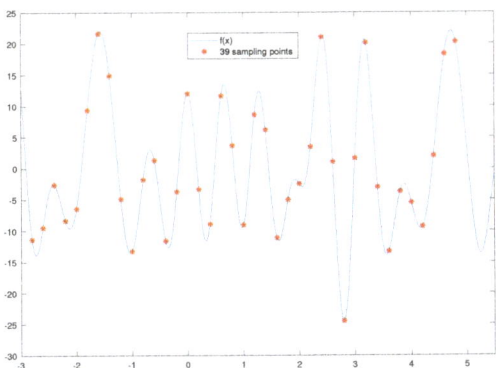

Figure 1. The signal $f(x)$ in (58) with 39 equispaced sampling values.

There are 20 original parameters in two sets: $\{c_1, \ldots, c_5, \phi_1, \ldots, \phi_5\}$ and $\{d_1, \ldots, d_5, \beta_1, \ldots, \beta_5\}$ corresponding to the two generators, respectively. To recover them, first we solve the following linear system for the coefficients of the Prony polynomial $\{\lambda_0, \ldots, \lambda_9\}$ based on the Equation (29)

$$H\lambda = -G,$$

where

$$H = \begin{bmatrix} -36.3064 & 24.4399 & 35.5543 & -40.2183 & -16.8633 & 18.9503 & -0.7573 & 18.8146 & 4.4668 & -52.7171 \\ 24.4399 & -0.3760 & -7.8892 & 9.3455 & -10.6340 & -8.8103 & 18.8824 & 1.8548 & -16.9513 & -9.4925 \\ 35.5543 & -7.8892 & -26.5849 & 21.6951 & 17.3985 & -10.7019 & -6.1982 & -16.8834 & -12.1046 & 24.6062 \\ -40.218 & 39.3455 & 21.6951 & -18.5319 & 21.6273 & 20.0106 & -46.4677 & -20.1576 & 24.6741 & 11.4779 \\ -16.8633 & -10.6340 & 17.3985 & 21.6273 & -15.9198 & -14.1386 & 6.0512 & -4.9102 & 3.4249 & 17.8445 \\ 18.9503 & -8.8103 & -10.7019 & 20.0106 & -14.1386 & -29.8792 & 27.4189 & 29.6337 & -11.7398 & -2.7658 \\ -0.7573 & 18.8824 & -6.1982 & -46.4677 & 6.0512 & 27.4189 & -6.2967 & 20.5893 & 23.4431 & -32.0445 \\ 18.8146 & 1.8548 & -16.8834 & -20.1576 & -4.9102 & 29.6337 & 20.5893 & -12.4873 & 0.2846 & 0.1035 \\ 4.4668 & -16.9513 & -12.1046 & 24.6741 & 3.4249 & -11.7398 & 23.4431 & 0.2846 & -35.8269 & 11.5787 \\ -52.7171 & -9.4925 & 24.6062 & 11.4779 & 17.8445 & -2.7658 & -32.0445 & 0.1035 & 11.5787 & -9.3042 \end{bmatrix}$$

and

$$G = \begin{bmatrix} 0.0458 & 0.0218 & -0.0275 & -0.0347 & -0.0103 & 0.0048 & 0.0510 & 0.0405 & -0.0520 & -0.0412 \end{bmatrix}^T.$$

We obtain

$$\lambda = \begin{bmatrix} -0.0088 & 0.0275 & -0.0639 & 0.1254 & -0.2113 & 0.3180 & -0.4300 & 0.5316 & -0.6010 & 0.3135 \end{bmatrix}^T,$$

which corresponds to the following Prony polynomial

$$\Lambda(z) = z^{10} + 0.3135z^9 - 0.6010z^8 + 0.5316z^7 - 0.4300z^6 + 0.3180z^5 - 0.2113z^4 + 0.1254z^3 - -0.0639z^2 + 0.0275z - 0.0088.$$

From the 10 zeros of this polynomial, we obtain 10 parameter values:

$$\{11.0000, 2.0000, 3.0000, 10.0000, 4.0000, 5.0000, 9.0000, 6.0000, 7.0000, 8.0000\}, \quad (59)$$

which correspond to $\{\phi_1, \ldots, \phi_5, \beta_1, \ldots, \beta_5\}$, but the explicit order is unknown. We must resolve the ambiguity: What five parameter values are for $\{\phi_1, \ldots, \phi_5\}$ (with the remaining five parameter values for $\{\beta_1, \ldots, \beta_5\}$)?

To separate the ϕ_j's from β_l's, we consider the following overdetermined linear system:

$$\begin{bmatrix} \cos(\phi_1 x_0) & \cdots & \cos(\phi_5 x_0) & \sin(\beta_1 x_0) & \cdots & \sin(\beta_5 x_0) \\ \cos(\phi_1 x_1) & \cdots & \cos(\phi_5 x_1) & \sin(\beta_1 x_1) & \cdots & \sin(\beta_5 x_1) \\ \vdots & \vdots & \vdots & \vdots & \cdots & \vdots \\ \cos(\phi_1 x_{19}) & \cdots & \cos(\phi_5 x_{19}) & \sin(\beta_1 x_{19}) & \cdots & \sin(\beta_5 x_{19}) \end{bmatrix} \begin{bmatrix} c_1 \\ \vdots \\ c_5 \\ d_1 \\ \vdots \\ d_5 \end{bmatrix} = \begin{bmatrix} f_0 \\ f_1 \\ \vdots \\ f_{19} \end{bmatrix}, \quad (60)$$

where we use the shorthand notations

$$x_n = x_0 + nh \quad \text{and} \quad f_n = f(x_0 + nh)$$

for $n = 0, 1, \ldots, 19$. Note: In this linear system, we only use 20 out of 39 original sampling values, which is adequate for this particular example. It is a trade-off issue between the accuracy of computation and the cost of computation (in time). In general, the more redundant equations we use, the more accuracy we can achieve in searching for the true solution. In other words, if we can obtain adequate accuracy, we focus on cutting the computation cost to the minimum. We do not solve this overdetermined linear system by the *least-square* method directly. We split these 20 equations into two parts: In the first part, we approximate the coefficients $\{c_1, \ldots, c_5, d_1, \ldots, d_5\}$ in (60) by the least-square method. Then we apply these derived coefficients to the equations in the second part so as to filter out the true solution.

Among the 10 values in (59), every time we select 5 of them for $\{\phi_1, \ldots, \phi_5\}$, the remaining 5 numbers are automatically for $\{\beta_1, \ldots \beta_5\}$. We will have total 252 possible choices (which is the combinatorial number $\binom{10}{5}$) as the candidates for the solution. Notice that this combinatorial number is a relatively big number. In order to speed up the processing, we reduce the redundant computation to the minimum. Let us use the notations $\{\phi_1^i, \ldots, \phi_5^i, \beta_1^i, \ldots \beta_5^i\}$ with $i = 1, 2, \ldots, 252$ representing those 252 candidates. Our method is based on the property that the information given in the sampling values has a lot of redundancy for selecting the true solution, and we only use just enough information from the given sampling values so as to save the computation time.

First, when we calculate the coefficients $\{c_1, \ldots, c_5, d_1, \ldots, d_5\}$ by the least-square method, we use exactly 10 equations (the same number of the coefficients) out of the 20 equations in (60). Based on our experiments, we do not have to use an overdetermined system for a good approximation by the least-square method. A determined system can give us excellent approximation for the least-square problem, while any underdetermined system usually does not approximate the data well through the least-square solution. For convenience, we select 10 consecutive equations in (60) somewhere in the middle, which we call the *least-square block* in our discussion, to approximate the coefficients $\{c_1, \ldots, c_5, d_1, \ldots, d_5\}$. Specifically, our least-square block takes the subscripts from 6 through 15, and the corresponding sampling values $\{f_6, f_7, \ldots, f_{15}\}$ should be selected as a reduced linear system given below,

$$\begin{bmatrix} \cos(\phi_1 x_6) & \cdots & \cos(\phi_5 x_6) & \sin(\beta_1 x_6) & \cdots & \sin(\beta_5 x_6) \\ \cos(\phi_1 x_7) & \cdots & \cos(\phi_5 x_7) & \sin(\beta_1 x_7) & \cdots & \sin(\beta_5 x_7) \\ \vdots & & \vdots & \vdots & & \vdots \\ \cos(\phi_1 x_{15}) & \cdots & \cos(\phi_5 x_{15}) & \sin(\beta_1 x_{15}) & \cdots & \sin(\beta_5 x_{15}) \end{bmatrix} \begin{bmatrix} c_1 \\ \vdots \\ c_5 \\ d_1 \\ \vdots \\ d_5 \end{bmatrix} = \begin{bmatrix} f_6 \\ f_7 \\ \vdots \\ f_{15} \end{bmatrix}. \quad (61)$$

Even if our new linear system (61) is a determined system, we still solve it for a least-square solution, because the determinant of the square matrix in (61) could be very close to zero. Then the remaining equations in (60) together with the coefficients derived from (61) will be used to detect which candidate is the true solution based on the error information.

For each set of values $\{\phi_1^i, \ldots, \phi_5^i, \beta_1^i, \ldots, \beta_5^i\}$ among the 252 candidates, the least-square solution for the linear system (61) would produce the 10 coefficients $[c_1^i, \ldots, c_5^i, d_1^i, \ldots, d_5^i]^T$, and we evaluate the following vector

$$\begin{bmatrix} f_0^i \\ f_1^i \\ \vdots \\ f_{19}^i \end{bmatrix} := \begin{bmatrix} \cos(\phi_1^i x_0) & \cdots & \cos(\phi_5^i x_0) & \sin(\beta_1^i x_0) & \cdots & \sin(\beta_5^i x_0) \\ \cos(\phi_1^i x_1) & \cdots & \cos(\phi_5^i x_1) & \sin(\beta_1^i x_1) & \cdots & \sin(\beta_5^i x_1) \\ \vdots & \ddots & \vdots & \vdots & \ddots & \vdots \\ \cos(\phi_1^i x_{19}) & \cdots & \cos(\phi_5^i x_{19}) & \sin(\beta_1^i x_{19}) & \cdots & \sin(\beta_5^i x_{19}) \end{bmatrix} \begin{bmatrix} c_1^i \\ \vdots \\ c_5^i \\ d_1^i \\ \vdots \\ d_5^i \end{bmatrix},$$

which is in general different from the original sampling vector $[f_0, f_1, \ldots, f_{19}]^T$. Then we will calculate the difference of these two vectors, and see how close they are. We define the error vector as follows:

$$\begin{bmatrix} \epsilon_0^i \\ \epsilon_1^i \\ \vdots \\ \epsilon_{19}^i \end{bmatrix} := \begin{bmatrix} |f_0^i - f_0| \\ |f_1^i - f_1| \\ \vdots \\ |f_{19}^i - f_{19}| \end{bmatrix}. \quad (62)$$

To search for the true solution among the 252 candidates, we discover an intrinsic property, shown in Figures 2 and 3, that can clearly separate the true solution from other candidates.

In Figure 2, we plot the error vector for one of the 252 candidates to view its typical behavior. The error values in the least-square block (with subscripts from 6 to 15) are very close to zero for a typical candidate; however, the error values that are out of the least-square block (with subscripts from 0 to 5 and from 16 to 19) are not close to zero in general for a candidate that is not the true solution.

This behavior can be explained in this way: The errors in the least-square block are usually very small due to the fact that the least-square solution of the determined system approximates the targeting sampling values $\{f_6, f_7, \ldots, f_{15}\}$ quite well; however, when we consider an error for a sampling value out of the least-square block, since the corresponding equation is not involved in the least-square approximation, there is no reason for this equation to generate a value that is very close to the targeting sampling value.

While for the true solution case, the behavior is different in the sense that the errors for all the equations in the linear system (60) are very close to zero (see Figure 3, and ignore the two reference points at the ends). Let us summarize the key property that helps us to find out the true solution among all the candidates: *For a candidate, if the coefficients generated from the determined linear system (61) by the least-square method cannot approximate just one sampling value out of the least-square block well, then it cannot be the true solution.*

However, if the coefficients for one candidate can approximate one particular sampling value out of the least-square block well, we can only say that it is *highly likely* that this candidate could be the true solution, because the probability for a *non-solution* candidate to approximate some sampling value out of the least-square block well is very small. Based on this observation from our experiments, we design the following strategy for the solution search.

Strategy: Eliminate as many as possible candidates in the first round filtering in two steps: *Step 1*. Select a determined linear system from the overdetermined linear system in (60) (as the least-square block), and approximate the coefficients $\{c_1, \ldots, c_5, d_1, \ldots, d_5\}$ by the least-square method for each of the 252 candidates. *Step 2*. Apply the derived coefficients in *Step 1* on one of the linear equations out of the least-square block to approximate the targeting sampling value and calculate the error with the targeting sampling value. If the error is greater than certain threshold (we use 0.1 as our threshold), we drop this candidate from the consideration; otherwise, this candidate passes the first round filtering. If only one candidate survives the first round filtering, it must be the true solution. If more than one candidates pass the first round filtering, we need to do the second round filtering. In the second round filtering, we simply apply the derived coefficients on another linear equation out of the least-square block, and calculate the error for the targeting sampling value. If the error is greater than the threshold, we eliminate this candidate. We keep doing these cycles until we identify the true solution. Since we have plenty of redundant equations out of the least-square block, we should be able to determine the true solution without going through too many cycles in general. Furthermore, those linear equations corresponding to the original sampling values that are not included in the linear system (60) can still be used for the above steps when necessary, but the probability to use those equations out of the linear system (60) will be extremely small. This simple strategy is designed to allow us to detect the true solution without unnecessary computation, while we still preserve the option to use the redundant information when necessary.

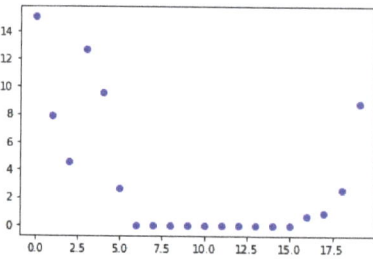

Figure 2. Display the error vector for one of the 252 candidates.

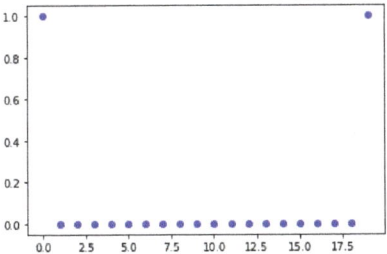

Figure 3. Display the error vector for the true solution with two reference points at the ends.

Here we would like to point out that as soon as we select values in ϕ-group or β-group, the order of those values in each group is not important, because their corresponding coefficients (c_j's or d_l's) will also be aligned with them accordingly when we solve the determined linear system (61) using the least-square method.

Example 2. *Our second function to be recovered has the following form*

$$f(\omega) = \frac{c_1}{2}\left(e^{-\frac{1}{2}(\phi_1-\omega)^2} + e^{-\frac{1}{2}(\phi_1+\omega)^2}\right)$$
$$+ \frac{c_2}{2}\left(e^{-\frac{1}{2}(\phi_2-\omega)^2} + e^{-\frac{1}{2}(\phi_2+\omega)^2}\right) \qquad (63)$$
$$+ \frac{c_3}{2}\left(e^{-\frac{1}{2}(\phi_3-\omega)^2} + e^{-\frac{1}{2}(\phi_3+\omega)^2}\right),$$

which is derived by applying the STFT on the following function

$$g(x) = \sum_{j=1}^{3} c_j \cos(\phi_j x), \qquad (64)$$

with the parameters of (64) listed in the following Table 2.

Table 2. Parameters of the function $f(x)$ in (64).

j	c_j	ϕ_j
1	0.5000	1.0000
2	0.2500	3.0000
3	1.0000	4.0000

To solve this problem, we need to use 12 (i.e., $4M$) sampling values. After we applied the method described in Section 4, we solved a linear system with 6 unknowns, and derived the Prony polynomial of degree 6 as follows

$$\Lambda(z) = 1.0000(z^6 + 1) - 14.4845(z^5 + z) + 65.9809(z^4 + z^2) + 108.8070z^3.$$

The symmetric structure of this polynomial tells us that its zeros appear in (z_j, z_j^{-1}) pairs for $j = 1, 2, 3$, which correspond to three pairs of parameters: $(1.0000, -1.0000)$, $(3.0000, -3.0000)$, and $(4.0000, -4.0000)$ for $(\phi_j, -\phi_j), j = 1, 2, 3$. Finally, we can solve another linear system for the coefficients c_j's with the errors listed in the Table 3.

Table 3. Parameters of the function $f(x)$ in (64) and approximate errors using 12 sampling values with $h = 0.5$.

j	c_j	ϕ_j	$|c_j - c_j^*|$	$|\phi_j - \phi_j^*|$
1	0.5000	1.0000	$3.7970.10^{-2}$	$5.2824.10^{-13}$
2	0.2500	3.0000	$5.0987.10^{-14}$	$4.5652.10^{-13}$
3	1.0000	4.0000	$5.8065.10^{-14}$	$1.4211.10^{-14}$

6. Conclusions

In this paper, we introduce a method that extends the Prony method to solve the *two-generator sparse expansion problem*. This method relies on the existence of a special linear operator for which the two generators must be the eigenfunctions. This two-generator problem has a special property: The zeros of its Prony polynomial correspond to two sets of parameters, and there is no straightforward way to separate them. We propose a *two-stage least-square detection method* on an overdetermined linear system for each candidate to extract the true solution, which relies on an intrinsic property for the true solution: *Only the true solution can use the coefficients derived from the least-square block to approximate the targeting sampling values out of the least-square block well*. Our method is designed to minimize the computation cost, while still maintain the computation accuracy.

It seems that the idea can be extended to the k-generator sparse expansion problem for $k > 2$; however, for the general k-generator case, the requirement that there exists a linear operator such that all the generators must be its eigenfunctions becomes *extremely* hard to achieve. For example, in the following sparse expansion problem,

$$f(x) = \sum_{j=1}^{M_1} c_j \cos(\phi_j x) + \sum_{l=1}^{M_2} d_l e^{\beta_l x}, \qquad (65)$$

it is not easy to find a linear operator, such that both $\cos(\phi x)$ and $e^{\beta x}$ are its eigenfunctions. One may argue that the problem could be solved by converting $\cos(\phi x)$ to $\frac{1}{2}(e^{i\phi x} + e^{-i\phi x})$, and then it becomes a one-generator problem. Notice that converting a two-generator problem to a one-generator problem may not work most of the time. We are interested in developing a general method that can solve the two-generator sparse expansion problem including the one in (65). We can see that there are many difficult problems to be solved in this multi-generator sparse expansion problem, and we would like to see more researchers contribute in this direction.

Our method for the two-generator sparse expansion problem can handle certain degree of uncertainty. For example, in problem (23), if we know the total number of terms (i.e., the value of $M_1 + M_2$), but we do not know the number of terms in each summation (i.e., the individual values of M_1 and M_2), we can still solve the problem using our *two-stage least-square detection* method described in Sections 3 and 5. If we increase the uncertainty a little more, can we still solve the problem?

For example, in the problem we considered in Section 4, if we do not know the exact number of terms (it is referred to *unknown order of sparsity M* in [1]) in the following expansion,

$$f(x) = \sum_{j=1}^{M} c_j e^{-\beta(\phi_j - x)^2} + \sum_{j=1}^{M} c_j e^{-\beta(\phi_j + x)^2},$$

and we are given K equispaced sampling values for some positive integer K. If we are told that these sampling values are sufficient to recover the signal, how do we recover it? In other words, we know that the number of terms M is in the range $1 \leq M \leq \lfloor K/4 \rfloor$, but we do not know the exact number M, can we solve the problem? The answer is *yes*, because we can try all the possible cases: $M = 1, 2, \ldots, \lfloor K/4 \rfloor$, and for each case, we apply our *two-stage least-square detection* method to tell us if the true solution can be extracted.

However, we are not satisfied with this kind of *exhaustive search type* solution due to its high cost. We plan to develop an efficient *term number detection* method, so that when we make a term number prediction, this method can tell us if it is correct or not immediately. In [1], two methods are proposed: One is based on the rank of the H matrix, and the other is based on the singular values of the H matrix. The main issue is: How to obtain a *reliable* method to determine the M value in the sparse expansion? Only after we obtain the correct term number we will pay the computation cost to go through all the necessary details to find the solution.

Author Contributions: Conceptualization, A.H. and W.H.; methodology, A.H. and W.H; software, A.H. and W.H.; validation, A.H. and W.H.; formal analysis, A.H. and W.H.; investigation, A.H. and W.H.; resources, A.H. and W.H.; data curation, A.H. and W.H.; writing—original draft preparation, A.H. and W.H.; writing—review and editing, A.H. and W.H.; visualization, A.H. and W.H. All authors have read and agreed to the published version of the manuscript.

Funding: This research received no external funding.

Conflicts of Interest: The authors declare no conflict of interest.

References

1. Peter, T. Generalized Prony Method. Ph.D. Thesis, University of Gottingen, Gottingen, Germany, 2014.
2. Hussen, A.M. Recover Data in Sparse Expansion Forms Modeled by Special Basis Functions. Ph.D. Thesis, University of Missouri, St. Louis, MO, USA, 2019. Available online: https://irl.umsl.edu/dissertation/896 (accessed on 22 March 2022).
3. Plonka, G.; Stampfer, K.; Keller, I. Reconstruction of Stationary and Non-stationary Signals by the Generalized Prony Method. *Anal. Appl.* **2019**, *17*, 179–210. [CrossRef]
4. Plonka, G.; Wischerhoff, M. How many Fourier samples are needed for real function reconstruction? *J. Appl. Math. Comput.* **2013**, *42*, 117–137. [CrossRef]
5. Beinert, R.; Plonka, G. Sparse phase retrieval of one-dimensional signals by Prony's method. *Front. Appl. Math. Stat.* **2017**, *3*, 5. [CrossRef]
6. Coluccio, L.; Eisinberg, A.; Fedele, G. A Prony-like method for non-uniform sampling. *Signal Process.* **2007**, *87*, 2484–2490. [CrossRef]
7. Peter, T.; Plonka, G.; Schaback, R. Prony's Method for Multivariate Signals. *Proc. Appl. Math. Mech.* **2015**, *15*, 665–666. [CrossRef]
8. Peter, T.; Plonka, G.; Roşca, D. Representation of sparse Legendre expansions. *J. Symb. Comput.* **2013**, *50*, 159–169. [CrossRef]
9. Peter, T.; Plonka, G. A generalized Prony method for reconstruction of sparse sums of eigenfunctions of linear operators. *Inverse Prob.* **2013**, *29*, 025001. [CrossRef]
10. Peter, T.; Potts, D.; Tasche, M. Nonlinear approximation by sums of exponentials and translates. *SIAM J. Sci. Comput.* **2011**, *33*, 1920–1947. [CrossRef]
11. Plonka, G.; Tasche, M. Prony methods for recovery of structured functions. *GAMM Mitt.* **2014**, *37*, 239–258. [CrossRef]
12. Wischerhof, M. Reconstruction of Structured Functions From Sparse Fourier Data. Ph.D. Thesis, Niedersächsische Staats- und Universitätsbibliothek Göttingen, Göttingen, Germany, 2015.

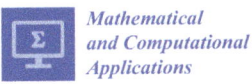

Article

A Bivariate Beta from Gamma Ratios for Determining a Potential Variance Change Point: Inspired from a Process Control Scenario

Schalk W. Human [1], Andriette Bekker [1,2,*], Johannes T. Ferreira [1,2] and Philip Albert Mijburgh [1]

[1] Department of Statistics, University of Pretoria, Pretoria 0002, South Africa; schalk.human@up.ac.za (S.W.H.); johan.ferreira@up.ac.za (J.T.F.); symstat@up.ac.za (P.A.M.)
[2] Centre of Excellence in Mathematical and Statistical Sciences, Johannesburg 2000, South Africa
* Correspondence: andriette.bekker@up.ac.za

Abstract: Within statistical process control (SPC), normality is often assumed as the underlying probabilistic generator where the process variance is assumed equal for all rational subgroups. The parameters of the underlying process are usually assumed to be known—if this is not the case, some challenges arise in the estimation of unknown parameters in the SPC environment especially in the case of few observations. This paper proposes a bivariate beta type distribution to guide the user in the detection of a permanent upward or downward step shift in the process' variance that does not directly rely on parameter estimates, and as such presents itself as an attractive and intuitive approach for not only potentially identifying the magnitude of the shift, but also the position in time where this shift is most likely to occur. Certain statistical properties of this distribution are derived and simulation illustrates the theoretical results. In particular, some insights are gained by comparing the newly proposed model's performance with an existing approach. A multivariate extension is described, and useful relationships between the derived model and other bivariate beta distributions are also included.

Keywords: bivariate beta; gamma; hypergeometric function; sequential; shift in process variance

1. Introduction

1.1. Problem Contextualisation

The monitoring of the variance of independent and identically distributed (i.i.d) normal random variables over time by taking successive, independent samples of measurements over time remains an interesting and valuable research consideration within quality control environment. In this case, when the variance σ^2 changes to $\sigma_1^2 = \lambda \sigma^2$ for some $\lambda \neq 1$, the practitioner needs to investigate the scope of such a change (a value of λ), and ideally, the position within the successive measurements where such a change could've taken place. Suppose that X_{ij} are i.i.d. $N(\mu, \sigma^2)$, $i = 0, 1, 2, \cdots, \kappa - 1$ and $X_{ij} \sim$ i.i.d. $N(\mu, \sigma_1^2 = \lambda \sigma^2)$, $i = \kappa, \kappa + 1, \cdots, m$ where $j = 1, 2, \cdots, n_i \geq 2$ and $\lambda > 0$, as outlined in Figure 1. The values of κ and λ are assumed to be unknown, but deterministic in nature. The order of these samples is important and cannot be re-ordered; in other words, the samples have a set sequence corresponding to the order in which they were obtained.

Thus, inspired by a practical objective, this paper aims to present a theoretically motivated framework to

1. present and contextualise this problem within the quality control environment;
2. follow a systematic approach to build up the distributional foundations from this practical perspective;
3. exploratively focus on the development of the (new) resulting bivariate beta distribution;
4. compare this model with an existing approach; and

5. determine whether $\lambda \neq 1$, and if this is indeed the case, to determine κ, the location of where the shift in the variance occurred.

Therefore, from sample κ onwards, the process would be considered out-of-control. Note that it is assumed that the shift occurs between two samples.

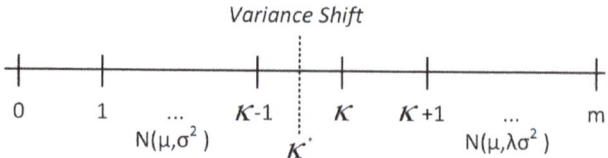

Figure 1. Process shift.

Assume that both the process mean (μ) and variance (σ^2) are unknown, and that they are estimated by their respective minimum variance unbiased estimators (MVUE), given by:

$$\bar{X}_i = \frac{\sum_{j=1}^{n_i} X_{ij}}{n_i}, i = 0, 1, 2, \cdots, m, \tag{1}$$

$$S_i^2 = \frac{1}{n_i - 1} \sum_{j=1}^{n_i} (X_{ij} - \bar{X}_i)^2, i = 0, 1, 2, \cdots, m, \tag{2}$$

where $\bar{X}_i$ and S_i^2 denote the mean and variance of sample i respectively. Some particular notes on Equation (1) and (2) include:

- The index variable i ranges from 0 to m: a total of $m + 1$ independent rational subgroups or samples.
- At least two samples are needed for a potential shift between them to be possible, therefore we assume $m \geq 1$.
- The sample size n_i can vary between different samples.
- $n_i \geq 2$ is necessary since the process mean and variance are both assumed to be unknown and have to be estimated.
- The pooling approach here is to use $m - r + 1$ and r sample means and variances in the construction of the test statistic in Section 1.2. Alternatively one can consider a single mean/variance in this construction, which would result in additional information $\sum_{i=r}^{m} n_i - 1$ and $\sum_{i=0}^{r-1} n_i - 1$ such that $n_i \geq 1$ and probability density functions are valid. In this case, the approach would reduce to a two sample comparison testing for a change in the variance.

The problem of determining if a shift in the process variance has occurred can be divided into two stages, namely before the potential shift and after, as indicated below ($Gamma(\cdot, \cdot)$ denotes the usual gamma distribution with suitable shape and scale parameters [1]).

Before the shift	
Samples:	$i = 0, 1, 2, \cdots, \kappa - 1$ $j = 1, 2, \cdots, n_i$
Distribution:	$X_{ij} \sim N(\mu, \sigma^2)$

$$W_i = \frac{(n_i - 1)S_i^2}{\sigma^2} \stackrel{d}{=} \text{Gamma}\left(\frac{n_i - 1}{2}, 2\right)$$
$$\stackrel{d}{=} \chi^2(n_i - 1). \quad (3)$$

After the shift	
Samples:	$i = \kappa, \kappa + 1, \cdots, m$ $j = 1, 2, \cdots, n_i$
Distribution:	$X_{ij} \sim N(\mu, \sigma_1^2 = \lambda \sigma^2)$

$$W_i = \frac{(n_i - 1)S_i^2}{\lambda \sigma^2} \stackrel{d}{=} \text{Gamma}\left(\frac{n_i - 1}{2}, 2\lambda\right). \quad (4)$$

1.2. A Solution: Sequential Statistic Framework

The proposed distribution compares all the sample variances before a certain point (where the potential shift occurs), with all sample variances after the time of the shift. In essence, the multi-sample hypothesis testing problem is approached by using m sequential two sample tests; described as:

$$\begin{array}{lll} S_0^2 & \text{is compared with} & S_1^2, S_2^2, \cdots, S_m^2 \\ S_0^2, S_1^2 & \text{is compared with} & S_2^2, S_3^2, \cdots, S_m^2 \\ & \text{and so forth until} & \\ S_0^2, S_1^2, \cdots, S_{m-1}^2 & \text{is compared with} & S_m^2. \end{array} \quad (5)$$

Note that, due to the assumption that we have $m + 1$ samples, the procedure to identify the possible change in the variance requires m comparisons. Of these m comparisons the one that leads to the largest disparity between the sample variables on the left and the right of (5) will indicate the most likely position where the process experienced a change in variance. In essence, what is then needed is to quantify the difference between the sample variances of the left and right of (5), and then determine which of the m different comparisons has the largest difference between the sample variances, and finally to use some measure to determine if this maximum difference is within some set tolerance range. Our proposed mathematical construct on how to achieve this is discussed for the remained of this introduction. Assuming that no shift in the process variance has occurred, it is possible to construct a series of two sample statistics that correspond to the general procedure described in (5). Each statistic corresponds to whether at sample $r = \kappa$ the two independent samples (the sample variances before time r and the sample variances after and including time r) are from normal distributions with the same unknown variance σ^2. This can alternatively be viewed as testing whether $\sigma^2 = \sigma_1^2$, which is similar to testing $\lambda = 1$. As such, it follows that detecting a shift in the process variance can be reduced to the following hypothesis test:

$$H_0 : \sigma^2 = \sigma_1^2 \text{ vs } H_A : \sigma^2 \neq \sigma_1^2 \text{ or alternatively } H_0 : \lambda = 1 \text{ vs } H_A : \lambda \neq 1.$$

Suppose that a shift of size λ occurs in the process variance, then the corresponding random variables after the shift would be $W_i \sim \text{Gamma}(\frac{n_i-1}{2}, 2\lambda), i = \kappa, \kappa+1, \cdots, m$ distributed (see (4)). If there is no shift in the variance, $\lambda = 1$, and it follows that $W_i \sim \text{Gamma}(\frac{n_i-1}{2}, 2), i = 0, 1, \cdots, m$. Thus the hypothesis being investigated can be changed depending on the choice of the scale parameter of the specific gamma random variables. From (5) it follows that the series of statistics that forms the building blocks of the process are given by

$$U_r^* = \frac{\left(\frac{\sum_{i=r}^m (n_i-1) S_i^2}{\lambda \sigma^2 \sum_{i=r}^m (n_i-1)}\right)}{\left(\frac{\sum_{i=0}^{r-1}(n_i-1) S_i^2}{\sigma^2 \sum_{i=0}^{r-1}(n_i-1)}\right)} \equiv \frac{\frac{\sum_{i=r}^m W_i}{\sum_{i=r}^m (n_i-1)}}{\frac{\sum_{i=0}^{r-1} W_i}{\sum_{i=0}^{r-1}(n_i-1)}}, r = 1, 2, \cdots, m-1, m, \quad (6)$$

where $W_i \sim Gamma(\frac{n_i-1}{2}, 2)$ for $i = 1, 2, \cdots, r-1$ and $W_i \sim Gamma(\frac{n_i-1}{2}, 2\lambda)$ for $i = r, r+1, \cdots, m$.

In essence, (6) implies that the sample variances before the potential shift are pooled together, and the sample variances after the potential shift are pooled together. The numerator of the statistic at time r is the average, weighted by each statistic's degrees of freedom, of all the sample variances between and including samples r and m, while the denominator is the corresponding weighted average of all the sample variances between and including samples 0 and $r-1$; this is graphically presented in Figure 2. Thus U_r^* is the test statistic that is typically used to test the equality of two population variances from a normal distribution.

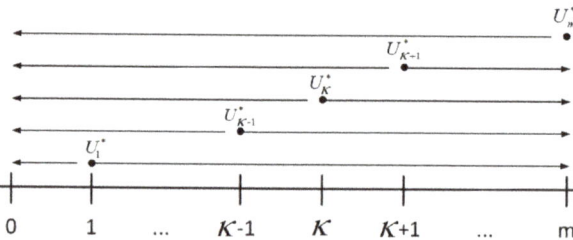

Figure 2. Building blocks of new distribution.

From (3), (4) and (6), it follows that if no shift has occurred in the process variance, i.e., $\lambda = 1$, each statistic U_r^* is univariate F distributed with $\sum_{i=r}^m (n_i - 1)$ and $\sum_{i=0}^{r-1}(n_i - 1)$ degrees of freedom, respectively.

The reasoning behind the critical values that would indicate whether $\lambda \neq 1$ is justified by inspecting the sequence of statistics in (6). Suppose that an increase ($\lambda > 1$) in the process variance has occurred from sample $r = \kappa$ onward, then:

- The statistic U_r^*'s numerator will contain only sample variances that come from a $N(\mu, \lambda \sigma^2), \lambda > 1$ distribution, whereas the denominator will contain only sample variances that come from a $N(\mu, \sigma^2)$ distribution.
- If k_1 is some integer value such that $1 \leq r - k_1 < r$, then the statistic $U_{r-k_1}^*$ will contain k_1 sample variances in its numerator that are from a $N(\mu, \sigma^2)$ distribution. This will reduce the weighted average of the sample variances of $U_{r-k_1}^*$'s numerator in comparison to the numerator of U_r^*.
- Similarly, if k_2 is some integer value such that $r < r + k_2 \leq m$, then the statistic $U_{r+k_2}^*$ will contain k_2 sample variances in its denominator that are from a $N(\mu, \lambda \sigma^2)$ distribution. This will increase the weighted average of sample variances of $U_{r+k_2}^*$'s denominator in comparison to the denominator of U_r^*.
- Thus, any statistic other than the one immediately following the shift in the process variance, will contain either smaller sample variances in its numerator, or larger sample variances in its denominator (on average). Either of these scenarios result in a high probability that all other statistics are smaller relative to U_r^*.
- This leads to the conclusion that the most probable place where an upwards shift in the process variance will be detected is at the statistic immediately following the shift.

- The value that this statistic assumes also has a high likelihood of being the maximum value of all the $U_r^*, r = 1, 2, \cdots, m-1, m$ statistics.
- As such, the most reasonable method of calculating the critical value to detect an upwards shift in the process variance is to calculate the maximum order statistic of the charting statistics $U_r^*, r = 1, 2, \cdots, m-1, m$, (under the null hypothesis) and to set the critical value equal to some percentile of the distribution of the maximum order statistic.

Using a similar but inverted argument, it can be justified that the critical value of the control chart should be set equal to some percentile of the minimum order statistic of the charting statistics, under the null hypothesis of no shift having occurred, if the detection of a downward shift in the process variance is of concern. Due to space limitations, the minimum order statistics are not presented in this article.

To simplify matters going forward and for notational purposes we omit the factors $\sum_{i=r}^{m}(n_i - 1)$ and $\sum_{i=0}^{r-1}(n_i - 1)$ in (6), and drop the superscript *, and therefore the statistics of interest become

$$U_r = \frac{\sum_{i=r}^{m} W_i}{\sum_{i=0}^{r-1} W_i}, r = 1, 2, \cdots, m-1, m, \quad (7)$$

where $W_i \sim Gamma(\alpha_i > 0, \beta_i > 0), i = 0, 1, \cdots, m$ and independent. Since $\alpha_i = \frac{n_i - 1}{2}$ and $\beta_i = 2\lambda$, the shape parameter is related to the sample size of the i_{th} sample, and the scale parameter is related to the underlying distribution's variance. The theoretical focus here is based on the statistics in (7), whereas the statistics in (6) are those that are practically applicable and is the basis in the simulation study in Section 3.

Constructing statistics using ratios of random variables as in (7) is of practical interest in many areas of science. Ref. [2] studied and derived the joint density functions of ratios of Rayleigh, Rician, Nakagami-*m*, and Weibull random variables; [3] approached the ratios of generalised gamma variables via exact- and near exact solutions, and [4] derived closed-form expressions for the ratio of independent non-identically distributed variables from an *α-μ* distribution which have applications in the performance analysis of wireless communication systems.

The proposed model in this paper will be compared to the model of [5] in Section 3, and is described here for the convenience of the reader. If $r = 2$, that the bivariate joint probability density function of the statistics in (10) is given by

$$\begin{aligned}g(t_1, t_2) &= \frac{\left(\beta_0^{\alpha_1+\alpha_2}\beta_1^{\alpha_0+\alpha_2}\beta_2^{\alpha_0+\alpha_1}\right)\Gamma(\alpha_0+\alpha_1+\alpha_2)}{\Gamma(\alpha_0)\Gamma(\alpha_1)\Gamma(\alpha_2)}(t_1)^{\alpha_1-1}(t_2)^{\alpha_2-1}(1+t_1)^{\alpha_2} \\ &\times (\beta_1\beta_2 + \beta_0\beta_2 t_1 + \beta_0\beta_1(1+t_1)t_2)^{-\alpha_0-\alpha_1-\alpha_2}\end{aligned} \quad (8)$$

where $t_1, t_2 > 0$, $\alpha_i, \beta_i > 0$ for $i = 0, 1, 2,$, $t_1 = \frac{w_1}{w_0}$ and $t_2 = \frac{w_2}{(w_0+w_1)}$, and $\Gamma(\cdot)$ is the gamma function [6].

Refs. [5,7] proposed a beta distribution that is used to detect a shift in the process variance which is based on the Q chart that was developed by [8], and as such the series of comparisons of sample statistics they made were

$$\begin{array}{ll} S_1^2 \text{ is compared with} & S_0^2 \\ S_2^2 \text{ is compared with} & S_0^2, S_1^2 \\ \text{and so forth until} & \\ S_m^2 \text{ is compared with} & S_0^2, S_1^2, \cdots, S_{m-1}^2. \end{array} \quad (9)$$

Using a similar approach to the method described earlier the statistics can be given as

$$T_r = \frac{W_r}{\sum_{i=0}^{r-1} W_i}, r = 1, 2, \cdots, m-1, m, \quad (10)$$

which is graphically presented in Figure 3—with W_i defined in (3) and (4).

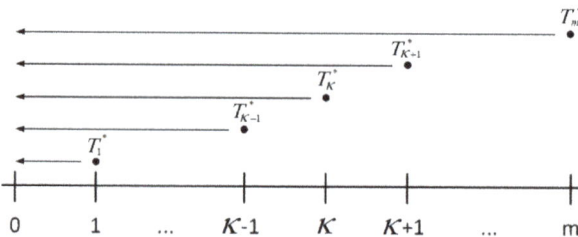

Figure 3. Building blocks of [7].

Based on the work of [5,9] provided insight for detecting the change in the parameter structure if the underlying process is multivariate normal.

1.3. Outline of Paper

In Section 2, the bivariate joint probability density function which emanates from (7) is derived, accompanied by an exploratory shape analysis. This is followed by the derivation of the marginal probability density functions, the product moment as well as the maximum order statistic of the distribution. In Section 3 the performance of the model that this article proposes is compared to the Q chart studied by [5], this will be conducted through a simulation study. Tables of simulated values for the 95th percentiles of the maximum order statistics of the sets of random variables in (6) and (7) are provided in Section 3, for varying parameter choices, to enable practical application of the proposed model. The values in these tables are corroborated through numerical integration of the derived expressions for the maximum order statistics. The Appendices A and B contains proofs of the obtained results as well as the positioning of this distribution among other often considered bivariate beta models.

2. Proposed Model

In this section, the joint probability density function of the random variables U_1 and U_2 (see (7) when $r = 2$) is derived, followed by a shape analysis, the derivation of the marginal probability density functions, the product moment as well as the maximum order statistic. A brief review of this new candidate with respect to its partners is provided in the Appendices A and B—which provides additional insight for modelling as well as expressions with closed form.

2.1. Bivariate Probability Density Function

Theorem 1. *Let W_i be independent gamma random variables with parameters $\alpha_i = \frac{n_i - 1}{2} > 0$, $\beta_i = 2\lambda > 0$ for $i = 0, 1, 2$. Let $U_1 = \frac{W_1 + W_2}{W_0}$ and $U_2 = \frac{W_2}{W_0 + W_1}$ then the joint probability density function of U_1 and U_2 is given by*

$$f(u_1, u_2) = \frac{\left(\beta_0^{\alpha_1 + \alpha_2} \beta_1^{\alpha_0 + \alpha_2} \beta_2^{\alpha_0 + \alpha_1}\right) \Gamma(\alpha_0 + \alpha_1 + \alpha_2)}{\Gamma(\alpha_0) \Gamma(\alpha_1) \Gamma(\alpha_2)} (u_1 - u_2)^{\alpha_1 - 1} u_2^{\alpha_2 - 1} (1 + u_1)^{\alpha_2} (1 + u_2)^{\alpha_0}$$
$$\times \left(\beta_1 \beta_2 (1 + u_2) + \beta_0 \beta_2 (u_1 - u_2) + \beta_0 \beta_1 u_2 (1 + u_1)\right)^{-\alpha_0 - \alpha_1 - \alpha_2} \quad (11)$$

where $u_1 > u_2 > 0$.

2.2. Shape Analysis

In this section a shape analysis is conducted for the joint probability density function (11). A standard set of parameters has been chosen as a baseline. The parameters are chosen to be $\alpha_0 = \alpha_1 = \alpha_2 = 5$ and $\beta_0 = \beta_1 = \beta_2 = 2$, in other words, a process where all three samples consist of $5 \times 2 + 1 = 11$ observations, and where no shift has occurred in the process variance. Some of the parameters will then be varied from this baseline in order to investigate the effect that a change in the specific parameters will have on the general shape of the joint probability density function. Note that the change in some parameters will be large - so large that they lose practical realism; this is conducted to emphasise and investigate the general change in the shape, and is not meant to be an indication of the practical applications of the joint probability density function. The functions will only be plotted for $u_1 \in [0,3]$, $u_2 \in [0,3]$.

In Figure 4 it can be seen what effect increasing the sample sizes (while keeping them equal) has on the joint probability density function. It is seen that increasing the sample sizes also increases the height of the peak of the probability density function. Larger sample sizes also shrink the length and width of the "tails" of the joint probability density function. In essence, the larger the sample sizes ($n_i = 11, n_i = 51, n_i = 101$), the smaller the domain on which the function has non-trivial values.

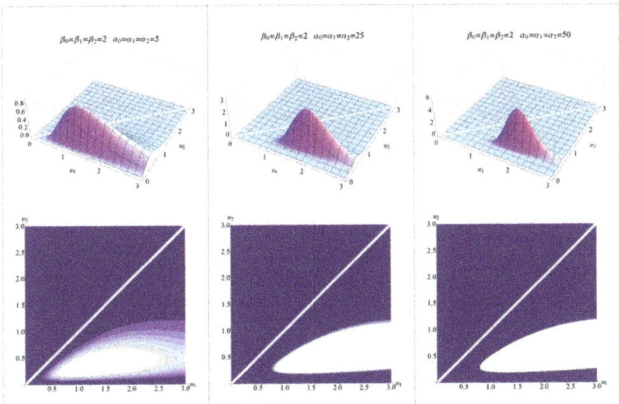

Figure 4. Equal sample sizes, no shift in the variance.

Figure 5 below demonstrates that a sustained increase in the process variance, irrespective of size, minimally affects the general shape and location of the joint probability density function, but does affect the height of the probability density function's peak. In the below example, the shift in the process variance occurs at time 1, and as one would hope and expect, the joint probability density function relies heavily on the value of the statistic at time 1, U_1. A similar effect, where the joint probability density function relies heavily on the value of U_2 is seen when the shift occurs at time 2.

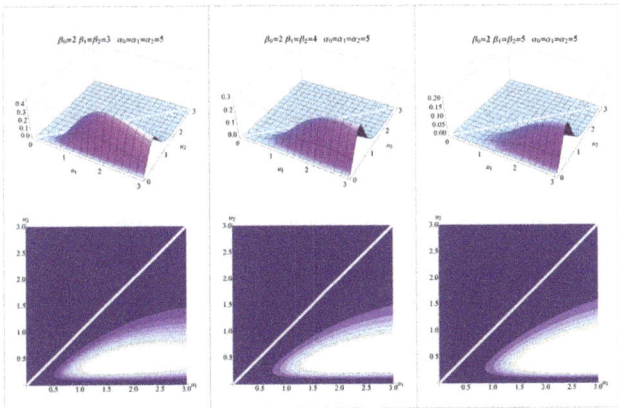

Figure 5. Equal sample sizes, increase in variance at time 1.

2.3. Marginal Probability Density Functions

Theorem 2. Assume that (U_1, U_2) has the joint probability density function in (11), then the marginal probability density function of U_1 is given by

$$
\begin{aligned}
f(u_1) &= \frac{\left(\beta_0^{\alpha_1+\alpha_2} \beta_1^{\alpha_0+\alpha_2} \beta_2^{\alpha_0+\alpha_1}\right) \Gamma(\alpha_0+\alpha_1+\alpha_2)}{\Gamma(\alpha_0)\Gamma(\alpha_1)\Gamma(\alpha_2)} (\beta_1\beta_2 + \beta_0\beta_2 u_1)^{-\alpha_0-\alpha_1-\alpha_2} \\
&\times (1+u_1)^{\alpha_2} \sum_{k=0}^{\infty} \left[\left(\frac{\alpha_0!}{k!(\alpha_0-k)!} \right) (u_1)^{k+\alpha_1+\alpha_2-1} \frac{\Gamma(\alpha_1)\Gamma(k+\alpha_2)}{\Gamma(k+\alpha_1+\alpha_2)} \right. \\
&\times \left. {}_2F_1\left(\alpha_0+\alpha_1+\alpha_2, k+\alpha_2; k+\alpha_1+\alpha_2; -\frac{u_1(\beta_1\beta_2-\beta_0\beta_2+\beta_0\beta_1+\beta_0\beta_1 u_1)}{\beta_1\beta_2+\beta_0\beta_2 u_1}\right) \right]
\end{aligned}
\tag{12}
$$

where $u_1 > 0$, $\alpha_0, \alpha_1, \alpha_2 \in \mathbb{Z}$, $\beta_1(\beta_2+\beta_0(1+u_1)) > \beta_0\beta_2$, $\left|-\frac{u_1(\beta_1\beta_2-\beta_0\beta_2+\beta_0\beta_1+\beta_0\beta_1 u_1)}{\beta_1\beta_2+\beta_0\beta_2 u_1}\right| < 1$, and the marginal probability density function of U_2 is given by

$$
\begin{aligned}
f(u_2) &= \frac{\left(\beta_0^{\alpha_1+\alpha_2}\beta_1^{\alpha_0+\alpha_2}\beta_2^{\alpha_0+\alpha_1}\right)\Gamma(\alpha_0+\alpha_1+\alpha_2)}{\Gamma(\alpha_0)\Gamma(\alpha_2)}(1+u_2)^{\alpha_0}\sum_{k=0}^{\infty}\left[\frac{\alpha_2!}{k!(\alpha_2-k)!}\right. \\
&\times \sum_{l=0}^{\infty}\left[(-1)^l \binom{\alpha_0+\alpha_1+\alpha_2+l-1}{l}(\beta_0\beta_2+\beta_0\beta_1 u_2)^{-\alpha_0-\alpha_1-\alpha_2-l}\right. \\
&\times \left.\left.(\beta_1\beta_2+\beta_1\beta_2 u_2 - \beta_0\beta_2 u_2 + \beta_0\beta_1 u_2)^l u_2^{k-\alpha_0-l-1} \frac{\Gamma(\alpha_0+\alpha_2+l-k)}{\Gamma(\alpha_0+\alpha_1+\alpha_2+l-k)}\right]\right]
\end{aligned}
\tag{13}
$$

where $u_2 > 0$, $\alpha_0, \alpha_1, \alpha_2 \in \mathbb{Z}$, $\alpha_i, \beta_i > 0$ for $i = 0, 1, 2$ and ${}_2F_1(,)$ is the Gauss hypergeometric function ([6] p. 1005). Note that in (12) if $\alpha_0 \in \{1, 2, \cdots\}$ the sum changes from $\sum_{k=0}^{\infty}$ to $\sum_{k=0}^{\alpha_0}$ (See [6] p. 25, Equations (1).110 and 1.111). Similarly if $\alpha_2 \in \{1, 2, \cdots\}$ the sum changes from $\sum_{k=0}^{\infty}$ to $\sum_{k=0}^{\alpha_2}$ in (13).

If $\beta_0 = \beta_1 = \beta_2$ it follows that:

$$
\begin{aligned}
f(u_1) &= \frac{\Gamma(\alpha_0+\alpha_1+\alpha_2)}{\Gamma(\alpha_0)\Gamma(\alpha_1)\Gamma(\alpha_2)}(1+u_1)^{-\alpha_0-\alpha_1}\sum_{k=0}^{\infty}\left[\left(\frac{\alpha_0!}{k!(\alpha_0-k)!}\right)(u_1)^{k+\alpha_1+\alpha_2-1}\right. \\
&\times \left.\frac{\Gamma(\alpha_1)\Gamma(k+\alpha_2)}{\Gamma(k+\alpha_1+\alpha_2)} {}_2F_1(\alpha_0+\alpha_1+\alpha_2, k+\alpha_2; k+\alpha_1+\alpha_2; -u_1)\right]
\end{aligned}
$$

where $0 < u_1 < 1$, and

$$
\begin{aligned}
f(u_2) &= \frac{\Gamma(\alpha_0+\alpha_1+\alpha_2)}{\Gamma(\alpha_0)\Gamma(\alpha_2)}(1+u_2)^{\alpha_0}\sum_{k=0}^{\infty}\left[\frac{\alpha_2!}{k!(\alpha_2-k)!}\sum_{j=0}^{\infty}\left[(-1)^j\binom{\alpha_0+\alpha_1+\alpha_2+j-1}{j}\right.\right. \\
&\times \left.\left.(1+u_2)^{-\alpha_0-\alpha_1-\alpha_2}u_2^{k-\alpha_0-j-1}\frac{\Gamma(\alpha_0+\alpha_2+j-k)}{\Gamma(\alpha_0+\alpha_1+\alpha_2+j-k)}\right]\right]
\end{aligned}
$$

where $u_2 > 0$.

2.4. Product Moment and Order Statistics

Theorem 3. Assume that (U_1, U_2) has the joint probability density function (11), then the product moment of U_1 and U_2 is given by

$$
\begin{aligned}
&E\left(U_1^r U_2^s\right) \\
&= \sum_{p=0}^{r}\binom{r}{p}\frac{\left(\beta_0^{\alpha_1-p-s}\beta_1^{-\alpha_1}\beta_2^{p+s}\right)\Gamma(\alpha_2+p+s)\Gamma(\alpha_0+\alpha_1-p-s)\Gamma(\alpha_1+r-p)\Gamma(\alpha_0-r)}{\Gamma(\alpha_0)\Gamma(\alpha_1)\Gamma(\alpha_2)\Gamma(\alpha_0+\alpha_1-p)} \\
&\times {}_2F_1\left(\alpha_0+\alpha_1-p-s, \alpha_1+r-p; \alpha_0+\alpha_1-p; 1-\frac{\beta_0\beta_2}{\beta_1\beta_2}\right)
\end{aligned}
\tag{14}
$$

where $r, s \in \{0, 1, 2, \cdots\}$, $\alpha_0+\alpha_1 > r+s$, $\alpha_0 > \alpha_1$, $\alpha_0+\alpha_1 > p$, and $\left|1-\frac{\beta_0\beta_2}{\beta_1\beta_2}\right| < 1$.

The maximum order statistics are of importance in detecting whether a shift in the process variance does indeed occur, as was discussed in Section 1 and will be demonstrated in Section 3. Although a closed form expression for the maximum order statistic is not tractable in a closed form, an expression is provided that can be implemented to calculate numerical values.

Assume that (U_1, U_2) has the joint probability density function (11), then the maximum order statistic of U_1 and U_2 can be determined either by

$$\begin{aligned} P(max(U_1, U_2) < z) &= P(U_1 < z) \\ &= \int_0^z f(u_1) du_1 \end{aligned} \quad (15)$$

when $\alpha_0, \alpha_1, \alpha_2 \in \Re$ (since $u_1 > u_2$).

3. Comparison Study and Discussion

Deriving the order statistics of the series of statistics in (6) and (7) is a complex task since they are neither independent nor identically distributed, (see [10]). Hence values for the 95th percentiles of the maximum order statistics are simulated for varying numbers of samples (m equal to 1, 4, 9, 14, 19, 24, 29, 49, 99 and 499) and sample sizes (n equal to 2, 5, 10, 15, 20, 25, 30, 50, 100 and 500) in Section 3.1.

In Section 3.1, properties of (7) are also studied when no shift in the variance occurs, this is imperative since the control limits of a control chart are constructed under the null hypothesis that no shift has occurred, or alternatively given that the process is in control. The in control properties that are investigated for both the Q chart and the new proposed model (6) are where the maximum order statistic is most likely to occur when no shift has occurred in the process variance. Practically, this is a very important question since, if the maximum order statistic consistently occurs at roughly the same place in the sequence of samples, it implies that the distribution should be treated with added suspicion if it indicates that a shift in the process occurs at this location.

In Section 3.2, the out of control performance of the newly proposed distribution is compared with the Q chart model investigated by [5]. The probability that the respective control charts will detect a shift in the process variance is investigated for differing sizes of shifts in the process variance. The comparison is made for varying numbers of samples as well as sample sizes.

3.1. Comparison When the Process Is in Control

The location of the maximum order statistic when the process is in control is investigated using graphs. For brevity's sake only one possible combination ($m = 29$, $n = 15$) of the number of samples and sample sizes mentioned above is included in this paper. All of them, however, lead to the same general conclusion.

As can be seen from Figure 6 the [5] distribution's maximum order statistic occurs most often at the first statistic, with the probability of the maximum occurring at subsequent statistics steadily decreasing. This implies that the Q chart becomes more stable as the process progresses, which makes practical sense since each subsequent statistic includes more of the sample data. The newly proposed distribution's maximum order statistic occurs most often at the first statistic, and second-most often at the last statistic. This due to the way in which the statistics of the distribution are constructed (see (6)). This implies that while our proposed model may detect shifts at the ends of the samples, signals received at these locations should be treated with a bit of skepticism.

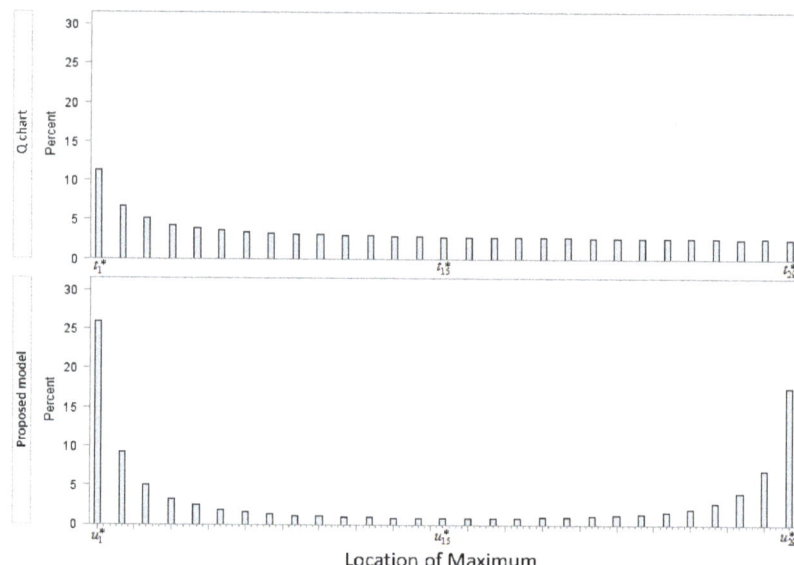

Figure 6. Location of maximum order statistics: $m = 29$ and $n = 15$.

In Table 1 the 95th percentiles of the maximum order statistics of the newly proposed distributions are simulated (to the third decimal) using Monte Carlo simulation. In other words z in the equation $P\left(max\left(U_1^*, U_2^*, \cdots, U_m^*\right) < z\right) = 0.95$. Similarly the 95th percentiles of the maximum order statistics of (6) are simulated in Table 2.

Table 1. 95th percentiles of the maximum order statistics of $U_1^*, U_2^*, \cdots, U_m^*$.

m \ n	2	5	10	15	20	25	30	50	100	500
1	214.286	6.652	3.230	2.512	2.189	1.998	1.872	1.615	1.399	1.160
4	242.046	6.229	3.047	2.385	2.093	1.919	1.805	1.569	1.371	1.150
9	254.592	5.992	2.908	2.301	2.023	1.863	1.755	1.534	1.349	1.142
14	251.747	5.874	2.871	2.266	1.997	1.842	1.739	1.522	1.341	1.139
19	255.413	5.836	2.851	2.254	1.987	1.831	1.730	1.517	1.337	1.137
24	257.880	5.813	2.837	2.243	1.980	1.827	1.723	1.514	1.335	1.137
29	259.105	5.815	2.824	2.238	1.977	1.822	1.720	1.512	1.334	1.136
49	259.963	5.799	2.818	2.228	1.968	1.815	1.715	1.508	1.331	1.135
99	258.810	5.797	2.803	2.222	1.959	1.811	1.710	1.504	1.329	1.134
499	263.017	5.772	2.801	2.215	1.957	1.808	1.707	1.501	1.328	1.133

Table 2. 95th percentiles of the maximum order statistics of $U_1, U_2, \cdots, U_m$.

m \ n	2	5	10	15	20	25	30	50	100	500
1	399.762	12.091	5.933	4.638	4.072	3.735	3.513	3.062	2.688	2.275
4	1151.103	29.015	14.061	11.045	9.708	8.958	8.434	7.411	6.553	5.618
9	2413.470	57.222	27.649	21.699	19.101	17.623	16.615	14.615	12.984	11.184
14	3691.861	85.496	41.144	32.376	28.469	26.292	24.837	21.864	19.419	16.754
19	4944.966	112.744	54.656	43.071	37.870	34.913	33.020	29.081	25.816	22.318
24	6187.147	141.315	68.325	53.653	47.313	43.549	41.182	36.293	32.248	27.887
29	7528.303	170.174	81.753	64.408	56.621	52.313	49.397	43.531	38.678	33.457
49	12560.904	281.708	135.824	107.163	94.200	86.900	82.168	72.395	64.349	55.728
99	25774.144	565.834	270.688	213.065	187.991	173.740	164.160	144.638	128.625	111.365
499	126434.246	2815.186	1353.458	1064.841	939.259	868.035	819.248	722.038	642.752	556.808

Note that:

- Since the critical values of the distribution are derived under the assumption of the null hypothesis, the equations for the maximum order statistic can be simplified to be constructed out of chi-square random variables instead of the more complex gamma case.
- When the sample sizes of all the samples are equal ($n_i = n$ for $i = 1, \cdots, m$) the removal of the constant terms in (7) is superfluous since the series of statistics in (6) and (7) are merely scalar multiples of each other. This can be seen in Tables 1 and 2 where $P\left(max\left(U_1^*, U_2^*, \cdots, U_m^*\right) < z\right) \approx P(max(U_1, U_2, \cdots, U_m) < mz)$.
- Instead of using simulation, the maximum order statistics in Table 2 could have been calculated using (15). Table 3 demonstrates their equivalence for the bivariate case by numerically integrating (15).

Table 3. Simulated and theoretical 95th percentiles of the maximum order statistics of U_1, U_2.

Method \ n	2	5	10	15	20	25	30	50	100	500
Simulated values	399.762	12.091	5.933	4.638	4.072	3.735	3.513	3.062	2.688	2.275
Theoretical values	399.000	12.083	5.920	4.640	4.068	3.735	3.515	3.064	2.688	2.276

3.2. Comparisons When the Process is Out of Control

In this section, the proposed model's potential to detect shifts in compared with that of the Q chart form investigated by [5]. The probability of signaling that a shift in the process variance has occurred depends on a few variables. In this paper, these variables are: the number of samples, the sample size of the samples, where in the process the shift in the process variance occurs, and the size of the shift. The figures in this section take all of these parameters into account.

In each of the following figures, the probability of signaling a shift in the variance is displayed as a function of the size of the shift, where the shift size λ, ranges from $\lambda = 1$ (no shift) to $\lambda = 5$ (a 500% increase in the process variance). Many different combinations of the number of samples, the sample sizes, as well as the locations of the shift were tested; however, only the graphs that illustrate key findings are included in this paper. The chosen parameters that were simulated are: number of samples (m) equal to 10, 20 and 30, the sample sizes (n) equal to 2, 5, 10 and 20, and the location of the shift in the process variance (κ) occurring at (roughly, due to integer sample numbers) 25%, 50% and 75% of the way through the samples.

By simulating graphs such as those in Figures 7–9, certain conclusions can be reached about the properties and efficacy of the two competing models ((5) and (9)):

- When $n = 2$, irrespective of the number of samples, the newly proposed model outperforms the Q chart. There are some caveats however that should be noted:
    1. In the above graphs, each plotted point was simulated 500,000 times during the Monte Carlo process. For all the $n = 2$ graphs, the points vary erratically between each 0.05 increases in the shift size, and thus even for a large number of simulations the process cannot be described as "stable".
    2. The Q chart seems to be completely incapable of detecting an increase in the process variance when $n = 2$, with the probability of detecting a shift remaining at roughly 5%, irrespective of the size of the shift.
    3. While the new model's probability of detecting a shift does increase as the size of the shift increases, it remains relatively low, at roughly 7% to 10%, just marginally higher than the 5% chance when the process is actually IC. This implies that while it might be theoretically possible to implement the new model for samples sizes of 2, it would likely not be a practically useful technique.
    4. The new model's probability of detecting a shift does not increase as the number of samples increases, as would be expected (and as is the case for the other choices of n)
- From these points above, it can be concluded that using a sample size of 2 does not lead to an effective control chart.
- For a small numbers of samples ($m = 9$), the newly proposed model outperforms the Q chart for all simulated sample sizes as well as locations of shifts (for all shift sizes).
- When there are 20 samples ($m = 19$), the newly proposed model outperforms the Q chart in nearly all situations. The Q chart does have a higher probability of detecting a shift in the process variance only when the sample sizes are small ($n = 5$), and the shift occurs relatively late in the process ($\kappa = 15$), for shifts in the process variance between $\lambda = 3$ and $\lambda = 4.75$. Since a 300% to 475% increase in the process variance is unlikely to occur in practice, the newly proposed model would likely be more effective for $m = 19$.
- For $m = 29$, sweeping statements about the performances of the two methods are more difficult to make since the plotted percentage lines cross often. However it can be said that:
    1. For small sample sizes ($n = 5$), the proposed model outperforms the Q chart for small shifts in the process variance, whereas the Q chart performs better for larger shifts.
    2. The Q chart performs at its best when the shift in the process variance occurs late in the series of samples.
    3. For larger sample sizes ($n = 20$) the proposed model outperforms the Q chart when the shift in the process variance occurs early, but when the shift occurs roughly half way through the series of samples, or further, the performance of the two methods are very similar.

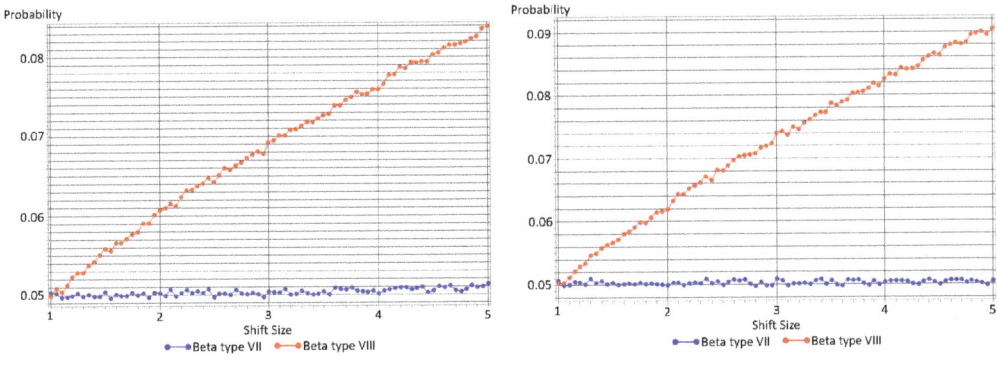

$m = 9, n = 2, \kappa = 7$ $m = 29, n = 2, \kappa = 15$

Figure 7. Probability of detecting a shift when $n = 2$.

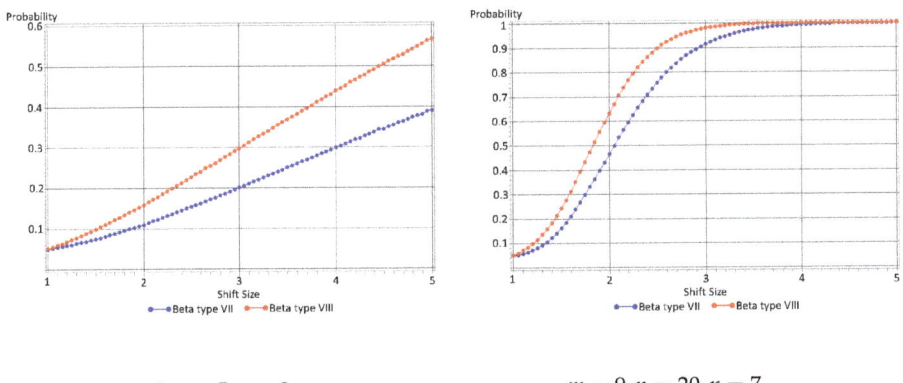

$m = 9, n = 5, \kappa = 3$ $m = 9, n = 20, \kappa = 7$

Figure 8. Probability of detecting a shift when $m = 9$.

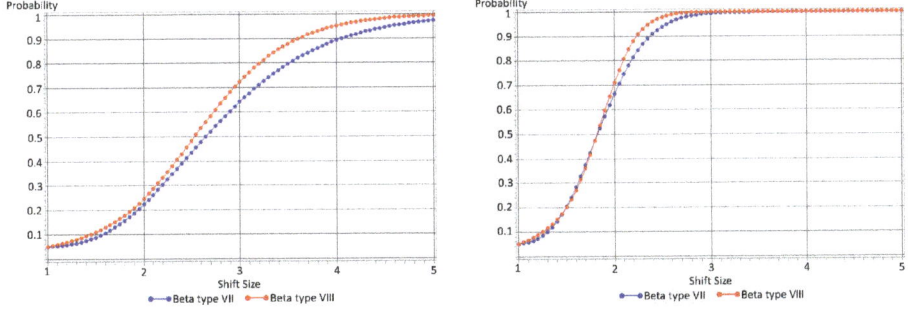

$m = 19, n = 10, \kappa = 15$ $m = 29, n = 20, \kappa = 22$

Figure 9. Probability of detecting a shift when $\kappa \approx 0.75\, m$.

4. Concluding Remarks

In this paper a new bivariate beta type distribution is proposed that can be utilised to detect a shift in a process's variance when the underlying process follows a normal distribution. The proposed model compares favourably with the Q chart model studied by [5] in most situations; especially when the number of samples is small, and when the process variance experiences a change early on in the series of samples. Future work can include focus on (i) when a shift occurs within a sample, (ii) expanding underlying distributional assumptions to that of the class of scale mixtures to consider departures from normality (see [11]), and (iii) the multivariate setup, of which we propose the probability density function in the following theorem and a proof outlined in the Appendices A and B.

Theorem 4. *Let W_i be independent gamma random variables with parameters $(\alpha_i > 0, \beta_i > 0)$ for $i = 0, 1, 2, \cdots, m$. Let $U_r = \frac{\sum_{i=r}^{m} W_i}{\sum_{i=0}^{r-1} W_i}, r = 1, 2, \cdots, m-1, m$, then the joint probability density function of $U_1, U_2, \cdots, U_m$ is given by*

$$
\begin{aligned}
& f(u_1, u_2, \cdots, u_m) \\
&= \frac{\prod_{i=1}^{m-1}\left[(u_i - u_{i+1})^{\alpha_i - 1}\right](u_m)^{\alpha_m - 1}\Gamma\left(\sum_{i=0}^{m}\alpha_i\right)}{\prod_{i=0}^{m}\left[\beta_i^{\alpha_i}\Gamma(\alpha_i)\right]} \\
& \times (1+u_1)^{\sum_{i=2}^{m}\alpha_i} \prod_{i=2}^{m}\left[(1+u_i)^{-\alpha_{i-1}-\alpha_i}\right] \\
& \times \left(\frac{1}{\beta_0} + \frac{(u_1 - u_2)}{\beta_1(1+u_2)} + \sum_{i=2}^{m-1}\left[\frac{(1+u_1)(u_i - u_{i+1})}{\beta_j(1+u_i)(1+u_{i+1})}\right] + \frac{(1+u_1)u_m}{\beta_m(1+u_m)}\right)^{-\sum_{i=0}^{m}\alpha_i}
\end{aligned} \tag{16}
$$

where $u_1 > u_2 > \cdots > u_m > 0$.

Author Contributions: Conceptualization, S.W.H. and A.B.; methodology, S.W.H., A.B. and P.A.M.; software, P.A.M.; validation, S.W.H., A.B., J.T.F. and P.A.M.; formal analysis, P.A.M.; investigation, A.B., J.T.F. and P.A.M.; writing—original draft preparation, P.A.M.; writing—review and editing, A.B., J.T.F. and P.A.M.; supervision, S.W.H. and A.B.; funding acquisition, A.B. and J.T.F. All authors have read and agreed to the published version of the manuscript.

Funding: This work was based upon research supported in part by the National Research Foundation (NRF) of South Africa, grant RA201125576565, nr 145681; NRF ref. SRUG190308422768 nr. 120839; the RDP296/2022 grant from the University of Pretoria, South Africa, as well as the Centre of Excellence in Mathematical and Statistical Sciences, based at the University of the Witwatersrand, South Africa. The opinions expressed and conclusions arrived at are those of the authors and are not necessarily to be attributed to the NRF.

Acknowledgments: The authors acknowledge the StatDisT group based at the University of Pretoria, Pretoria, South Africa. Furthermore, the authors thank the editor as well as three anonymous reviewers for their insightful contributions which led to an improved manuscript.

Conflicts of Interest: The authors declare no conflict of interest.

Appendix A. Positioning

The aim of this section is to show the relationships between some of the most commonly used bivariate beta distributions, and to relate these distributions to the bivariate distribution derived by [5] and the one proposed by this article. In this section the statistics in the relationships are constructed out of chi-square random variables, not gamma, as is conducted throughout the rest of this article. This is conducted for the sake of simplicity; however, the relationships hold regardless.

Let $Y_1 \sim \chi^2(\alpha)$, $Y_2 \sim \chi^2(\beta)$ and $Y_3 \sim \chi^2(\gamma)$. Then:

- If $Q_1 = \frac{Y_1}{Y_1 + Y_2 + Y_3}$ and $Q_2 = \frac{Y_2}{Y_1 + Y_2 + Y_3}$. Then the joint distribution of Q_1 and Q_2 is called a bivariate beta type I distribution. In Figure A1 this will be denoted alternatively as $(Q_1, Q_2) \sim B_I(\alpha, \beta, \gamma)$. The multivariate generalisation of this distribution is called the Dirichlet type I distribution (see [12,13]).

- If $V_1 = \frac{X_1}{X_3}$ and $V_2 = \frac{X_2}{X_3}$ then (V_1, V_2) has a bivariate beta type II distribution [14].
- If $W_1 = \frac{Y_1}{Y_1+Y_2+2Y_3}$ and $W_2 = \frac{Y_2}{Y_1+Y_2+2Y_3}$ then (W_1, W_2) has a bivariate beta type III distribution. The multivariate generalisation was derived and studied by [15,16] considered the case when $W_1 = \frac{Y_1}{Y_1+Y_2+cY_3}$ and $W_2 = \frac{Y_2}{Y_1+Y_2+cY_3}$, which is a generalisation of the above type III distribution.
- If $X_1 = \frac{Y_1}{Y_1+Y_3}$ and $X_2 = \frac{Y_2}{Y_2+Y_3}$ then (X_1, X_2) has a bivariate beta type IV distribution, ([17,18]).
- If $C_1 = \frac{aY_1}{aY_1+bY_2+cY_3}$ and $C_2 = \frac{bY_2}{aY_1+bY_2+cY_3}$, $a, b, c > 0$. Then (C_1, C_2) has a bivariate beta type V distribution. If $a = 1, b = 1$ and $c = 1$, the bivariate beta type V reduces to the bivariate beta type I [19].
- If $Z_1 = \frac{Y_1}{Y_2+Y_3}$ and $Z_2 = \frac{Y_2}{Y_1+Y_3}$ then (Z_1, Z_2) has a bivariate beta type VI distribution. This joint probability density function has not yet been derived in the literature and could potentially be applied to detecting shifts in a process variance.
- If $T_1 = \frac{Y_2}{Y_1}$ and $T_2 = \frac{Y_3}{Y_1+Y_2}$ then (T_1, T_2) be referred to as the bivariate beta type VII distribution, ([5]).
- If $U_1 = \frac{Y_2+Y_3}{Y_1}$ and $U_2 = \frac{Y_3}{Y_1+Y_2}$ then (U_1, U_2) has a bivariate beta type VIII distribution. This is the model that this article proposes in terms of gamma variables in Section 2, but in its special case it can be reduced to be constructed from chi-square variables.

Relationships between these models are graphically represented in Figure A1.

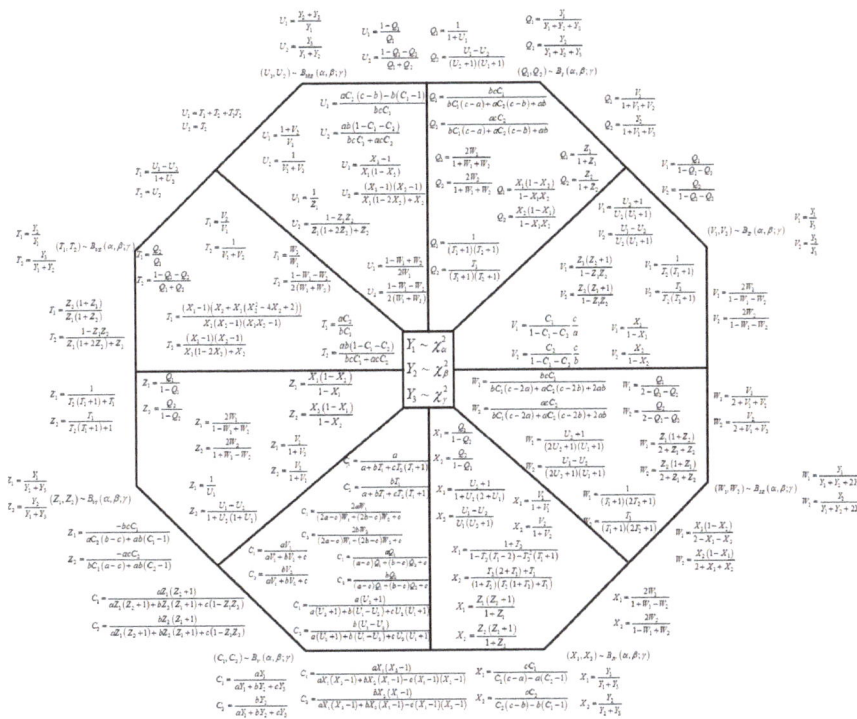

Figure A1. Relationships between several bivariate beta distributions.

Appendix B. Proofs

Proof of Theorem 1. The joint probability density function of W_0, W_1, W_2 is given by

$$f(w_0, w_1, w_2) = \frac{1}{\beta_0^{\alpha_0} \beta_1^{\alpha_1} \beta_2^{\alpha_2} \Gamma(\alpha_0) \Gamma(\alpha_1) \Gamma(\alpha_2)} \left(w_0^{\alpha_0-1} e^{-\frac{w_0}{\beta_0}} \right) \left(w_1^{\alpha_1-1} e^{-\frac{w_1}{\beta_1}} \right) \left(w_2^{\alpha_2-1} e^{-\frac{w_2}{\beta_2}} \right) \quad \text{(A1)}$$

where $w_0, w_1, w_2 > 0$. Let $U_0 = W_0$, then $W_1 = \frac{U_0(U_1-U_2)}{(1+U_2)}$ and $W_2 = \frac{U_0 U_2(1+U_1)}{(1+U_2)}$, and

$$\begin{aligned}
f(u_0, u_1, u_2) &= \frac{1}{\beta_0^{\alpha_0} \beta_1^{\alpha_1} \beta_2^{\alpha_2} \Gamma(\alpha_0)\Gamma(\alpha_1)\Gamma(\alpha_2)} (u_1 - u_2)^{\alpha_1-1} u_2^{\alpha_2-1} (1+u_1)^{\alpha_2}(1+u_2)^{-\alpha_1-\alpha_2} \\
&\times u_0^{(\alpha_0+\alpha_1+\alpha_2)-1} \exp\left(\frac{-u_0(\beta_1\beta_2(1+u_2)+\beta_0\beta_2(u_1-u_2)+\beta_0\beta_1 u_2(1+u_1))}{\beta_0\beta_1\beta_2(1+u_2)}\right).
\end{aligned} \quad \text{(A2)}$$

By integrating (A2) with respect to u_0, and rearranging the terms it follows that

$$\begin{aligned}
f(u_1, u_2) &= \frac{1}{\beta_0^{\alpha_0} \beta_1^{\alpha_1} \beta_2^{\alpha_2} \Gamma(\alpha_0)\Gamma(\alpha_1)\Gamma(\alpha_2)} (u_1 - u_2)^{\alpha_1-1} u_2^{\alpha_2-1} (1+u_1)^{\alpha_2}(1+u_2)^{-\alpha_1-\alpha_2} \\
&\times \int_0^\infty u_0^{(\alpha_0+\alpha_1+\alpha_2)-1} \exp\left(\frac{-u_0(\beta_1\beta_2(1+u_2)+\beta_0\beta_2(u_1-u_2)+\beta_0\beta_1 u_2(1+u_1))}{\beta_0\beta_1\beta_2(1+u_2)}\right) du.
\end{aligned} \quad \text{(A3)}$$

By applying [6] p. 346, Equation (3).381.4, to (A3), the result in (11) follows. □

Proof of Theorem 2. From (11), by rearranging the terms, and applying [6] p. 25, Equations (1).110 and 1.111, it follows that

$$\begin{aligned}
f(u_1) &= \frac{\left(\beta_0^{\alpha_1+\alpha_2}\beta_1^{\alpha_0+\alpha_2}\beta_2^{\alpha_0+\alpha_1}\right)\Gamma(\alpha_0+\alpha_1+\alpha_2)}{\Gamma(\alpha_0)\Gamma(\alpha_1)\Gamma(\alpha_2)}(1+u_1)^{\alpha_2} \\
&\times (\beta_1\beta_2 - \beta_0\beta_2 + \beta_0\beta_1 + \beta_0\beta_1 u_1)^{-\alpha_0-\alpha_1-\alpha_2} \sum_{k=0}^{\infty}\left(\frac{\alpha_0!}{k!(\alpha_0-k)!}\right) \\
&\times \int_0^{u_1} (u_1-u_2)^{\alpha_1-1} u_2^{k+\alpha_2-1} \left(\frac{\beta_1\beta_2+\beta_0\beta_2 u_1}{\beta_1\beta_2-\beta_0\beta_2+\beta_0\beta_1+\beta_0\beta_1 u_1} + u_2\right)^{-\alpha_0-\alpha_1-\alpha_2} du_2.
\end{aligned} \quad \text{(A4)}$$

By applying [6] p. 317, Equation (3). 197.8, to (A4), the result in (12) follows. From (11), by rearranging the terms, and applying [6] p. 25, Equations (1).110 and 1.111 twice, it follows that

$$\begin{aligned}
f(u_2) &= A(u_1, u_2) \int_{u_2}^\infty (u_1-u_2)^{\alpha_1-1}(1+u_1)^{\alpha_2}\left(1+\frac{u_1(\beta_0\beta_2+\beta_0\beta_1 u_2)}{\beta_1\beta_2+\beta_1\beta_2 u_2-\beta_0\beta_2 u_2+\beta_0\beta_1 u_2}\right)^{-\alpha_0-\alpha_1-\alpha_2} du_1 \\
&= A(u_1, u_2) \sum_{k=0}^\infty \left[\frac{\alpha_2!}{k!(\alpha_2-k)!}\sum_{l=0}^\infty \left[(-1)^l \binom{\alpha_0+\alpha_1+\alpha_2+l-1}{l}\right.\right.\\
&\left.\left.\times \left(\frac{(\beta_0\beta_2+\beta_0\beta_1 u_2)}{\beta_1\beta_2+\beta_1\beta_2 u_2-\beta_0\beta_2 u_2+\beta_0\beta_1 u_2}\right)^{-\alpha_0-\alpha_1-\alpha_2-l}\int_{u_2}^\infty (u_1-u_2)^{\alpha_1-1} u_1^{k-\alpha_0-\alpha_1-\alpha_2-l} du_1\right]\right]
\end{aligned} \quad \text{(A5)}$$

where

$$\begin{aligned}
A(u_1, u_2) &= \frac{\left(\beta_0^{\alpha_1+\alpha_2}\beta_1^{\alpha_0+\alpha_2}\beta_2^{\alpha_0+\alpha_1}\right)\Gamma(\alpha_0+\alpha_1+\alpha_2)}{\Gamma(\alpha_0)\Gamma(\alpha_1)\Gamma(\alpha_2)} u_2^{\alpha_2-1}(1+u_2)^{\alpha_0} \\
&\times (\beta_1\beta_2+\beta_1\beta_2 u_2 - \beta_0\beta_2 u_2+\beta_0\beta_1 u_2)^{-\alpha_0-\alpha_1-\alpha_2}.
\end{aligned} \quad \text{(A6)}$$

By applying [6] p. 315, Equation (3).191.2 to (A5), the result in (13) follows. □

Proof of Theorem 3. By using the relationships in Figure A1, and reordering the terms it follows that

$$\begin{aligned}
E(U_1^r U_2^s) &= E\big((T_1+T_2+T_1 T_2)^r (T_2)^s\big) \\
&= \int_0^\infty \int_0^\infty (t_1+t_2+t_1 t_2)^r (t_2)^s g(t_1, t_2) dt_2 dt_1 \\
&= \frac{\left(\beta_0^{-\alpha_0}\beta_1^{-\alpha_1}\beta_2^{\alpha_0+\alpha_1}\right)\Gamma(\alpha_0+\alpha_1+\alpha_2)}{\Gamma(\alpha_0)\Gamma(\alpha_1)\Gamma(\alpha_2)} \int_0^\infty (1+t_1)^{r-\alpha_0-\alpha_1}(t_1)^{\alpha_1-1} \\
&\times \int_0^\infty \left(\frac{t_1}{1+t_1}+t_2\right)^r (t_2)^{\alpha_2+s-1}\left(\frac{\beta_1\beta_2+\beta_0\beta_2 t_1}{\beta_0\beta_1(1+t_1)}+t_2\right)^{-\alpha_0-\alpha_1-\alpha_2} dt_2 dt_1.
\end{aligned} \quad \text{(A7)}$$

By applying [6] p. 25, Equations (1).110 and 1.111 to (A8), and reordering the terms, it follows that

$$
\begin{aligned}
E(U_1^r U_2^s) &= \sum_{p=0}^{r} \binom{r}{p} \frac{\left(\beta_0^{-\alpha_0} \beta_1^{-\alpha_1} \beta_2^{\alpha_0+\alpha_1}\right) \Gamma(\alpha_0+\alpha_1+\alpha_2)}{\Gamma(\alpha_0)\Gamma(\alpha_1)\Gamma(\alpha_2)} \int_0^{\infty} (1+t_1)^{-\alpha_0-\alpha_1+p} (t_1)^{\alpha_1+r-p-1} \\
&\quad \times \left(\frac{\beta_1\beta_2+\beta_0\beta_2 t_1}{\beta_0\beta_1(1+t_1)}\right)^{-\alpha_0-\alpha_1-\alpha_2} \int_0^{\infty} (t_2)^{\alpha_2+p+s-1} \left(1+\frac{\beta_0\beta_1(1+t_1)}{\beta_1\beta_2+\beta_0\beta_2 t_1} t_2\right)^{-\alpha_0-\alpha_1-\alpha_2} dt_2 dt_1
\end{aligned}
\tag{A8}
$$

From [6] p. 315, Equation (3).194.3, it follows that (A8) may be expressed as

$$
\begin{aligned}
E(U_1^r U_2^s) &= \sum_{p=0}^{r} \binom{r}{p} \frac{\left(\beta_0^{-\alpha_0} \beta_1^{-\alpha_1} \beta_2^{\alpha_0+\alpha_1}\right) \Gamma(\alpha_0+\alpha_1+\alpha_2)}{\Gamma(\alpha_0)\Gamma(\alpha_1)\Gamma(\alpha_2)} \int_0^{\infty} (1+t_1)^{-\alpha_0-\alpha_1+p} (t_1)^{\alpha_1+r-p-1} \\
&\quad \times \left(\frac{\beta_1\beta_2+\beta_0\beta_2 t_1}{\beta_0\beta_1(1+t_1)}\right)^{-\alpha_0-\alpha_1-\alpha_2} \left(\frac{\beta_0\beta_1(1+t_1)}{\beta_1\beta_2+\beta_0\beta_2 t_1}\right)^{-\alpha_2-p-s} \\
&\quad \times B(\alpha_2+p+s, \alpha_0+\alpha_1+\alpha_2-\alpha_2-p-s) dt_1 \\
&= \sum_{p=0}^{r} \binom{r}{p} \frac{\left(\beta_0^{\alpha_1-p-s} \beta_1^{-\alpha_0-p-s} \beta_2^{\alpha_0+\alpha_1}\right) \Gamma(\alpha_2+p+s)\Gamma(\alpha_0+\alpha_1-p-s)}{\Gamma(\alpha_0)\Gamma(\alpha_1)\Gamma(\alpha_2)} \\
&\quad \times (\beta_1\beta_2)^{-\alpha_0-\alpha_1+p+s} \int_0^{\infty} (1+t_1)^{-s} (t_1)^{\alpha_1+r-p-1} \left(1+\frac{\beta_0\beta_2}{\beta_1\beta_2} t_1\right)^{-\alpha_0-\alpha_1+p+s} dt_1
\end{aligned}
\tag{A9}
$$

By applying [6] p. 317, Equation (3). 197.5 to (A9), the result in (14) follows. □

Proof of Theorem 4. The joint probability density function of W_i, $i = 0, 1, 2, \cdots, m$ is given by

$$
f(w_0, w_1, \cdots, w_m) = \prod_{i=0}^{m} \frac{\left(w_i^{\alpha_i-1} e^{-\frac{w_i}{\beta_i}}\right)}{\beta_i^{\alpha_i} \Gamma(\alpha_i)}, \quad w_0, w_1, \cdots, w_m > 0.
\tag{A10}
$$

Let $U_0 = W_0$, $U_r = \frac{\sum_{i=r}^{m} W_i}{\sum_{i=0}^{r-1} W_i}$, $r = 1, 2, \cdots, m-1, m$. It follows that

$$
\begin{aligned}
W_0 &= U_0 \\
W_1 &= \frac{U_0(U_1-U_2)}{(1+U_2)} \\
W_r &= \frac{U_0(1+U_1)(U_r - U_{r+1})}{(1+U_r)(1+U_{r+1})}, \quad r = 2, 3, \cdots, m-1 \\
W_m &= \frac{U_0(1+U_1)U_m}{(1+U_m)},
\end{aligned}
$$

with $J(w_0, \cdots, w_m \to u_0, u_1, \cdots, u_m) = \frac{u^m (1+u_1)^{m-1}}{\prod_{j=2}^{m}(1+u_j)^2}$. Subsequently the joint probability density function of $U_0, U_1, U_2, \cdots, U_m$ is

$$
\begin{aligned}
&f(u_0, u_1, u_2, \cdots, u_m) \\
&= \frac{\prod_{i=1}^{m-1}\left[(u_i - u_{i+1})^{\alpha_i-1}\right](u_m)^{\alpha_m-1}}{\prod_{i=0}^{m}\left[\beta_i^{\alpha_i}\Gamma(\alpha_i)\right]} u^{\sum_{i=0}^{m}[\alpha_i]-1} (1+u_1)^{\sum_{i=2}^{m}[\alpha_i]} \prod_{i=2}^{m}\left[(1+u_i)^{-\alpha_{i-1}-\alpha_i}\right] \\
&\quad \times e^{-u_0\left(\frac{1}{\beta_0} + \frac{(u_1-u_2)}{\beta_1(1+u_2)} + \sum_{i=2}^{m-1}\left[\frac{(1+u_1)(u_i-u_{i+1})}{\beta_i(1+u_i)(1+u_{i+1})}\right] + \frac{(1+u_1)u_m}{\beta_m(1+u_m)}\right)}.
\end{aligned}
\tag{A11}
$$

By applying [6] p. 346, Equation (3). 381.4 to (A11), the result in (16) is proved. □

References

1. Bain, L.J.; Engelhardt, M. *Introduction to Probability and Mathematical Statistics*; Duxbury Press: Belmont, CA, USA, 1992; Volume 4.
2. Pavlović, D.Č.; Sekulović, N.M.; Milovanović, G.V.; Panajotović, A.S.; Stefanović, M.Č.; Popović, Z.J. Statistics for ratios of Rayleigh, Rician, Nakagami-, and Weibull distributed random variables. *Math. Probl. Eng.* **2013**, *2013*, 252804. [CrossRef] [PubMed]
3. Bilankulu, V.; Bekker, A.; Marques, F. The ratio of independent generalized gamma random variables with applications. *Comput. Math. Methods* **2021**, *3*, e1061. [CrossRef]
4. Leonardo, E.J.; da Costa, D.B.; Dias, U.S.; Yacoub, M.D. The ratio of independent arbitrary α-μ random variables and its application in the capacity analysis of spectrum sharing systems. *IEEE Commun. Lett.* **2012**, *16*, 1776–1779. [CrossRef]

5. Adamski, K. Generalised Beta Type II Distributions-Emanating from a Sequential Process. Ph.D. Thesis, University of Pretoria, Pretoria, South Africa, 2013.
6. Gradshteyn, I.S.; Ryzhik, I.M. *Table of Integrals, Series, and Products*; Academic Press: Cambridge, MA, USA, 2014.
7. Adamski, K.; Human, S.W.; Bekker, A. A generalized multivariate beta distribution: Control charting when the measurements are from an exponential distribution. *Stat. Pap.* **2012**, *53*, 1045–1064. [CrossRef]
8. Quesenberry, C.P. SPC Q charts for start-up processes and short or long runs. *J. Qual. Technol.* **1991**, *23*, 213–224. [CrossRef]
9. Bekker, A.; Ferreira, J.T.; Human, S.W.; Adamski, K. Capturing a Change in the Covariance Structure of a Multivariate Process. *Symmetry* **2022**, *14*, 156. [CrossRef]
10. David, H.A.; Nagaraja, H.N. *Order Statistics*; John Wiley & Sons: Hoboken, NJ, USA, 2004.
11. Ferreira, J.T.; Bekker, A.; Marques, F.; Laidlaw, M. An enriched α-μ model as fading candidate. *Math. Probl. Eng.* **2020**, *2020*, 5879413. [CrossRef]
12. Gupta, R.D.; Richards, D.S.P. The history of the Dirichlet and Liouville distributions. *Int. Stat. Rev.* **2001**, *69*, 433–446. [CrossRef]
13. Balakrishnan, N.; Lai, C.D. *Continuous Bivariate Distributions*; Springer Science & Business Media: Berlin/Heidelberg, Germany, 2009.
14. Tiao, G.G.; Cuttman, I. The inverted Dirichlet distribution with applications. *J. Am. Stat. Assoc.* **1965**, *60*, 793–805. [CrossRef]
15. Cardeno, L.; Nagar, D.K.; Sánchez, L.E. Beta type 3 distribution and its multivariate generalization. *Tamsui Oxf. J. Math. Sci.* **2005**, *21*, 225–242.
16. Ehlers, R. Bimatrix Variate Distributions of Wishart Ratios with Application. Ph.D. Thesis, University of Pretoria, Pretoria, South Africa, 2011.
17. Jones, M. Multivariate t and beta distributions associated with the multivariate F distribution. *Metrika* **2002**, *54*, 215–231. [CrossRef]
18. Olkin, I.; Liu, R. A bivariate beta distribution. *Stat. Probab. Lett.* **2003**, *62*, 407–412. [CrossRef]
19. Ehlers, R.; Bekker, A.; Roux, J.J. Triply noncentral bivariate beta type V distribution. *S. Afr. Stat. J.* **2012**, *46*, 221–246.

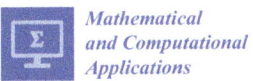

Mathematical and Computational Applications

Article

The Binomial–Natural Discrete Lindley Distribution: Properties and Application to Count Data

Shakaiba Shafiq [1], Sadaf Khan [1], Waleed Marzouk [2], Jiju Gillariose [3] and Farrukh Jamal [1,*]

[1] Department of Statistics, The Islamia University of Bahawalpur, Bahawalpur 63100, Pakistan; shakaiba.hashmi@gmail.com (S.S.); smkhan6022@gmail.com (S.K.)
[2] Faculty of Graduate Studies for Statistical Research, Department of Mathematical Statistics, Cairo University, Giza 12613, Egypt; wgm-74@hotmail.com
[3] Department of Statistics, CHRIST (Deemed to be University), Hosur Road, Bangalore 560029, India; jijugillariose@yahoo.com
* Correspondence: farrukh.jamal@iub.edu.pk

Abstract: In this paper, a new discrete distribution called Binomial–Natural Discrete Lindley distribution is proposed by compounding the binomial and natural discrete Lindley distributions. Some properties of the distribution are discussed including the moment-generating function, moments and hazard rate function. Estimation of the distribution's parameter is studied by methods of moments, proportions and maximum likelihood. A simulation study is performed to compare the performance of the different estimates in terms of bias and mean square error. SO_2 data applications are also presented to see that the new distribution is useful in modeling data.

Keywords: discretizing; natural discrete Lindley distribution; over dispersion; maximum likelihood estimation

1. Introduction

Count data modeling is a challenging task in many areas, including, but not limited to, public health, medicine, epidemiology, applied science, sociology, and agriculture. In many situations, the life length of a device cannot be measured on a continuous scale and the survival function is assumed to be a function of a count random variable instead of being a function of a continuous-time random variable. Therefore, discrete distributions are somewhat meaningful to model lifetime data in situations where output may be of a discrete nature. The traditional discrete distributions have limited applicability as models for reliability, failure times, aggregate loss, etc., especially with the count data with over-dispersion in which the variance is greater than the mean. This has led to the development of some discrete distributions based on popular continuous models in reliability analysis, actuarial sciences survival analysis, etc. The discretization of continuous distributions has produced many discrete distributions in the last few decades in the statistical literature. However, the quest for a quintessential model remains the crux of the matter in the diverse scientific paradigm.

One of the many approaches to define new models is the discretization of distributions. Until recently, the majority of discrete lifetime distributions have been proposed in the statistical literature by discretizing the survival function $S(x)$ of continuous lifetime distributions (see the work of authors, for example, in references [1–12]).

The probability mass function (pmf) $P(X = x)$ is defined as follows

$$P(X = x) = S(x) - S(x+1) \qquad x = 0, 1, 2, \ldots$$

Away from this method, Afify [12] have introduced and studied a new discrete Lindley distribution by constructing a mixture of discrete analogs to the continuous components used in creating the continuous Lindley distribution.

In this paper, we propose and study a new probability mass function (pmf), denoted by p_x, by compounding the binomial and the NDL distributions. The basic principle of this method is stated as if N(input) and X(output) are two random variables denoting the number of particles entering and leaving an attenuator, then the probability functions $p(n)$ and $f(x)$ of these two random variables are connected by the binomial decay transformation

$$P(X = x) = \sum_{n=x}^{\infty} \binom{n}{x} p^x (1-p)^{n-x} p(n); \quad x = 0, 1, \ldots, \infty \quad (1)$$

where $0 \leq p \leq 1$ is the attenuating coefficient which is discussed by Hu et al. [7]. They considered $p(n)$ as a Poisson distribution with the parameter $\lambda > 0$, and then they showed that $\Pr(X = x)$ is the Poisson distribution with the parameter λp. For clarity, attenuators are electrical devices built to lower the amount of voltage flowing through them without severely compromising the signal's integrity. They serve as a safeguard against systems being exposed to signals with power levels that are too high to be decoded. Déniz [13] introduced uniform Poisson distribution using the idea of Hu et al. [7] by interchanging in Equation (1) the binomial distribution and the discrete uniform distribution and maintaining $P(n)$ as the Poisson distribution. Some new discrete distributions also are proposed in the literature using the methodology of [7]. Akdoğan et al. [14] proposed uniform-geometric distribution and Coşkun et al. [15] constructed binomial–discrete Lindley distribution.

The rest of the paper is arranged as follows: Section 2 defines the natural discrete Lindley distribution and proposes the new binomial–natural discrete Lindley distribution with important properties, subsequently. In Section 3, various parameter estimation and simulation studies are given. Section 4 concerns the real data illustration of the findings. In Section 5, some conclusions are provided.

2. Natural Discrete Lindley Distribution

Recently, Al-Babtain et al. [16] proposed and studied a new natural discrete analog of the continuous Lindley distribution as a mixture of geometric and negative binomial distributions. The new distribution is called natural discrete Lindley (NDL) distribution and it has many interesting properties that make it superior to many other discrete distributions, particularly in analyzing over-dispersed count data. The NDL can be applied in the collective risk models and is competitive with the Poisson distribution to fit automobile-claim-frequency data. Let N be a non-negative random variable obtained as a finite mixture of geometric (p) and negative binomial $(2, p)$ with mixing probabilities $\frac{p}{p+1}$ and $\frac{1}{p+1}$, respectively, then the probability mass function of the NDL distribution is defined as

$$P(N = n) = \frac{p^2}{p+1}(2+n)(1-p)^n \; ; \quad n = 0, 1, 2, \ldots \text{ and } p \in (0, 1) \quad (2)$$

One of the most important features of this distribution is that it has a single parameter and it has attractive properties, which makes it suitable for applications not only in insurance settings but also in other fields where over-dispersions are observed. For more details about this distribution, see Al-Babtain et al. [16]. Given the usefulness of NDL, the discrete analogue due to NDL known as the binomial NDL (BNDL) seems to be naturally interesting to explore.

2.1. The Proposed Discrete Analog

The probability mass function (1) can be expressed as

$$P(X = x) = \sum_{n=x}^{\infty} P(X = x | N = n) P(N = n),$$

where $P(X|N = n)$ has the binomial $b(n, p)$ distribution. Suppose that N is the random variable from NDL with parameter p given in (2); then, the probability mass function of the discrete random variable X is obtained as

$$p_x(x;p) = P(X = x) = \sum_{n=x}^{\infty} P(X = x|N = n)P(N = n) = \sum_{n=x}^{\infty} \binom{n}{x} p^x(1-p)^{n-x} \frac{p^2}{p+1}(2+n)(1-p)^n$$
$$= \sum_{n-x=0}^{\infty} \binom{n}{x} p^x(1-p)^{n-x} \frac{p^2}{p+1}(2+n)(1-p)^n = \sum_{k=0}^{\infty} \binom{x+k}{x} p^x(1-p)^k \frac{p^2}{p+1}(2+x+k)(1-p)^{x+k}$$
$$= \frac{p^2}{p+1} \sum_{k=0}^{\infty} \binom{x+k}{x} p^x(2+x+k)(1-p)^{x+2k} = \frac{(1-p)^x(1+x+2p-p^2)}{(p+1)(2-p)^{x+2}}; \quad x$$
$$= 0, 1, 2 \dots \text{ and } p \in (0, 1)$$

(3)

If X has the pmf (3), then it is called a binomial natural discrete Lindley (BNDL) random variable and it is denoted by $X \sim \text{BNDL}(p)$. For $n = 0$, this means that no particles enter into the attenuator and it will be termed as failure. Consequently, the corresponding cumulative distribution function (cdf) of BNDL distribution is given by

$$F(x;p) = P(X \leq x) = \sum_{t=0}^{x} p_x(t) = \sum_{t=0}^{x} \frac{(1-p)^t(1+t+2p-p^2)}{(p+1)(2-p)^{t+2}} = 1 - \frac{(1-p)^{x+1}(3+x+p-p^2)}{(p+1)(2-p)^{x+2}}.$$

(4)

Figure 1 shows the probability mass function (pmf) plots of the proposed distribution for various values of parameter p. Thus, the pmf is always a decreasing function, and the new discrete random variable tends to take small values when p increases. The stochastic process tends to happen very quickly once the parameter value grows, which is implied quite strongly by the model's behavior. Therefore, the BNDL model is a logical substitute for the traditional exponential distribution to characterize such phenomena. Additionally, the flexibility of the proposed BNDL can be tested for varied count data sources. For example, this model may be helpful for simulating aggregate losses that are typically limited to actuarial data by maximizing the overall garment fit for a particular number of sizes and accommodation rate, crucial to assessing the goodness of the scaling system. Furthermore, it may be helpful to overcome the problem of over-dispersed data in social sciences, as in anthropology where civilizations grew near the existence of a consistent water source, which is necessary for human survival. Figure 2 complements the results of Figure 1.

2.2. Statistical Properties of the BNDL Distribution

Primarily in this section, we provide some explicit results based on the mathematical properties of the BNDL distribution.

2.2.1. Moment-Generating Function

If $X \sim \text{BNDL}(p)$ distribution, then the moment-generating function of X is given as

$$M_X(t) = E\left(e^{tX}\right) = \sum_{x=0}^{\infty} e^{tx} \frac{(1-p)^x(1+x+2p-p^2)}{(p+1)(2-p)^{x+2}} = \frac{1-p(e^t-2)+p^2(e^t-1)}{(2-e^t+pe^t-p)^2(p+1)}.$$

For more on generating functions, see Yalcin and Simsek [17], Yalcin and Simsek [18] and Simsek [19].

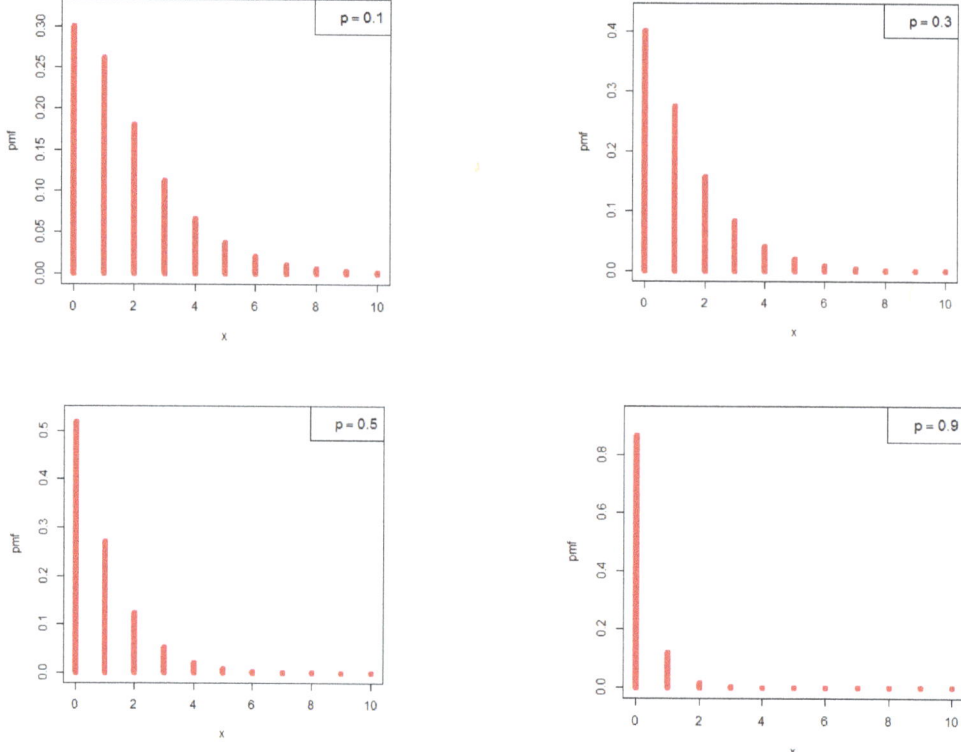

Figure 1. Pmf of BNDL distribution for some choices of p.

2.2.2. Probability-Generating Function

The probability-generating function of the random variable $X \sim \text{BNDL}(p)$ can be obtained using its moment-generating function which is equivalent to calculating $E(t^X)$; therefore, the probability-generating function of the random variable X is

$$G_X(t) = E\left(t^X\right) = M_X(\log(t)) = \frac{1 - p(t-2) + p^2(t-1)}{(2 - t + p(t-1))^2 (p+1)}.$$

Since,

$$G_X^{(k)}(t) = \frac{d^k G_X(t)}{dt^k} = E\left\{X(X-1)(X-2)\ldots(X-k+1)t^{X-k}\right\}.$$

Therefore, at $t = 1$, we can obatin

$$G_X^{(k)}(1) = \left.\frac{d^k G_X(t)}{dt^k}\right|_{t=1} = E\{X(X-1)(X-2)\ldots(X-k+1)\},$$

where $\mu_{(k)} = E\{X(X-1)(X-2)\ldots(X-k+1)\}$ is the kth factorial moment of X.

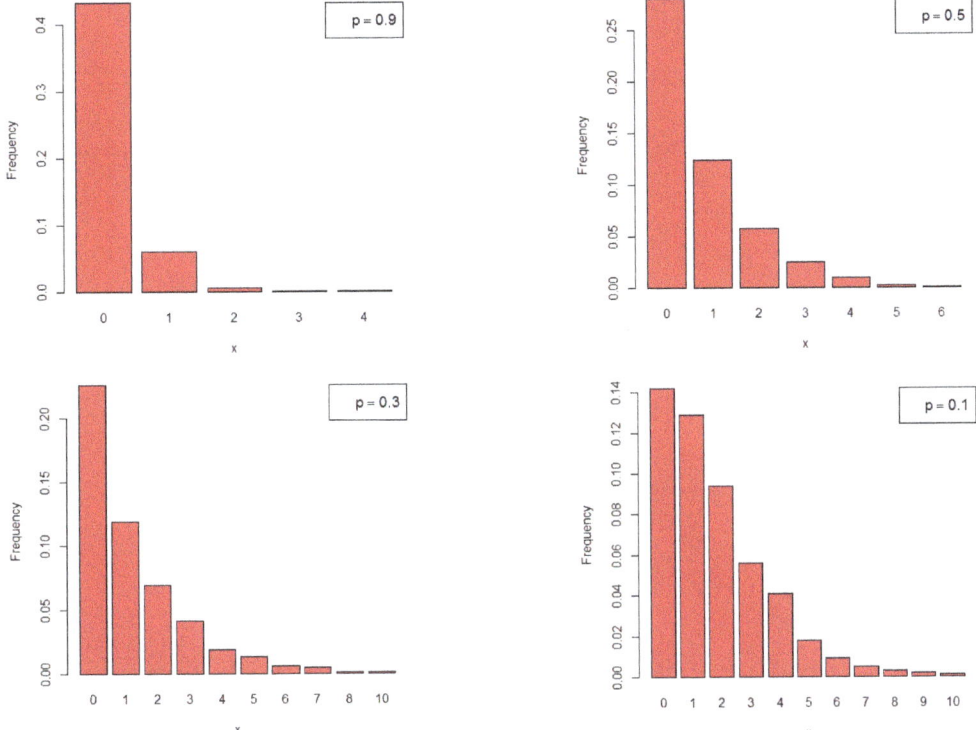

Figure 2. Histograms of the BNDL model for simulated data.

2.2.3. Non-Central Moments and Variance

If $X \sim \text{BNDL}(p)$ distribution, then the kth moment about zero of X is given by

$$\mu'_k = E(X^r) = \sum_{x=0}^{\infty} x^k p_x = \sum_{x=0}^{\infty} x^k \frac{(1-p)^x (1+x+2p-p^2)}{(p+1)(2-p)^{x+2}}.$$

The first four raw moments can be obtained as follows

$$\mu'_1 = E(X) = \frac{(p+2)(1-p)}{p+1},$$

$$\mu'_2 = E(X^2) = \frac{(1-p)(8-3p-2p^2)}{p+1},$$

$$\mu'_3 = E(X^3) = \frac{(1-p)(44-53p+6p^2+6p^3)}{p+1},$$

and

$$\mu'_4 = E(X^4) = \frac{(1-p)(308-516p+346p^2-12p^3-24p^4)}{p+1}.$$

The variance in the random variable X is

$$Var(X) = E(X^2) - [E(X)]^2 = \frac{(1-p)(4+5p-2p^2-p^3)}{(p+1)^2}.$$

2.2.4. Central Moments

The kth moment about the mean of X is

$$\mu_r = E\left[(X - \mu_1')^k\right] = \sum_{x=0}^{\infty} (x - \mu_1')^k p_x(x) = \sum_{x=0}^{\infty} (x - \mu_1')^k \frac{(1-p)^x (1 + x + 2p - p^2)}{(p+1)(2-p)^{x+2}}.$$

Therefore, the second, third and fourth central moments of the random variable X are

$$\mu_2 = \frac{(1-p)(4 + 5p - 2p^2 - p^3)}{(p+1)^2},$$

$$\mu_3 = \frac{(1-p)(12 + 21p - 7p^2 - 21p^3 + 5p^4 + 2p^5)}{(p+1)^3},$$

and

$$\mu_4 = \frac{(1-p)(100 + 181p - 132p^2 - 285p^3 + 50p^4 + 137p^5 - 27p^6 - 9p^7)}{(p+1)^4}.$$

2.2.5. Skewness and Kurtosis

The coefficient of skewness and the coefficient of kurtosis of the of BNDL distribution are, respectively,

$$\beta_1 = \frac{\mu_3}{\sqrt{\mu_2^3}} = \frac{(1-p)(12 + 21p - 7p^2 - 21p^3 + 5p^4 + 2p^5)}{(4 + p - 7p^2 + p^3 + p^4)^{3/2}}.$$

$$\beta_2 = \frac{\mu_4}{\mu_2^2} = \frac{100 + 181p - 132p^2 - 285p^3 + 50p^4 + 137p^5 - 27p^6 - 9p^7}{(1-p)(4 + 5p - 2p^2 - p^3)^2}.$$

2.2.6. Index of Dispersion

The index of dispersion (ID) indicates whether a certain distribution is suitable for under- or over-dispersed datasets. For example, ID $= 1$ for the Poisson distribution where the variance is equal to the mean, for the geometric distribution and the negative binomial distribution ID > 1, while the binomial distribution has ID < 1.

Theorem 1. *If $X \sim BNDL(p)$, then $Var(X) > E(X)$ for all $p \in (0,1)$.*

Proof. We have

$$ID(X) = \frac{Var(X)}{E(X)} = \frac{4 + 5p - 2p^2 - p^3}{p^2 + 3p + 2}.$$

This function is a monotonic decreasing function as $p \in (0,1)$ increases. It converges to 2 when $p \to 0$, while it tends to 1 as $p \to 1$; therefore, $ID(X) \in (1,2)$, which means that $ID(X) > 1$, and hence, $Var(X) > E(X)$. □

From Theorem 1, BNDL distribution should only be used in the count data analysis with over-dispersion. In Table 1, some of the empirical findings of these measured are due for considerations.

Table 1. Mean, Variance, Skewness, kurtosis and ID of the BNDL distribution for different values of the parameter p.

p	0.1	0.2	0.3	0.4	0.5	0.6	0.7	0.8	0.9
Mean	1.71818	1.4666	1.2384	1.0285	0.8333	0.6500	0.4764	0.3111	0.1526
Variance	3.3314	2.7288	2.1923	1.7191	1.3055	0.9475	0.6412	0.3832	0.1703
Skewness	1.5578	1.6186	1.6831	1.7542	1.8372	1.9427	2.0935	2.3522	2.9813
Kurtosis	7.7069	9.4991	11.8378	15.0902	19.9488	27.8656	42.3746	74.4447	180.1786
ID	1.9389	1.8606	1.770268	1.6714	1.5666	1.4576	1.3459	1.2317	1.1159

2.2.7. Log-Concavity

A necessary and sufficient condition that p_x be strongly unimodal is that it has to be log-concave, i.e., $p_{x+1}^2 \geq p_x p_{x+2}$ for all x (see Keilson and Gerber [20]).

Theorem 2. *The pmf of the BNDL distribution in (3) is log-concave.*

Proof. From (3), we can directly reach

$$p_{x+1}^2 = \frac{(1-p)^{2x+2}(2+x+2p-p^2)^2}{(p+1)^2(2-p)^{2x+6}},$$

and

$$p_x p_{x+2} = \frac{(1-p)^{2x+2}(1+x+2p-p^2)(3+x+2p-p^2)}{(p+1)^2(2-p)^{2x+6}}.$$

After some algebraic operations, we find that

$$p_{x+1}^2 - p_x p_{x+2} = \frac{(1-p)^{2x+2}}{(p+1)^2(2-p)^{2x+6}} > 0,$$

for all x and for all choices $p \in (0,1)$.

Theorem 2 confirms that the BNDL distribution is strongly unimodal. □

2.3. Reliability Properties of the BNDL Distribution

2.3.1. Survival Function

If $X \sim \mathrm{BNDL}(p)$ distribution, then from (4), the survival function of X is

$$S(x;p) = P(X \geq x) = \frac{(1-p)^{x+1}(3+x+p-p^2)}{(p+1)(2-p)^{x+2}}.$$

2.3.2. Hazard Rate and Mean Residual Life Functions

The hazard (failure) rate function is the probability that an item has survived time x, given that it has survived to at least time x. If $X \sim \mathrm{BNDL}(p)$ distribution, then its hazard rate (failure rate) function is given as

$$r(x;p) = P(X = x | X > x) = \frac{p_x(x;p)}{S(x;p)} = \frac{1+x+2p-p^2}{(1-p)(3+x+p-p^2)}.$$

Obviously, the upper limit of the failure rate function is $\frac{1}{1-p}$, i.e., $\lim_{x \to \infty} r(x;p) = \frac{1}{1-p}$. Graphical illustrations of hazard rate function are presented in Figure 3 while descriptive measures are presented in Figure 4.

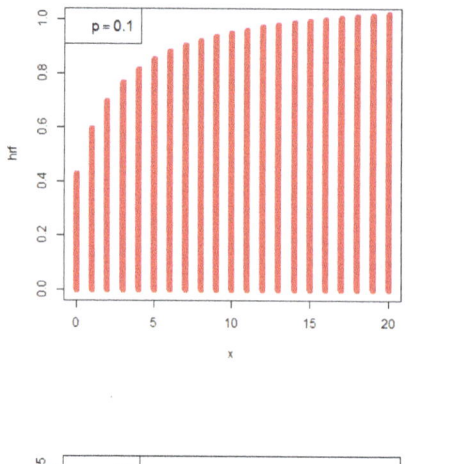
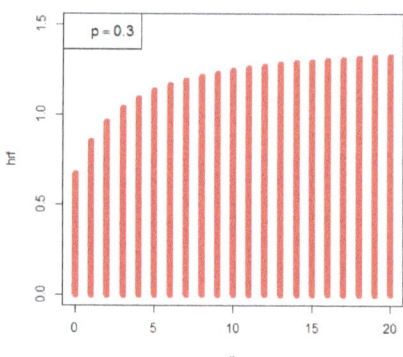
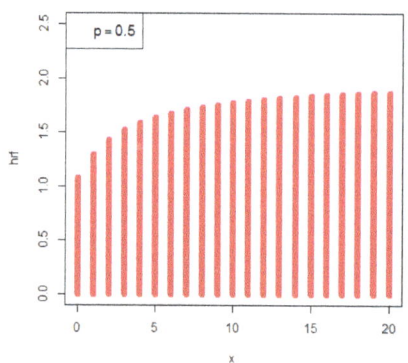
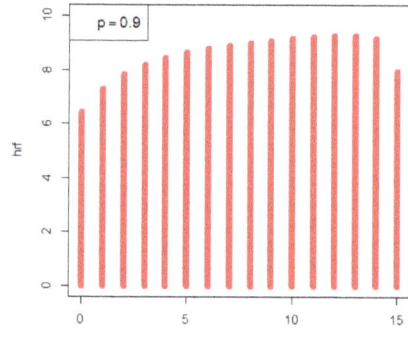

Figure 3. Plots of hazard rate of BNDL distribution for some choices of p.

The mean residual life function of X is given by

$$m(x;p) = P(X - x | X > x) = \frac{\sum_{t=x+1}^{\infty} S(t;p)}{S(x;p)} = \frac{(p-1)(p^2 - x - 5)}{3 + p - p^2 + x}.$$

Corollary 1. *If $X \sim BNDL(p)$ distribution, then it has an increasing failure rate and decreasing mean residual life.*

As we explained through Theorem 2, the BNDL distribution has a property of log-concavity; therefore, according to Gupta et al. [21], the BNDL distribution has an IFR property. According to Kemp [22], the next chain is verified

$$\text{IFR} \Rightarrow \text{IFRA} \Rightarrow \text{NBU} \Rightarrow \text{NBUE} \Rightarrow \text{DMRL}.$$

So, the BNDL distribution is

1. **IFR** (increasing failure rate).
2. **IFRA** (increasing failure rate average).
3. **NBU** (new better than used).
4. **NBUE** (new better than used in expectation).
5. **DMRL** (decreasing mean residual lifetime).

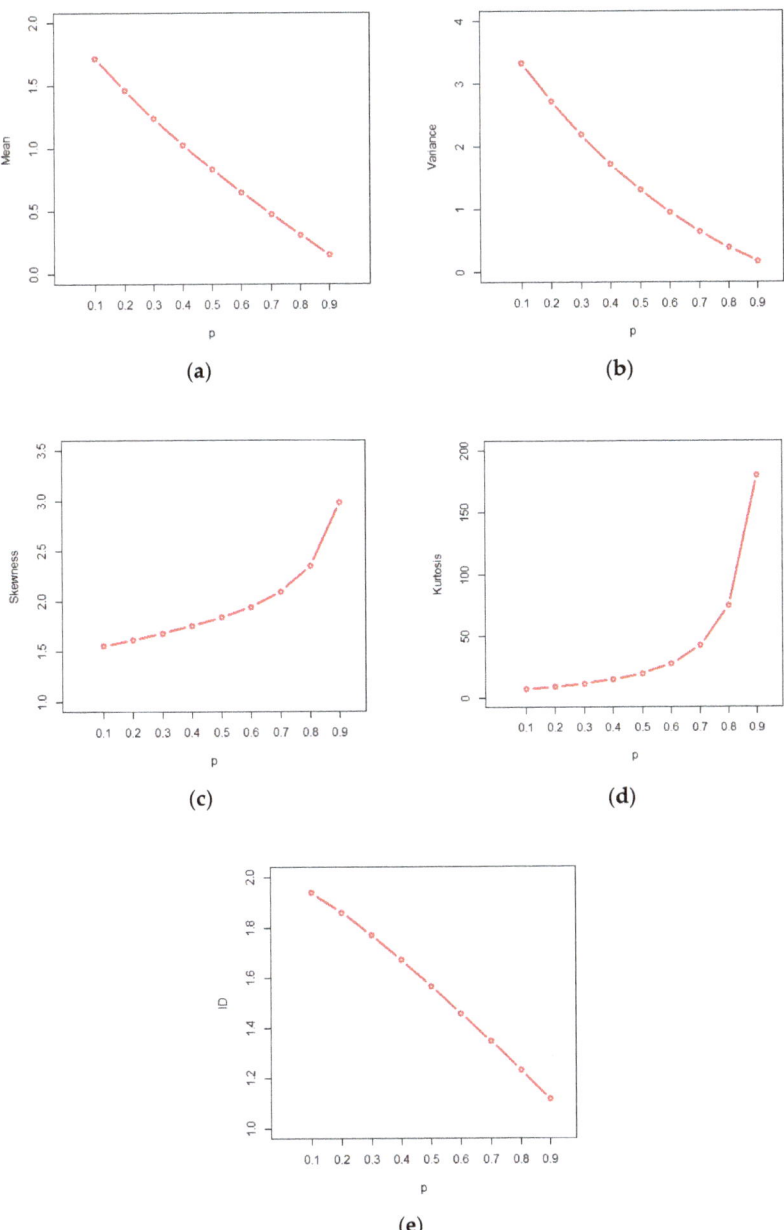

Figure 4. Plots of the BNDL model for (**a**) Mean, (**b**) Variance, (**c**) Skewness, (**d**) Kurtosis and (**e**) ID.

2.4. Stochastic Orderings

Stochastic orders are important measures to judge comparative behaviors of random variables. Shaked and Shanthikumar [8] showed that many stochastic orders exist and have various applications. Given two random variables X and Y, we say that X is smaller than Y in the

1. Usual stochastic order, denoted by $X \leq_{st} Y$, if $F_X(x) \geq F_Y(x)$, for all x.
2. Hazard rate order, denoted by $X \leq_{hr} Y$, if $h_X(x) \geq h_Y(x)$, for all x.
3. Reversed hazard rate order, denoted by $X \leq_{rh} Y$, if $F_X(x)/F_Y(x)$ decreases in x.
4. Mean residual life order, denoted by $X \leq_{mrl} Y$, if $m_X(x) \leq m_Y(x)$, for all x.
5. Likelihood ratio order, denoted by $X \leq_{lr} Y$, if $f_X(x)/f_Y(x)$ decreases in x.

For all the previous orders, we have the following chains of implications:

$$X \leq_{lr} Y \Rightarrow X \leq_{hr} Y \Rightarrow X \leq_{st} Y,$$

and

$$X \leq_{lr} Y \Rightarrow X \leq_{rh} Y \Rightarrow X \leq_{st} Y$$

also,

$$X \leq_{hr} Y \Rightarrow X \leq_{mrl} Y.$$

Theorem 3. Let $X \sim BNDL(p_1)$ and $Y \sim BNDL(p_2)$; then, $X \leq_{lr} Y$ for all $p_1 > p_2$.

Proof. Let

$$L(x; p_1, p_2) = \frac{p_X(x; p_1)}{p_Y(x; p_2)}.$$

Now,

$$L(x; p_1, p_2) = \frac{(p_2+1)(2-p_2)^{x+2}(1-p_1)^x(1+x+2p_1-p_1^2)}{(p_1+1)(2-p_1)^{x+2}(1-p_2)^x(1+x+2p_2-p_2^2)},$$

and

$$L(x+1; p_1, p_2) = \frac{(p_2+1)(2-p_2)^{x+3}(1-p_1)^{x+1}(2+x+2p_1-p_1^2)}{(p_1+1)(2-p_1)^{x+3}(1-p_2)^{x+1}(2+x+2p_2-p_2^2)}.$$

Therefore,

$$\frac{L(x+1; p_1, p_2)}{L(x; p_1, p_2)} = \frac{(2-p_2)(1-p_1)(2+x+2p_1-p_1^2)(1+x+2p_2-p_2^2)}{(2-p_1)(1-p_2)(2+x+2p_2-p_2^2)(1+x+2p_1-p_1^2)} \quad (5)$$

Let $p_1 = 1 - \delta$ and $p_2 = 1 - \delta - \varepsilon$, where $0 < \delta < 1$ and $0 < \varepsilon < 1 - \delta$.
After substitution of the values p_1 and p_2 in (5), we obtain

$$\frac{L(x+1; p_1, p_2)}{L(x; p_1, p_2)} = \frac{\eta_1(\delta + \delta^2 + \delta\varepsilon)}{\eta_2(\delta + \delta\varepsilon + \delta^2 + \varepsilon)},$$

where

$$\eta_1 = \left(3 + x - \delta^2\right)\left(2 + x - (\delta + \varepsilon)^2\right),$$

and

$$\eta_2 = \left(3 + x - (\delta + \varepsilon)^2\right)\left(2 + x - (\delta)^2\right).$$

After some algebraic operations, we find that

$$\eta_1 - \eta_2 = -\varepsilon(2\delta + \varepsilon) < 0 \Rightarrow \eta_1 < \eta_2.$$

Therefore,

$$\eta_1\left(\delta + \delta^2 + \delta\varepsilon\right) < \eta_2\left(\delta + \delta\varepsilon + \delta^2 + \varepsilon\right).$$

This implies that

$$\frac{L(x+1;\ p_1,p_2)}{L(x;p_1,p_2)} < 1 \Rightarrow L(x+1;\ p_1,p_2) < L(x;p_1,p_2).$$

□

2.5. Entropy

Entropy is a measure of uncertainty of a random variable. The entropy of a discrete random variable X with pmf $p(x)$ and alphabet $\mathcal{X}$ is given by

$$\mathbb{H}(X) = -E(\log p(X)) = -\sum_{x \in \mathcal{X}} p(x)\log(p(x)).$$

Entropy can be interpreted as the measure of average uncertainty in X or the average number of bits needed to describe X. For more details on entropy and information theory, we refer the reader to Gray [23].

Now, if $X \sim \text{BNDL}(p)$, then the entropy of the random variable X can be calculated by the following formula

$$\mathbb{H}(X) = \frac{1}{(2-p)^2(1+p)}\Big\{(2-p)^2\big[(-2+p+p^2)\log(1-p) + (4+p-p^2)\log(2-p) + (1+p)\log(1+p)\big] + \text{LerchPhi}^{(0,1,0)}\Big[\frac{1-p}{2-p}, -1, 1+2p-p^2\Big]\Big\},$$

where $\text{LerchPhi}^{(0,1,0)}[z,s,a]$ gives the Lerch transcendent $\Phi(z,s,a) = \sum_{k=0}^{\infty}\frac{z^k}{(a+k)^s}$. Table 2 presents some numerical values of the entropy of $X \sim \text{BNDL}(p)$ for different choices of p. From Table 2, one can observe that $\mathbb{H}(X)$ is monotonically decreasing in $p \in (0,1)$ with its limits tending to be 1.88 as p tends to 0 as $p \to 1$.

Table 2. Numerical results of $\mathbb{H}(X)$ for different values of the parameter p.

p	$\mathbb{H}(X)$	p	$\mathbb{H}(X)$
0.0001	1.87934	0.5	1.25943
0.01	1.86852	0.55	1.18391
0.03	1.84654	0.6	1.10402
0.05	1.82437	0.65	1.01888
0.07	1.80201	0.7	0.927315
0.09	1.77948	0.75	0.827736
0.11	1.75675	0.8	0.717861
0.14	1.72231	0.85	0.594157
0.17	1.6874	0.9	0.450497
0.2	1.652	0.95	0.273684
0.25	1.59181	0.96	0.231718
0.3	1.52994	0.97	0.186252
0.35	1.46611	0.98	0.135994
0.4	1.40002	0.99	0.078212
0.45	1.33128	0.999	0.0112562

Figure 5 relates the $\mathbb{H}(X)$ to the values of parameter p. One may note that (X) is monotonically decreasing in $p \in (0,1)$ with its limit inclining to zero as p tends to 1.

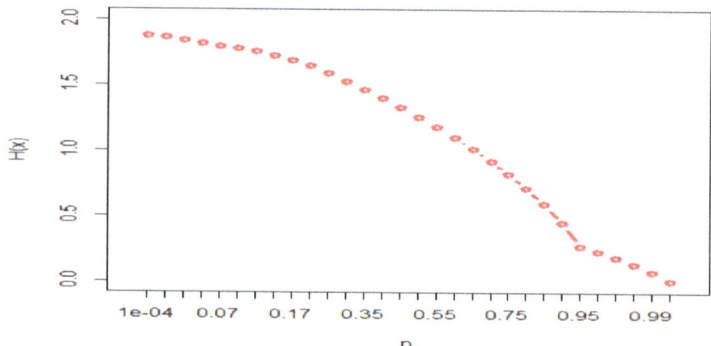

Figure 5. $\mathbb{H}(X)$ of X versus p.

3. Estimation and Simulation

In this section, we determine the estimation of unknown parameter p by the maximum likelihood, moment and proportion methods.

3.1. Method of Maximum Likelihood Estimation

Let $x_1, x_2, \ldots, x_n$ be the observed values from the BNDL distribution with parameter p. The likelihood and log-likelihood function are given, respectively, as

$$L(p) = \prod_{i=1}^{n} f(x_i) = \prod_{i=1}^{n} \frac{(1-p)^{x_i}(1+x_i+2p-p^2)}{(p+1)(2-p)^{x_i+2}},$$

and

$$l(p) = \log(1-p)\sum_{i=1}^{n} x_i + \sum_{i=1}^{n} \log\left(1+x_i+2p-p^2\right) - n\log(p+1) - 2n\log(2-p) - \log(2-p)\sum_{i=1}^{n} x_i.$$

The maximum likelihood estimate (MLE) of the parameter p can be obtained by solving the following equation using some numerical procedures.

$$\frac{\partial l(p)}{\partial p} = \frac{3pn}{2+p-p^2} - \frac{\sum_{i=1}^{n} x_i}{2-3p+p^2} + 2\sum_{i=1}^{n} \frac{1-p}{1+2p-p^2+x_i} = 0$$

3.2. Method of Moments Estimation

Let $X_1, X_2, \ldots, X_n$ be a random sample from the BNDL distribution with parameter p. The moment estimate (ME) of the parameter p can be obtained by solving the following equation.

$$\frac{(p+2)(1-p)}{p+1} = \frac{1}{n}\sum_{i=1}^{n} X_i.$$

3.3. Method of Proportions Estimation

Let $X_1, X_2, \ldots, X_n$ be a random sample from the BNDL distribution with parameter p. For $i = 1, 2, \ldots, n$, we define the indicator functions

$$I(X_i) = \begin{cases} 1 & \text{if } X_i = 0 \\ 0 & \text{if } X_i > 0 \end{cases}$$

Therefore, the proportion of 0s in the sample $\Pi = \frac{1}{n}\sum_{i=1}^{n} I(X_i)$. The proportion estimate (PE) of the parameter p can be obtained by solving the following equation with respect to p

$$\Pi = \frac{1 + 2p - p^2}{(p+1)(2-p)^2}.$$

3.4. Simulation Study

In this section, we assess the behavior of the maximum likelihood estimators for a finite sample of size n. Based on BNDL distribution, a simulation study is carried out. The simulation study is based on the following steps: firstly, generate $N = 1000$ samples of sizes $n = 25, 50, \ldots, 500$ from the BNDL distribution. Then, compute the maximum likelihood estimators for the model parameters. Lastly, compute the MSEs given by

$$\text{MSE}(p) = \frac{1}{1000}\sum_{i=1}^{1000}(\hat{p} - p)^2$$

For various parameters' values, the simulation's results provided in Figure 6 indicate that the estimated MSEs fall off toward zero when the sample size n increases. Hence, we have conclusive evidence to claim that the maximum likelihood estimation of p satisfies the asymptotic convergence of normality. The asymptotic normality of the MLE is a very well-known classic property given as follows. In a parametric model, we say that an estimator $\hat{p}$ based on $X_1, X_2, X_3, \ldots, X_n$ is consistent if $\hat{p} \to p$ in probability as $n \to \infty$. We say that it is asymptotically normal if $\sqrt{n}(\hat{p} - p)$ converges in distribution to a normal distribution. So $\hat{p}$ above is consistent and asymptotically normal.

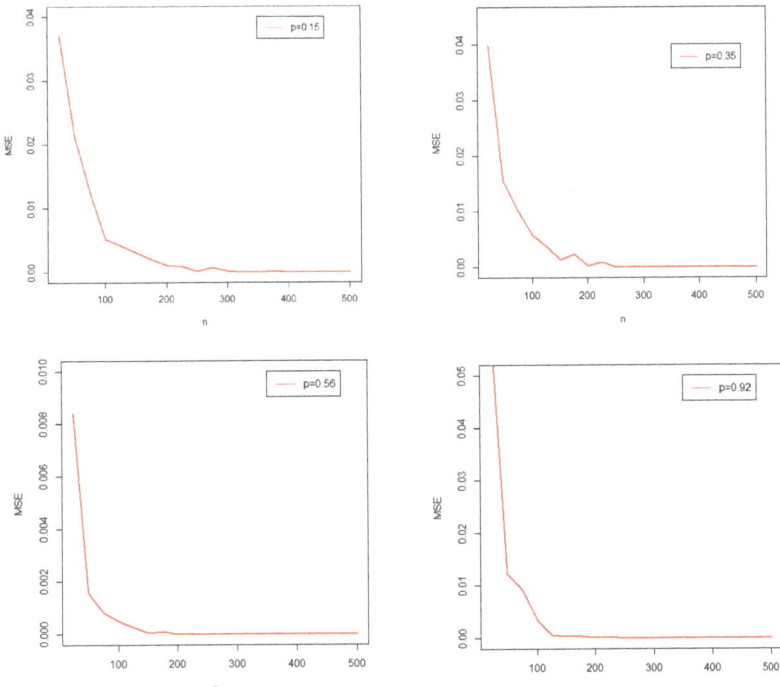

Figure 6. Plots of the estimated parameter and MSEs for various values of p.

4. Applications to Count Data

In this section, to show the application, we used a real-life data set to examine the efficiency and superiority of the BNDL distribution in modeling real data practice, recently studied by Balakarishnan et al. [24], consisting of 744 discrete observations. Santiago, Chile is recognized as one of the most environmentally contaminated cities in the world. In order to obtain the level of air pollution and its associated adverse effects on humans in Santiago, the National Commission of Environment (CONAMA) of the government of Chile collects data on sulfur dioxide (SO_2) concentrations in the air. The data corresponding to the hourly SO_2 concentrations (in ppm) observed at a monitoring station located in Santiago city are:

x	1	2	3	4	5	6	7	8	9	10 and above
f	86	235	120	119	35	15	11	9	4	10

The descriptive statistics of the data sets are, Mean = 2.93, Median = 2, Mode = 3, SD = 2.02, Coefficient of Variation = 0.69, Skewness = 4.32, Kurtosis = 34.57, Range = 24, Min value = 1 and Max value = 25.

We compare BNDL to Binomial–Discrete Lindley Distribution (BDLD) by Kuş et al. [15] and Negative Binomial distribution. The pmf of BDLD is given as

$$p_x(x;p) = \frac{p^{2x}\left[\{p^3 - (1-p)(1-p-x)\}log(p) + (1-p)\{1-p(1-p)\}\right]}{\{1-log(p)\}\{1-p(1-p)\}^{x+2}}$$

We considered the AIC (Akaike Information Criterion), CAIC (Consistent Akaike Information Criterion), BIC (Bayesian Information Criterion) and HQIC (Hannan–Quinn Information Criterion). The model with minimum values for these statistics could be chosen as the best model to fit the data. All results in Table 3 were obtained using the R PROGRAM.

Table 3. MLEs and their standard errors (in parentheses) with statistics AIC, BIC, HQIC and CAIC values for given data.

Distribution	MLE (SE)	MEASURES			
		AIC	CAIC	BIC	HQIC
BNDL (p)	0.6283 (0.0129)	2681.839	2681.844	2686.451	2683.616
BDLD (p)	0.6922 (0.0055)	3092.3700	3092.3760	3096.9820	3094.1480
Negative Binomial (n, k)	17.2957, 2.9262 (4.7378, 0.0678)	2824.156	2849.44	2833.38	2818.69

Figure 7 gives the quantile–quantile plot (Q-Q plot) and box plot and Figure 8 gives TTT plot versus the EHRF for the given data set. Total Time on Test (TTT plots) showed that the data set has an increasing hazard rate shape which is confirmed by EHRF. Figures 9 and 10 show the fitted model against its comparative distributions. These plots clearly show that the BNDL model is superior to well-known BDLD and Negative Binomial models.

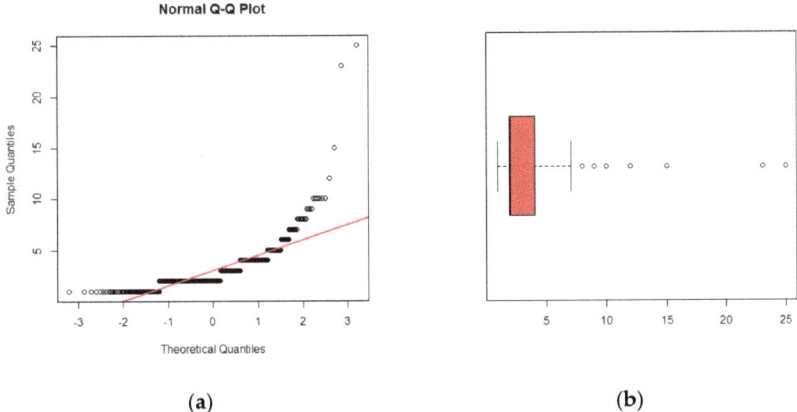

Figure 7. (**a**) QQ plot and (**b**) box for the given data.

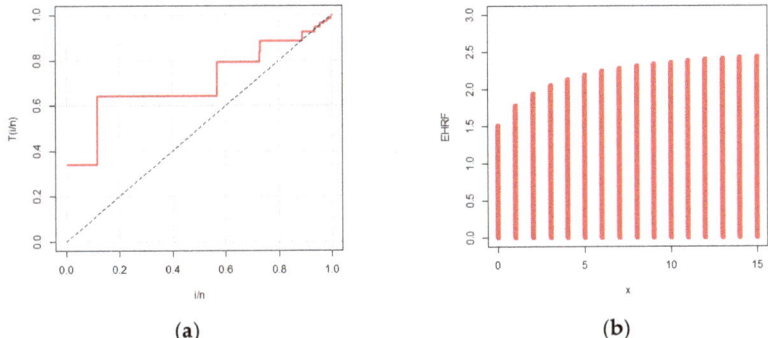

Figure 8. (**a**) TTT plot and (**b**) Expected Hazard Rate Function (EHRF) for the BDLD model for the dataset.

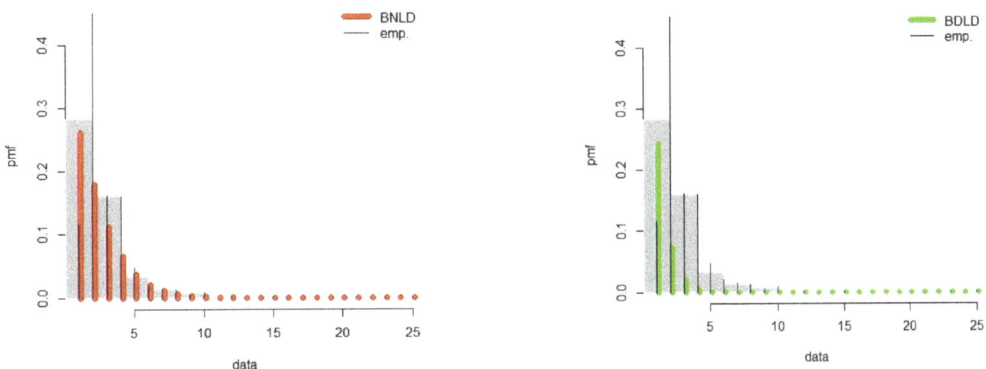

Figure 9. Fitted plots of BNDL and BDLD distribution for given data set.

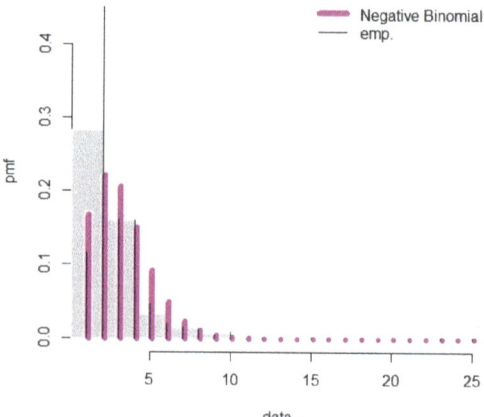

Figure 10. Fitted plot of Negative Binomial distributions for given data set.

5. Concluding Remarks

A new one-parameter discrete distribution was proposed and its important distributional, monotonic, and reliability characteristics were explored. Some statistical and reliability properties of the proposed discrete model were derived. Various estimating approaches were discussed. A simulation study was conducted to determine the MLEs' accuracy and precision. The applicability of the proposed distribution in modeling a real-life discrete data set was demonstrated. It is clear from the comparison that the new distribution is the best distribution for fitting the data sets from among the all-tested distributions and it will be a useful contribution to the field of count data modeling.

Author Contributions: Conceptualization, S.S. and S.K.; methodology, W.M.; software, J.G.; validation, S.S. and S.K.; formal analysis, W.M.; investigation, S.S.; resources, F.J.; data curation, W.M.; writing—original draft preparation, S.S. and W.M.; writing—review and editing, S.K.; visualization, J.G.; supervision, S.K.; project administration, F.J. All authors have read and agreed to the published version of the manuscript.

Funding: This research received no external funding.

Conflicts of Interest: The authors declare no conflict of interest.

References

1. Aryuyuen, S.; Bodhisuwan, W.; Volodin, A. Discrete Generalized Odd Lindley—Weibull Distribution with Applications. *Lobachevskii J. Math.* **2020**, *41*, 945–955. [CrossRef]
2. Chakraborty, S. A New Discrete Distribution Related to Generalized Gamma Distribution and Its Properties. *Commun. Stat. Theory Methods* **2015**, *44*, 1691–1705. [CrossRef]
3. Chakraborty, S.; Chakravarty, D. Discrete Gamma Distributions: Properties and Parameter Estimations. *Commun. Stat. Theory Methods* **2012**, *41*, 3301–3324. [CrossRef]
4. Chakraborty, S.; Dhrubajyoti, C. A Discrete Gumbel Distribution. *arXiv* **2014**. Available online: https://arxiv.org/abs/1410.7568 (accessed on 8 June 2022).
5. El-Morshedy, M.; Eliwa, M.S.; Nagy, H. A New Two-Parameter Exponentiated Discrete Lindley Distribution: Properties, Estimation and Applications. *J. Appl. Stat.* **2018**, *47*, 354–375. [CrossRef]
6. Gómez-Déniz, E.; Calderín-Ojeda, E. The Discrete Lindley Distribution: Properties and Applications. *J. Stat. Comput. Simul.* **2011**, *81*, 1405–1416. [CrossRef]
7. Hu, Y.; Peng, X.; Li, T.; Guo, H. On the Poisson Approximation to Photon Distribution for Faint Lasers. *Phys. Lett. A* **2007**, *367*, 173–176. [CrossRef]
8. Shaked, M.; Shanthikumar, J.G. *Stochastic Orders*; Springer: New York, NY, USA, 2007. [CrossRef]
9. Nekoukhou, V.; Alamatsaz, M.H.; Bidram, H. Discrete Generalized Exponential Distribution of a Second Type. *Statistics* **2013**, *47*, 876–887. [CrossRef]

10. Para, B.A.; Jan, T.R. Discrete Generalized Weibull Distribution: Properties and Applications in Medical Sciences. *Pak. J. Stat.* **2017**, *33*, 337–354.
11. Roy, D. The Discrete Normal Distribution. *Commun. Stat.-Theory Methods* **2003**, *32*, 1871–1883. [CrossRef]
12. Afify, A.Z.; Elmorshedy, M.; Eliwa, M.S. A New Skewed Discrete Model: Properties, Inference, and Applications. *Pak. J. Stat. Oper. Res.* **2021**, *17*, 799–816. [CrossRef]
13. Déniz, E.G. A New Discrete Distribution: Properties and Applications in Medical Care. *J. Appl. Stat.* **2013**, *40*, 2760–2770. [CrossRef]
14. Akdoğan, Y.; Kuş, C.; Asgharzadeh, A.; Kinaci, I.; Sharafi, F. Uniform-Geometric Distribution. *J. Stat. Comput. Simul.* **2016**, *86*, 1754–1770. [CrossRef]
15. Kuş, C.; Akdoğan, Y.; Asgharzadeh, A.; Kınacı, I.; Karakaya, K. Binomial-Discrete Lindley Distribution. *Commun. Fac. Sci. Univ. Ank. Ser. A1 Math. Stat.* **2019**, *68*, 401–411. [CrossRef]
16. Al-Babtain, A.A.; Ahmed, A.H.N.; Afify, A.Z. A New Discrete Analog of the Continuous Lindley Distribution, with Reliability Applications. *Entropy* **2020**, *22*, 603. [CrossRef]
17. Yalcin, F.; Simsek, Y. Formulas for characteristic function and moment generating functions of beta type distribution. *Rev. Real Acad. Cienc. Exactas Físicas Y Naturales. Ser. A Matemáticas* **2022**, *116*, 86. [CrossRef]
18. Yalcin, F.; Simsek, Y. A new class of symmetric beta type distributions constructed by means of symmetric Bernstein type basis functions. *Symmetry* **2020**, *12*, 779. [CrossRef]
19. Simsek, B. Formulas derived from moment generating functions and Bernstein polynomials. *Appl. Anal. Discret. Math.* **2019**, *13*, 839–848. [CrossRef]
20. Keilson, J.; Gerber, H. Some Results for Discrete Unimodality. *J. Am. Stat. Assoc.* **1971**, *66*, 386–389. [CrossRef]
21. Gupta, P.L.; Gupta, R.C.; Tripathi, R.C. On the monotonic properties of discrete failure rates. *J. Stat. Plan. Inference* **1997**, *65*, 255–268. [CrossRef]
22. Kemp, A.W. Classes of discrete lifetime distributions. *Commun. Stat. Theory Methods* **2004**, *33*, 3069–3093. [CrossRef]
23. Gray, R.M. *Entropy and Information Theory*; Springer: New York, NY, USA, 2011. [CrossRef]
24. Balakrishnan, N.; Leiva, V.; Sanhueza, A.; Cabrera, E. Mixture inverse Gaussian distributions and its transformations, moments and applications. *Statistics* **2009**, *431*, 91–104. [CrossRef]

MDPI
St. Alban-Anlage 66
4052 Basel
Switzerland
Tel. +41 61 683 77 34
Fax +41 61 302 89 18
www.mdpi.com

Mathematical and Computational Applications Editorial Office
E-mail: mca@mdpi.com
www.mdpi.com/journal/mca

www.ingramcontent.com/pod-product-compliance
Lightning Source LLC
LaVergne TN
LVHW070216100526
838202LV00015B/2055